solid state
electronic circuits:
for engineering
technology

INTERNATIONAL TECHNOLOGICAL UNIVERSITY
This Book is Donated by:
PROF. WAI-KAI CHEN

Date:

# solid state electronic circuits: for engineering technology

Anthony S. Manera

*President*
*Niagara College of Applied Arts and Technology*

McGRAW-HILL BOOK COMPANY

New York
St. Louis
San Francisco
Düsseldorf
Johannesburg

Kuala Lumpur
London
Mexico
Montreal
New Delhi

Panama
Rio de Janeiro
Singapore
Sydney
Toronto

Library of Congress Cataloging in Publication Data

Manera, Anthony S.
  Solid state electronic circuits.
  Includes bibliographical references.
  1. Transistor circuits. I. Title.
TK7871.9.M324  621.3815'3'0422  72-13156
ISBN  0-07-039871-2

## SOLID STATE ELECTRONIC CIRCUITS: FOR ENGINEERING TECHNOLOGY

Copyright © 1973 by McGraw-Hill, Inc. All rights reserved. Printed in the United States of America. No part of this publication may be reproduced, stored in a retrieval system, or transmitted, in any form or by any means, electronic, mechanical, photocopying, recording, or otherwise, without the prior written permission of the publisher.

1234567890KPKP79876543

*The editors for this book were Alan W. Lowe and Cynthia Newby, the designer was Marsha Cohen, and its production was supervised by James E. Lee. It was set in Modern by Progressive Typographers. It was printed and bound by Kingsport Press, Inc.*

*to my mother and father*

# contents

PREFACE     *xiii*

1. BASIC PROPERTIES OF SEMICONDUCTORS     1

    1.1    Introduction     1
    1.2    Intrinsic Conduction     6
    1.3    Doped Semiconductors     9
    1.4    Drift and Diffusion Currents     10
    1.5    PN Junction with No External Voltage     11
    1.6    PN Junction with Forward Bias     14
    1.7    PN Junction with Reverse Bias     15
    1.8    Junction Capacitance     16
    1.9    Breakdown     18

2. THE DIODE     21

    2.1    The Ideal Diode     21
    2.2    The Real Diode: Large-signal Operation     26
    2.3    The Real Diode: Small-signal Operation     36

3. CIRCUIT APPLICATIONS OF DIODES     47

    3.1    Introduction     47
    3.2    Power Supplies     47
           3.2a    Half-wave Rectification     49
           3.2b    Full-wave Rectification     54
           3.2c    The Half-wave Rectifier with RC Load     60
           3.2d    The Full-wave Rectifier with RC Load     67
           3.2e    Voltage Regulation     70
    3.3    Clipping Circuits     76
    3.4    Clamping Circuits     85

4. PERFORMANCE MEASURES FOR AMPLIFYING CIRCUITS     93

    4.1    The Black Box Approach     93
    4.2    The Decibel     100

| | | |
|---|---|---|
| 4.3 | *Transducers* | 103 |
| 4.4 | *Ideal Amplifiers* | 109 |
| 4.5 | *Frequency Response* | 114 |
| | 4.5a  *Introduction* | 114 |
| | 4.5b  *Bode Plots* | 118 |
| | 4.5c  *Cascaded Stages* | 136 |
| 4.6 | *Distortion* | 140 |
| 4.7 | *Noise* | 148 |

## 5. INTRODUCTION TO JUNCTION TRANSISTORS — 151

| | | |
|---|---|---|
| 5.1 | *The Junction Transistor* | 151 |
| 5.2 | *Static Characteristics* | 158 |
| 5.3 | *Graphic Analysis* | 162 |
| | 5.3a  *Locating the Q Point* | 162 |
| | 5.3b  *Determining Amplification* | 168 |
| | 5.3c  *AC Loading* | 173 |
| 5.4 | *DC Circuit Analysis* | 176 |
| 5.5 | *Cutoff Current* | 187 |
| 5.6 | *Breakdown Voltage* | 191 |
| 5.7 | *The Transistor as a Switch* | 192 |
| 5.8 | *Manufacturing Techniques* | 199 |
| | 5.8a  *Junction-forming Processes* | 200 |
| | 5.8b  *Transistor Structures* | 202 |
| | 5.8c  *Integrated Circuits* | 204 |

## 6. JUNCTION TRANSISTOR SMALL-SIGNAL MODELS — 207

| | | |
|---|---|---|
| 6.1 | *Junction Transistor Models* | 207 |
| 6.2 | *The Hybrid (h) Parameter Model* | 208 |
| | 6.2a  *General Properties of Hybrid Parameters* | 213 |
| | 6.2b  *Experimental Determination of h Parameters* | 214 |
| | 6.2c  *Graphic Evaluation of h Parameters* | 224 |
| | 6.2d  *Use of Data Sheets to Obtain h Parameters* | 231 |
| | 6.2e  *Conversion of h Parameters from One Configuration to Another* | 232 |
| 6.3 | *The Hybrid-Pi Model* | 237 |

## 7. JUNCTION TRANSISTOR SMALL-SIGNAL ANALYSIS — 245

| | | |
|---|---|---|
| 7.1 | *The Common Emitter Amplifier* | 245 |
| 7.2 | *The Common Base Amplifier* | 249 |

|  |  |  |
|---|---|---|
| 7.3 | The Common Collector Amplifier | 253 |
| 7.4 | Transistor Amplifier with Unbypassed Emitter Resistance | 265 |
| 7.5 | Low-frequency Response | 285 |
| 7.6 | High-frequency Response | 296 |

## 8. JUNCTION TRANSISTOR BIASING — 301

|  |  |  |
|---|---|---|
| 8.1 | General Considerations | 301 |
| 8.2 | Constant Base Current Bias | 303 |
| 8.3 | Series (Emitter) Feedback Bias | 305 |
| 8.4 | Stability Factors | 308 |
| 8.5 | Shunt Feedback Biasing | 314 |
| 8.6 | Nonlinear Bias Stabilization | 316 |

## 9. FIELD EFFECT TRANSISTORS AND CIRCUITS — 319

|  |  |  |
|---|---|---|
| 9.1 | The Junction Field Effect Transistor (JFET) | 320 |
| 9.2 | The Metal Oxide Semiconductor Field Effect Transistor (MOSFET) | 322 |
| 9.3 | Graphic Analysis of FET Amplifier | 325 |
| 9.4 | Static Characteristics and Ratings | 328 |
| | 9.4a Leakage Current | 328 |
| | 9.4b Voltage Breakdown in JFETs | 329 |
| | 9.4c Voltage Breakdown in MOSFETs | 330 |
| | 9.4d Drain Current Specifications | 332 |
| 9.5 | Biasing the FET | 333 |
| 9.6 | Dynamic Characteristics | 345 |
| 9.7 | Analysis of FET Amplifier Circuits | 349 |
| | 9.7a The Common Source Configuration | 349 |
| | 9.7b The Common Drain Configuration | 351 |
| | 9.7c The Common Gate Configuration | 354 |
| | 9.7d Maximizing the Input Resistance | 357 |
| | 9.7e Frequency Response | 362 |

## 10. FEEDBACK PRINCIPLES AND APPLICATIONS — 367

|  |  |  |
|---|---|---|
| 10.1 | Basic Definitions | 368 |
| 10.2 | Effects of Negative Feedback on Amplifier Performance | 375 |
| 10.3 | General Feedback Connections | 379 |
| 10.4 | Multistage Feedback Circuits | 396 |
| 10.5 | Stability of Feedback Systems | 401 |

## 11. DC CIRCUITS — 413

- 11.1 Direct-coupled Amplifiers — 413
- 11.2 The Differential Amplifier — 418
- 11.3 Chopper-stabilized Amplifiers — 429
- 11.4 Operational Amplifiers — 431
  - 11.4a Basic Relationships — 433
  - 11.4b The Inverter Connection — 435
  - 11.4c The Adder Circuit — 435
  - 11.4d The Subtractor Circuit — 437
  - 11.4e Constant Voltage and Constant Current Sources — 438
  - 11.4f The Noninverter Connection — 439
  - 11.4g The Voltage Follower — 439
  - 11.4h Offset Voltage — 440
  - 11.4i Offset Current — 441
  - 11.4j Common-mode Rejection Ratio — 442
  - 11.4k Frequency-response Considerations — 444
- 11.5 Voltage Regulation — 445
  - 11.5a Shunt Regulator — 446
  - 11.5b Series Regulator — 448
  - 11.5c Series Regulator with Current Preregulator — 453

## 12. LARGE-SIGNAL AMPLIFICATION — 457

- 12.1 General Considerations — 457
  - 12.1a Output Power — 458
  - 12.1b Transistor Parameters — 460
  - 12.1c Distortion — 461
  - 12.1d Efficiency — 463
- 12.2 Operating Region — 465
- 12.3 Thermal Considerations — 467
- 12.4 Amplifier Circuits — 474
  - 12.4a Class A Circuits — 474
  - 12.4b Class B Circuits — 488
- 12.5 Complementary Symmetry Circuits — 508
- 12.6 Other Power Circuits — 511

## 13. RF CIRCUITS — 515

- 13.1 Small-signal RF Amplifiers — 515
  - 13.1a Performance Measures — 516

|  |  |  |  |
|---|---|---|---|
|  | 13.1b | Basic Circuits | 517 |
|  | 13.1c | Tuned Circuits | 520 |
|  | 13.1d | Circuits Employing Bipolar Transistors | 545 |
|  | 13.1e | Analysis Using Admittance Parameters | 556 |
|  | 13.1f | Automatic Gain Control (AGC) | 571 |
| 13.2 | Nonlinear Circuits | | 574 |
|  | 13.2a | Class C Amplification | 576 |
|  | 13.2b | Frequency Multiplication | 578 |
|  | 13.2c | Mixing | 580 |

## 14. OSCILLATORS — 585

|  |  |  |  |
|---|---|---|---|
| 14.1 | Basic Concepts and Definitions | | 585 |
| 14.2 | Harmonic Oscillators | | 592 |
|  | 14.2a | The RC Phase Shift Oscillator | 595 |
|  | 14.2b | The Colpitts Oscillator | 602 |
|  | 14.2c | The Hartley Oscillator | 608 |
|  | 14.2d | Crystal Oscillators | 611 |
|  | 14.2e | Tunnel Diode Oscillators | 614 |
| 14.3 | Relaxation Oscillators | | 623 |
|  | 14.3a | The Unijunction Transistor Oscillator | 623 |
|  | 14.3b | The Astable Multivibrator | 633 |

## 15. PULSE CIRCUITS — 639

|  |  |  |  |
|---|---|---|---|
| 15.1 | Properties of Pulses | | 639 |
| 15.2 | Linear Wave Shaping | | 644 |
|  | 15.2a | Attenuation Networks | 644 |
|  | 15.2b | Amplification and Phase Inversion | 653 |
|  | 15.2c | Differentiation and Integration | 658 |
| 15.3 | Nonlinear Wave Shaping | | 664 |
|  | 15.3a | The Bipolar Transistor | 666 |
|  | 15.3b | The Field Effect Transistor | 668 |
|  | 15.3c | The SCR | 672 |
|  | 15.3d | Sweep Generation | 680 |
|  | 15.3e | Staircase Generation | 684 |
|  | 15.3f | The One Shot (Monostable Multivibrator) | 688 |
|  | 15.3g | The Schmitt Trigger | 693 |
|  | 15.3h | The Flip-flop | 700 |

## APPENDIX

| | | |
|---|---|---|
| 1. | Abbreviations of Units and Prefixes | 709 |
| 2. | Definition of Symbols | 711 |
| 3. | Voltage and Current Dividers | 713 |
| 4. | Determinants | 719 |
| 5. | Loop and Node Analysis | 723 |
| 6. | Complex Algebra | 729 |
| 7. | Thevenin and Norton Equivalent Circuits | 737 |
| 8. | Power Exchange between Source and Load | 741 |
| 9. | Superposition | 747 |
| 10. | Semiconductor Data Sheets | 751 |
| | 2N2137–46 | |
| | 2N2386 | |
| | 2N3199–201 | |
| | 2N3392–4 | |
| | 2N3796–7 | |
| | 2N3823 | |
| 11. | Derivations | 777 |
| | 11.1 Average Value of a Waveform | 777 |
| | 11.2 RMS Value of a Waveform | 780 |
| | 11.3 Derivation of Eq. (9.15) | 783 |
| | 11.4 Calculation of Power Dissipation for Each Transistor in an Ideal Class B Amplifier | 784 |
| | 11.5 Derivation of Eq. (13.13) | 785 |

## ANSWERS TO SELECTED PROBLEMS    787

## INDEX    801

# preface

This book discusses basic semiconductor circuits, including the physical operation and electrical characteristics of the more common solid state devices.

The level should be suitable for engineering technology curricula in junior and community colleges as well as training programs in industry and government. The practicing technician or engineer may also find the book useful as a refresher or reference.

Abbreviations for units and prefixes conform to IEEE (Institute of Electrical and Electronics Engineers) Standard No. 260. A list is given in Appendix 1. Letter symbols for electrical quantities and parameters generally conform to IEEE Standard No. 255. Conventional current flow is used throughout the book; this convention, as well as graphic symbols for voltage and current sources, is illustrated in Appendix 2.

The content is both qualitative and quantitative; simple design is used wherever feasible. The mathematical level is limited to basic algebra and trigonometry. Although the student should have had or be taking concurrently a course in electric circuits (passive), a review is provided, along with illustrative examples, in the appendices. This review covers voltage and current dividers, loop and node analysis, determinants, complex numbers, Thevenin and Norton equivalent circuits, power transfer between source and load, and superposition. Some derivations requiring the use of calculus are carried out in Appendix 11.

Information from manufacturers' data sheets is used throughout the book; a number of representative data sheets are reproduced in Appendix 10. References are given at the end of each chapter to other books and publications dealing with the specific content of that chapter in greater depth. The book includes a large number of examples and problems. Problem sets generally appear at the end of individual sections in each chapter. Thus the student knows when he can tackle the problems and obtains the benefits of immediate reinforcement.

Chapter 1 develops properties of the PN junction as a starting point for the understanding of diodes and transistors.

The diode as an ideal and real device is treated in Chap. 2. In addition to diode familiarization, the purpose of this chapter is to introduce analytic techniques for later application. Characteristic curves, model construction, load line analysis, and large-signal versus small-signal operation are some of the topics treated. The diode's relative simplicity makes the introduction of these concepts simpler than with active devices.

Applications of diodes are discussed in Chap. 3. These include basic rectifier and power supply circuits, as well as Zener diode voltage regulators and clipping and clamping circuits. Diode models from the previous chapter are utilized to solve problems.

Chapter 4 introduces the amplifier as a black box. Performance measures are defined for amplifiers regardless of the active device used. The concepts and analysis techniques therefore apply to any type of amplifier, using discrete components or integrated circuits. The material should be of interest to anyone who needs to know about the external performance of amplifiers but is not concerned with their internal operation. In addition to basic performance measures, such as amplification, gain, and impedance levels, the topics of frequency response, distortion, and noise are also introduced. Bode plots are presented and extensively used. The concepts of *source* and *load* are treated under "Transducers," giving typical examples. The material in this chapter can be covered entirely before active devices are discussed, or it can be used as a reference while studying active circuits. The advantage of this approach is that, when circuits are later introduced, the student is already familiar with performance criteria and can better appreciate the reasons for using various circuit configurations. The student who is not concerned with the internal behavior of amplifiers can bypass Chaps. 5 through 9 and go directly to Chaps. 10 and 11 ("Feedback" and "DC Circuits").

Chapters 5 through 8 deal with junction (bipolar) transistors. Applications here are limited to small signals and low frequencies; however, static switching behavior is also discussed. The operation, characteristics, ratings, and manufacturing techniques, including the use of data sheets, are presented along with graphic and equivalent circuit analysis methods.

Chapter 9 discusses field effect transistors (FETs). The physical structure, operation, and electrical characteristics (static and dynamic) are treated. The various amplifier configurations are analyzed, including high-input impedance techniques and frequency response.

Feedback as a tool for improving amplifier performance and for system control is presented in Chap. 10. Stability considerations are discussed, including Nyquist's criterion. Many of these concepts are applied in Chap. 11 ("DC Circuits") in conjunction with differential and operational amplifiers as well as with transistorized voltage regulators.

Large-signal amplifiers are covered in Chap. 12. Output power, distortion, efficiency, and thermal considerations, including heat sinks, are emphasized. Class B circuits are treated in some detail, leading to complementary symmetry configurations.

Chapter 13 considers RF circuits, including some background indicating their application in communication systems. $y$ parameters are developed and applied to the analysis of field effect and bipolar transistor circuits. Many RF circuits require extensive alignment "on the bench"; hence a lot of emphasis is placed on tuned circuits and impedance transformation techniques. Stability through neutralization, unilateralization, and mismatching is discussed. Frequency spectra, as well as filtering concepts, are developed in conjunction with class C amplifiers, frequency multipliers, and mixers. AGC techniques for both FETs and bipolar transistors are also covered.

Oscillators of the harmonic and relaxation type are treated in Chap. 14. Concepts and techniques from Chaps. 10 and 13 are used to develop the theory and methods of analysis. Active devices used are the bipolar, field effect, and unijunction transistors, as well as tunnel diodes. The basic $RC$, Colpitts, Hartley, and Clapp configurations are analyzed using techniques appropriate to each specific circuit. The problems of frequency stability are introduced, leading to a discussion of crystal-controlled oscillators.

Chapter 15 deals with pulse circuits. Pulse parameters are defined, and pulse shaping in terms of linear and nonlinear processing is discussed. Compensated voltage dividers, linear inverters, and differentiating and integrating circuits represent the first part. The switching characteristics of bipolar and field effect transistors as well as SCRs are then presented, leading to nonlinear circuits. These include the monostable multi, the Schmitt trigger, step and sweep generators, SCR control circuits, and the flip-flop, including triggering techniques.

Many persons and organizations have contributed to the development of this book. The author is grateful to the manufacturers and publishers who have granted permission to use various illustrations and data sheets. The suggestions of many reviewers have been incorporated throughout the book. Special thanks go to D. Roddy, J. Coolen, and D. McLean (Lakehead University); G. M. Mitchell (Spokane Community College); G. Schickman (Miami Dade Junior College); N. Wipond (Sir Sandford Fleming College); I. Morgulis, G. Martinson, M. Ghorab, and C. Barsony (Ryerson Polytechnical Institute); J. Kendall (University of Alberta, Calgary); M. Anderson, S. Stewart, D. Ropchan, D. Driscoll, and L. Ozbolt (Confederation College). Last, but not least, I should like to acknowledge my wife's extreme patience in enduring for so long the hardships inherent in the preparation of this work.

*A. S. Manera*

solid state
electronic circuits:
   for engineering
   technology

# chapter 1

# basic properties of semi-conductors

## 1.1 INTRODUCTION

Solid state devices, whose study we propose to undertake, are the basic building blocks of modern electronics. In addition to increased reliability, smaller size, lower power consumption, and other advantages over vacuum tubes, solid state devices have often been responsible for the simplification of complex circuitry needed to perform various functions.

Before we attempt to discuss the characteristics of these devices, however, a brief review of the fundamentals of atomic theory will be worthwhile.

The existence of matter is relatively obvious; matter is the stuff around us, it takes up space. The concept of *charge*, however, is often a stumbling block to an understanding of certain physical phenomena. Yet matter and charge are very real quantities, intimately working together to make up the world around us.

Charge is related to force; it is, in fact, a concept invented to explain the existence of certain forces. The fact that solids actually exist implies the presence of forces. Since great pressure is required to compress a solid block of steel, there must be repulsive forces acting inside the steel; similarly, if we try to tear the block apart, the existence of strong attractive forces becomes evident. The theory which uses the concept of charge goes a long way toward explaining these forces of attraction and repulsion.

Although there are different kinds of matter, the basic particles are the same. These are electrons, protons, and neutrons. All three are needed to make up the atom, which is the smallest distinguishable unit of different kinds of matter. We know of about 100 different kinds of atoms, each involving a different arrangement of electrons, protons, and neutrons. Each different atom represents a different element, unique in its characteristics as observed by external experiments. Iron, copper, hydrogen, silver, and gold are examples of elements. Each of these elements contains only one kind of atom. An atom of iron differs from an atom of copper only in the number and arrangement of the basic particles. Different elements can also be chemically combined; in this case the resulting substance is called a compound.

An atom resembles our solar system in that it appears to be made up of a central core, or *nucleus*, around which electrons continuously rotate. The nucleus contains protons and neutrons. Neutrons have no electrical property of interest to us, but electrons and protons do. Protons have about 2,000 times the mass of electrons; neutrons have essentially the same mass as protons. The distinguishing feature of greatest importance, however, is the electric *charge* associated with each of these particles. Electrons appear to be repelled by other electrons and attracted to protons. The physical property responsible for these effects of attraction and repulsion is what we call charge. The unit of measurement for charge is the coulomb, abbreviated as C.[1] The charge on electrons is arbitrarily called negative, while that on protons is positive. The magnitude of charge on either the electron or proton is approximately $1.6 \times 10^{-19}$C. A neutron has no charge.

Electrons rotate around the nucleus of their atoms in specific energy levels. Energy is required to move electrons away from the nucleus; thus electrons closest to the nucleus have less energy than those farthest from the nucleus, as shown graphically in Fig. 1.1. Energy is measured in electron volts (eV), where one electron volt is the energy required to move one electron across a potential difference of one volt.

Substances are formed by the *bonding* of many atoms. When atoms are bonded, the individual energy levels of electrons merge into *bands* of allowable energy states. Separating these bands are energy gaps, or forbidden energy bands. These are energy levels in which no electron can exist. Exciting the atom—that is, adding energy to it—may result in electrons overcoming the energy gap and moving into a band further from the nucleus.

The outermost energy band of an unexcited atom is the valence band. If energy is added to the atom, electrons may be boosted to the conduction band, where they are relatively free and can move about the material without being associated with any particular atom. Figure 1.2 depicts the valence and

---

[1] See Appendix 1 for a list of units and their standard abbreviations.

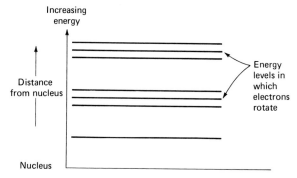

*Fig. 1.1 Electrons in a single atom orbit the nucleus in specific energy levels; energy increases as we move further away from the nucleus.*

conduction bands, as well as the energy gap separating them. An atom that has more or fewer electrons than its normal complement is said to be ionized. In the case of an electron loss, the atom is a positive ion; if electrons have been gained, it is a negative ion.

The energy gap between valence and conduction bands is an important measure of how materials behave electrically. Certain materials are characterized by a wide energy gap. This implies that valence electrons cannot be boosted to the conduction band unless a relatively large amount of energy is applied. These materials are electric insulators. On the other hand, elements for which the conduction and valence bands overlap are electric conductors. Here conduction electrons are readily available.

Somewhere between these extremes lie elements for which the energy gap between valence and conduction bands is intermediate to insulators and

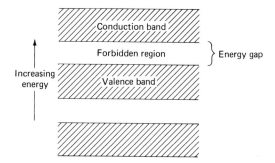

*Fig. 1.2 The forbidden region represents a discrete energy gap that must be overcome before valence electrons can reach the conduction band.*

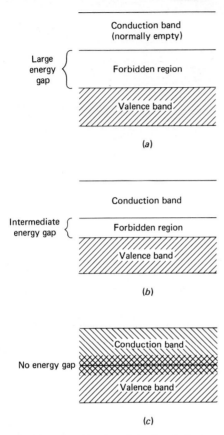

Fig. 1.3 *Classification of the electrical properties of materials according to the width of energy gap between valence and conduction bands:* (a) *insulators;* (b) *semiconductors;* (c) *conductors.*

conductors. These elements are semiconductors. Figure 1.3 illustrates the relationship between energy gap and electrical properties of materials.

Although several different materials behave as semiconductors, there are two which have found the greatest application. These are the elements silicon and germanium (Si and Ge), both of which have four valence electrons. Elements such as carbon and compounds such as gallium arsenide (GaAs) may also be used, but due to practical difficulties silicon and germanium continue to dominate.

Both Si and Ge atoms are arranged in a crystalline structure. This is a three-dimensional structure, characteristic of most solids, that involves an orderly and repetitive arrangement of atoms. Practical semiconductor devices are manufactured using single continuous crystals; that is, throughout the sample, the orderly arrangement of atoms keeps repeating itself without any abrupt changes. When more than one crystal is involved (polycrystalline materials), irregularities at crystal boundaries affect electrical performance in a manner that is difficult to predict and control.

The four valence electrons associated with each Si or Ge atom form pairs of electrons with neighboring atoms, as shown in the simplified two-dimensional view of Fig. 1.4. Here valence electrons are "shared" among neighboring atoms, so that each atom appears to have not four, but eight valence electrons. The motion of these valence electrons produces attractive forces that bind the various atoms in a specific crystal structure. These attractive forces represent a bond, appropriately called a covalent bond.

Covalent bonds involve electrons from the valence band which are not free charge carriers. If one were to apply enough energy, however, the covalent bond could be broken, and the electrons would move to the conduction band, where they would be free to carry an electric current. The energy required to break the bond could come from heat, light, or an electric field.

A pure semiconductor in which all valence electrons form covalent bonds is an insulator. Here energy levels in the conduction band are empty, or unfilled. At room temperature, however, enough heat energy is present to break some covalent bonds. Thus many free electrons are usually available. If, however, the semiconductor is cooled to a temperature of absolute zero, there is no heat energy, and all electrons form covalent bonds (unless some other form of energy is supplied). In this case, the material does not conduct.

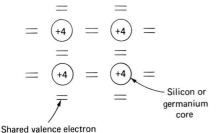

Shared valence electron

*Fig. 1.4 Covalent bond structure for silicon or germanium; simplified two-dimensional representation. Each electron has a negative charge; the shared electrons form a covalent bond. All 32 protons and 28 electrons are included in the Si or Ge core. The remaining four valence electrons are shown separately to illustrate covalent bonding. The core has a net charge of $+4$, while the valence electrons have a net charge of $-4$. Therefore the total charge for each atom is zero.*

## Problem Set 1.1

1.1.1. Why is the concept of charge useful?
1.1.2. Is an electron which is far from the nucleus of its atom at a higher or lower energy level than an electron which is closer to the nucleus?
1.1.3. What two energy bands in atoms are of greatest importance to the study of semiconductors?
1.1.4. When is an atom ionized?
1.1.5. Are electrons in the valence band normally "free"?
1.1.6. When would an electron from the valence band move to the conduction band?
1.1.7. How does the energy gap between valence and conduction bands determine the electrical properties of a material?
1.1.8. What is a covalent bond?
1.1.9. If no external source of energy is available, will a pure semiconductor such as germanium or silicon be capable of electric conduction?
1.1.10. Does the electric resistance of a semiconductor bar depend on temperature? If so, in what way?

## 1.2 INTRINSIC CONDUCTION

A semiconductor without any impurities is an *intrinsic* semiconductor. Any electric conduction that may take place in such a material is called intrinsic conduction.

You will recall that, at room temperature, some covalent bonds are broken because of the presence of heat. As this happens, electrons from the broken covalent bond move into the conduction band, where they move randomly around the crystalline structure. If an external voltage is applied, these electrons flow, resulting in current.

When an atom loses a valence electron, an incomplete covalent bond results. The missing valence electron is called a *hole*. The hole is a positive-charge carrier, since it is due to the absence of a negatively charged particle (valence electron). Even though a hole is not a particle, its positive charge can move, resulting in an electric current.

Suppose that a valence electron manages to move into a hole. Since it has been filled by an electron, the hole is no longer there. In this case recombination has taken place—that is, the hole has been neutralized. The electron which fills this hole, however, has left a vacancy in its original covalent bond. In effect, a hole has now appeared there. We can say either that the valence electron moved to fill the hole or that the hole moved to where the valence electron used to be. Figure 1.5 illustrates the sequence of events.

Why must we consider the hole concept at all when we could talk only

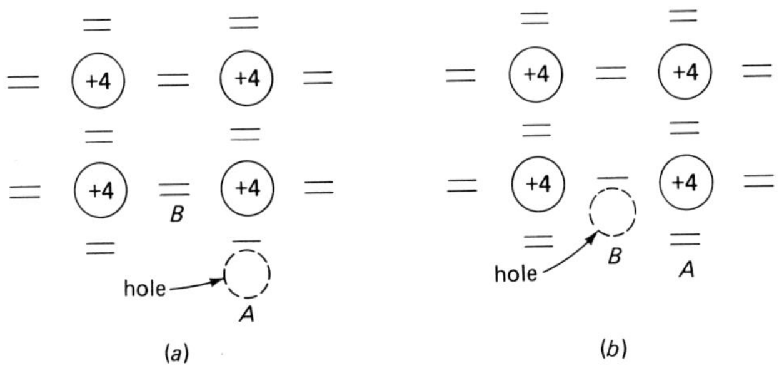

Fig. 1.5  Hole motion: (a) a missing valence electron at A represents a hole; (b) valence electron from B moves into hole at A or hole from A moves to B.

in terms of electrons? To see why we differentiate between electron and hole motion, note that electrons associated with hole motion are valence electrons, while electrons in the conduction band do not result in hole movement. For example, consider the motion of a conduction electron when an external voltage is applied. The conduction electron is not under the influence of any one atom; thus it can move freely in response to the applied potential. For valence electrons, however, the situation is different. Their energy level is not sufficient to place them in the conduction band, but they can move into nearby holes. Their motion is thus more restricted than for conduction electrons. In fact, for any given applied voltage, a valence electron will move a shorter average distance than a conduction electron in the same period of time, as shown in Fig. 1.6. Hence, (conduction) electrons are more "mobile" than holes.

Note that at room temperature many covalent bonds are broken; each time this happens an electron-hole pair is generated. The hole and electron are charge carriers, and if an electric potential is applied, they move and result in current. This is intrinsic conduction—that is, conduction occurs as a result of

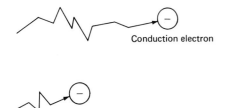

Fig. 1.6  A conduction electron travels further than a valence electron for any applied voltage in the same period of time.

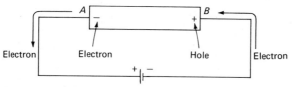

*Fig. 1.7  Intrinsic conduction in a semiconductor.*

properties of the pure material and is not due to the addition of any external impurities.

As an example of intrinsic conduction, consider the semiconductor bar shown in Fig. 1.7. Although there are many electron-hole pairs in such a sample, only one pair is shown in the figure for simplicity. Here the following process takes place:

1. Conduction electrons within the bar are attracted toward $A$ and flow through the connecting wire into the positive terminal of the battery. Additional electrons leave the negative terminal of the battery and enter the semiconductor at $B$. Thus there is a continuous flow of electrons in a counterclockwise direction around the circuit.
2. Holes in the bar are attracted toward $B$. Holes that reach $B$ are neutralized by electrons entering the semiconductor from the wire near $B$. This is a *recombination* process. At the same time, some valence electrons near $A$ break their covalent bond and enter the wire leading toward the positive terminal of the battery. The breaking of covalent bonds is a charge *generation* process in which both electrons and holes are produced. We therefore have holes "appearing" at $A$, moving from $A$ toward $B$, and "disappearing" at $B$. The net effect is a flow of holes in a clockwise direction around the circuit.
3. There is continuous generation and recombination of electrons and holes. There are no holes in the circuit external to the semiconductor. Holes "disappearing" at $B$ represent electrons entering $B$, while holes "appearing" at $A$ represent electrons leaving $A$. As far as the external circuit is concerned, the total current is due to a counterclockwise flow of electrons, as if the semiconductor bar were a simple resistor. It is only inside the semiconductor that the total current can be broken down into its two basic components, electron and hole flow.

If more energy is supplied to the semiconductor, additional electron-hole pairs are generated, resulting in a decrease of the bar's ohmic resistance. Typical room-temperature values of resistance for a cubic centimeter of semiconductor are in the order of 50 Ω for Ge and 50,000 Ω for Si. The higher value for Si is due to the greater energy gap between its valence and conduction bands (1.1 eV) relative to Ge (0.7 eV).

## Problem Set 1.2

1.2.1. What is the difference between a hole and a free electron?
1.2.2. In which band does hole motion take place?
1.2.3. In which band does conduction electron motion take place?
1.2.4. What is the meaning of mobility?
1.2.5. Why are holes less mobile than electrons?
1.2.6. What does intrinsic conduction mean?
1.2.7. Would you expect any difference in the resistivity of a sample of intrinsic Si compared to intrinsic Ge at a temperature of absolute zero? Assume no other forms of energy are present.
1.2.8. Why is the resistance of Si greater than that of Ge, when all other factors (purity of sample, dimensions, presence of energy) are equal?
1.2.9. Under what conditions do you think a bar of Si might have a lower electric resistance than a bar of Ge?

## 1.3 DOPED SEMICONDUCTORS

Until now, our discussion has been limited to intrinsic conduction. This type of conduction is due to the presence of electron-hole pairs which have been generated by heat or some other source of energy. This is not a very useful process, since we have no practical way of controlling the conductivity of our semiconductor. Because of this, impurities in carefully measured amounts are added to semiconductors to alter their electrical properties in a predictable manner. This process of adding impurities is called *doping*.

Useful impurities are elements with five or three valence electrons (pentavalent or trivalent). If an impurity with five valence electrons is used, four of these form covalent bonds with Ge or Si atoms as shown in Fig. 1.8. The remaining electron is not needed to form a covalent bond. This electron assumes an energy level very close to the conduction band. We have therefore produced a

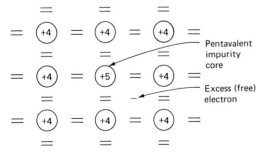

*Fig. 1.8* The addition of a pentavalent impurity (5 valence electrons) to a pure semiconductor results in excess (free) electrons.

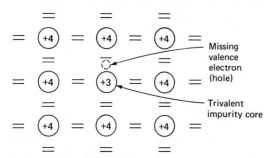

*Fig. 1.9* The addition of a trivalent impurity (3 valence electrons) to a pure semiconductor results in a deficiency of valence electrons (holes).

free negative-charge carrier by the introduction of an impurity. This electron is different from an electron produced by the rupture of a covalent bond in that there is no hole associated with it.

Elements with five valence electrons are called *donor impurities*, since they "donate" an extra electron. Typical pentavalent elements are antimony, arsenic, and phosphorus. A semiconductor doped with a donor impurity has an excess of negative-charge carriers and is referred to as N-type material.

If we dope with a trivalent impurity, the three valence electrons form covalent bonds with neighboring electrons from the Ge or Si atoms. One of the bonds, however, is not complete, since four valence electrons would be required and only three are available. This is shown in Fig. 1.9. The missing valence electron is, in effect, a hole. The hole is a fully qualified charge carrier and no different from an electron except as outlined earlier.

Trivalent elements are called *acceptors*, and when they are used to dope semiconductors, the resulting material is called P-type, because of the positive-charge carriers. There are no free electrons associated with holes produced by doping.

## 1.4 DRIFT AND DIFFUSION CURRENTS

There are two types of current involved in semiconductors: drift and diffusion. Drift current is the flow of charge due to a potential difference. This is what takes place in a resistive material when a voltage is applied.

Diffusion current is not due to a potential difference, but results from the random motion of charged particles due to thermal energy. For example, if a large number of free electrons are concentrated in one area of a sample and there is another area in that same sample with very few free electrons, electrons from the higher-density region will flow toward the lower-density region until an equilibrium is established. This phenomenon is called diffusion, and it characterizes a natural tendency toward equilibrium; particles move from where there are lots of them to where there are few.

Diffusion is a slower process than drift, but it may be speeded up through the addition of thermal energy.

## Problem Set 1.3

1.3.1. What is the difference between an intrinsic and a doped semiconductor?
1.3.2. What is the difference between a trivalent impurity and a pentavalent impurity?
1.3.3. What is different about an electron produced by the addition of impurities in a semiconductor and one which came about through the rupture of a covalent bond?
1.3.4. What is a donor impurity?
1.3.5. What is an acceptor impurity?
1.3.6. What is N-type semiconductor?
1.3.7. What is P-type semiconductor?
1.3.8. Is a potential difference required to produce diffusion current?
1.3.9. What current results when charges move because of an electric potential?
1.3.10. Why do you think diffusion can be speeded up by the addition of heat?

## 1.5 PN JUNCTION WITH NO EXTERNAL VOLTAGE

Consider a semiconductor sample which is doped P-type on one side and N-type on the other. The area where the P- and N-type regions meet is called a PN *junction*, and the resulting device is a diode. A useful PN junction cannot be produced by connecting P-type to N-type material by welding or similar processes; this would result in discontinuities across the crystalline structure. As stated earlier, semiconductor devices must be made from single crystals in order to be of practical use. Impurities may be introduced into a semiconductor by a variety of processes, some of which are discussed in Sec. 5.8.

Figure 1.10 shows a simplified two-dimensional view of such a semicon-

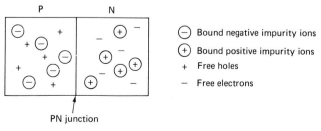

Fig. 1.10  *Electric charges in a semiconductor which is oppositely doped on each side.*

ductor. Only holes and negative impurity ions are shown in the P side. On the N side, free electrons and positive impurity ions are indicated. Ions represent atoms that have gained or lost electrons; these atoms are not mobile and should not be confused with mobile charges. The whole sample is electrically neutral; therefore, there must be a net positive charge for each free electron and a net negative charge for each hole.

When the junction is formed, the following processes are initiated:

1. Holes from the P side diffuse into the N side, where they combine with free electrons.
2. Free electrons from the N side diffuse into the P side, where they combine with holes.
3. The diffusion of holes and electrons is an electric current, referred to as a recombination current. The recombination process decays exponentially with both time and distance from the junction. Thus most of the recombination occurs very soon after the junction is formed and in the region very near the junction. A measure of the rate of recombination is the lifetime $\tau$, defined as the time required for the density of carriers to decrease to 37 percent of the original concentration.
4. Electrons have been removed and holes added to the N side; similarly, the P side has lost holes and acquired electrons. This exchange of charge occurs mainly in a small region on each side of the junction which becomes depleted of free charge carriers, leaving only positive ions on the N side and negative ions on the P side. Since there are few free electrons on the N side near the junction, the positive ions are said to be uncovered; similarly, there are uncovered negative ions on the P side near the junction, as shown in Fig. 1.11. Remember that ions are fixed and cannot move.

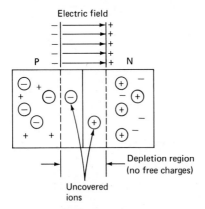

Fig. 1.11 *Electric field due to presence of uncovered ions in the depletion region near a PN junction.*

5. Uncovered ions within the depletion region produce an electric field as shown in the figure. The electric field represents a potential difference between the two regions, also called barrier or space charge potential.
6. The space charge potential discourages diffusion of charges across the junction. For example, an electron trying to diffuse from the N to the P side is repelled by the negative charge on the P side. The same applies to holes trying to diffuse from the P to the N side. Therefore, the diffusion process does not continue indefinitely but continues only as long as it takes to build up the potential barrier, after which time further diffusion is substantially reduced.
7. The physical width of the depletion region depends on the doping level. If very heavy doping is used, the depletion region is physically thin because a diffusing charge need not travel far across the junction before recombining (lifetime is short). If light doping is used, however, the depletion region is wider (lifetime is long). Figure 1.12 compares depletion-region widths for relatively heavy and light doping. The widths shown are exaggerated; in practice they can be as low as $10^{-6}$ cm or less.

So far we have discussed the flow of electrons from the N side and holes from the P side. Free electrons in N-type material and free holes in P-type material are called majority carriers.

We must now consider minority carriers. Due to the presence of heat and to some extent other forms of energy, a number of covalent bonds are broken in all semiconductors, whether they are doped or not. Each time a covalent bond is broken, two charge carriers result, an electron and a hole. An N-type semiconductor, for example, has many free electrons because of doping; there are also some free electrons and holes due to the rupture of covalent bonds. Free electrons represent the desirable majority carriers; holes are the minority carriers.

When the PN junction was formed, there was a recombination current, caused by the diffusion of majority carriers across the junction. Minority carriers produce an additional component of current. The polarity of the electric field

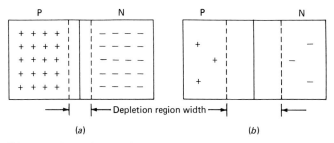

Fig. 1.12  Dependence of depletion-region width on doping level (only free charges are shown for simplicity): (a) heavy doping; (b) light doping.

(potential barrier) across the junction discourages diffusion of majority carriers. Minority carriers (holes in the N-type and electrons in the P-type), however, are not opposed; in fact, the electric field helps them drift across the junction. Each time a minority carrier drifts across the junction, the potential barrier is reduced and more diffusion of majority carriers takes place, which raises the barrier. The process continues, in effect, forever. A balance is reached, and a continuous diffusion of majority carriers and drift of minority carriers maintain that balance.

The generation of minority carriers is primarily a function of temperature. This can be a serious problem for both Si and Ge devices operating at elevated temperatures, but it is less of a problem in Si because of the larger energy gap between valence and conduction bands.

### Problem Set 1.4

1.4.1. What causes diffusion of holes and electrons across a PN junction?
1.4.2. Why does most recombination take place in the region near a PN junction?
1.4.3. What causes the barrier potential?
1.4.4. Why does the area in the immediate vicinity of a PN junction have relatively few free charges?
1.4.5. Is there a net charge within the depletion region? Explain.
1.4.6. How does the width of the depletion region depend on the degree of doping?
1.4.7. What happens to the potential barrier when a majority carrier diffuses across the junction?
1.4.8. What happens to the potential barrier when a minority carrier drifts across the junction?
1.4.9. Which type of carrier is hindered by the potential barrier as it tries to cross the PN junction? Which type of carrier is aided by the potential barrier at the junction?
1.4.10. Is there a continuous flow of charges across a PN junction when no external potential is applied?
1.4.11. What is the main factor that causes minority carriers?
1.4.12. Assuming all other things equal, which element (Si or Ge) do you think has the greatest concentration of minority carriers?
1.4.13. Which of the two elements Si and Ge do you think would behave more reliably at higher temperatures? Explain.

## 1.6 PN JUNCTION WITH FORWARD BIAS

Suppose that an external voltage is applied to a PN junction as shown in Fig. 1.13a. The P side is connected to the positive terminal and the N side to the negative terminal of the battery. In this case, the PN junction is said to be

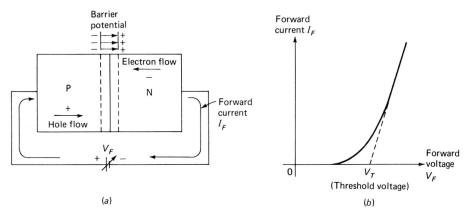

*Fig. 1.13* (a) *PN junction with forward bias;* (b) *voltage-current characteristic for the forward-biased PN junction.*

forward-biased. The polarity of external voltage causes free electrons and holes to move across the junction as indicated in the figure. These majority carriers cover some of the previously uncovered ions, thereby reducing the potential barrier. Majority carriers now find it easier to diffuse across the junction, since more of them have the energy required to overcome the reduced barrier. If we increase the magnitude of the forward-biasing voltage $V_F$, the barrier is further reduced; thus the effective resistance of the junction decreases with the amount of forward bias. This results in a larger number of majority carriers crossing the junction. The flow of these carriers represents forward current $I_F$ and is characterized by holes and electrons flowing toward and across the junction, as shown in Fig. 1.13a. Since no holes exist in the external wiring, only electrons flow in the outside circuit. The direction of $I_F$ is shown in the figure; this is the current that would be measured by an ammeter connected in the circuit for that purpose.

If the forward bias is further increased, the current may rise to an excessive value, causing junction overheating and device destruction. Ge devices can usually stand junction temperatures around 100°C; Si units can function up to 175°C or higher.

When the forward voltage and current are measured, it is found that current rises with an approximately exponential relationship to voltage, as shown on the characteristic of Fig. 1.13b. The threshold voltage $V_T$ is an approximate value of $V_F$ below which current flow is usually negligible. $V_T$ is around 0.2 V for Ge and 0.6 V for Si units.

## 1.7 PN JUNCTION WITH REVERSE BIAS

Consider now a reverse-biased PN junction as in Fig. 1.14a. The negative terminal of the battery is connected to the P side and the positive terminal to the

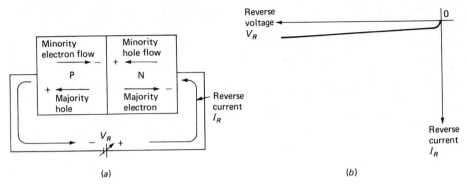

*Fig. 1.14* (a) *PN junction with reverse bias;* (b) *voltage-current characteristic for the reverse-biased PN junction. Note that the total reverse current is slightly voltage-dependent (it is primarily temperature-dependent).*

N side; this reinforces the potential barrier. As the magnitude of reverse-biasing voltage $V_R$ is increased, the potential barrier also increases, and majority carriers find it more difficult to diffuse across the junction. In fact, majority carriers are pulled away from the junction; this results in very little current, and the junction appears as a relatively large resistance.

Unfortunately, in addition to majority carriers, we also have minority carriers, which, although not significantly influenced by the magnitude of reverse voltage (their existence is due to temperature), still represent a definite component of reverse current, called the saturation current.

The voltage-current characteristic of Fig. 1.14b shows that as the reverse voltage is increased in magnitude, the total current also increases slightly. This is due to impurities on the surface of the semiconductor, which behave as an effective resistance, obeying Ohm's law. This additional component of current, which is dependent on voltage rather than on temperature, is called the surface-leakage component.

In practice the total reverse current $I_R$ is temperature-dependent to a greater degree than it is voltage-dependent. In Ge, it doubles for approximately every 10°C rise, while in Si it doubles about every 6°C. Generally $I_R$ is of the order of μA for Ge and nA for Si.[1]

## 1.8 JUNCTION CAPACITANCE

Capacitive effects are exhibited by PN junctions whether they are forward- or reverse-biased.

---

[1] μA is the abbreviation for microampere ($10^{-6}$ ampere); similarly nA is the abbreviation for nanoampere ($10^{-9}$ ampere).

You will recall that there are few free charge carriers in the depletion region. Hence this region behaves as the dielectric material used for making capacitors. The P- and N-type conducting regions on each side of the semiconductor act as the *plates*. We therefore have all the components needed for a capacitive effect. This junction capacitance $C_T$ is also called *transition* or *space charge* capacitance. It can be objectionable in certain applications while useful in others.

Space charge capacitance is voltage-dependent. If the reverse voltage is increased, the depletion region extends further into the conducting areas away from the junction, increasing the separation between the plates and causing the capacitance to decrease.

Capacitive effects are also present in forward-biased junctions. Here the term used is *diffusion capacitance* $C_D$ to account for the time delay in moving charges across the junction by the diffusion process. Therefore, diffusion capacitance cannot be identified in terms of a dielectric and plates as space charge capacitance. Instead it must be considered as a fictitious but convenient element that allows us to predict actual behavior, such as time delay. Because the time-delay effect is greater if the amount of charge to be moved across the junction is increased, it follows that diffusion capacitance varies directly with the magnitude of forward current.

## Problem Set 1.5

1.5.1. How does forward biasing a PN junction affect its potential barrier?
1.5.2. How does reverse bias affect the potential barrier across a PN junction?
1.5.3. If excessive current is allowed, what can happen to a PN junction?
1.5.4. What does threshold voltage mean?
1.5.5. Why do you think the threshold voltage of Si units is higher than that of Ge units?
1.5.6. What is the basic mechanism of charge flow through a PN junction under forward-bias conditions? Under reverse-bias conditions?
1.5.7. What does the saturation current primarily depend on?
1.5.8. Why does the total reverse current also depend to some degree on the amount of reverse voltage applied?
1.5.9. When does a PN junction exhibit an effective low resistance? An effective high resistance?
1.5.10. At what rate does the reverse current increase with temperature?
1.5.11. What two factors affect the depletion region width and in what way?
1.5.12. What is space charge capacitance?
1.5.13. What is diffusion capacitance?
1.5.14. Under what bias conditions (forward or reverse) is diode capacitance voltage-dependent? Current-dependent?

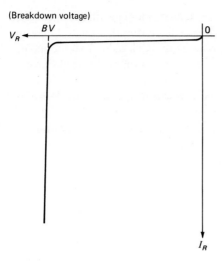

*Fig. 1.15  PN junction reverse characteristic, including breakdown region.*

## 1.9  BREAKDOWN

If the reverse bias on a PN junction is increased beyond a level known as the reverse breakdown voltage $BV$, the reverse current begins to rise abruptly, as shown in Fig. 1.15. Although breakdown is not necessarily destructive, the current must be limited to a safe value; otherwise the device will self-destroy from the heat produced.

There are two basic processes that can cause junction breakdown. One is Zener breakdown; the other is avalanche breakdown. Zener breakdown is caused by excessive electric field strength in the depletion region. A strong electric field can provide enough energy to break electron-pair bonds. Thus, as the external voltage increases, the field strength increases, many charge carriers are generated, and the current rises very rapidly.

Another breakdown process is avalanche. This phenomenon generally occurs in wider depletion regions, where the field is not strong enough to produce Zener breakdown. Instead, free electrons accelerated by the field collide with electrons in covalent bonds. Upon collision, the covalent bonds are broken, and electron-hole pairs are generated. The freed electrons are accelerated by the field, resulting in more collisions, creating more charge carriers. The process quickly avalanches to produce the abrupt rise in current which is known as breakdown.

All diodes break down at some voltage, some as low as 2 or 3 V, others as high as 20,000 V or more. The voltage at which breakdown occurs can be controlled in the manufacturing process, since it depends on the width of the depletion region which in turn depends on the doping level.

## Problem Set 1.6

1.6.1. Is the junction breakdown caused by forward or reverse bias?
1.6.2. Is the voltage required to break down a PN junction always the same?
1.6.3. What is the difference between Zener and avalanche breakdown?
1.6.4. Is breakdown always destructive?

## REFERENCES

1. Brophy, J. J.: "Semiconductor Devices," McGraw-Hill Book Company, New York, 1964.
2. Cutler, P.: "Semiconductor Circuit Analysis," McGraw-Hill Book Company, New York, 1964.
3. Gibbons, J. F.: "Semiconductor Electronics," McGraw-Hill Book Company, New York, 1966.
4. Ristenbatt, M. P., and R. L. Riddle: "Transistor Physics and Circuits," Prentice-Hall, Inc., Englewood Cliffs, N.J., 1966.
5. Seidman, A. H.: Solid State Principles, *Electro-Technology*, December 1964, pp. 59–78.
6. Kendall, E. J. M.: "Transistors," Pergamon Press, New York, 1969.
7. Brazee, J. G.: "Semiconductor and Tube Electronics: An Introduction," Holt, Rinehart and Winston, Inc., New York, 1968.
8. Basic Theory and Application of Transistors, TM 11-690, Headquarters, Department of the Army, 1959.
9. "General Electric Transistor Manual," 7th ed., Syracuse, 1964.
10. Millman, J., and C. C. Halkias: "Electronic Devices and Circuits," McGraw-Hill Book Company, New York, 1967.
11. Harris, D. F., and P. N. Robson: "Vacuum and Solid State Electronics: An Introductory Course," The MacMillan Company, New York, 1963.
12. Phillips, A. B.: "Transistor Engineering and Introduction to Integrated Semiconductor Circuits," McGraw-Hill Book Company, New York, 1962.

# chapter 2

# the diode

It has been shown that a PN junction diode exhibits some interesting characteristics. We shall discuss the diode first as an ideal device; the practical limitations will be taken up later.

## 2.1 THE IDEAL DIODE

There is no such thing as an ideal or perfect diode. The existence of such a device is postulated only as an aid in learning to analyze circuits that use diodes.

An ideal diode may be described as a two-terminal device that conducts with zero resistance when forward-biased and appears as an infinite resistance when reverse-biased.

Figure 2.1a shows the schematic symbol for a diode. The terms *anode* and *cathode* are used to identify the terminals connected to the P and N regions, respectively. When forward-biased, the diode conducts, and current flows as shown in Fig. 2.1b. The direction shown is for conventional current flow. Since the ideal diode has zero resistance when forward-biased, it follows that the voltage is also zero. This information is displayed on the graph in Fig. 2.1d as a solid line along the positive vertical axis. The voltage corresponding to this vertical line is always zero, no matter how much current flows.

When reverse-biased, as in Fig. 2.1c, the ideal diode behaves as an infinite resistance, and hence no current flows, whatever the reverse voltage. This

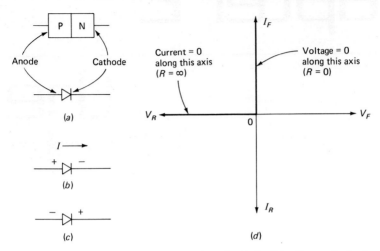

Fig. 2.1  (a) *Diode symbol, showing anode and cathode;* (b) *voltage polarity for forward bias;* (c) *voltage polarity for reverse bias;* (d) *voltage-current characteristic of ideal diode.*

relationship is displayed on the graph of Fig. 2.1d along the negative horizontal axis. Anywhere along this axis, the current is zero.

Figure 2.2 illustrates the analogy between an ideal diode and a switch. In Fig. 2.2a the diode is forward-biased, thus behaving as a closed switch. In this case, the full voltage $E$ is developed across $R$, resulting in current flow. There is no voltage across the closed switch.

In Fig. 2.2b the diode behaves as an open switch. Since there is no current, no voltage can be developed across $R$, and the full value of applied voltage $E$ is developed across the switch contacts.

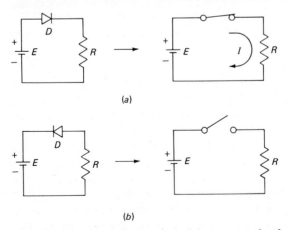

Fig. 2.2  (a) *Forward-biased diode behaves as a closed switch;* (b) *reverse-biased diode behaves as an open switch.*

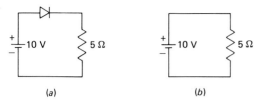

Fig. 2.3  Circuits for Example 2.1.

Because power is the product of voltage and current ($P = VI$), there can be no power if either the voltage or the current is zero. For the ideal diode, only two states are possible. When it is conducting, there is current, but no voltage; when it is not conducting, there is voltage, but no current. Hence, the ideal diode never dissipates power.

### Example 2.1

Calculate the current in the circuit of Fig. 2.3a.

### Solution

The diode is forward-biased, and hence it may be replaced with a short circuit (closed switch), as shown in Fig. 2.3b. We now have

$$I = \frac{E}{R} = \frac{10}{5} = 2 \text{ A}$$

### Example 2.2

Calculate the current in the circuit of Fig. 2.4a.

### Solution

The diode is reverse-biased, and hence it may be replaced with an open circuit (infinite resistance), as in Fig. 2.4b. No current can exist in this circuit; therefore $I = 0$.

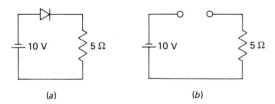

Fig. 2.4  Circuits for Example 2.2.

**Example 2.3**

Calculate the current in the circuit of Fig. 2.5a.

**Solution**

$D_1$ is reverse-biased, and hence there is no current through the branch involving $D_1$ and the 4 Ω resistor; $D_2$ is forward-biased, and it may be replaced with a short circuit. This is shown in Fig. 2.5b. The current is

$$I = \frac{E}{R} = \frac{10}{12 + 8} = \frac{10}{20} = 0.5 \text{ A}$$

**Example 2.4**

Determine the current waveform in the circuit of Fig. 2.6a.

**Solution**

The generator produces a sinusoidally alternating voltage $e$ whose peak value is 100 V. During the portion of the cycle that the top end of the generator is positive, the diode is forward-biased. Current flows during this time. Since the voltage is sinusoidal, so is the current. The peak current must be

$$I_m = \frac{E_m}{R} = \frac{100}{100} = 1 \text{ A}$$

During the next half-cycle, the polarity of the generator voltage is reversed—that is, the top end of the generator is negative, which reverse-biases the diode. Since a reverse-biased diode acts as an infinite resistance, there is no current. A complete time graph of the current is shown in Fig. 2.6c. This graph indicates that current is allowed to flow in one direction only; it is blocked in the other. This process is called rectification. Diodes, in fact, are often called rectifiers, because they allow current flow in only one direction.

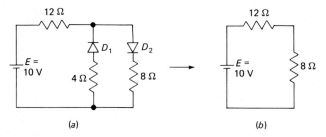

**Fig. 2.5** Circuits for Example 2.3.

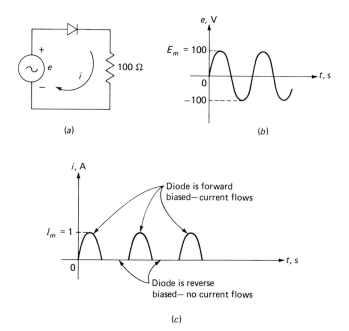

*Fig. 2.6* (a) *Circuit for Example 2.4;* (b) *applied voltage for the circuit of* (a); (c) *resulting current.*

## Problem Set 2.1

2.1.1. Draw the circuit schematic for a diode; label its cathode and anode. Indicate the polarity of external voltage required to
   a. Forward-bias the diode
   b. Reverse-bias the diode

2.1.2. For an ideal diode which is forward-biased, what are the possible values of voltage, resistance, current?

2.1.3. Repeat Prob. 2.1.2 for a reverse-biased ideal diode.

2.1.4. Why does an ideal diode never dissipate power?

2.1.5. Indicate, for each circuit diagram in Fig. 2.7, whether the diode conducts (ON) or does not conduct (OFF).

2.1.6. Calculate all currents in each circuit of Fig. 2.7.

2.1.7. For the circuit of Fig. 2.6a, plot, as a function of time, a graph of
   a. Voltage across the 100-$\Omega$ resistance
   b. Current through diode
   c. Voltage across diode
   d. Power dissipated by the diode

2.1.8. In Prob. 2.1.7 what must the instantaneous values of (a) and (c) add up to?

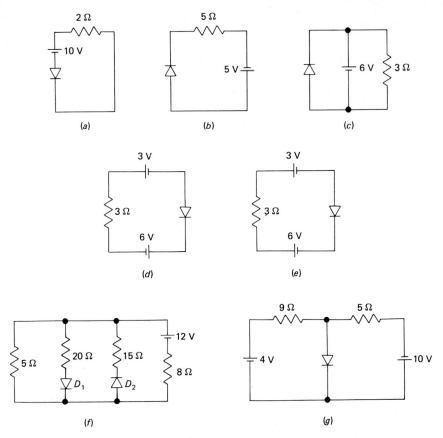

Fig. 2.7  Circuits for Probs. 2.1.5 and 2.1.6.

## 2.2  THE REAL DIODE: LARGE-SIGNAL OPERATION

Real diodes are different from the ideal device just discussed. Often it is possible to neglect these differences and treat all diodes as if they were ideal. To decide when this simplification is justified, however, one must understand the practical limitations involved.

These limitations can be explained with the help of voltage-current characteristics. These characteristics, for both forward and reverse bias, are shown in Fig. 2.8a. Note the change in scale between forward and reverse regions. In the forward direction relatively small voltages produce relatively large currents, while in the reverse direction the current is very small, even with large voltages. It is therefore necessary to use different scales to display both forward and reverse characteristics on the same graph.

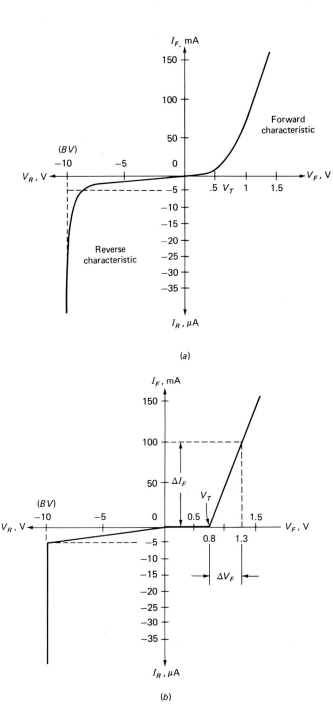

Fig. 2.8 (a) Voltage-current characteristic of real diode; (b) straight-line approximation to (a).

In the forward region, the threshold voltage $V_T$ must be overcome before significant conduction takes place. Even after $V_F$ exceeds $V_T$, however, current does not flow completely unopposed. This implies the existence of some resistance, which we will refer to as the forward diode resistance. In the reverse region some current also flows, although it is usually small. The phenomenon of breakdown, which was introduced earlier, must also be considered.

These departures from the ideal make it necessary to develop some specialized analysis techniques. Due to the nonlinearity of the characteristic, the analysis must often be of an approximate nature, since an exact analysis might involve complexities not justified by the ultimate application.

One such approach is the straight-line method, in which the actual characteristic is approximated with the straight line segments, as shown in Fig. 2.8b. Such an approximation is bound to be inexact near the knees of the curve ($V_T$ or $BV$) but yields fairly reliable results when "large signals" are involved—that is, voltages and currents whose magnitudes are substantially greater than $V_T$ or $BV$.

Once a diode's characteristic has been approximated with straight lines, some important parameters can be quickly obtained. In the forward region, for example, the threshold voltage $V_T$ can be read directly from the graph; in this case it is 0.8 V. The large-signal forward resistance $r_F$ is obtained by determining the voltage increment ($\Delta V_F$) required to produce a given current increment ($\Delta I_F$), as follows:

$$r_F = \frac{\Delta V_F}{\Delta I_F} \tag{2.1}$$

The magnitudes of current and voltage increments selected to compute $r_F$ are immaterial with a straight-line approximation. If a 100-mA current increment is selected in Fig. 2.8b, we can compute $r_F$ as follows:

$$r_F = \frac{1.3 - 0.8}{(100 - 0)(10^{-3})} = 5 \, \Omega$$

Both $V_T$ and $r_F$ can be used to formulate a model for the real diode. By *model* we mean a circuit diagram whose function is to describe, within a given accuracy, the performance of a real device. To construct such a model, we begin with the ideal device and add various circuit components to account for actual behavior. Figure 2.9a shows the ideal diode. To simulate the threshold voltage, we add a fictitious DC voltage source in Fig. 2.9b. To show that this actually takes into account the threshold voltage effect, assume an external voltage between $X$ and $Y$ of such a polarity as to forward-bias the diode. Thus $X$ is positive with respect to $Y$, as indicated in the figure. The voltage across the ideal diode is the difference between $V_{XY}$ (the externally applied voltage) and $V_T$. If $V_{XY} < V_T$, the anode is negative with respect to the cathode, reverse-biasing the ideal diode and resulting in zero current. If $V_{XY} > V_T$, the anode is positive with respect to the cathode, and the diode conducts.

The model of Fig. 2.9b does not include the diode's forward resistance $r_F$; this is added in Fig. 2.9c. The combination of the elements $V_T$, $r_F$, and the ideal diode allows us to approximately predict the behavior of the real diode. Note that this model is only valid when the diode is forward-biased.

When the real diode is reverse-biased, we are concerned with reverse current and breakdown voltage. Reverse current is heavily dependent on temperature, but contains a voltage-dependent component as well. Since the diode characteristic is for a fixed temperature, any model we develop is valid only at the temperature for which the information is supplied.

A reverse-biased real diode is shown in Fig. 2.10a. The polarity of $V_{XY}$ indicates that $X$ is negative with respect to $Y$. The straight-line approximation of Fig. 2.8b indicates that as the reverse voltage changes from, say, 0 to $-10$ V, the corresponding change in reverse current is from 0 to $-5$ μA. This change of reverse current with reverse voltage can be simulated with a reverse resistance $r_R$, as follows:

$$r_R = \frac{\Delta V_R}{\Delta I_R} \tag{2.2}$$

$$r_R = \frac{10}{5 \times 10^{-6}} = 2 \text{ M}\Omega$$

Hence, an adequate model for a real diode which has a reverse bias less than the breakdown voltage is a simple resistance, 2 MΩ in our case, as shown in Fig. 2.10b.

When the reverse voltage exceeds the breakdown level, small increases

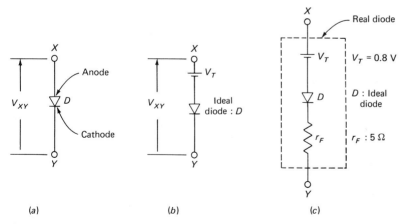

Fig. 2.9 (a) Ideal diode; (b) ideal diode plus threshold voltage; (c) ideal diode with threshold voltage and forward resistance simulates the real diode when X is positive with respect to Y.

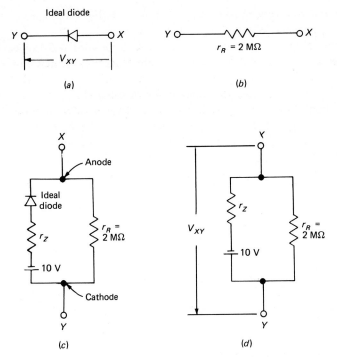

*Fig. 2.10* (a) *Reverse-biased ideal diode (Y positive with respect to X);* (b) *real diode model when reverse-biased with less than the breakdown voltage;* (c) *real diode model valid for any reverse voltage;* (d) *real diode model valid only in breakdown region.*

in voltage result in large current increases. A diode in the breakdown stage, therefore, exhibits a low resistance. This can be simulated by adding a few more components to our model, as in Fig. 2.10c. Assume that the reverse bias is less than the breakdown voltage, 10 V in this case. This means that the 2-MΩ resistance is the only element in the circuit, since the 10-V source prevents the ideal diode from conducting until the anode-to-cathode voltage exceeds 10 V.

Once the breakdown voltage is exceeded, the model of Fig. 2.10c simplifies to that of Fig. 2.10d, which is valid only in the breakdown region. The Zener resistance $r_Z$ is the only resistance to limit diode current and may be determined as follows:

$$r_Z = \frac{\Delta V_Z}{\Delta I_Z} \tag{2.3}$$

where $\Delta V_Z$ and $\Delta I_Z$ are voltage and current increments in the breakdown region. The graphic technique for determining $r_Z$ is illustrated in Fig. 2.11; $r_Z$ can be quite low, from a few ohms down to 0.01 Ω or less for specially designed units.

The real diode models for forward and reverse bias can be lumped to

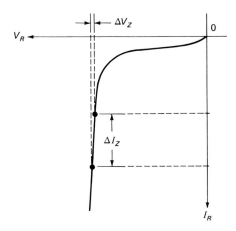

*Fig. 2.11  Technique for determining Zener resistance of diode.*

yield a composite model valid in all cases. This is shown in Fig. 2.12. At any given time the diode can be in only one of its three possible regions. If it is forward-biased, only the branch involving $V_T$, $D_1$, and $r_F$ is significant. Although $r_R$ is also in the model, it is large compared to $r_F$, and it can be neglected. When the real diode is reverse-biased (but not in breakdown), neither $D_1$ nor $D_2$ conducts, and hence the only significant branch is that involving $r_R$.

If the breakdown voltage is exceeded, $D_2$ conducts, and the branch involving $D_2$, $r_Z$, and $BV$ applies. The branch with $D_1$ is out of the circuit ($D_1$ cannot conduct), while $r_R$ is too large to be significant.

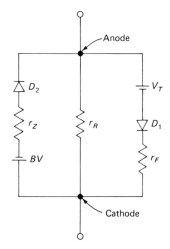

*Fig. 2.12  Real diode model, valid in all regions of operation.*

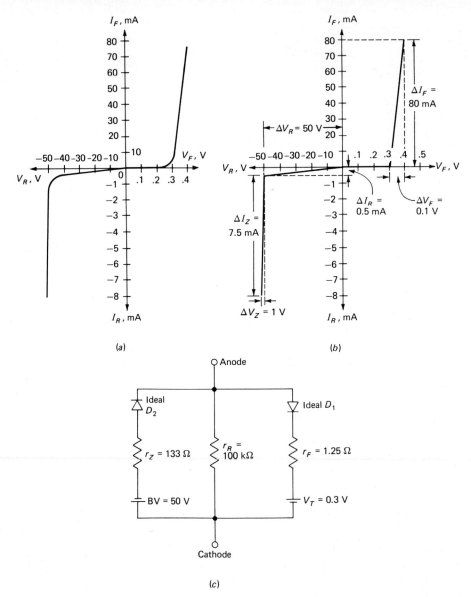

Fig. 2.13 (a) *Diode characteristic;* (b) *straight-line approximation to* (a); (c) *model for real diode.*

## Example 2.5

Using the diode characteristic of Fig. 2.13a, develop a model by straight-line approximation methods.

## Solution

The diode characteristic is first approximated by straight line segments, as shown in Fig. 2.13b. Both $V_T$ and $BV$ may be read directly from the curve:

$V_T = 0.3$ V
$BV = -50$ V

The forward resistance is the ratio of appropriate voltage to current increments, as shown in the diagram:

$$r_F = \frac{\Delta V_F}{\Delta I_F} \qquad (2.1)$$

$$r_F = \frac{0.1}{80 \times 10^{-3}} = 1.25 \; \Omega$$

Similarly, $r_R$ can be found as

$$r_R = \frac{\Delta V_R}{\Delta I_R} \qquad (2.2)$$

$$r_R = \frac{50}{0.5 \times 10^{-3}} = 100 \; \text{k}\Omega$$

In the breakdown region

$$r_Z = \frac{\Delta V_Z}{\Delta I_Z} \qquad (2.3)$$

$$r_Z = \frac{1}{7.5 \times 10^{-3}} = 133$$

These results are used to construct the model shown in Fig. 2.13c.

### Example 2.6

Using the models developed in Example 2.5, find the current in the circuit of Fig. 2.14a.

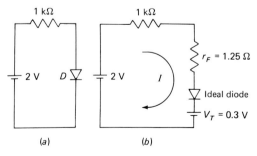

Fig. 2.14  Circuits for Example 2.6.

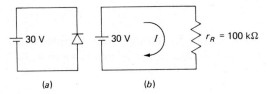

Fig. 2.15 Circuits for Example 2.7.

**Solution**

The diode is forward-biased, and hence only the branch involving $V_T$ and $r_F$ applies. The complete equivalent circuit, including the diode model, is shown in Fig. 2.14b. Here the ideal diode clearly conducts, and the current is

$$I = \frac{2 - 0.3}{1,001.25} = 1.7 \text{ mA}$$

**Example 2.7**

Using the models of Example 2.5, find the current in the circuit of Fig. 2.15a.

**Solution**

The diode is reverse-biased, but the voltage is not sufficient to cause breakdown. The diode's reverse resistance of 100 kΩ is an adequate model, as shown in Fig. 2.15b.

$$I = \frac{30}{100 \times 10^3} = 0.3 \text{ mA}$$

**Example 2.8**

Using an appropriate model from Example 2.5, determine the current in the circuit of Fig. 2.16a.

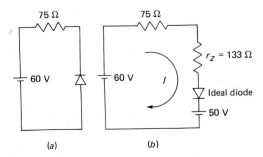

Fig. 2.16 Circuits for Example 2.8.

## Solution

The applied voltage is large enough to cause breakdown, and the equivalent circuit of Fig. 2.16b applies.

$$I = \frac{60 - 50}{75 + 133} = 48 \text{ mA}$$

### Problem Set 2.2

2.2.1. For the diode characteristic of Fig. 2.17, using straight-line approximations, estimate

    a. Threshold voltage
    b. Reverse resistance
    c. Breakdown voltage
    d. Forward resistance
    e. Resistance in breakdown region

2.2.2. Using the information from Prob. 2.2.1, construct a model valid for

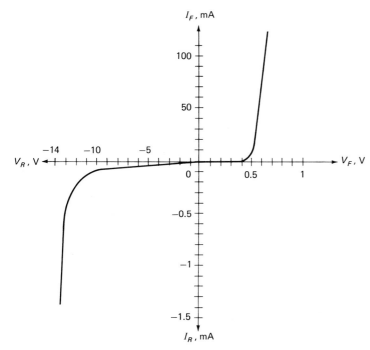

Fig. 2.17 Characteristic for Prob. 2.2.1.

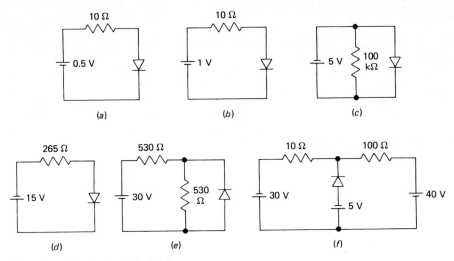

Fig. 2.18  Circuits for Prob. 2.2.3.

    a. Forward region only
    b. Reverse region (excluding breakdown) only
    c. Breakdown region only
    d. All three regions

2.2.3.  Estimate the diode current in each of the circuits of Fig. 2.18, using the models of Prob. 2.2.2. *Hint:* In the circuits of Figs. 2.18e and f replace the circuit seen by each diode with a Thevenin equivalent circuit.[1]

## 2.3 THE REAL DIODE: SMALL-SIGNAL OPERATION

An examination of the diode characteristic and its straight-line approximation (Fig. 2.8) reveals that the straight-line approximation is quite good in the large $V\text{-}I$ areas (both forward and reverse), but somewhat inaccurate if small signals are used, especially in the regions around the knees of the curve ($BV$ and $V_T$), where the discrepancy between actual and approximate characteristics is greatest.

    To analyze small-signal operation, consider the circuit of Fig. 2.19a. A forward diode characteristic is shown in Fig. 2.19b. The polarity of the 1-V DC source forward-biases the diode. The resulting current, however, causes some voltage drop across the 10-Ω resistance, and therefore the diode has less than 1 V across its terminals. This means that operation is in the region around the knee of the forward characteristic, where large-signal models (using the straight-line approximation) are inaccurate. In this case, it is convenient to use a graphic

---

[1] Thevenin's theorem is reviewed in Appendix 7.

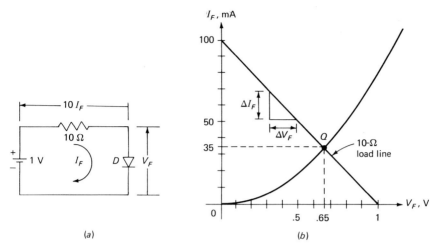

Fig. 2.19  (a) *Circuit to illustrate small-signal operation;* (b) *diode characteristic for* (a).

technique to determine the actual voltage and current for the diode. Writing Kirchhoff's voltage law around the loop of our circuit yields

$$1 = 10 I_F + V_F$$

The above equation involves two unknowns, and hence it cannot be solved unless a relationship between the unknowns is established. Such a relationship is provided by the characteristic of Fig. 2.19b which displays $V_F$ versus $I_F$ for our diode.

Note that the above equation describes a straight line, and hence it can be plotted by solving for and joining the horizontal and vertical intercepts. The solution of the equation is the point where the straight line crosses the diode characteristic. To locate the two intercepts we set each of the variables $V_F$ and $I_F$ in turn equal to zero and solve for the other:

a.  Set $V_F = 0$:  $1 = 10 I_F$

$$I_F = \frac{1}{10} = 100 \text{ mA}$$

b.  Set $I_F = 0$:  $1 = 0 + V_F$
$$V_F = 1 \text{ V}$$

The straight line obtained by joining these two intercepts crosses the diode characteristic at $V_F = 0.65$ V and $I_F = 35$ mA. The point ($V_F = 0.65$ V, $I_F = 35$ mA) is called $Q$ point (quiescent point) to denote the DC operating point for the diode. The straight line used to locate the $Q$ point is called a *load line*; in our case it is a 10-Ω load line. Note that a right triangle drawn along the load line, as in Fig. 2.19b, has sides $\Delta V_F$ and $\Delta I_F$ in the ratio of 10 Ω.

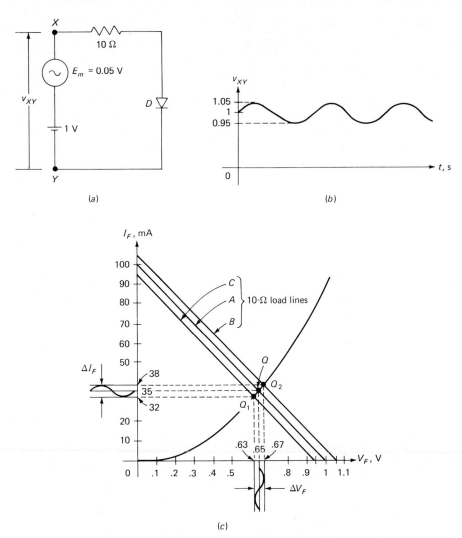

**Fig. 2.20** (a) Diode circuit with AC signal superimposed on DC level; (b) graph of $V_{XY}$ as a function of time; (c) graphic construction for (a).

The load line method can also be used to solve circuits in which the signals are large; it is usually preferred, however, to use large-signal models whenever they apply, since no graphic construction is then necessary.

The circuit of Fig. 2.20 will now be discussed. It is similar to the circuit in Fig. 2.19a except for the addition of a sinusoidal voltage generator in series with the 1-V DC source. The AC generator produces an alternating voltage whose peak value is 0.05 V. Hence, when the top end of the generator is at its maximum positive swing, the voltage between $X$ and $Y$ is $0.05 + 1 = 1.05$ V, while during the opposite part of the cycle, $v_{XY}$ swings down to $-0.05 + 1 = 0.95$ V. A graph of $v_{XY}$ is shown in Fig. 2.20b.

At the instant that the AC generator reverses polarity, the value of the AC voltage is 0, and hence $v_{XY} = 1$ V. At this time load line $A$ applies, as shown in Fig. 2.20c, which is exactly the situation we had for the circuit of Fig. 2.19. When $v_{XY}$ reaches its most positive value, however, load line $B$ applies, since the horizontal axis intercept is 1.05 V. Similarly, when $v_{XY}$ drops to 0.95 V, load line $C$ applies. The load line moves sinusoidally between the extremes of $C$ and $B$; it is always a 10-Ω load line, though, because the 10-Ω resistor, which determines the slope of the load line, is constant.

Following the generator voltage variations, the diode voltage and current also vary sinusoidally from 0.63 to 0.67 V and from 32 to 38 mA, respectively. The peak diode voltage is therefore 0.02 V, and the peak current is 3 mA.

There are a number of applications in which only the changing (AC) components of voltage and current are of interest. For example, the 1-V DC source in our circuit is required to set the $Q$ point at 0.65 V and 35 mA. Once the DC levels have been established, it is possible to consider only the AC components.

Shown in Fig. 2.21a is the AC equivalent for the circuit of Fig. 2.20. Since only the AC response of the circuit is being considered, the DC source is

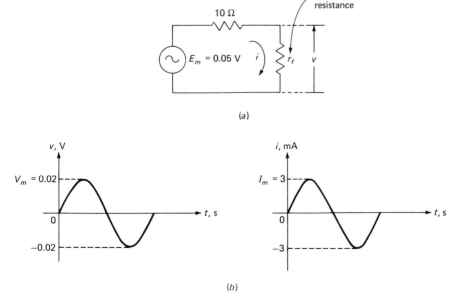

Fig. 2.21 (a) AC equivalent for the circuit of Fig. 2.20a; (b) diode voltage and current waveforms for (a).

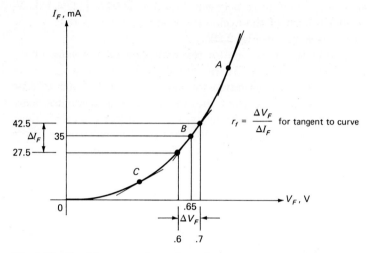

Fig. 2.22 Graphic construction to determine $r_f$.

not shown. The resistance $r_f$ represents the dynamic, or AC resistance, of the diode. It is obtained by taking the ratio of $\Delta V_F/\Delta I_F$ along a tangent to the curve at the $Q$ point, as shown in the graphic construction of Fig. 2.22:

$$r_f = \left.\frac{\Delta V_F}{\Delta I_F}\right|_Q \tag{2.4}$$

where $Q$ is the specific operating point. Note here that a 0.05-V change on each side of operating point $B$ results in a total current change of 15 mA, as follows:

$$\left.r_f\right|_{\text{at }B} = \left.\frac{\Delta V_F}{\Delta I_F}\right|_{\text{at }B} \tag{2.4}$$

$$\left.r_f\right|_{\text{at }B} = \frac{0.7 - 0.6}{(42.5 - 27.5)(10)} = 6.66\ \Omega$$

The difference between $r_f$ and $r_F$ is that the former is a small-signal or AC resistance valid for small changes about the operating point, while the latter is obtained using a straight-line approximation to the curve and hence is valid only for relatively large voltage and current levels.

The peak diode current and voltage in the circuit of Fig. 2.21a are

$$I_m = \frac{E_m}{10 + r_f} = \frac{0.05}{16.66} = 3\ \text{mA}$$

and $V_m = I_m r_f = (3 \times 10^{-3})(6.66) = 0.02$ V, as determined before. Both $v$ and $i$ are plotted in Fig. 2.21b.

The characteristic of Fig. 2.22 reveals that $r_f$ is relatively high near point $C$, while quite low near point $A$. Beyond $A$, in fact, the characteristic becomes essentially a straight line, and large-signal models apply.

Small-signal reverse and breakdown resistances ($r_r$ and $r_z$) may be defined in a manner similar to $r_f$. In general, though, provided operation is in the linear portion of the curve (away from the $BV$ point), the small-signal resistances ($r_r$ and $r_z$) are essentially the same as their large-signal counterparts ($r_R$ and $r_Z$).

### Example 2.9

Determine the $Q$ point and the AC voltage across the diode in the circuit of Fig. 2.23a, using the curve of Fig. 2.23b.

### Solution

The diode is forward-biased. The $Q$ point is located by constructing a 20-Ω load line on the characteristic, which yields

$$V_F = 0.14 \text{ V} \quad \text{and} \quad I_F = 3 \text{ mA}$$

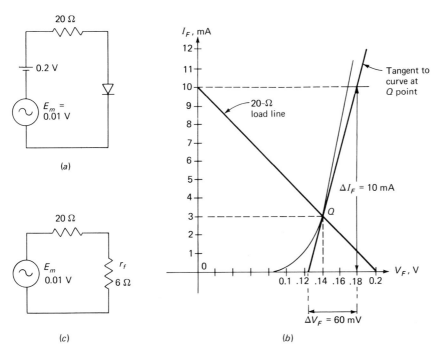

Fig. 2.23 (a) Circuit for Example 2.9; (b) graphic construction on diode forward characteristic; (c) AC equivalent for the circuit in (a).

**42** Solid State Electronic Circuits

By drawing a tangent to the curve at the $Q$ point, $\Delta V_F$ and $\Delta I_F$ may be taken as 60 mV and 10 mA, respectively; therefore $r_f = 60/10 = 6\ \Omega$. Hence, as far as the AC signal is concerned, we can work with equivalent circuit shown in Fig. 2.23c. The AC voltage across the diode is obtained by voltage division:

Peak voltage across diode: $\quad V_m = 0.01\,\dfrac{6}{26} = 2.31\text{ mV}$

### Example 2.10

Determine the $Q$ point and AC diode voltage in the circuit of Fig. 2.24a using the characteristic of Fig. 2.24b.

### Solution

Here the diode is reverse-biased. Since the total applied voltage is 8 V, we can safely assume that the diode is not in breakdown. The actual operating point is found by drawing a load line between $(-8\text{ V}, 0\ \mu\text{A})$ and $(0\text{ V}, -8\ \mu\text{A})$ in the reverse region. This yields a $Q$ point of around $-4$ V and $-4\ \mu$A.

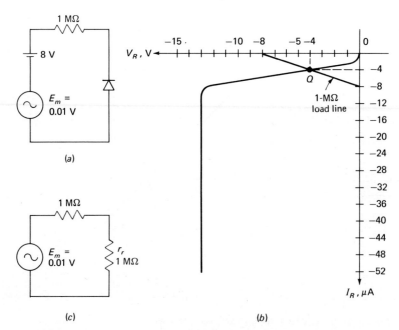

Fig. 2.24 (a) Circuit for Example 2.10; (b) graphic construction on diode reverse characteristic; (c) AC equivalent for the circuit in (a).

The AC voltage across the diode is found using the equivalent circuit of Fig. 2.24c. The small-signal reverse resistance $r_r = 1$ M$\Omega$, and hence the diode develops one-half of the AC voltage: $0.01/2 = 5$ mV peak.

## Example 2.11

Determine the $Q$ point and AC diode voltage in the circuit of Fig. 2.25 using the characteristic of Fig. 2.25b.

### Solution

The load line extends from $-20$ V on the reverse voltage axis to a point on the reverse current axis given by $-20/200 = -100$ mA. The $Q$ point is at $-12.4$ V

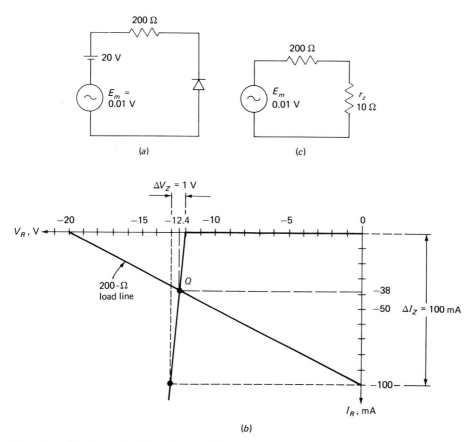

Fig. 2.25  (a) *Circuit for Example 2.11;* (b) *graphic construction on diode reverse characteristic;* (c) *AC equivalent for the circuit in* (a).

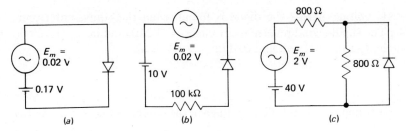

Fig. 2.26  Circuits for Problem Set 2.3.

and $-38$ mA. The dynamic (small-signal) breakdown resistance is the ratio of a small $\Delta V_Z$ to the corresponding $\Delta I_Z$ about the $Q$ point. Since the curve is fairly linear, large-signal increments can be used; these are more easily read from the graph:

$$r_z \simeq r_Z = \frac{\Delta V_Z}{\Delta I_Z} \tag{2.3}$$

$$r_z \simeq \frac{1}{100 \times 10^{-3}} = 10 \ \Omega$$

The equivalent circuit of Fig. 2.25c is used to calculate the AC voltage across the diode:

$$V_m = 0.01 \frac{10}{210} = 475 \ \mu V$$

## Problem Set 2.3

2.3.1. For each of the circuits in Fig. 2.26, determine the $Q$ point and the AC voltage and current for the diode. Use the appropriate characteristic from Figs. 2.23b, 2.24b, or 2.25b.

## REFERENCES

1. Cutler, P.: "Semiconductor Circuit Analysis," McGraw-Hill Book Company, New York, 1964.
2. Brazee, J. G.: "Semiconductor and Tube Electronics: An Introduction," Holt, Rinehart and Winston, Inc., New York, 1968.
3. Malvino, A. P.: "Transistor Circuit Approximations," McGraw-Hill Book Company, New York, 1968.
4. Mulvey, J.: "Semiconductor Device Measurements," Tektronix, Inc., Beaverton, Ore., 1968.
5. "Silicon Rectifier Handbook," Motorola, Inc., Phoenix, 1966.

6. Phillips, A. B.: "Transistor Engineering and Introduction to Integrated Semiconductor Circuits," McGraw-Hill Book Company, New York, 1962.
7. Millman, J., and C. C. Halkias: "Electronic Devices and Circuits," McGraw-Hill Book Company, New York, 1967.
8. Angelo, E. J.: "Electronic Circuits," McGraw-Hill Book Company, New York, 1964.

# chapter 3

# circuit applications of diodes

## 3.1 INTRODUCTION

Some common applications of diodes are discussed in this chapter. This is intended to achieve two basic purposes: familiarization with some important circuits and development of analytical skills required to treat new circuits later on.

## 3.2 POWER SUPPLIES

Electric energy is available in homes and industrial plants in the form of alternating voltage with a standard frequency of 60 Hz (in North America) and an effective (RMS) value of 115 V.

Figure 3.1 is a plot of the normal household voltage waveform. The instantaneous voltage $e$ varies sinusoidally with time $t$. From this graph we can obtain various parameters as follows:

Peak value: $E_m = 163$ V
Effective (RMS) value:[1] $E = 0.707\, E_m$ (3.1)
$E = 0.707\,(163) = 115$ V

[1] See Appendix 11.2, Eq. (A.45).

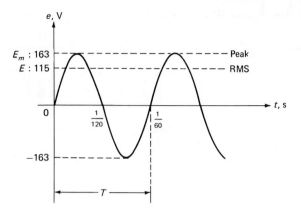

*Fig. 3.1   Plot of instantaneous voltage e versus time t.*

$$\text{Period:} \quad T = \frac{1}{60} \text{ s}$$

$$\text{Frequency:} \quad f = \frac{1}{T} \tag{3.2}$$

$$f = \frac{1}{1/60} = 60 \text{ Hz}[1]$$

Usually a DC voltage is required for the operation of most devices used in electronic equipment; for example, a transistor amplifier in a portable radio may require a 9-V DC power supply for proper operation. Other types of equipment may require higher or lower voltages, but the basic need for an unchanging voltage (as opposed to an alternating voltage) is almost universal in the field of electronics. For this reason we will discuss in some detail circuits whose function, very simply, is to change AC to DC. Such circuits are referred to as power supplies.

There are four functions that must be considered when dealing with power supply circuits. They are

a.  Voltage transformation
b.  Rectification
c.  Filtering
d.  Regulation

Voltage transformation is achieved through the use of transformers. It is a necessary function because the power supply must accept the standard 115-V AC and change it to the desired DC level. The transformer itself does not change

---

[1] The older unit *cycles per second* has been replaced with *hertz*, abbreviated Hz.

AC to DC; it merely steps AC voltage up or down, depending on the particular requirements.

Rectification is the conversion of AC voltage and alternating current (whose polarity keeps changing) to voltage and current of constant polarity. This is achieved by using diodes in various circuit configurations, since diodes allow current to flow in only one direction.

Filtering is a process of smoothing rectified voltage to remove level fluctuations (ripple) and provide a "smooth" DC voltage.

Regulation is a measure of the ability of a power supply to maintain an established DC voltage as the load varies. Voltage regulator circuits are usually designed to yield the best possible regulation for the particular application.

Each of the above processes is treated in the following sections.

### 3.2a  Half-wave Rectification

The simplified circuit of Fig. 3.2 can be used to illustrate both voltage transformation and rectification.

The generator $e_p$ represents the 115-V AC applied across the transformer primary; its time graph is shown in Fig. 3.3a. The peak primary voltage is $E_{pm} = 163$ V; the step-down action of the 10:1 transformer results in a secondary voltage $e_s$ whose peak value is $E_{sm} = {}^{163}\!/_{10} = 16.3$ V. The secondary voltage is identical to the primary voltage except for the level reduction by a factor of 10, as shown in Fig. 3.3b. The dots on primary and secondary windings indicate points in the transformer whose voltages are in phase. In this case, $e_p$ is in phase with $e_s$ (when $e_p$ rises, $e_s$ rises). It is, of course, possible to reverse one of the windings, in which case $e_p$ and $e_s$ are in opposite phase (when $e_p$ rises, $e_s$ drops, and vice versa).

When the dotted end of the transformer secondary is positive (the first half-cycle in this case), the diode conducts. Assuming an ideal diode (behaves as a short circuit), the voltage $v_L$ is identical to $e_s$. When the polarity of the AC changes, however, the diode cannot conduct; this results in no current and $v_L = 0$, as indicated by the graph of Fig. 3.3c.

The peak value of $v_L$ is $V_{LM} = 16.3$ V. $R_L$ is the load; it represents any-

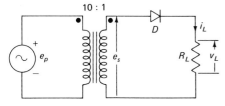

Fig. 3.2  Circuit to illustrate voltage transformation and half-wave rectification.

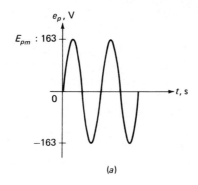

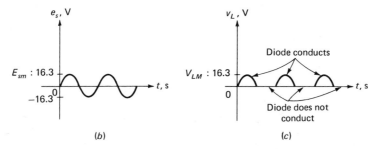

Fig. 3.3  Waveforms for the circuit of Fig. 3.2.

thing that uses electric power. In many applications, a resistor satisfactorily simulates the characteristics of whatever equipment is drawing power from the circuit. There are cases, however, when reactive loads must be considered. Motors, for example, behave, in part, inductively.

The function of the transformer, so far, is to alter the level of the AC voltage. In some cases a step-up, rather than step-down, transformer is used, depending on the ultimate application. The diode simply "rectifies" the flow of current, and hence the voltage across the load. By blocking each negative half-cycle, the diode prevents the load current $i_L$ from reversing its direction. The net result is a pulsating voltage $v_L$ across $R_L$ whose magnitude changes, but whose polarity is always the same (top end of $R_L$ can only be positive). The load voltage waveform is said to be half-wave rectified because only one half of the AC cycle is reproduced; similarly, the circuit is called a *half-wave rectifier*.

In power supply circuits such as the one of Fig. 3.2, care must be taken to ensure that the various components are not damaged by voltages and currents in excess of their maximum ratings.

Consider the load. The peak load voltage $V_{LM} = 16.3$ V; assuming $R_L = 100\ \Omega$, the peak load current is

$$I_{LM} = \frac{V_{LM}}{R_L} = \frac{16.3}{100} = 0.163 \text{ A} \tag{3.3}$$

The current through $R_L$ must follow the variations of $v_L$; thus the $i_L$ waveform is also half-wave rectified. Usually we are interested in the power absorbed by the load; for a resistive load, power can be computed as follows:

$$P_L = V_L I_L \tag{3.4}$$

where both $V_L$ and $I_L$ are RMS quantities. At this moment, however, we know only the peak values of $V_L$ and $I_L$; before power can be computed, these peak values must be converted to effective (RMS) values. For a full sine wave (current or voltage), the relationship between peak and RMS is given by Eq. (3.1), but this does not apply to half-wave rectified sine waves. Instead, the following applies:[1]

$$\text{RMS for half-wave sinusoid} = \frac{1}{2}\text{peak} \tag{3.5}$$

Using the relationships of Eqs. (3.5) and (3.4) we have

RMS load voltage: $V_L = \dfrac{V_{LM}}{2} = \dfrac{16.3}{2} = 8.15 \text{ V}$

RMS load current: $I_L = \dfrac{I_{LM}}{2} = \dfrac{0.163}{2} = 81.5 \text{ mA}$

Power absorbed by $R_L$: $P_L = V_L I_L = 8.15\,(81.5 \times 10^{-3})$
$P_L = 664 \text{ mW}$

$R_L$ must have a power rating greater than 664 mW if it is to safely dissipate the heat generated by the electric power.

We now come to the diode. You will recall that there is an upper limit of reverse voltage (often called PIV, for peak inverse voltage) that can be applied to any diode before voltage breakdown results. Our diode is forward-biased during each positive half-cycle and reverse-biased during the negative half-cycle. While the diode is reverse-biased, there is no voltage across $R_L$, and hence the transformer's full secondary voltage is across the diode. The maximum value that this voltage can reach is 16.3 V; thus a diode with a breakdown voltage rating greater than 16.3 V must be used.

Having assumed an ideal diode (zero forward resistance and infinite reverse resistance), we need not concern ourselves with diode power dissipation. In practice, however, a real diode is used and we must be more careful, especially if large currents are involved. In this case, the manufacturer's specifications of

---

[1] See Appendix 11.2, Eq. (A.47).

maximum allowable current for any particular diode must be observed. Diode specifications are generally given for peak and average current.

The transformer has maximum ratings too. Usually a kva (kilo-volt-ampere) rating is given, which is simply the product of RMS voltage and current handled by the transformer.

## Problem Set 3.1

3.1.1. For the circuit of Fig. 3.2, assume $E_{pm} = 163$ V, the transformer's turn ratio is 1.15:1, and the load is 10 Ω. Determine

    a. The transformer's secondary RMS voltage
    b. The reverse voltage rating for the diode
    c. The peak and RMS load current
    d. The power absorbed by the load

3.1.2. What effect does a real diode's reverse resistance (less than infinite Ω) have on the load voltage waveform of Fig. 3.3c? Under what conditions would this effect be most noticeable?

3.1.3. What effect does a real diode's forward voltage drop have on the load voltage waveform of Fig. 3.3c? Under what conditions would this effect be most noticeable?

We are now ready to discuss some basic properties of power supply waveforms. Since the objective of a power supply is to change AC to DC, a measure of performance is a comparison of the DC level to the AC component in the output.

As an example, consider the half-wave rectified voltage $v_L$ of Fig. 3.3c. The DC component of $v_L$ is the average value of this waveform. A mathematical derivation carried out in Appendix 11.1, Eq. (A.41), shows that for an unfiltered half-wave rectified waveform (such as that in Fig. 3.3c), the relationship between average (DC) and peak is

Average for half-wave sinusoid = 0.318 peak                     (3.6)

This means that the DC component of the load voltage in Fig. 3.3c is 0.318 (16.3) = 5.18 V. To calculate the AC component recall that the RMS value of a half-wave rectified sinusoid is one-half of the peak value. In our case, this is 8.15 V, which represents the RMS value of the total voltage (AC and DC components). Although beyond the scope of this textbook, it can be shown that when a signal is made up of more than one frequency component, the RMS value of the whole is related to the RMS values of each different frequency component by the following equation:[1]

RMS value of whole waveform: $\quad V = \sqrt{V_1^2 + V_2^2 + \cdots}\quad$ (3.7)

where $V_1$, $V_2$, etc., are the RMS values of each respective component. Our waveform (whose RMS value is 8.15 V) may be considered to be made up of an AC and a DC component. Hence

$$V_L = \sqrt{V_{L(AC)}^2 + V_{L(DC)}^2} \quad (3.7a)$$

where $V_{L(AC)}$ and $V_{L(DC)}$ represent the RMS values of AC and DC components, respectively. But we already know the value of the DC component (5.18 V), and since the RMS value of any DC component is equal to the DC component itself, the result is

$$8.15 = \sqrt{V_{L(AC)}^2 + 5.18^2} \quad (3.7a)$$
$$8.15^2 = V_{L(AC)}^2 + 26.8$$
$$V_{L(AC)}^2 = 66.4 - 26.8 = 39.6$$
$$V_{L(AC)} = \sqrt{39.6} = 6.3 \text{ V}$$

The 6.3-V AC component is called *ripple;* good power supplies are designed to minimize ripple.

Having computed the numerical values associated with AC and DC components as well as the whole waveform, we are now ready to use them in specifying certain figures of merit for power supply circuits. First of these is the ripple factor; this is the ratio of ripple to DC component:

$$RF = \frac{V_{L(AC)}}{V_{L(DC)}} \quad (3.8)$$

For our example, $RF = 6.3/5.18 = 1.21$, or 121 percent.

An additional measure of performance is the form factor—that is, the ratio of RMS load voltage to the DC component:

$$F = \frac{V_L}{V_{L(DC)}} \quad (3.9)$$

Ideally, if the load voltage contained only a DC component, $V_{L(AC)}$ would equal zero and the form factor $F = 1$. In our case,

$$F = \frac{8.15}{5.18} = 1.57$$

---

[1] See pp. 437–438 of Reference 7 at the end of this chapter.

54  Solid State Electronic Circuits

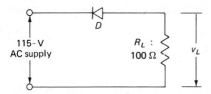

Fig. 3.4  Circuit for Problem Set 3.2.

## Problem Set 3.2

For the circuit of Fig. 3.4, assume an ideal diode and determine:

3.2.1.  Peak value of load voltage, load current, diode current, and diode voltage
3.2.2.  Average value of load voltage, load current, and diode current
3.2.3.  RMS value of load voltage, load current, and diode current
3.2.4.  AC component of load voltage
3.2.5.  Ripple factor
3.2.6.  Form factor
3.2.7.  Total load power
3.2.8.  Power dissipated by the diode

### 3.2b  Full-wave Rectification

The half-wave rectifier circuit produces too much ripple for some purposes. This is due to the utilization of only one half of the AC cycle, leaving a large gap between cycles.

Full-wave rectification is a process in which both halves of the AC cycle are utilized. This produces a waveform with much less ripple. We will discuss two different full-wave rectifier circuits, the center-tapped transformer type and the bridge type.

Consider the circuit of Fig. 3.5. The primary voltage is the 115-V supply. Assume a step-down transformer with a total RMS secondary voltage (from $B$ to

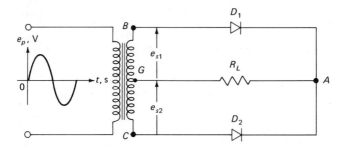

Fig. 3.5  Full-wave rectifier circuit.

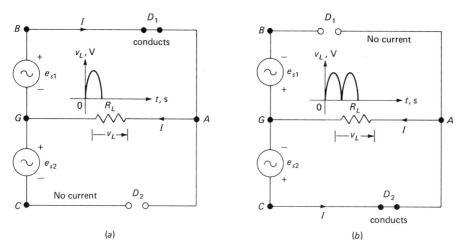

Fig. 3.6  Equivalent circuits for the full-wave rectifier; (a) first half-cycle; (b) second half-cycle.

$C$) of 40 V. This means that each half of the secondary winding develops 20 V. The peak value, of course, must be $20\sqrt{2} = 28.8$ V. To simplify the analysis, let us replace the transformer secondary with two fictitious generators, as shown in Fig. 3.6.

During the first half of each cycle, $B$ is positive with respect to $G$; this forces $D_1$ to conduct, as shown in Fig. 3.6a. At the same time, $C$ is negative with respect to $G$, which prevents $D_2$ from conducting. Hence during this first half of each cycle, $D_1$ conducts while $D_2$ stays off. Current flows only in the circuit with $e_{s1}$, $D_1$, and $R_L$; the direction of current makes $A$ positive with respect to $G$. The resultant voltage $V_L$ across $R_L$ is also shown.

During the second half of each cycle, $B$ is negative with respect to $G$; this keeps $D_1$ off, and hence no current flows in the upper circuit. At the same time, though, $C$ is positive with respect to $G$, which turns $D_2$ on, allowing current to flow, as shown in Fig. 3.6b. Note that the direction of current through $R_L$ is the same in both cases, and hence $v_L$ does not change polarity; that is, point $A$ is always positive with respect to $G$.

The process just described continues yielding a series of positive-going half sine waves, as shown in Fig. 3.7.

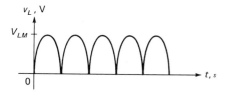

Fig. 3.7  Full-wave rectified voltage.

In comparing it with the half-wave rectifier, note that the full-wave rectifier requires two diodes instead of one; a center-tapped transformer is also required. The reverse voltage rating for each diode must be considered more closely. You will recall that, with the half-wave circuit, the diode was required to have a reverse voltage rating equal to the peak load voltage; for the example of Fig. 3.2 this was 16.3 V. To achieve the same 16.3-V peak across the load in the circuit of Fig. 3.6, however, both $e_{s1}$ and $e_{s2}$ must have peak values of 16.3 V. The maximum instantaneous voltage across each diode occurs halfway across each half-cycle (every 90°) when $e_{s1} = e_{s2} = V_{LM} = 16.3$ V. At that instant the voltage across the nonconducting diode is the sum of $V_{LM}$ and $e_{s2}$ (for $D_2$ during each first half-cycle) or $V_{LM}$ and $e_{s1}$ (for $D_1$ during each second half-cycle). Thus, for either diode, a voltage rating of $16.3 + 16.3 = 32.6$ V is necessary.

It can therefore be stated that for a full-wave circuit of the type in Figs. 3.5 and 3.6, each diode must have a reverse voltage rating equal to at least twice the peak load voltage.

### Example 3.1

The following data apply to the circuit of Fig. 3.5: AC voltage input to transformer primary is 115 V; transformer secondary voltage with 115 V across the primary is 50 V between $B$ and $C$; $R_L = 25\ \Omega$. Determine

a. Peak, DC component, RMS, and AC component of load voltage
b. Peak, DC component, RMS, and AC component of load current

### Solution

a. Since the transformer develops 50 V between $B$ and $C$, there must be 25 V across each half of the secondary winding. The secondary voltage is also sinusoidal, and hence the peak value across each half of the secondary is $25\sqrt{2} = 35.4$ V. Assuming ideal diodes, the rectified voltage across $R_L$ also has a peak value $V_{LM}$ of 35.4 V.

The average (DC) value of a full-wave rectified sinusoid is shown in Appendix 11.1, Eq. (A.42), to be

$$\text{Average for full-wave sinusoid} = 0.636 \text{ peak} \tag{3.10}$$

The DC load voltage is therefore

$$V_{L(\text{DC})} = 0.636 V_{LM} = 0.636(35.4) = 22.5 \text{ V}$$

The RMS value of a full-wave rectified sinusoid is shown in Appendix 11.2, Eq. (A.46), to be

$$\text{RMS for full-wave sinusoid} = 0.707 \text{ peak} \tag{3.1}$$

Therefore,

$$V_L = 0.707 V_{LM} = 0.707(35.4) = 25 \text{ V}$$

The load voltage waveform is shown in Fig. 3.8a. The AC component of load voltage is

$$V_{L(AC)} = \sqrt{V_L{}^2 - V_{L(DC)}{}^2} \qquad (3.7a)$$
$$V_{L(AC)} = \sqrt{625 - 507} = 10.9 \text{ V}$$

b. The peak load current is

$$I_{LM} = \frac{V_{LM}}{R_L} = \frac{35.4}{25} = 1.41 \text{ A} \qquad (3.3)$$

The DC component is

$$I_{L(DC)} = 0.636 I_{LM} \qquad (3.10)$$
$$I_{L(DC)} = 0.636(1.41) = 0.897 \text{ A}$$

The RMS load current is

$$I_L = 0.707 I_{LM} \qquad (3.1)$$
$$I_L = 0.707(1.41) = 1 \text{ A}$$

The load current waveform is shown in Fig. 3.8b. By direct analogy with Eq. (3.7a), the AC component of load current is

$$I_{L(AC)} = \sqrt{I_L{}^2 - I_{L(DC)}{}^2} \qquad (3.7b)$$
$$I_{L(AC)} = \sqrt{1 - 0.897^2} = 0.441 \text{ A}$$

## Example 3.2

Using the data of Example 3.1, determine

a. The ripple factor
b. The form factor

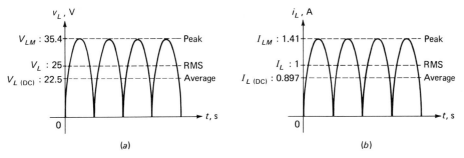

Fig. 3.8  Full-wave rectified load voltage and current waveforms.

c. Total power to the load
d. Peak diode current
e. Average diode current
f. Reverse voltage rating for each diode

**Solution**

a. The ripple factor is

$$RF = \frac{V_{L(AC)}}{V_{L(DC)}} \qquad (3.8)$$

$$RF = \frac{10.9}{22.5} = 0.487 \text{ or } 48.7 \text{ percent}$$

It is interesting to compare $RF$ for full-wave (48.7 percent) and half-wave (121 percent) circuits. The lower ripple factor in full-wave circuits is a direct consequence of the utilization of both half-cycles.

b. The form factor is

$$F = \frac{V_L}{V_{L(DC)}} \qquad (3.9)$$

$$F = \frac{25}{22.5} = 1.11$$

which is markedly better than the value of 1.57 for the half-wave circuit.

c. Total power to load is

$$P_L = V_L I_L \qquad (3.4)$$
$$P_L = 25(1) = 25 \text{ W}$$

d. The peak diode current is the same as the peak load current, that is, 1.41 A.

e. To calculate the average diode current $I_{D(AV)}$, we must bear in mind that each diode conducts for only one half-cycle. Hence,

$$I_{D(AV)} = 0.318(1.41) = 0.448 \text{ A}$$

f. Each diode must have a reverse voltage rating equal to twice the peak load voltage: $2(35.4) = 70.8$ V.

Another widely used full-wave rectifier is the bridge circuit shown in Fig. 3.9a. It consists of four diodes in a "bridge" configuration.

When the input causes $A$ to be positive with respect to $D$, diodes $D_2$ and $D_3$ conduct while $D_1$ and $D_4$ do not. The resulting current $i_L$ develops the load voltage $v_L$, as shown in Fig. 3.9b.

When the input polarity is reversed, the role of each diode pair also

Circuit Applications of Diodes 59

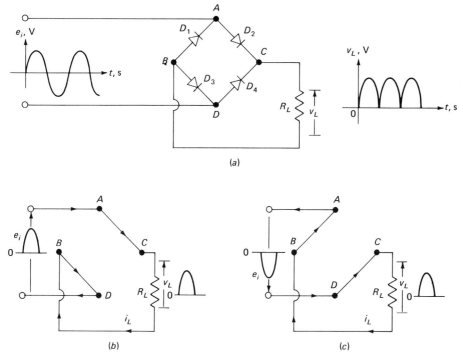

Fig. 3.9  (a) Bridge rectifier circuit; (b) equivalent circuit with positive-going input; (c) equivalent circuit with negative-going input.

reverses. The current $i_L$, however, develops a load voltage whose polarity is the same as during the previous half-cycle. Thus, regardless of the input voltage polarity, the load voltage is always positive-going. This is full-wave rectification, the same as that produced by the center-tapped transformer circuit.

Although the bridge circuit of Fig. 3.9a does not include a transformer, certain problems may arise unless a transformer is used. Suppose that the bottom end of $R_L$, for safety reasons, must be grounded. This means that $B$ (the junction of $D_1$ and $D_3$) is grounded. Now, if $e_i$ is the line voltage, also grounded at one end, the situation shown in Fig. 3.10a results, in which $D_3$ is continuously shorted out. Obviously, the circuit cannot function properly; in fact, when the input is negative and $D_1$ conducts, a direct short exists across the line. A technique for getting around this problem is to use an isolation transformer. The turns ratio of 1:1 provides no voltage-level change, but the lack of a direct connection between primary and secondary circuits "isolates" the two grounds, as shown in Fig. 3.10b. Naturally, a turns ratio other than 1:1 may be used if voltage transformation is also desired.

What about the reverse voltage on our diodes? An examination of Fig. 3.9b

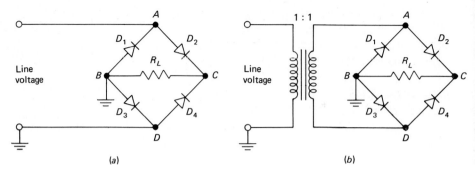

*Fig. 3.10  Use of isolation transformer in bridge rectifier;* (a) $D_3$ *is short-circuited if one end of the line voltage and* $R_L$ *are both grounded;* (b) *isolation transformer eliminates grounding problem.*

or $c$ indicates that the magnitude of voltage across the OFF diodes is identical to the load voltage; hence each diode must have a reverse voltage rating equal to the peak load voltage.

In summary, we may conclude that the bridge circuit has the disadvantage of requiring four instead of two diodes. The reverse voltage rating on each diode, however, is one-half that required by the center-tapped transformer circuit. Other factors, such as ripple and average diode current, are the same in both circuits. The transformer used with the bridge circuit does not need a center tap. It is required to handle only one-half the voltage of an equivalent center-tapped transformer circuit, thus resulting in a less costly transformer.

## Problem Set 3.3

For Probs. 3.3.1 through 3.3.6 the circuit of Fig. 3.9a applies. Assume ideal diodes; RMS input voltage $E_i = 115$ V; $R_L = 1$ kΩ. Determine

3.3.1. Peak, average, RMS, and AC component of load voltage
3.3.2. Peak, average, RMS, and AC component of load current
3.3.3. Reverse voltage rating for each diode
3.3.4. Ripple factor
3.3.5. Total power to load
3.3.6. Peak and average value of diode current
3.3.7. If a circuit of the type shown in Fig. 3.5 is used to achieve the same load voltage as in the previous problem, what secondary voltage rating would be required for the transformer?
3.3.8. Give two applications of transformers in power supply circuits.

### 3.2c  The Half-wave Rectifier with RC Load

Having discussed the basic principles of rectification, we now proceed with the third function of a power supply: filtering.

The half-wave rectified waveform for $v_L$ in Fig. 3.3c is far from the desired DC voltage that a good power supply must produce. To achieve a reasonably smooth DC voltage across $R_L$ we will use an electric component that opposes voltage changes; such a component is the capacitor. The modified circuit is shown in Fig. 3.11a. Waveforms of $v_L$ for various values of capacitance are shown in Fig. 3.11b.

Waveform 1 applies if $C = 0$; this is the situation in the circuit of Fig. 3.2 where no capacitance at all was used. Waveform 2 applies if some intermediate value of $C$ is connected across $R_L$, while with a relatively large capacitance, the much smoother waveform 3 results.

How does a capacitor succeed in smoothing a voltage waveform? The answer lies in understanding the charge-discharge mechanism of capacitors in general. In our case, assume that the transformer secondary voltage in Fig. 3.11a is just starting to go positive from 0 V and that the capacitor is initially discharged. As the diode conducts, it charges $C$, and $v_L$ rises as shown in the waveform of Fig. 3.12b (from 0 to $t_1$). Beyond $t_1$, $e_s$ (the voltage at the anode of $D$) starts to drop, but the voltage across $R_L$ is prevented by the capacitor from following the same variations as $e_s$. This is because the only way voltage across a capacitor can drop is for the capacitor to discharge; the diode prevents the capacitor from

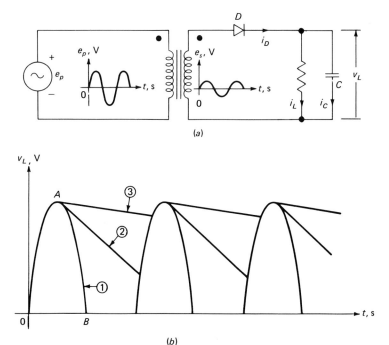

Fig. 3.11  Circuit and waveforms for a capacitively loaded half-wave rectifier.

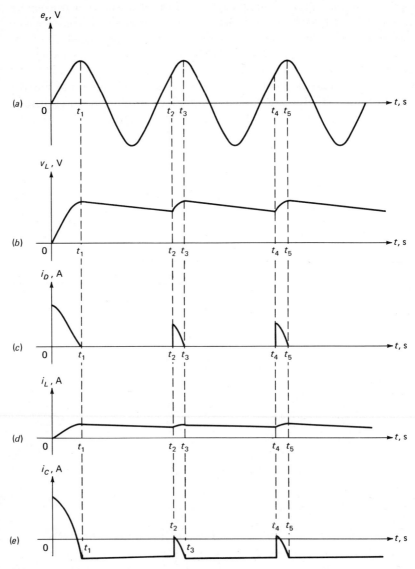

*Fig. 3.12  Waveforms for the circuit of Fig. 3.11a.*

discharging through the transformer secondary circuit, and hence $C$ can discharge only through $R_L$. To keep $C$ from appreciably discharging during the time between $t_1$ and $t_2$, both $C$ and $R_L$ should be large; this results in a long time constant. Usually $R_L$ is fixed in value (it is the load that we must feed), and hence only $C$ can be chosen at will. By using a large value of $C$, the $v_L$ waveform can be made extremely smooth.

Further insight into the "smoothing" operation can be gained by discussing the diode current, whose waveform is shown in Fig. 3.12c. During the period from 0 to $t_1$, diode current flows to deposit the initial charge on the capacitor. Around $t_1$, $e_s$ drops below $v_L$, reverse-biasing $D$, and hence $i_D$ becomes zero. This lasts until $t_2$, when $e_s$ rises above $v_L$. During the period from $t_1$ to $t_2$, $v_L$ drops as the capacitor slowly discharges through $R_L$. Obviously, the larger the $CR_L$ product (time constant), the less $v_L$ decays during this part of the cycle.

The diode current pulse from $t_2$ to $t_3$ and each subsequent pulse are smaller in magnitude than the very first. This is because at $t = 0$ the capacitor is uncharged ($v_L = 0$), while at $t_2$, $v_L$ is not much below the crest value, thus requiring a smaller current pulse to replenish the charge on the capacitor. It follows that for relatively large products of $R_L$ and $C$, the diode current pulse is fairly narrow and of small magnitude. If, for example, no load at all is used ($R_L = \infty$), the capacitor has nothing to discharge into except its own leakage resistance. If this leakage resistance can be neglected, $v_L$ remains perfectly constant at the crest value; it never drops below $e_s$, and the diode is reverse-biased at all times; therefore $i_D$ is always zero. No charging current is required if the charge on the capacitor is never depleted.

The remaining waveforms of Fig. 3.12 are easily explained. Current through a resistor always follows the voltage, and hence $i_L$ is identical in shape to $v_L$. The $i_C$ waveform is obtained by noting that $i_D = i_L + i_C$; hence $i_C = i_D - i_L$. Thus by graphically subtracting $i_L$ from $i_D$, the $i_C$ waveform results.

Consider now the load voltage waveform of Fig. 3.12b. The ripple has been purposely exaggerated; it is much less obvious in good power supplies. The waveform follows a sinusoidal curvature during the time period between $t_2$ and $t_3$; this is to be expected, since the diode is conducting and $v_L$ is essentially equal to $e_s$. From $t_3$ to $t_4$ the load voltage decays linearly with time. The decay is actually exponential, but it appears linear because only the initial portion of the exponential curve is utilized. As an example, let $R_L = 1$ k$\Omega$ and $C = 1{,}000$ $\mu$F; the time constant is

$$\tau = CR_L = (10^3)(10^{-6})(10^3) = 1 \text{ s}$$

which suggests that full discharge requires about 5 s (full charge or discharge takes five time constants). In our example, the capacitor is allowed to discharge from $t_3$ to $t_4$, which is a bit less than a full cycle. Since a full cycle lasts $\frac{1}{60}$ s, the capacitor is discharging for $\frac{1}{60}$ of a time constant, well within the initially linear region of the exponential decay. This effect is further demonstrated with the aid of Fig. 3.13.

The waveform of Fig. 3.12b is approximated with straight line segments in

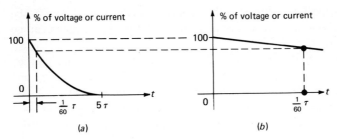

**Fig. 3.13** (a) *Exponential decay curve;* (b) *first $\frac{1}{60}\tau$ of exponential decay, expanded to emphasize linear shape of curve in this region.*

Fig. 3.14; the time between $t_2$ and $t_3$ is neglected (it is usually small compared to the time from $t_3$ to $t_4$). The result is a *sawtooth* waveform, slightly less accurate than the actual waveform of Fig. 3.12b, but definitely simpler to treat mathematically.

In Fig. 3.14, $V_{LM}$ is called the *crest* voltage, while $V_{R(PP)}$ is the peak-to-peak ripple voltage. If $V_{LM} = 10.1$ V and $V_{R(PP)}$ is 0.2 V, then the *trough* or minimum voltage is $V_{LM} - V_{R(PP)} = 9.9$ V.

The DC component $V_{L(DC)}$ of the total waveform is located halfway between the maximum and minimum levels:

$$V_{L(DC)} = V_{LM} - \tfrac{1}{2} V_{R(PP)} \tag{3.11}$$

To calculate the RMS value, the relationship between RMS and peak already established for sine waves cannot be used. Instead, an expression applicable to sawtooth waveforms must be developed; this is done in Appendix 11.2, Eq. (A.44), and the results are used here:

$$\text{RMS value of sawtooth waveform} = \frac{\text{peak}}{\sqrt{3}}$$

$$V_{L(AC)} = \frac{\tfrac{1}{2} V_{R(PP)}}{\sqrt{3}} \tag{3.12}$$

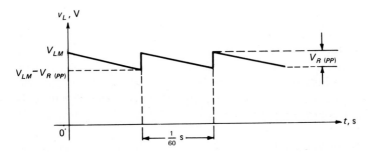

**Fig. 3.14** *Load voltage waveform with sawtooth approximation to ripple.*

In our example, $V_{R(PP)} = 0.2$ V; therefore,

$$V_{L(AC)} = \frac{0.1}{\sqrt{3}} = 57.7 \text{ mV} \tag{3.12}$$

The ripple factor is

$$RF = \frac{V_{L(AC)}}{V_{L(DC)}} \tag{3.8}$$

$$RF = \frac{57.7 \times 10^{-3}}{10} = 5.77 \times 10^{-5}$$

We will now develop an expression for predicting the ripple for any combination of $R_L$ and $C$. To do this note that, if the initially linear discharge of a capacitor were to continue, full discharge would occur in one time constant instead of five. This means that the ratio of $V_{R(PP)}$ to $\frac{1}{60}$ s is the same as the ratio of full decay ($V_{LM}$) to one time constant ($R_L C$):

$$\frac{V_{R(PP)}}{\frac{1}{60}} = \frac{V_{LM}}{\tau} = \frac{V_{LM}}{CR_L}$$

$$V_{R(PP)} = \frac{V_{LM}}{60 CR_L} \tag{3.13}$$

Equation (3.13) gives the peak-to-peak ripple in terms of crest (maximum) voltage and time constant. Substituting Eq. (3.13) into Eq. (3.12) yields the RMS ripple voltage:

RMS ripple voltage: $\quad V_{L(AC)} = \dfrac{V_{R(PP)}}{2\sqrt{3}} \tag{3.12}$

$$V_{L(AC)} = \frac{V_{LM}}{120 \sqrt{3} \, CR_L} \tag{3.14}$$

Substitution of Eqs. (3.14) and (3.11) into Eq. (3.8) yields an expression for the ripple factor, as follows:

$$RF = \frac{V_{LM}/(120 \sqrt{3} \, CR_L)}{V_{LM} - \frac{1}{2} V_{R(PP)}} = \frac{V_{LM}/(120 \sqrt{3} \, CR_L)}{V_{LM} - [V_{LM}/(120 CR_L)]}$$

$$RF = \frac{1}{120 \sqrt{3} \, CR_L - 3} \tag{3.15}$$

### Example 3.3

Given the load voltage waveform of Fig. 3.14, assume a crest value of 20 V, $R_L = 100$ Ω, and $C = 1{,}000$ μF. Determine

a. The RMS ripple voltage
b. Maximum and minimum load voltage

c. DC component of load voltage
d. Ripple factor
e. Power to load

**Solution**

a. The RMS ripple voltage is given by Eq. (3.14):

$$V_{L(AC)} = \frac{V_{LM}}{120\sqrt{3}\,CR_L} = \frac{20}{(120\sqrt{3})(10^{-3})(100)} = 0.962 \text{ V} \qquad (3.14)$$

b. The maximum load voltage is equal to the crest value of 20 V. The minimum value is obtained by subtracting the peak-to-peak ripple from the crest, as follows:

$$\text{Peak-to-peak ripple voltage:} \quad V_{R(PP)} = 2\sqrt{3}\,V_{L(AC)} \qquad (3.12)$$
$$V_{R(PP)} = 2\sqrt{3}\,(0.962) = 3.34 \text{ V}$$

Minimum $V_L = 20 - 3.34 = 16.66$ V

c. The DC voltage is halfway between maximum and minimum levels, as follows:

$$V_{L(DC)} = V_{LM} - \tfrac{1}{2}V_{R(PP)} \qquad (3.11)$$
$$V_{L(DC)} = 20 - \tfrac{1}{2}(3.34) = 18.33 \text{ V}$$

d. The ripple factor can be calculated using either Eq. (3.8) or Eq. (3.15). Since the necessary data are available, Eq. (3.8) is simpler to use:

$$RF = \frac{V_{L(AC)}}{V_{L(DC)}} = \frac{0.962}{18.33} = 0.0525 = 5.25 \text{ percent} \qquad (3.8)$$

e. Power to the load

$$P_L = V_L I_L \qquad (3.4)$$
$$V_L = \sqrt{V_{L(AC)}^2 + V_{L(DC)}^2} = \sqrt{0.962^2 + 18.33^2} \qquad (3.7a)$$
$$V_L = 18.35 \text{ V}$$

Note that the AC component is quite insignificant compared with the DC component here; this is true for low-ripple circuits in general.

$$P_L = 18.35 \left[\frac{18.35}{100}\right] = 3.37 \text{ W}$$

## Problem Set 3.4

3.4.1. For the circuit of Fig. 3.11a, what is the relationship between the value of capacitance and the resulting ripple? Explain your answer.

3.4.2. In what way does the diode in the circuit of Fig. 3.11a help the capacitor maintain a constant output voltage?

3.4.3. What would be the form factor for a power supply capable of producing a perfectly smooth DC output?

3.4.4. For the circuit of Fig. 3.11a, assume the peak transformer secondary voltage is 50 V and the diode is ideal. Determine the DC component of load voltage ($V_{L(DC)}$) and the ripple ($V_{L(AC)}$) when $C = 0$.

3.4.5. Repeat Prob. 3.4.4 when $R_L = 1$ k$\Omega$, $C = 2,000$ $\mu$F. Compare the DC level and ripple obtained in each case.

3.4.6. Compute the load power for Prob. 3.4.5.

3.4.7. What capacitance would be required in Prob. 3.4.5 to achieve a ripple factor of 0.1 percent? What are the DC and AC components of load voltage in this case?

3.4.8. How does average diode current depend on ripple voltage for the circuit of Fig. 3.11a?

3.4.9. Why is load voltage decay linear during the nonconducting half-cycle for the diode in the circuit of Fig. 3.11a?

## 3.2d  The Full-wave Rectifier with RC Load

Figure 3.15a shows a full-wave rectifier feeding an $RC$ load. The function of the capacitor is again to smooth the load voltage waveform; since both halves

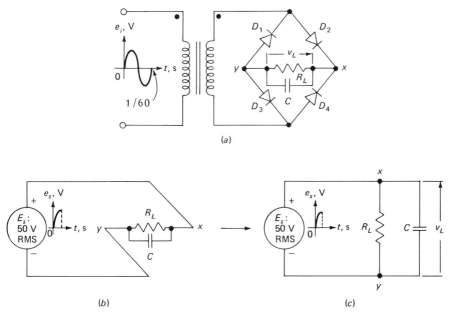

Fig. 3.15 (a) Bridge rectifier with RC load; (b) and (c) equivalent circuits for (a) during the first 90° of positive half-cycle.

of the AC cycle are used, however, we should expect this circuit to produce less ripple (all other things being equal) than a half-wave rectifier.

In the circuit of Fig. 3.15a, both $D_2$ and $D_3$ conduct during the first 90° of the positive half-cycle; meanwhile $D_1$ and $D_4$ are OFF. This yields the equivalent circuit of Fig. 3.15b, redrawn in simplified form in Fig. 3.15c. During this time the voltage across the $R_L$-$C$ combination follows the input voltage.

During the second 90° of the positive half-cycle, the input to the bridge falls from the crest value toward zero. The voltage across the $R_L$-$C$ combination, however, cannot follow the input. This is because voltage across a capacitor can only decrease if current flows in the opposite direction to that which deposited the initial charge on the capacitor. Here this cannot happen because $D_2$ and $D_3$ allow current only in one direction. Thus $v_L$ remains close to the crest, while the input voltage to the bridge continues to drop. During this time (second 90° of positive half-cycle) both $D_2$ and $D_3$ are reverse-biased in addition to $D_1$ and $D_4$. Since none of the diodes conduct, $C$ can discharge only through $R_L$.

The load voltage waveform and its approximation are shown in Fig. 3.16. The approximation in Fig. 3.16b involves a sawtooth waveform similar to that for a half-wave circuit. The only significant difference is that, in a full-wave circuit, the capacitor discharges for only $1/120$ s as opposed to $1/60$ s in the half-wave rectifier.

An expression for the ripple similar to Eq. (3.13) can now be developed:

$$\frac{V_{R(PP)}}{1/120} = \frac{V_{LM}}{\tau} = \frac{V_{LM}}{CR_L}$$

$$V_{R(PP)} = \frac{V_{LM}}{120 R_L C} \tag{3.16}$$

### Example 3.4

Assume the following data for the circuit of Fig. 3.15a: ideal diodes; RMS input voltage to transformer primary is $E_i = 115$ V; transformer turns ratio of 2.3:1; $R_L = 500\ \Omega$ and $C = 1{,}000\ \mu\mathrm{F}$. Determine

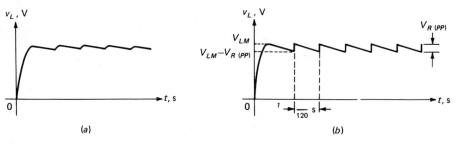

Fig. 3.16 (a) Load voltage waveform for the circuit of Fig. 3.15a; (b) approximation to (a).

a. RMS ripple voltage
b. DC component of load voltage
c. Total load power

**Solution**

Since the transformer's turns ratio is 2.3:1, the secondary voltage is $115/2.3 = 50$ V. The peak load voltage is therefore $50\sqrt{2} = 70.7$ V.

a. The peak-to-peak ripple is

$$V_{R(PP)} = \frac{V_{LM}}{120R_LC} \qquad (3.16)$$

$$V_{R(PP)} = \frac{70.7}{120(500 \times 10^{-3})} = 1.18 \text{ V}$$

The RMS ripple is

$$V_{L(AC)} = \frac{V_{R(PP)}}{2\sqrt{3}} \qquad (3.12)$$

$$V_{L(AC)} = \frac{1.18}{2\sqrt{3}} = 0.341 \text{ V}$$

b. The DC component is

$$V_{L(DC)} = V_{LM} - \tfrac{1}{2}V_{R(PP)} \qquad (3.11)$$
$$V_{L(DC)} = 70.7 - 0.59 \simeq 70.1 \text{ V}$$

c. Since the ripple is relatively small, we can neglect it and assume $V_L \simeq V_{L(DC)}$:

$$P_L = V_L I_L \qquad (3.4)$$

$$P_L = 70.1 \left[\frac{70.1}{500}\right] = 9.85 \text{ W}$$

## Problem Set 3.5

3.5.1. In Example 3.4 determine the capacitance required for a ripple factor of 0.5 percent. Assume all other parameters remain unchanged. Calculate the new DC components of load voltage and load current.

3.5.2. Derive an expression for the ripple factor similar to Eq. (3.15) that applies to the circuit of Fig. 3.15a.

### 3.2e Voltage Regulation

Voltage regulation is a measure of a circuit's ability to maintain a constant output voltage. The output voltage of a good power supply should remain reasonably constant under a number of varying conditions, such as line voltage fluctuations and changes in current drawn by the load.

Line voltage fluctuations are usually reflected across the load; the exact amount depends on the degree of filtering between the rectification circuitry and the load. As current drawn from the power supply increases, more voltage is dropped across the various internal resistances (diodes and transformers have DC resistances through which the load current must flow), thus leaving less voltage for the load itself.

Shown in Fig. 3.17 is the diagram of a power supply. Voltage transformation, rectification, and filtering have already been discussed. The $LC$ filter shown here is capable of providing much more of a "smoothing" action on the rectified waveform than the simple $RC$ load discussed earlier.

In this section, a simple voltage regulator circuit is discussed. Discussion of more complex circuits must be delayed until transistors are understood. Such a

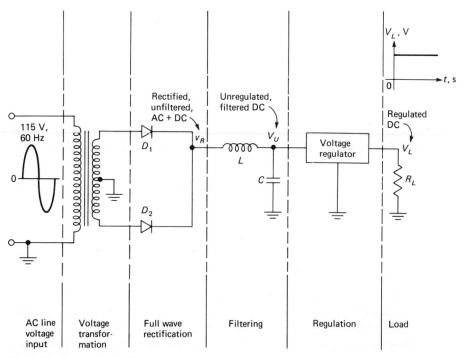

Fig. 3.17  Power supply diagram.

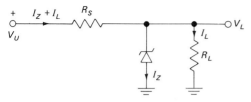

Fig. 3.18  Zener diode regulator.

circuit is shown in Fig. 3.18; it represents the block labeled "voltage regulator" in Fig. 3.17. The input is $V_U$, a filtered but unregulated DC; the output is the load voltage $V_L$, regulated, we hope. The circuit works this way:

1. A diode with a very sharp reverse breakdown characteristic (called a *Zener diode*) is connected across the load.
2. When a diode is in breakdown, its voltage remains extremely close to the breakdown value, even if the current varies over a large range. This can be seen by examining the reverse diode characteristic in Fig. 3.19.
3. If the unregulated DC voltage ($V_U$) rises, for example, the current through $R_S$ increases; this extra current, however, is diverted to the diode instead of flowing through the load. The diode voltage is virtually unaffected by the increase in current, and hence the load voltage (which is the same as the diode voltage) remains practically the same.
4. If the load requires more current (when $R_L$ is reduced), the Zener can supply the extra current without affecting the load voltage.

A numerical example will be used to further discuss the circuit. Suppose that in Fig. 3.18 $V_U$ is constant at 15 V DC, $R_S = 2.5\ \Omega$, the diode's breakdown voltage is 10 V, and $R_L$ can be anything from an open circuit (no load) to 10 $\Omega$. With an ideal Zener (the slope of the breakdown characteristic perfectly vertical), $r_Z = 0$ and $V_L$ can only be 10 V. Since $V_U$ is 15 V, $R_S$ has a constant 5 V. This yields

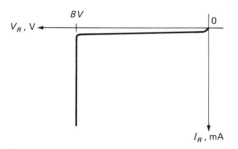

Fig. 3.19  Reverse characteristic for Zener diode.

Current through $R_S$: $I_Z + I_L = \dfrac{5}{2.5} = 2$ A

It follows that $R_S$ must dissipate $5(2) = 10$ W.

When we are dealing with quantities that can range between a minimum value and a maximum value, minimum values will be denoted with an underscore (minimum load current $\underline{I_L}$) and maximum values with an overscore (maximum load current $\overline{I_L}$). In our case, the load current can take on any value between two extremes:

$\underline{I_L} = 0 \quad$ when there is no load

$\overline{I_L} = \dfrac{10}{10} = 1$ A

When the load draws 1 A, the Zener current is also 1 A, since the total must equal the 2 A through $R_S$. Under these conditions, the diode dissipates $1(10) = 10$ W. With no load, the Zener current is 2 A, which generates $10(2) = 20$ W. We therefore require a 20-W Zener (10-V breakdown) and a 10-W resistor ($R_S$). These rather large power ratings are a basic disadvantage of the circuit, although when we are dealing with lower current swings, the power dissipation may be substantially lower.

In general, maximum $I_Z$ flows when $I_L$ is minimum; $I_Z$ is minimum when $I_L$ is maximum. This means that the range of load current variation ($\overline{I_L} - \underline{I_L}$) is limited by the maximum permissible Zener current. For example, if a Zener diode is rated at 5 A, the load current range must be less than 5 A.

So far we have assumed a perfect Zener, which, of course, yields perfect regulation. Practical Zeners do not have resistances of 0 $\Omega$, although a few milliohms are possible. Let us assume $r_Z = 0.1\ \Omega$ and compute voltage regulation for our example:

$$\text{Percent regulation} = \left[\dfrac{\text{no load voltage} - \text{full load voltage}}{\text{full load voltage}}\right] 100 \qquad (3.17)$$

Figure 3.20 shows the complete equivalent circuit, including $r_Z$. With no load,

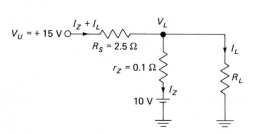

Fig. 3.20  Circuit of Fig. 3.18 with large-signal model substituted for Zener diode.

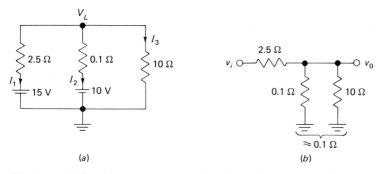

*Fig. 3.21* (a) General circuit to calculate $V_L$ in Zener diode regulator; (b) equivalent circuit for calculating the changing component of output voltage ($v_o$) due to a changing component of input voltage ($v_i$).

$$I_z + I_L = I_z = \frac{15 - 10}{2.6} = 1.925 \text{ A}$$
$$V_L = 15 - 2.5(1.925) = 10.19 \text{ V}$$

With full load ($R_L = 10 \, \Omega$), the circuit of Fig. 3.21a results, which can be solved by means of node equations:[1]

$$I_1 + I_2 + I_3 = 0$$
$$\frac{V_L - 15}{2.5} + \frac{V_L - 10}{0.1} + \frac{V_L}{10} = 0$$
$$4V_L - 60 + 100V_L - 1{,}000 + V_L = 0$$
$$105V_L = 1{,}060$$
$$V_L = 10.10 \text{ V}$$

$$\text{Percent regulation} = \left[ \frac{10.19 - 10.10}{10.10} \right] 100 = 0.89 \quad (3.17)$$

A measure of how well the circuit attenuates input voltage fluctuations is the ripple reduction factor (*RRF*):

$$RRF = \frac{\text{change in input voltage}}{\text{change in output voltage}} \quad (3.18)$$

We can illustrate the *RRF* by assuming in our example that the unregulated DC can swing anywhere between extremes of 13 and 17 V. To compute the changes in output voltage resulting from such fluctuations, we assume a full load and compute $V_L$ for both extremes:

a. $V_U = 17 \text{ V}$: Using the circuit of Fig. 3.21a (except for the 17-V instead of 15-V source), we have

$$I_1 + I_2 + I_3 = 0$$

---

[1] See Appendix 5 for a discussion of node analysis.

$$\frac{V_L - 17}{2.5} + \frac{V_L - 10}{0.1} + \frac{V_L}{10} = 0$$
$$4V_L - 68 + 100V_L - 1{,}000 + V_L = 0$$
$$105 V_L = 1{,}068$$
$$V_L = 10.171 \text{ V}$$

b. $V_U = 13$ V: The procedure in (a) is followed once more:

$$I_1 + I_2 + I_3 = 0$$
$$\frac{V_L - 13}{2.5} + \frac{V_L - 10}{0.1} + \frac{V_L}{10} = 0$$
$$4V_L - 52 + 100V_L - 1{,}000 + V_L = 0$$
$$105 V_L = 1{,}052$$
$$V_L = 10.019 \text{ V}$$

The $RRF$ is

$$RRF = \frac{17 - 13}{10.171 - 10.019} = \frac{4}{0.152} \simeq 26 \tag{3.18}$$

A $RRF$ of 26 implies that any change in voltage at the input to the regulator is attenuated by a factor of 26 when appearing across the load.

The $RRF$ can also be obtained using the principle of superposition. Since $RRF$ has to do with changes in the output due to changes in the input, its value may be determined by considering only the changing component of input voltage and disregarding the constant (DC) component. Equation (3.18) may be rewritten as

$$RRF = \frac{v_i}{v_o} \tag{3.18a}$$

where $v_i$ and $v_o$ are the changing components of input and output voltage, respectively. In the equivalent circuit of Fig. 3.21b, all DC voltages are considered to be zero and

$$v_o \simeq v_i \left[ \frac{0.1}{2.5 + 0.1} \right] = \frac{v_i}{26}$$

therefore,

$$RRF \simeq 26$$

as calculated earlier.

**Example 3.5**

The following data apply to the circuit of Fig. 3.18: $\overline{V_U} = 30$ V; $\underline{V_U} = 20$ V; $\underline{R_L} = 10 \ \Omega$; $\overline{R_L} = 100 \ \Omega$; desired $V_L = 10$ V. A 10-V Zener diode will be used, which has a maximum power dissipation rating of 50 W and needs a minimum current of 10 mA to ensure that it operates always in the breakdown region, well below the knee of the curve. Determine a range of values for $R_S$.

## Solution

As implied by the problem, a range of values for $R_S$, rather than a specific value, is required. This will be dictated by two constraints:

a. The current through the Zener must never drop below 10 mA, or the Zener will come out of breakdown, and regulation will be lost.
b. The current through the Zener must never be so high that the maximum power dissipation rating of 50 W is exceeded.

The maximum allowable Zener current is

$$\overline{I_Z} = \frac{P}{V} = \frac{50}{10} = 5 \text{ A}$$

In general, the current through $R_S$ is

$$I_Z + I_L = \frac{V_U - V_L}{R_S}$$

from which

$$I_Z = \frac{V_U - V_L}{R_S} - I_L \qquad (a)$$

We want $I_Z$ to exceed 10 mA under all operating conditions:

$$I_Z > 10 \text{ mA}$$

$I_Z$ is minimum when $V_U$ is minimum and $I_L$ is maximum. Therefore Eq. (a) can be rewritten as

$$\underline{I_Z} = \frac{\underline{V_U} - V_L}{R_S} - \overline{I_L}$$

But $\underline{V_U} = 20$ V and $\overline{I_L} = V_L/\underline{R_L} = {}^{10}\!/_{10} = 1$ A. We therefore have

$$\underline{I_Z} = \frac{20 - 10}{R_S} - 1 > 10 \text{ mA}$$

$$\frac{10}{R_S} - 1 > 0.01$$

$$\frac{10}{R_S} > 1.01$$

$R_S < 9.9\ \Omega$

$I_Z$ is maximum when $I_L$ is minimum and $V_U$ is maximum. Therefore Eq. (a) can be rewritten as

$$\overline{I_Z} = \frac{\overline{V_U} - V_L}{R_S} - \underline{I_L}$$

But $\overline{V_U} = 30$ V, $\underline{I_L} = V_L/\overline{R_L} = {}^{10}\!/_{100} = 0.1$, and $\overline{I_Z} = 5$ A. Therefore we have

$$\frac{30 - 10}{R_S} - 0.1 < 5$$

$R_S > 3.92\ \Omega$

A value of $R_S$ between 3.92 and 9.9 $\Omega$ is required. It seems logical to choose $R_S$ around 7 $\Omega$ to provide an equal margin of safety.

### Problem Set 3.6

3.6.1. For the circuit of Fig. 3.18, $V_L = 15$ V; maximum load current is 100 mA; minimum load current is 0; dynamic Zener resistance is 0.05 $\Omega$; input voltage range is 18 to 20 V; $R_S = 10\ \Omega$. Determine

    a. Maximum power dissipated by $R_S$
    b. Maximum power dissipated by diode
    c. Minimum diode current
    d. Voltage regulation for an input voltage of 20 V
    e. Ripple reduction factor under full load conditions
    f. The power that must be dissipated by $R_S$ if the output is accidentally short-circuited

3.6.2. It is desired to regulate at 10 V, using the circuit of Fig. 3.18.

$\underline{R_L} = 100\ \Omega$    $\overline{R_L} = 1{,}000\ \Omega$    $I_Z = 5$ mA
$\underline{V_U} = 13$ V    $\overline{V_U} = 15$ V    $\overline{P_Z} = 20$ W

Determine a suitable value for $R_S$.

## 3.3 CLIPPING CIRCUITS

Clipping circuits utilize nonlinear properties of diodes to limit the amplitude of a particular voltage at some desired level. For example, in radio receivers designed to receive weak signals, random bursts of noise may be present which tend to be quite a nuisance; these bursts may be voltage spikes whose large amplitude completely masks the low-level signal during their brief duration. A diode limiter can be used to *clip* all voltage levels in excess of some preset value, thus improving the intelligibility of whatever communication is being received.

To illustrate clipping, the circuit of Fig. 3.22a is used. The circuit is polarity-sensitive (because of the diode); we will apply both negative and positive inputs. The results are easier to see when displayed on a graph, as in Fig. 3.22b. This graph is the transfer characteristic, and it tells us what the output is going to be for any given input. For example, if $v_1 = 0$ (input terminal grounded), the diode is reverse-biased, no current flows, and $v_2 = 0$. This is one point on the graph, namely ($v_1 = 0$, $v_2 = 0$). Similarly, for any negative input, the diode is reverse-biased, no current flows, there is no voltage across $R$, and $v_2 = v_1$, as shown graphically by the 45° plot in the region to the left and below the origin. With positive inputs below 10 V the same circumstances prevail: the diode does not conduct, and hence $v_2 = v_1$, as indicated by the 45° plot in the positive region below 10 V.

So far, there is no current through the diode or the battery and $v_2 = v_1$ for any value of $v_1$ below 10 V. Now assume $v_1$ slightly greater than $+10$ V; this forces $D$ into conduction. With an ideal diode, $r_F = 0\,\Omega$, making $v_2 = 10$ V. Now no matter how large $v_1$ becomes, $v_2$ remains at 10 V because the diode is conducting. This is shown on the graph as a horizontal line beyond $v_1 = 10$ V.

The relationship between $v_1$ and $v_2$, which is displayed graphically in Fig. 3.22b, can also be expressed mathematically as follows:

$v_2 = v_1$     when $v_1 \leq 10$ V
$v_2 = 10$ V    when $v_1 > 10$ V

If clipping is desired for both positive and negative inputs, another branch

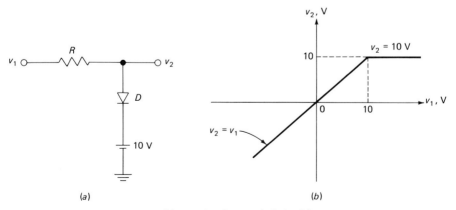

Fig. 3.22  (a) *Clipping circuit;* (b) *transfer characteristic for* (a).

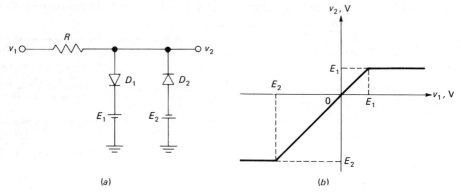

Fig. 3.23 (a) *Circuit that clips both positive and negative levels;* (b) *transfer characteristic for* (a).

is added, as in the circuit of Fig. 3.23a. Here the branch involving $D_1$ and $E_1$ becomes significant whenever the input reaches a positive level greater than $E_1$. Similarly, the other branch becomes significant whenever the input drops below $E_2$. The relationship between $v_1$ and $v_2$ is displayed graphically in Fig. 3.23b; it can also be expressed mathematically:

$v_2 = E_1$ when $v_1 > E_1$
$v_2 = v_1$ when $E_2 \leq v_1 \leq E_1$
$v_2 = E_2$ when $v_1 < E_2$

**Example 3.6**

Assume a 100-V peak sinusoidal input for the circuit of Fig. 3.22a. Determine the output.

**Solution**

All values of input voltage below 10 V are faithfully reproduced at the output, but the positive half is clipped at 10 V; the resulting waveform is shown in Fig. 3.24.

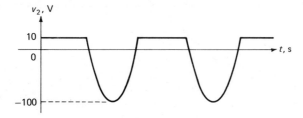

Fig. 3.24 *Output waveform for Example 3.6.*

Circuit Applications of Diodes 79

### Example 3.7

The triangular voltage of Fig. 3.25a is applied to the input of the circuit in Fig. 3.23a. Assuming $E_1 = 10$ V and $E_2 = 5$ V, determine the output.

### Solution

All levels greater than $+10$ V and below $-5$ V are clipped; the output waveform is shown in Fig. 3.25b.

### Example 3.8

Assume that the real diode in the circuit of Fig. 3.22a has the following large-signal parameters:

Forward threshold voltage: $V_T = 0.7$ V
Forward resistance: $r_F = 10\ \Omega$
Reverse resistance: $r_R = 1\ M\Omega$

Determine the new transfer characteristic if $R = 100\ \Omega$.

### Solution

Since it takes 0.7 V of forward bias to start the diode conducting, $D$ is OFF until $v_2$ exceeds 10.7 V. For values of $v_1$ that result in $v_2 < 10.7$ V, the diode is reverse-

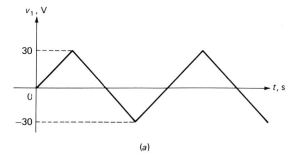

(a)

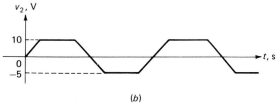

(b)

Fig. 3.25 Input (a) and output (b) waveforms for Example 3.7.

80   Solid State Electronic Circuits

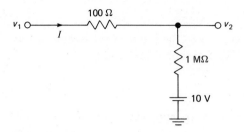

*Fig. 3.26   Equivalent circuit when diode is not conducting (Example 3.8).*

biased and the equivalent circuit of Fig. 3.26 applies. It is sufficient to solve this circuit for two different points only. When these points are joined with a straight line, they yield the transfer characteristic for the region in which the equivalent circuit is valid. In general,

$$v_2 = v_1 - 100I \quad \text{where } I \simeq \frac{v_1 - 10}{10^6}$$

Hence

$$v_2 = v_1 - \frac{100(v_1 - 10)}{10^6}$$

$$v_2 = v_1 - 10^{-4}v_1 + 10^{-3}$$

$$v_2 \simeq v_1 + 0.001 \tag{a}$$

Equation (a) is plotted in Fig. 3.27 by locating the following two points:

1. When $v_1 = 0$, $v_2 \simeq 0.001$ V
2. When $v_2 = 10.7$, $10.7 \simeq v_1 + 0.001$
$$v_1 \simeq 10.7 \text{ V}$$

For values of $v_1$ that result in $v_2 > 10.7$ V, the diode conducts and the equivalent circuit of Fig. 3.28 applies. An expression for $v_2$ is developed as follows:

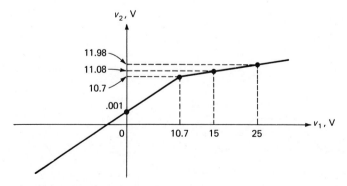

*Fig. 3.27   Transfer characteristic for the circuit of Fig. 3.22 when parameters of the real diode are used.*

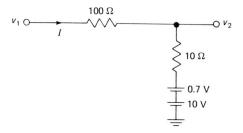

Fig. 3.28 Equivalent circuit when diode conducts (Example 3.8).

$v_2 = v_1 - 100I$

$I \simeq \dfrac{v_1 - 10.7}{110}$

$v_2 = v_1 - \dfrac{100(v_1 - 10.7)}{110}$

$v_2 = 0.09v_1 + 9.73$ \hfill (b)

Equation (b) allows us to solve for the output voltage whenever the equivalent circuit of Fig. 3.28 is valid (for inputs > 10.7 V). If, for example, $v_1 = 15$ V,

$v_2 = 0.09(15) + 9.73 = 11.08$ V

and if $v_1 = 25$ V,

$v_2 = 0.09(25) + 9.73 = 11.98$ V

Both of the above points allow us to complete the transfer characteristic, as shown in Fig. 3.27. Examination of this characteristic points out that perfect clipping of positive levels above 10 V is not possible; in fact, clipping does not occur until the input exceeds 10.7 V. Although the clipping level could be adjusted at 10 V by using a 9.3-V DC source (instead of the 10-V shown), there would still be a slight slope to our characteristic because of the diode's nonzero forward resistance.

### Example 3.9

Using the transfer characteristic of Fig. 3.27, determine the output voltage if the input of Fig. 3.25a is used.

### Solution

Both input and output waveforms are shown in Fig. 3.29. The output is obtained by locating several critical points, which are then joined with straight lines:

a. When $v_1 = 0$, $v_2 \simeq 0.001$ V
b. When $v_1 = 10.7$ V, $v_2 \simeq 10.7$ V

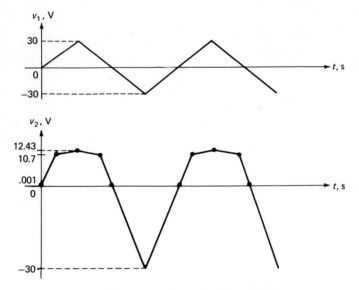

Fig. 3.29  Input and output waveforms for Example 3.9.

c. When $v_1 = 30$ V, Eq. (b) from Example 3.8 is used to yield $v_2 = 0.09(30) + 9.73 = 12.43$ V
d. When $v_1 = -30$ V, $v_2 \simeq -30$ V, as indicated by Eq. (a) from Example 3.8

## Problem Set 3.7

3.7.1. The circuit of Fig. 3.22a clips at 10.7 V instead of 10 V as might be expected from the value of DC source connected between diode cathode and ground. What causes the extra 0.7 V before clipping occurs?

3.7.2. It is desired to use a Zener diode that breaks down at 10 V to achieve clipping of positive levels above 10 V, while allowing any input below 10 V to appear undisturbed at the output. The circuit of Fig. 3.30 is

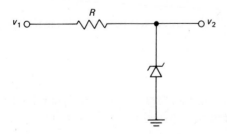

Fig. 3.30  Circuit for Prob. 3.7.2.

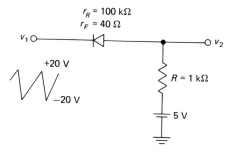

Fig. 3.31  Circuit for Prob. 3.7.3.

proposed. Do you see any problem in achieving the above objectives with this circuit, and if so, how would you solve the problems?

3.7.3. Draw the transfer characteristic and output voltage for the circuit of Fig. 3.31. Assume $V_T = 0$.

3.7.4. Sketch the output waveform for each circuit in Fig. 3.32. The input is a 50-V peak sinusoidal voltage for all circuits. Assume ideal diodes.

3.7.5. Match each circuit in Fig. 3.33 with its output waveform. The input is the same in all cases, a 100-V peak-to-peak sinusoidal voltage.

3.7.6. Sketch the diagram of a circuit that will

a. Clip all levels above 30 V
b. Clip all levels below $-30$ V
c. Clip all levels above 30 V and below $-30$ V

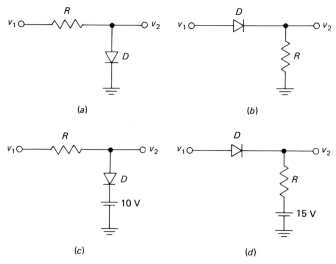

Fig. 3.32  Circuits for Prob. 3.7.4.

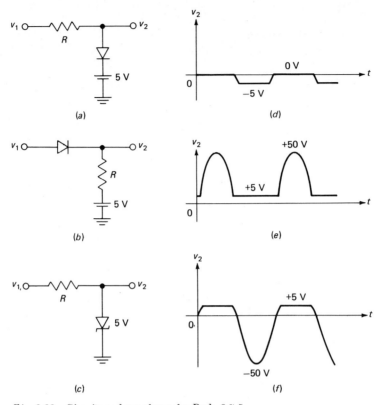

Fig. 3.33 Circuits and waveforms for Prob. 3.7.5.

    d. Clip all levels below 30 V
    e. Clip all levels above $-30$ V

3.7.7. Develop the transfer characteristic for the circuit of Fig. 3.23a if $R = 100$ k$\Omega$, $E_1 = 8$ V, $E_2 = -15$ V, and both diodes are ideal.

3.7.8. Repeat Prob. 3.7.7 if both diodes have the following large-signal parameters: $V_T = 0.3$ V, $r_F = 15$ $\Omega$, $r_R = 100$ k$\Omega$.

3.7.9. Under what conditions would a diode's breakdown voltage rating be significant in analyzing clipping circuits?

3.7.10. A 40-V peak sinusoidal voltage is applied to the circuit of Prob. 3.7.7. Determine the output.

3.7.11. Determine the transfer characteristic for the circuit of Fig. 3.22a if a 1,000-$\Omega$ load resistance is connected across the output. Assume $R = 1$ k$\Omega$ and the diode is ideal.

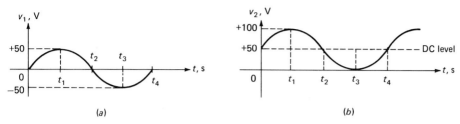

*Fig. 3.34* (a) *Waveform with no DC level;* (b) *waveform of* (a), *positively clamped at 0 V.*

## 3.4 CLAMPING CIRCUITS

Clamping is a process of introducing a DC level into a signal. For example, the waveform in Fig. 3.34a is a sine wave with equal positive and negative swings about 0 V, and hence its average value is 0 (it has no DC level). The waveform in Fig. 3.34b, however, has been "lifted" so as to just touch the horizontal axis; it has now acquired a DC level of 50 V. A circuit capable of accepting the input shown in Fig. 3.34a and delivering the output shown in Fig. 3.34b is called a *clamper;* the resulting output waveform is said to be positively clamped at 0 V.

A diode clamper is shown in Fig. 3.35a. The sinusoidal voltage of Fig. 3.34a is the input. As soon as $v_1$ rises from 0 toward 50 V, the diode conducts. Assuming an ideal diode, its voltage, which is also the output, must be zero during the time between 0 and $t_1$. During the time that $D$ is conducting, however, $C$ charges to 50 V; at $t_1$, the voltage across $C$ is exactly 50 V. At the same time, $v_1$ starts to drop, which means that the anode of $D$ is negative relative to its cathode, thus reverse-biasing the diode and preventing the capacitor from discharging. Since the capacitor is holding its charge, it behaves as a DC voltage source while the nonconducting diode appears as an open circuit, yielding the equivalent circuit of Fig. 3.36, for which we can write

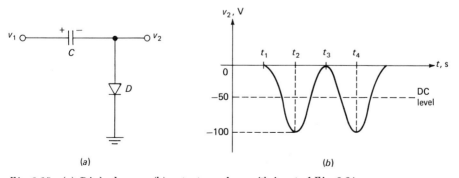

*Fig. 3.35* (a) *Diode clamper;* (b) *output waveform with input of Fig. 3.34a.*

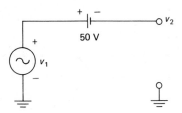

Fig. 3.36 Equivalent circuit for clamper of Fig. 3.35a.

$$v_2 = -50 + v_1$$

Thus our circuit behaves as if a 50-V DC source had been placed in series with the AC generator; this shifts everything by 50 V in the negative direction, as shown in the output waveform of Fig. 3.35b. Such a waveform is said to be negatively clamped. Positive clamping could be achieved by reversing the diode. The period from 0 to $t_1$ is required to deposit the initial charge on the capacitor, and hence this first segment of the waveform is lost.

How about clamping at levels other than 0 V? It can be achieved by connecting a DC source as shown in Fig. 3.37a. The output here is positively clamped at $-20$ V.

To study the performance of real diode clampers, consider the circuit of Fig. 3.38a. The circuit should negatively clamp the input at 5 V. The input is shown in Fig. 3.38b; it is 0 until $t = 0$; then it rises to 15 V and stays there for ½ ms. At $t = 0.5$ ms, $v_1$ drops to $-15$ V and is held there for 5 ms. The cycle keeps repeating from then on.

The sequence of events is as follows:

1. Just before $t = 0$, the capacitor is charged to 5 V due to the DC source in the circuit, and therefore $v_2 = 5$ V, as shown in Fig. 3.38d. At $t = 0$, $v_1$ rises from 0

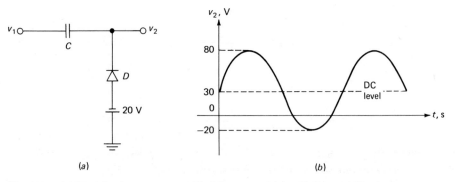

Fig. 3.37 (a) Positive clamper at $-20$ V; (b) output of (a) with input of Fig. 3.34a.

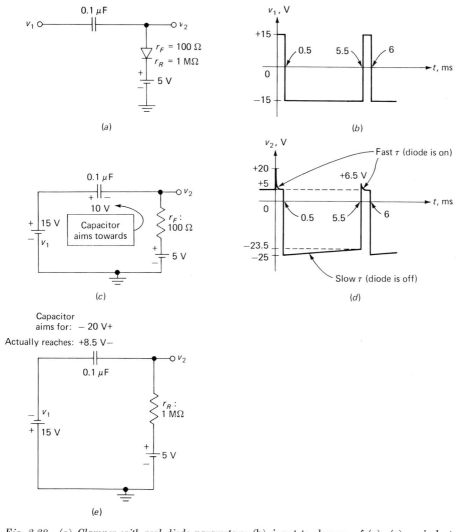

Fig. 3.38 (a) Clamper with real diode parameters; (b) input to clamper of (a); (c) equivalent circuit of clamper when $v_1 = +15$ V; (d) output; (e) equivalent circuit of clamper when $v_1 = -15$ V.

to 15 V, but $C$ has had no time to change its charge; therefore its voltage is still 5 V, and $v_2$ rises to $v_1 + 5 = 20$ V.

2. The equivalent circuit of Fig. 3.38c applies while the input is at 15 V. During this period the diode conducts and $C$ charges toward 10 V as shown. This takes about five time constants: $5RC = 5(100 \times 10^{-7}) = 0.5$ ms. As the

capacitor voltage rises from 0 to 10 V, $v_2$ decays from 20 to 5 V as shown in the output waveform of Fig. 3.38d.

3. At $t = 0.5$ ms, $v_1$ changes from $+15$ to $-15$ V, a total drop of 30 V. Since the capacitor's voltage cannot change instantaneously, the output must also drop by 30 V, which takes it from $+5$ down to $-25$ V.
4. With $v_2$ at $-25$ V, the diode cannot conduct and the circuit in Fig. 3.38e applies. The capacitor tries to charge toward 20 V, which requires five time constants: $5RC = (5 \times 10^6)(10^{-7}) = 500$ ms.
5. As the capacitor charges toward 20 V, $v_2$ aims toward $20 - 15 = 5$ V. This is a total change of 30 V (from the $-25$ V value at $t = 0.5$ ms), and requires 500 ms to complete. The $-15$ V input, however, lasts only 5 ms; it follows that there is not enough time for the capacitor to reach 20 V, and hence $v_2$ cannot reach $+5$ V either.
6. The initial portion of any exponential charge or discharge is linear, and hence the change in voltage $x$ taking place during 5 ms is related to the total voltage change (30 V) that would occur in one time constant (100 ms) as follows:

$$\frac{5 \times 10^{-3}}{100 \times 10^{-3}} = \frac{x}{30}$$
$$x = 1.5 \text{ V}$$

which means that at $t = 5.5$ ms, $v_2$ has changed from $-25$ to $-25 + 1.5 = -23.5$ V. Similarly, the voltage across the capacitor has changed from 10 to 8.5 V.

7. At $t = 5.5$ ms, $v_1$ rises by 30 V; since $C$ has had no time to alter its charge, $v_2$ must also rise by 30 V, bringing it to $-23.5 + 30 = +6.5$ V.
8. With $v_2 = 6.5$ V, the diode conducts and the equivalent circuit of Fig. 3.38c applies. The capacitor, of course, charges to 10 V as shown, with the fast time constant. As soon as $C$ reaches 10 V, $v_2$ equals 5 V. From here on, the sequence keeps repeating.

From the analysis just completed it is obvious that, although we have succeeded in clamping our signal negatively at 5 V, this has been achieved with some loss in faithful reproduction of the input waveform. There is some "overshoot" on the positive side to 6.5 V and decay on the negative side from $-25$ to $-23.5$ V. A diode with low forward and high reverse resistance would minimize both of these problems. In practice, however, the $v_1$ generator has some internal resistance which is in series with the diode's $r_F$, while any load resistance fed by $v_2$ reduces the benefits of a large diode reverse resistance. Therefore some waveform distortion is always to be expected.

An interesting application of clampers is shown in Fig. 3.39a. This is a voltage doubler circuit made up of a clamper and a peak rectifier. Steady state voltage waveforms are shown in Fig. 3.39b, c, and d. These waveforms apply after the initial "transient" period is over and $C_1$ has been charged at 10 V. The

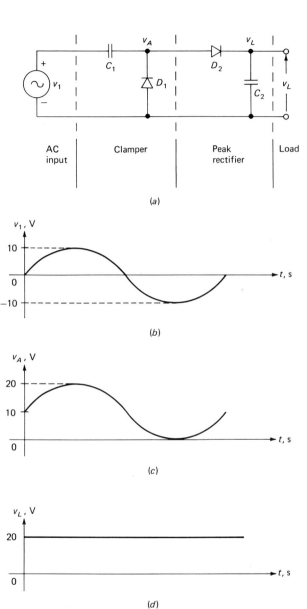

Fig. 3.39 (a) Voltage doubler circuit; (b), (c), and (d) waveforms.

circuit made up of $C_1$ and $D_1$ positively clamps $v_1$ at 0 V, yielding waveform $v_A$. $v_A$ causes $D_2$ to conduct only as long as it takes for $C_2$ to charge up to the peak value of $v_A$, 20 V. As soon as $v_L$ reaches 20 V, $D_2$ stops conducting, which prevents

$C_2$ from discharging. Thus $v_L$ remains at the *peak* value of $v_A$ (hence the name *peak rectifier*).

As long as there is no load, $C_2$ has no way of discharging, and $v_L$ is a smooth, ripple-free 20-V DC. If a small amount of current is drawn from the circuit, $v_L$ drops somewhat, but $D_2$ conducts during the period in which $v_A > v_L$, thus replenishing the charge lost by $C_2$. The net result is the introduction of ripple in $v_L$; obviously, the more current is drawn, the greater the ripple. In a number of applications for which the current requirements are low, however, the voltage doubler is a convenient and economical power supply.

## Problem Set 3.8

3.8.1. Sketch the output waveform and determine the DC level for each of the circuits shown in Fig. 3.40; the input, in all cases, is a 100-V peak sinusoidal voltage. Assume ideal diodes and no output loading.

3.8.2. A 100-V peak square wave whose period is 20 ms is to be positively clamped at 25 V.

    a. Sketch the ideal output waveform.
    b. Determine the DC level in the output waveform.
    c. Sketch the circuit diagram required.
    d. Determine the actual output waveshape if $C_2 = 0.2~\mu F$, there is no load, and the diode parameters are $r_F = 50~\Omega$, $r_R = 1~M\Omega$, $V_T = 0$. Give all critical values of $v_L$.

3.8.3. Sketch the diagram of a circuit that will

    a. Positively clamp at 0 V
    b. Positively clamp at $-50$ V
    c. Positively clamp at $+50$ V
    d. Negatively clamp at 0 V
    e. Negatively clamp at $-50$ V
    f. Negatively clamp at $+50$ V

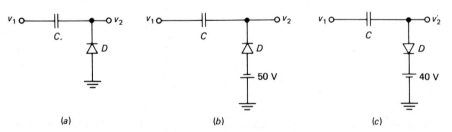

Fig. 3.40    Circuits for Prob. 3.8.1.

## REFERENCES

1. Brazee, J. G.: "Semiconductor and Tube Electronics: An Introduction," Holt, Rinehart and Winston, Inc., New York, 1968.
2. Malvino, A. P.: "Transistor Circuit Approximations," McGraw-Hill Book Company, New York, 1968.
3. Mulvey, J.: "Semiconductor Device Measurements," Tektronix, Inc., Beaverton, Ore., 1968.
4. Cowles, L. G.: "Transistor Circuits and Applications," Prentice-Hall, Inc., Englewood Cliffs, N.J., 1968.
5. Millman, J., and C. C. Halkias: "Electronic Devices and Circuits," McGraw-Hill Book Company, New York, 1967.
6. Angelo, E. J.: "Electronic Circuits," McGraw-Hill Book Company, New York, 1964.
7. Skilling, H. H.: "Electrical Engineering Circuits," John Wiley & Sons, Inc., New York, 1957.
8. Grabinski, J.: "Electronic Power Supplies," Holt, Rinehart and Winston, Inc., New York, 1969.

# chapter 4
# performance measures for amplifying

Basic terminology of amplifying circuits is presented in this chapter.

An electronic amplifier is a circuit capable of magnifying the amplitude of electrical signals. Such a circuit consists of passive components (resistors, inductors, capacitors, diodes, transformers) and active devices properly connected. Passive components are used to perform a variety of tasks such as setting DC levels and coupling signals into and out of the amplifier. Active devices are the key components of any amplifier. They can be vacuum tubes (triode, tetrode, or pentode) or semiconductor devices (bipolar transistor, field effect transistor, tunnel diode).

## 4.1  THE BLACK BOX APPROACH

In this section we consider general properties of amplifiers as opposed to specific analysis of individual components. As such, the amplifier can be characterized as a *black box*, meaning that only the behavior of electric signals across terminals external to the black box are considered. In Fig. 4.1a the triangular symbol represents an amplifier; there are two sets of terminals. The input terminals connect the signal to the amplifier, and the amplified signal appears across the output terminals. $E_g$ is a voltage generator that represents the signal source; $R_g$ is the generator's internal resistance. The load $R_L$ simulates whatever component absorbs the amplifier's output power.

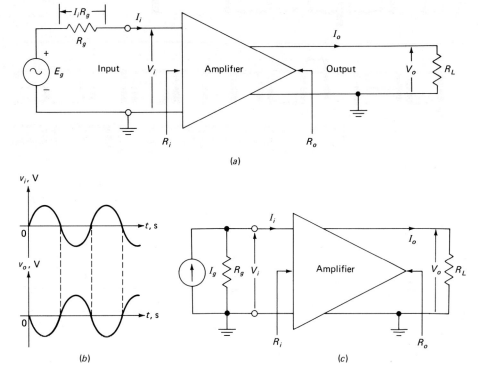

*Fig. 4.1* (a) *Amplifier driven from voltage source;* (b) *input and output voltage waveforms when amplifier introduces 180° phase shift;* (c) *amplifier driven from current source.*

As a result of $E_g$, there is current and voltage at the amplifier's input. The input current depends on the total resistance seen by $E_g$:

$$I_i = \frac{E_g}{R_g + R_i} \tag{4.1}$$

where $R_i$ is the resistance looking into the amplifier's input. The input voltage is

$$V_i = I_i R_i \tag{4.2}$$

or

$$V_i = E_g - I_i R_g \tag{4.3}$$

which can also be expressed as[1]

$$V_i = E_g \frac{R_i}{R_i + R_g} \tag{4.4}$$

[1] See Appendix 3 for a discussion of voltage and current dividers.

On the output side, the amplifier produces an output voltage $V_o$ and an output current $I_o$. For the resistive load shown,

$$I_o = \frac{V_o}{R_L} \tag{4.5}$$

The relationships between $V_o$ and $V_i$ as well as $I_o$ and $I_i$ are denoted by appropriate amplification factors, as follows:

Voltage amplification: $A_v = \dfrac{V_o}{V_i}$ (4.6)

Current amplification: $A_i = \dfrac{I_o}{I_i}$ (4.7)

Note that in the diagram of Fig. 4.1a, $V_i$ and $V_o$ are shown positive with respect to ground. This is simply a convention used when analyzing amplifier circuits. The same applies to $I_i$ and $I_o$; the assumed directions are arbitrary.

In many types of amplifiers a positive-going input voltage produces a negative-going output voltage, and vice versa. Here we say that the output is 180° out of phase with the input, or more simply, the amplifier produces phase inversion. Typical input and output waveforms for such an amplifier are shown in Fig. 4.1b. In this case, when $v_i$ is positive, $v_o$ is negative, and the voltage amplification $A_v$ acquires a negative sign.

## Example 4.1

A generator with an internal resistance of 600 Ω drives an amplifier whose input resistance is 400 Ω. The generator's open-circuit voltage is 1 mV; the amplifier develops 100 mV across an 8-kΩ load. Determine $A_v$ and $A_i$.

**Solution**

Using Fig. 4.1a, the input voltage to the amplifier is

$$V_i = E_g \frac{R_i}{R_g + R_i} = \frac{10^{-3} \times 400}{600 + 400} = 0.4 \text{ mV} \tag{4.4}$$

The input current is

$$I_i = \frac{E_g}{R_g + R_i} = \frac{10^{-3}}{1{,}000} = 1 \text{ μA} \tag{4.1}$$

The output current is

$$I_o = \frac{V_o}{R_L} = \frac{0.1}{8{,}000} = 12.5 \text{ μA} \tag{4.5}$$

The voltage amplification is

$$A_v = \frac{V_o}{V_i} = \frac{0.1}{4 \times 10^{-4}} = 250 \tag{4.6}$$

The current amplification is

$$A_i = \frac{I_o}{I_i} = \frac{12.5 \times 10^{-6}}{10^{-6}} = 12.5 \tag{4.7}$$

In the circuit of Fig. 4.1a, $E_g$ and $V_i$ are the same if $R_g = 0$. In this case $E_g$ would be an ideal voltage source. If $R_g$ is large compared with $R_i$, however, most of $E_g$ is dropped across $R_g$, and $V_i$ is much less than $E_g$.

In the circuit of Fig. 4.1c, $I_g$ and $I_i$ are equal if $R_g$ is infinite. In this case all $I_g$ must flow through the amplifier's input, and $I_i = I_g$. If $R_g$ is small compared with $R_i$, however, most of $I_g$ flows through $R_g$, and $I_i \ll I_g$. In general,

$$I_i = I_g \frac{R_g}{R_i + R_g} \tag{4.8}$$

### Example 4.2

In the circuit of Fig. 4.1a, $V_o = 5$ mV when $E_g = 100$ $\mu$V. If $R_g = 1{,}000$ $\Omega$ and $R_i = 1{,}000$ $\Omega$, determine $V_i$ and $A_v$.

### Solution

The amplifier's input voltage is

$$V_i = E_g \frac{R_i}{R_i + R_g} = \frac{10^{-4} \times 1{,}000}{2{,}000} = 50 \ \mu V \tag{4.4}$$

The voltage-amplification factor is

$$A_v = \frac{V_o}{V_i} = \frac{5 \times 10^{-3}}{50 \times 10^{-6}} = 100$$

### Example 4.3

In the circuit of Fig. 4.1c, $I_o = 0.15$ mA when $I_i = 3$ $\mu$A. If $R_g = 100$ k$\Omega$ and $R_i = 150$ k$\Omega$, determine $I_g$ and $A_i$.

### Solution

The generator current is

$$I_g = I_i \frac{R_i + R_g}{R_g} = \frac{(3 \times 10^{-6})(250 \times 10^3)}{100 \times 10^3} = 7.5 \ \mu A \tag{4.8}$$

The current-amplification factor is

$$A_i = \frac{I_o}{I_i} = \frac{15 \times 10^{-5}}{3 \times 10^{-6}} = 50 \qquad (4.7)$$

Note that both voltage and current are involved in the input and output circuits. Electric power is therefore supplied to the amplifier by the signal generator; similarly, the amplifier delivers power to the load. Generally the load power is larger than the input power, yielding a power gain $G$. For a resistive load and input, $G$ may be determined as follows:

Input power: $\quad P_i = |V_i I_i| \qquad (4.9)$
Load power: $\quad P_o = |V_o I_o| \qquad (4.10)$

Power gain: $\quad G = \dfrac{P_o}{P_i} \qquad (4.11)$

or $\qquad G = \dfrac{|V_o I_o|}{|V_i I_i|} = |A_v A_i| \qquad (4.12)$

Absolute values are used to remove any negative signs that may be present because of phase reversal through the amplifier. Note that $G$ is always a positive number.

### Example 4.4

Determine the power gain for Example 4.1.

### Solution

The input power is

$$P_i = |V_i I_i| = (0.4)(10^{-3})(10^{-6}) = 0.4 \times 10^{-9} \text{ W} \qquad (4.9)$$

The output power is

$$P_o = |V_o I_o| = (0.1)(12.5 \times 10^{-6}) = 1.25 \ \mu\text{W} \qquad (4.10)$$

The power gain is

$$G = \frac{P_o}{P_i} = \frac{1.25 \times 10^{-6}}{0.4 \times 10^{-9}} = 3{,}125$$

The circuits of Fig. 4.1a and c may be analyzed using an appropriate model. The amplifier delivers current to $R_L$, and therefore its behavior can be simulated with an output current generator, including an output resistance $R_o$, as shown in Fig. 4.2. The current produced by this generator is $\alpha I_i$, where $\alpha$ is a current-amplification factor for the device. Note that $\alpha I_i$ is proportional to the input current $I_i$; this is therefore a current-controlled current generator. There are two paths shunting $\alpha I_i$, $R_L$, and $R_o$. The current through $R_L$ is

$$I_o = \alpha I_i \frac{R_o}{R_o + R_L} \tag{4.13}$$

### Example 4.5

A 2,000-Ω microphone feeds an amplifier whose input resistance is also 2,000 Ω. $V_i$ is 10 mV; the load is a 16-Ω loudspeaker. There is no phase inversion. The current-amplification factor $\alpha$ in Fig. 4.2 is 100, and $R_o = 16$ Ω. Determine

a. Current amplification
b. Voltage amplification
c. Power gain

### Solution

a. $I_i = \dfrac{V_i}{R_i} = \dfrac{10 \times 10^{-3}}{2{,}000} = 5 \ \mu\text{A}$

$I_o = \dfrac{\alpha I_i R_o}{R_o + R_L} = \dfrac{(100)(5 \times 10^{-6})(16)}{32} = 250 \ \mu\text{A} \tag{4.13}$

$A_i = \dfrac{I_o}{I_i} = \dfrac{250 \times 10^{-6}}{5 \times 10^{-6}} = 50$

b. $V_o = I_o R_L = (250 \times 10^{-6})(16) = 4$ mV

$A_v = \dfrac{V_o}{V_i} = \dfrac{4 \times 10^{-3}}{10 \times 10^{-3}} = 0.4$

c. $G = |A_v A_i| = 0.4(50) = 20$

Additional measures of performance that may be used in conjunction

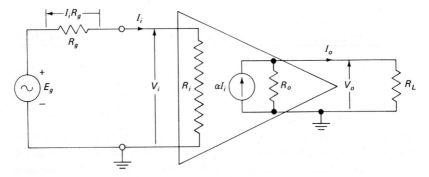

Fig. 4.2 Amplifier analysis using a current-controlled current generator model.

with power gain are the transducer gain $G_T$ and the maximum available gain $G_{MA}$, also called MAG. The transducer gain is defined as follows:

$$G_T = \frac{P_o}{P_{i(max)}} \qquad (4.14)$$

where $P_o$ is the actual output power and $P_{i(max)}$ is the maximum power available from the source. The actual power gain $G$ is equal to the transducer gain if the amplifier's input resistance is matched to the source resistance.[1] The maximum available gain is

$$G_{MA} = \frac{P_{o(max)}}{P_{i(max)}} \qquad (4.15)$$

where $P_{o(max)}$ is the maximum power available from the amplifier output. $P_{o(max)}$ can be achieved only if the load is matched to $R_o$; therefore, in order for $G$ to equal $G_{MA}$, a power match is required at both the input and output. The value of $G_{MA}$ for a given amplifying device is often given by manufacturers; it is a figure of merit useful for comparing different devices. Whether $G_{MA}$ is actually achieved depends on whether the source and load resistances are matched to $R_i$ and $R_o$, respectively.

## Problem Set 4.1

4.1.1. A 50-mV signal generator with an internal resistance of 75 Ω feeds an amplifier of the type in Fig. 4.2 with the following parameters: $R_i$ = 100 kΩ, $\alpha$ = 100, $R_o$ = 10 kΩ, $R_L$ = 10 kΩ. Determine $A_v$, $A_i$, and $G$.

4.1.2. An amplifier is driven from a 10-μA current generator with an internal resistance of 10 kΩ. If $R_i$ = 10 kΩ, $I_o$ = 0.5 mA, and $R_L$ = 2 kΩ, determine $A_i$ and $A_v$.

4.1.3. The following voltage measurements are taken on an amplifier circuit of the type in Fig. 4.2: $E_g$ = 1,000 μV, $V_i$ = 500 μV, $V_o$ = 1 mV, $R_g$ = 10 kΩ, $R_L$ = 100 Ω. Determine the power gain.

4.1.4. Under what conditions do you think $A_i = \alpha$?

4.1.5. An amplifier whose voltage amplification is 10,000 is used to deliver 10 μA to a 100-kΩ load. If the input current is 1 μA, determine $R_i$.

4.1.6. What values of $R_i$ would result in zero power input to an amplifier? What would be the input voltage and current in each case?

4.1.7. What is the output resistance of the amplifier in Prob. 4.1.5 if $\alpha$ = 100?

4.1.8. What is the voltage amplification for an amplifier with $R_i$ = 0?

4.1.9. What is the current amplification for an amplifier with $R_i$ = ∞?

4.1.10. What is the power gain in Probs. 4.1.8 and 4.1.9?

4.1.11. Determine the transducer power gain for the amplifier in Prob. 4.1.3.

---

[1] See Appendix 8 for a discussion of power exchange between source and load.

## 4.2 THE DECIBEL

A common method for indicating the ratio between two quantities whose units are the same is to express the ratio in terms of its logarithm. This approach roughly simulates the response of the human ear; for example, if the power level of a particular sound is increased by a factor of 100, the increase in sound intensity perceived by a human ear is not 100, but a smaller factor proportional to the logarithm of 100 and called the *decibel*. Although amplification is not exclusively limited to sound signals, use of the decibel is widespread because it tends to simplify many calculations involving large ratios.

The actual power gain of an amplifier was defined as

$$G = \frac{P_o}{P_i} \tag{4.11}$$

The same gain can be expressed as

$$G_{(\text{dB})} = 10 \log G \tag{4.16}$$

where log means $\log_{10}$ and dB is the abbreviation for decibel. Suppose, for example, that the power gain of a given amplifier is 1,000. Expressed in decibels, this becomes

$$G_{(\text{dB})} = 10 \log 1{,}000$$
$$G_{(\text{dB})} = 10(3) = 30 \text{ dB}$$

We can therefore say that this amplifier has a power gain of 30 dB or that the output power level is 30 dB above the input power level.

Since power gain is related to the amplification of voltage and current, it is possible to derive the following relationships:

1. $G = \dfrac{P_o}{P_i} = \dfrac{V_o^2/R_L}{V_i^2/R_i}$

   $G = A_v^2 \dfrac{R_i}{R_L}$

   $G_{(\text{dB})} = 10 \log \left( A_v^2 \dfrac{R_i}{R_L} \right) = 10 \log A_v^2 + 10 \log \dfrac{R_i}{R_L}$

   $$G_{(\text{dB})} = 20 \log A_v + 10 \log \dfrac{R_i}{R_L} \tag{4.17}$$

2. $G = \dfrac{P_o}{P_i} = \dfrac{I_o^2 R_L}{I_i^2 R_i}$

   $G = A_i^2 \dfrac{R_L}{R_i}$

   $G_{(\text{dB})} = 10 \log \left( A_i^2 \dfrac{R_L}{R_i} \right) = 10 \log A_i^2 + 10 \log \dfrac{R_L}{R_i}$

$$G_{(dB)} = 20 \log A_i + 10 \log \frac{R_L}{R_i} \tag{4.18}$$

For the special case when $R_L = R_i$, the second terms in Eqs. (4.17) and (4.18) become zero, and

$$G_{(dB)} = 20 \log A_v \tag{4.19}$$
$$G_{(dB)} = 20 \log A_i \tag{4.20}$$

Note that when the ratio of two quantities is less than 1, the decibel equivalent is negative. This would happen, for example, if we were to analyze an attenuation network whose function is to "attenuate" the input signal power by a factor of, say, 100. This means that $P_o = P_i/100$ and

$$G = \frac{P_o}{P_i} = \frac{1}{100}$$

$$G_{(dB)} = 10 \log G = 10 \log \frac{1}{100} = 10(\log 1 - \log 100)$$

$$G_{(dB)} = 10(0 - 2) = -20 \text{ dB}$$

Here the input power level is 20 dB above the output power level, or the output power level is 20 dB below the input power level; either statement is correct.

Another useful relationship to consider is the decibel equivalent for a power ratio of 2:

$$G_{(dB)} = 10 \log 2 = 3 \text{ dB}$$

This relationship may allow us to solve certain problems using quick mental calculations. For example, 100 W is 3 dB above 50 W; similarly 50 W is 3 dB below 100 W. Each time a power level is doubled, it increases by 3 dB; each time it is halved, it decreases by 3 dB. To determine what power level is 9 dB above 1 W, we can say that, since 9 dB is three times 3 dB, the unknown signal must be 1 W doubled three times, namely 8 W. By a similar process, a signal level 9 dB below 1 W corresponds to 1 W halved three times successively: $(1)(\frac{1}{2})(\frac{1}{2})(\frac{1}{2}) = \frac{1}{8}$ W.

We should not think of the decibel as a unit. It does not measure any physical quantity; it is simply a measure of the ratio between two physical quantities, such as output power and input power. In some cases, however, we may want to compare a specific power level $P_x$ with a reference power level. A common reference is 1 mW. The number of decibels that a given power level $P_x$ is above or below the reference power level of 1 mW is

$$P_{x(dBm)} = 10 \log \frac{P_x}{10^{-3}} \tag{4.21}$$

The notation dBm, instead of dB, is used to denote the 1-mW reference level.

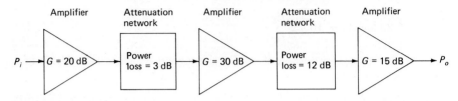

Fig. 4.3 Block diagram of several cascaded amplifiers with interstage attenuation networks.

This allows for the expression of specific power levels in terms of dBm; for example, 100 mW corresponds to $10 \log (100/1) = 20$ dBm. Similarly, $-3$ dBm is one-half of 1 mW, namely 0.5 mW.

An additional advantage of decibel notation results when various power gains $G_1$, $G_2$, etc., must be multiplied to yield an overall gain $G$. For example, if an amplifier stage whose power gain is 1,000 is followed by another stage whose power gain is also 1,000, the overall gain is

$$G = G_1 G_2 \tag{4.22}$$
$$G = 1{,}000(1{,}000) = 10^6$$

If the same operation is performed using decibels, we have

$$G_{(dB)} = 10 \log G_1 G_2 \tag{4.22a}$$
$$G_{(dB)} = 10(\log G_1 + \log G_2)$$
$$G_{(dB)} = 10(3 + 3) = 60 \text{ dB}$$

In general, the decibel notation is simpler because multiplication simplifies to addition and division simplifies to subtraction.

Consider now the block diagram in Fig. 4.3; here a number of amplifiers are *cascaded* to produce greater overall amplification. The attenuation between each cascaded stage may be due to a long cable or to networks used to *couple* one stage to the next, either of which exhibits a power loss. Since the power gains of amplifiers and power losses of attenuation networks are expressed in decibels, it is a simple matter to compute the overall power gain:

$$G = 20 - 3 + 30 - 12 + 15 = 50 \text{ dB}$$

## Problem Set 4.2

4.2.1. Express the following ratios in decibels:

a. 250

b. $\dfrac{1}{50}$

c. $10^4$

d. $4.5 \times 10^{-8}$

e. $\dfrac{350}{1{,}000}$

f. $\dfrac{800 \times 10^{-4}}{0.005}$

4.2.2. Determine the power ratio corresponding to the following:
  a. 80 dB
  b. 0 dB
  c. −40 dB
  d. −12 dB
  e. −24 dB
  f. 15 dB

4.2.3. An amplifier raises the power level of its 5-$\mu$W input signal by 40 dB. What is the output power?

4.2.4. An attenuation network yields an output of 1 $\mu$W with an input of 1 mW. What is the decibel loss of the network?

4.2.5. What is the decibel difference between a 50-kW and a 10-kW radio transmitter?

4.2.6. The noise level of a particular tape recording is 30 dB below the signal level. If the signal power is 10 mW, what is the noise power?

4.2.7. The input resistance to an amplifier with a power gain of 36 dB is 10 k$\Omega$. What input voltage (RMS) is required to produce an output power of 10 W?

4.2.8. The gain of all amplifiers drops at higher frequencies. At some frequency $f_H$ the power gain will have dropped to one-half of its original (low-frequency) value. What is the magnitude of $A_v$ at $f_H$, expressed as a percent of its low-frequency value?

4.2.9. A given radio-frequency transmission line has a power loss of 1 dB per 100 ft. Noise is also picked up along the line at the rate of 1 mW per 100 ft. Assuming a noiseless 10-mW signal at the input, how long can the line be before the noise level rises to 2.22 dB below the signal?

## 4.3 TRANSDUCERS

Although it is customary to analyze amplifier circuits assuming an input voltage or current generator, in practice the "signal" may originate from any one of a large number of possible "sources." Very often the original source is a device capable of converting small changes in some physical parameter to corresponding electrical variations. In this case, as in every case where energy is converted from one form to another, the device is referred to as a *transducer*.

A phonograph cartridge is a typical transducer. Because of the *piezoelectric* effect, certain crystals (such as Rochelle salts), when subjected to mechanical stress, develop a small voltage proportional to the mechanical stress; this process involves the conversion of mechanical to electric energy. For the phonograph cartridge, the lateral motion of a phonograph needle along the record groove results in corresponding voltage variations. Thus, the phonograph cartridge behaves as a signal generator, including an effective internal resistance.

Another example of a signal source is the dynamic or moving-coil microphone. Here, sound vibrations striking a diaphragm cause motion of a coil in a magnetic field, thus generating a proportional emf. The small emf, along with a relatively low internal resistance, (30 to 500 Ω) can serve as the input to an electronic amplifier.

Various photoelectric devices are also available whose electrical properties depend on the light intensity striking their photosensitive surface. Television cameras operate on this principle; changes in light intensity (from the image to be televised) cause corresponding variations in electric current. These electrical variations represent the "signal" which is then amplified, transmitted, and processed until it again becomes a visible image on a television screen. The television camera is therefore a transducer of light to electric energy.

*Strain gages* are often used for the measurement of mechanical strain. The strain gage is little more than fine wire mounted on thin paper backing which is then bonded to the structural member whose mechanical strain we wish to measure. If the member is in tension, the wire stretches, its cross section shrinks, and its electric resistance increases. Thus, if current is caused to flow through the strain gage, its magnitude decreases when the structural member is in tension. As a result of this action, changes in mechanical strain cause corresponding changes in electric current. Since these current variations are relatively small, some amplification is usually required before they can deflect the needle on a meter movement or are otherwise detected.

Other "signal sources" include receiving antennas, thermocouples, and so on. Whatever the nature of a signal source, there is a time-varying quantity involved (there is no point amplifying a signal whose magnitude never changes) and also an effective internal resistance. If the internal resistance is very small, the signal approaches a constant voltage source. In this case *the signal voltage is virtually unaffected by the value of resistance* connected across the signal source. If, on the other hand, the internal resistance is very large, a constant current source is approached. Such a source tends to deliver *the same current to any resistance* connected across its terminals.

Power transfer between source and load (the load may be the input to another amplifier stage) can be maximized if the source's internal resistance is equal to the load resistance it sees. Where there is significant capacitive or inductive reactance, the source has an effective internal *impedance*, rather than a simple resistance; similarly the input to the amplifier may involve an effective *impedance* as well. In cases such as this, power transfer can be maximized by a conjugate match between the source and its load, as follows:

Source impedance = conjugate of load impedance[1]

$$Z_S = Z_L^*  \quad (4.23)$$
$$R_S + jX_S = R_L - jX_L \quad (4.24)$$

[1] See Appendix 8, Eqs. (A.36) and (A.36a).

In the above equations $R_S$ and $X_S$ represent the equivalent resistance and reactance of the source, while $R_L$ and $X_L$ are equivalent resistance and reactance of the load seen by the source.

Just as signal sources can be transducers or other electronic circuits, the load to be fed by an amplifier's output can be another circuit or an appropriate transducer. For example, when more than one stage of amplification is required, the load for stage 1 is simply the input to stage 2, and so on. In any case, any component that absorbs electric power from an amplifier's output appears as an equivalent resistance. This is true of a loudspeaker, for example, which "transduces" an amplifier's electric power output to sound waves; a transmitting antenna is also a resistive load in that it absorbs electric energy from the transmitter's last amplifier stage and converts it to electromagnetic energy for radiation to the atmosphere.

## Example 4.6

The strain gage bridge of Fig. 4.4a is used to generate an electric signal proportional to the mechanical strain on strain gage $R_x$. $R_x$ is bonded to a structural member which is subjected to a sinusoidally time-varying strain; as a result of this strain, the value of $R_x$ alternates between extremes of 999 and 1,001 Ω. The bridge output is fed to an amplifier connected across terminals $XY$.

Determine the equivalent circuit of the signal source (level and internal resistance) feeding the amplifier.

## Solution

The output from the bridge is the signal that has to be amplified. An equivalent circuit for the "signal source" involves a Thevenin voltage and a Thevenin resistance. The Thevenin voltage is the value of $V_{XY}$ when no load is connected across terminals $XY$. The circuit of Fig. 4.4a may be used to determine $V_{XY}$ as follows:

$$V_{XY} = V_{XZ} - V_{YZ} \tag{a}$$

$$V_{YZ} = E \frac{R_2}{R_1 + R_2} \tag{b}$$

$$V_{YZ} = 10 \frac{10^4}{2 \times 10^4} = 5 \text{ V}$$

Note that $V_{YZ}$ is constant, since none of the quantities used to determine it ($E$, $R_1$, $R_2$) ever change.

$$V_{XZ} = E \frac{R_x}{R_3 + R_x} \tag{c}$$

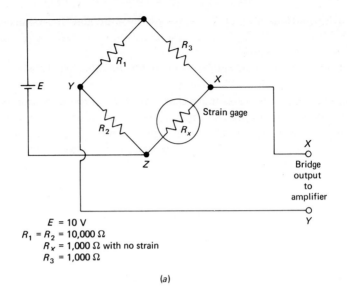

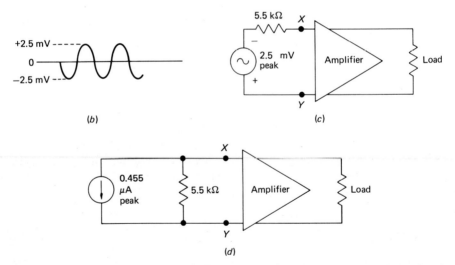

*Fig. 4.4* (a) Strain gage bridge circuit feeding amplifier; (b) open-circuit output voltage of strain gage bridge in (a); (c) Thevenin equivalent circuit of signal source; (d) Norton equivalent circuit of signal source.

Since $R_x$ depends on strain, $V_{XZ}$ is not constant but must be calculated for each value of strain:

a. No strain: $$R_x = 1,000 \ \Omega; \ V_{XZ} = E \frac{R_x}{R_3 + R_x} = \frac{10(1,000)}{2,000}$$
$$= 5 \text{ V}$$

b. Maximum compression: $R_x = 999 \ \Omega$; $V_{XZ} = \dfrac{10(999)}{1,000} = 4.9975 \text{ V}$

c. Maximum tension: $R_x = 1,001 \ \Omega$; $V_{XZ} = \dfrac{10(1,001)}{2,001} = 5.0025 \text{ V}$

The bridge's open-circuit output $V_{XY}$ can now be calculated using Eq. (a):

a. No strain: $V_{XY} = 5 - 5 = 0 \text{ V}$
b. Maximum compression: $V_{XY} = 4.9975 - 5 = -0.0025 \text{ V}$
c. Maximum tension: $V_{XY} = 5.0025 - 5 = +0.0025 \text{ V}$

The Thevenin equivalent voltage is therefore a 2.5-mV peak sine wave, as in Fig. 4.4b. The Thevenin (or Norton) resistance is seen by looking into terminals $XY$ with the battery shorted. This is the internal resistance of the signal source:

$$R_g = \dfrac{R_1 R_2}{R_1 + R_2} + \dfrac{R_3 R_x}{R_3 + R_x} = 5.5 \text{ k}\Omega$$

The nominal value of $R_x$ (1,000 $\Omega$) was used above because the variations in $R_x$ have little effect on the value of $R_g$. The equivalent circuit feeding the amplifier can be simulated using Thevenin parameters as in Fig. 4.4c (2.5-mV peak voltage source in series with a 5.5-k$\Omega$ internal resistance) or with Norton parameters as in Fig. 4.4d [$(2.5 \times 10^{-3})/(5.5 \times 10^3) = 0.455$-$\mu$A peak current source shunted by the 5.5-k$\Omega$ internal resistance]. Either model is correct and adequate in analyzing the amplifying circuit that follows.

### Example 4.7

The maximum RMS open-circuit output voltage for a particular amplifier, without significant distortion, is 1 V. Under matched conditions (output resistance of amplifier matched to load resistance), the maximum undistorted load power is 125 mW. Determine the output resistance of the amplifier and the effective load resistance.

### Solution

Under matched conditions, $R_o = R_L$, and the output voltage is

$$V_o = 1 \left( \dfrac{R_L}{R_o + R_L} \right) = 1 \left( \dfrac{R_L}{2R_L} \right) = 0.5 \text{ V}$$

If the load power is 125 mW,

$$P_o = \dfrac{V_o^2}{R_L} = \dfrac{V_o^2}{R_o}$$

$$0.125 = \frac{0.5^2}{R_o}$$

$$R_o = \frac{0.25}{0.125} = 2 \, \Omega$$

**Example 4.8**

The output voltage for a given audio amplifier is 10 V when the load power is 25 W and 4 V when the load power is 16 W. Determine the amplifier's output resistance.

**Solution**

a. $V_o = 10$ V   $P_o = 25$ W

$$P_o = \frac{V_o^2}{R_L}$$

$$R_L = \frac{V_o^2}{P_o} = \frac{100}{25} = 4 \, \Omega$$

b. $V_o = 4$ V   $P_o = 16$ W

$$R_L = \frac{V_o^2}{P_o} = \frac{16}{16} = 1 \, \Omega$$

c. Using the diagram of Fig. 4.5a,

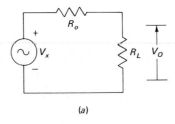

(a)

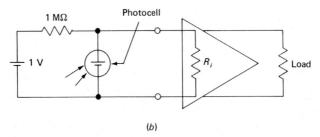

(b)

*Fig. 4.5* (a) *Output equivalent circuit for amplifier of Example 4.8;* (b) *circuit for Prob. 4.3.2.*

$$V_o = V_x \frac{R_L}{R_o + R_L}$$

When $V_o = 10$ V, $R_L = 4\ \Omega$,

$$10 = \frac{4V_x}{R_o + 4}$$

$10R_o + 40 = 4V_x$ \hfill (a)

When $V_o = 4$ V, $R_L = 1\ \Omega$,

$$4 = \frac{1V_x}{R_o + 1}$$

$4R_o + 4 = V_x$ \hfill (b)

Combining Eqs. (a) and (b) we get

$10R_o + 40 = 4(4R_o + 4)$

from which $R_o = 4\ \Omega$.

## Problem Set 4.3

For each problem that follows, draw an appropriate diagram.

4.3.1. A 1-M$\Omega$ crystal microphone delivers $-50$ dBm across 47 k$\Omega$. How much power can it deliver to the input of an amplifier if there is a power match?

4.3.2. The photocell in the circuit of Fig. 4.5b has a "dark" resistance of 200 k$\Omega$ and a "light" resistance of 5 k$\Omega$. The input resistance to the amplifier is 200 k$\Omega$. Determine the nature and magnitude of signal across the amplifier's input terminals if the photocell is alternately illuminated for 1 s and kept dark for 4 s.

4.3.3. A given ceramic cartridge delivers $-47$ dBm into a 500-k$\Omega$ load. What is its internal resistance if the open-circuit voltage is 200 mV?

4.3.4. The load for amplifier $A$ is the input to amplifier $B$. Amplifier $A$ has an output resistance of 10 k$\Omega$, and when properly matched, can deliver 500 $\mu$W to a load. If the input resistance to amplifier $B$ is 2 k$\Omega$, determine its input voltage.

4.3.5. The AC voltage across a radio transmitter's antenna terminals is 100 V. If the power being transmitted is 139 W, determine the load seen by the transmitter's output stage.

## 4.4 IDEAL AMPLIFIERS

As with our discussion of diodes, the ideal amplifier is a useful concept to develop, not because it actually exists, but because it provides a reference against which real amplifiers can be compared. We will define two types of ideal amplifiers: the ideal voltage amplifier and the ideal current amplifier.

The ideal voltage amplifier has infinite input resistance and zero output resistance. Such an amplifier, shown in Fig. 4.6a, produces an output voltage dependent only on the value of input voltage; hence it is said to be *voltage-controlled*. The internal resistance of the source need not be considered, because the "open circuit" which represents the amplifier's input forces $I_i$ to equal zero, and hence $E_g = V_i$ at all times. Since no current is drawn, it follows that the power extracted from the source is zero. On the output side, it is convenient to use a voltage source as opposed to a current source, because $R_o = 0$. The amplification factor $k$ is unitless; it determines by how much $V_i$ is amplified to produce $V_o$. Since $R_o = 0$, the voltage across $R_L$ is always equal to $kV_i$; hence $V_o$ is never affected by the value of $R_L$.

The graphic characteristics for the ideal voltage amplifier are given in Fig. 4.6b. Suppose that $V_i = 2$ V and $k = 20$; since $V_o = kV_i$, it follows that $V_o = 2(20) = 40$ V. The vertical line $V_i = 2$ V expresses this result graphically; it further points out that $V_o$ remains the same for all values of $I_o$. This means that once $V_i$ has been set, $V_o$ does not vary, even if the output current drawn by $R_L$ changes.

To summarize, the ideal voltage amplifier of Fig. 4.6a has an infinite input resistance, and therefore it does not load the signal source. This is often desirable because loading of the source may result in substantial signal loss (when $R_g$ is high) or excessive current drain. Additionally, the fact that $R_o = 0 \ \Omega$ allows the amplifier to drive any load, no matter how small the resistance,

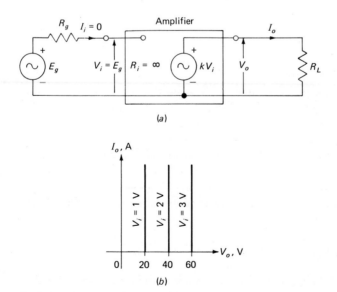

*Fig. 4.6  Ideal voltage amplifier;* (a) *circuit;* (b) *output characteristics when* $k = 20$.

without any reduction in $V_o$. Theoretically, at least, $V_o$ remains the same as $R_L$ varies from infinity to 0 Ω. Thus, the ideal voltage amplifier requires no input power and can deliver infinite output power (when $R_L = 0$ Ω). Practical devices whose characteristics approach those described above include the vacuum tube and field effect transistor, at least so far as the input resistance is concerned.

Consider now the ideal current amplifier of Fig. 4.7a. Ideal current amplifiers are characterized by zero input and infinite output resistance. On the input side, there is a signal generator with an internal resistance $R_g$. The Norton equivalent circuit is used for the signal source to facilitate the analysis. The amplifier's zero input resistance causes $I_g$ to equal $I_i$, no matter what the value of $R_g$. This amplifier is *current-controlled*, because the only input is the current $I_i$. There is no input voltage or input power, because $R_i = 0$. This condition may be approached by junction transistor amplifiers.

The output circuit consists of a current generator that produces an output current equal to $\alpha I_i$. Since $R_o = \infty$, it follows that $\alpha I_i = I_o$, regardless of the value of $R_L$. The load, in fact, is driven from a constant current source. Should $R_L$ change, for example, $I_o$ remains the same; only $V_o$ changes, since $V_o = I_o R_L$.

The output characteristics of Fig. 4.7b illustrate the amplifier's operation graphically. If, for example, $I_i = 2$ mA and $\alpha = 50$,

$$I_o = \alpha I_i = 50(2 \times 10^{-3}) = 100 \text{ mA}$$

The horizontal line corresponding to $I_i = 2$ mA and $I_o = 100$ mA implies that

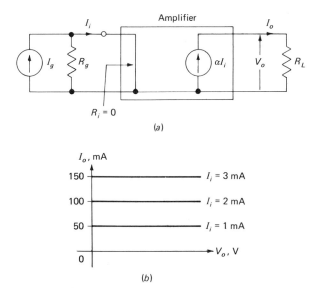

Fig. 4.7 Ideal current amplifier; (a) circuit; (b) output characteristics for $\alpha = 50$.

$V_o$ can take on any value. The actual value of $V_o$ is therefore controlled by $R_L$, and if $R_L = \infty$, an infinite voltage and hence infinite load power results. Thus the ideal current amplifier can produce infinite load power while requiring no input power whatever.

**Example 4.9**

An ideal voltage amplifier is driven from a crystal cartridge whose open-circuit voltage and internal resistance are 50 mV and 100 k$\Omega$, respectively. The voltage-amplification factor $k = 20$; $R_L$ can take on any of three different values: 50, 100, 500 $\Omega$. Plot $I_o$ versus $V_o$.

**Solution**

From the equivalent circuit of Fig. 4.6a, it is evident that the input voltage to the amplifier $V_i$ is equal to the generator's open-circuit voltage $E_g$ because $R_i = \infty$. Thus, the generator's internal resistance does not enter the picture at all.

Since $V_i = E_g = 50$ mV, it follows that $V_o = kV_i = 20(50 \times 10^{-3}) = 1$ V. The output current can be calculated for each specific value of $R_L$; using $I_o = V_o/R_L$, we get

a. $R_L = 50\ \Omega$: $I_o = \dfrac{1}{50} = 20$ mA

b. $R_L = 100\ \Omega$: $I_o = \dfrac{1}{100} = 10$ mA

c. $R_L = 500\ \Omega$: $I_o = \dfrac{1}{500} = 2$ mA

The above results are plotted on the graph of Fig. 4.8a. The three points

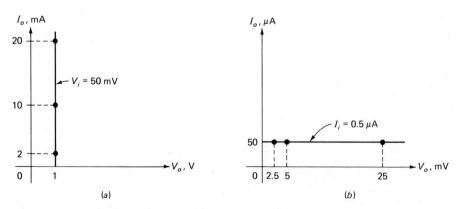

*Fig. 4.8* Plot of $V_o$ versus $I_o$ for (a) *Example 4.9;* (b) *Example 4.10.*

have been connected with a straight line to obtain a continuous plot that applies for any intermediate value of $R_L$. If additional values of $V_i$ were to be considered, a family of curves such as in Fig. 4.6b would result.

### Example 4.10

An ideal current amplifier is driven from the same signal source and drives the same load as in the previous example. The amplifier's current-amplification factor $\alpha = 100$. Plot $I_o$ versus $V_o$.

### Solution

As indicated in the circuit of Fig. 4.7a, $R_i$ for an ideal current amplifier is zero. The input current is the short-circuit current of the signal source (Norton equivalent circuit); it can be determined as follows:

$$I_g = I_i = \frac{E_g}{R_g} = \frac{50 \times 10^{-3}}{100 \times 10^3} = 0.5 \ \mu A$$

Since $\alpha = 100$, $\alpha I_i = 50 \ \mu A$; this is the output current for any value of $R_L$. The output voltage, however, depends on $R_L$. Using $V_o = I_o R_L$, we get

a. $R_L = 50 \ \Omega$: $V_o = (50 \times 10^{-6})(50) = 2.5$ mV
b. $R_L = 100 \ \Omega$: $V_o = (50 \times 10^{-6})(100) = 5$ mV
c. $R_L = 500 \ \Omega$: $V_o = (50 \times 10^{-6})(500) = 25$ mV

The above data are used to plot $V_o$ versus $I_o$ in Fig. 4.8b for the specific case $I_i = 0.5 \ \mu A$. If we consider additional values of $I_i$, a family of curves similar to that in Fig. 4.7b results.

In addition to the ideal voltage and current amplifiers, we may also define ideal *transresistance* and *transconductance* amplifiers. The ideal transconductance amplifier is characterized by infinite $R_i$ and $R_o$; hence it is *voltage-controlled* ($I_i = 0$) and delivers an *output current* independent of $R_L$. The performance measure for this type of amplifier is $I_o/V_i$, which has the units of mhos, and hence the term *transconductance*. The ideal transresistance amplifier is characterized by $R_i = R_o = 0$; hence it is *current-controlled* ($V_i = 0$) and delivers an *output voltage* independent of $R_L$. The performance measure is $V_o/I_i$, which has the units of ohms, and hence the term *transresistance*. See Table 4.1 for a summary.

### Problem Set 4.4

4.4.1. What is the input resistance to

    a. An ideal voltage amplifier?
    b. An ideal current amplifier?

Table 4.1. Characteristics of Ideal Amplifiers.

| Type of ideal amplifier | $R_i$ | $I_i$ | $V_i$ | $R_o$ | $V_o$ | $I_o$ | Applicable performance measure |
|---|---|---|---|---|---|---|---|
| Voltage | $\infty$ | 0 | $V_i$ | 0 | $V_o$ | $V_o/R_L$ | $V_o/V_i$ |
| Current | 0 | $I_i$ | 0 | $\infty$ | $I_o R_L$ | $I_o$ | $I_o/I_i$ |
| Transresistance | 0 | $I_i$ | 0 | 0 | $V_o$ | $V_o/R_L$ | $V_o/I_i$ |
| Transconductance | $\infty$ | 0 | $V_i$ | $\infty$ | $I_o R_L$ | $I_o$ | $I_o/V_i$ |

4.4.2. What effect does the internal resistance of a signal generator have on the input voltage to an ideal voltage amplifier?

4.4.3. What effect does the internal resistance of a signal generator have on the input current to an ideal current amplifier?

4.4.4. Why is the signal power input to an ideal amplifier (voltage or current) always zero?

4.4.5. Do you think it is possible to analyze an ideal voltage amplifier using a Norton equivalent circuit for the signal source?

4.4.6. Do you think it is possible to analyze an ideal current amplifier using a Thevenin equivalent circuit for the signal source?

4.4.7. Why is an ideal voltage amplifier said to be *voltage-controlled* while an ideal current amplifier is *current-controlled*?

4.4.8. What is the maximum load power for each of the ideal amplifiers of Figs. 4.6a and 4.7a?

4.4.9. Regardless of the value of $R_L$, what are the values of voltage amplification and power gain for the ideal amplifier of Fig. 4.6a?

4.4.10. Regardless of the value of $R_L$, what is $A_i$ and $G$ for the ideal amplifier of Fig. 4.7a?

4.4.11. Repeat Example 4.9 for a signal source whose short-circuit current and internal resistance are 10 μA and 10 kΩ, respectively.

4.4.12. Repeat Example 4.10 for a signal source whose open-circuit voltage and internal resistance are 1 V and 1 MΩ, respectively.

4.4.13. Are considerations involving maximum power transfer meaningful for the circuits of Figs. 4.6a and 4.7a? Explain your answer.

## 4.5 FREQUENCY RESPONSE

### 4.5a Introduction

You will recall that all signals to be amplified vary with time in some manner. Applications exist for signals whose period (time to complete a full

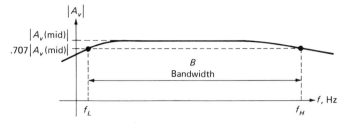

Fig. 4.9  Typical frequency response curve.

cycle) extends anywhere from a few tenths of a nanosecond to several hours; this corresponds to a frequency spread from several gigahertz, or GHz (G = giga = $10^9$), down to 0.001 Hz or less. The problems encountered in amplifying signals at extremely low frequencies are different from those at the opposite end of the frequency spectrum.

No amplifier can satisfactorily amplify signals of all frequencies. There are limitations in the amplifying devices themselves which preclude this ideal situation. There are also many cases where an amplifier is intentionally designed to be selective. This is done if a specific range of frequencies is of interest, as in a radio broadcast receiver, where we wish to receive only one station at a time. In such instances, the amplifier performs a *filtering* function in addition to amplifying. Its selective amplification results in the filtering out of signals whose frequencies lie outside the desired band.

In general, amplification drops at both low and high frequencies. A typical frequency response curve is shown in Fig. 4.9. Here the absolute value (magnitude) of voltage amplification $|A_v|$ is plotted as a function of frequency. The absolute value is used because, in general, $A_v$ is a complex number—that is, it involves magnitude and phase shift. The phase shift through the amplifier is due to the presence of reactive components;[1] it is frequency-dependent and may be plotted as a function of frequency on a graph similar to that of Fig. 4.9. Note that $|A_v|$ is reasonably flat over the midfrequency range but falls off on each side. The "roll-off" is not abrupt, but gradual; hence frequencies far from the flat response region are also amplified, but to a lesser degree. A measure of an amplifier's frequency response is the bandwidth $B$, which may be defined as the frequency band in which $|A_v|$ remains above 70.7 percent of the midfrequency amplification $|A_{v(\text{mid})}|$. In Fig. 4.9, the frequencies where $|A_v|$ drops below 0.707 $|A_{v(\text{mid})}|$ are indicated as $f_L$ and $f_H$; these represent the low- and high-frequency cutoff, respectively. The bandwidth is defined as

$$B = f_H - f_L \tag{4.25}$$

Since power is proportional to voltage squared, it follows that when $|A_v|$

---

[1] See Appendix 6 for a review of complex algebra.

drops to 0.707 of its midfrequency value, power gain drops to $.707^2 = \frac{1}{2}$ of the midfrequency value. Note that this is a 3-dB drop. The frequencies $f_H$ and $f_L$ are therefore called the *half-power* or 3-dB frequencies. Similarly, $B$ is referred to as the 3-dB bandwidth.

In addition to the amplifying device, frequency limitations arise because of *coupling* networks. A coupling network may involve nothing more than a capacitor or transformer, or it may be a more complex, tuned circuit. In the simplest of cases, the coupling network allows the desired signal to pass from source to amplifier input and from amplifier output to load while it prevents all other voltages and currents that are not part of the signal from doing the same. These nonsignal voltages and currents are usually DC levels required to operate the amplifying device.

In the diagram of Fig. 4.10a capacitors $C_1$ and $C_2$ are used to couple the signal from the generator to the amplifier's input and from the output of the amplifier to the load, respectively. Both capacitors are selected to have negligible reactance at the lowest signal frequency to be amplified so that signal current flow is unopposed by their presence. The direct currents required by the amplifier (they are not part of the signal), however, are prevented by the capacitors from flowing through either the signal generator or the load.

Transformer coupling in Fig. 4.10b performs the same basic function as

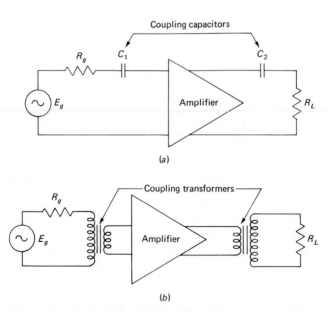

*Fig. 4.10* Techniques for allowing AC signal to pass from source to amplifier to load while blocking DC levels; (a) capacitor coupling; (b) transformer coupling.

the coupling capacitors. Since a transformer does not respond to DC, only AC signals are coupled from the generator to the amplifier's input and from the amplifier's output to the load. The transformer has an advantage in that its turns ratio can be specified to provide the impedance levels necessary to optimize circuit performance. In order to carry out its coupling function efficiently, however, the transformer's shunt reactance must be relatively high at the signal frequency.

Although coupling capacitors and transformers perform reasonably well in the frequency range for which they are selected, neither have an unlimited frequency response. As the signal frequency decreases, for example, the reactance of coupling capacitors increases; this reduces the input current and voltage to the amplifier. The net result is a reduction in the output level. When transformers are used, a similar effect takes place. At low frequencies, the shunt reactance of the transformer drops; this causes more current to be drawn from the signal generator, and therefore more voltage is dropped across $R_g$, leaving less voltage to be developed across the amplifier's input terminals. We can therefore conclude that both coupling capacitors and transformers degrade the low-frequency response.

At higher frequencies, the amplifying device exhibits a lower amplification factor due to transit-time effects (transit time refers to the ability of charge carriers to respond to rapid signal variations). There are also various capacitances in shunt with the signal path. Some of the capacitances are inherent in the amplifying device. Others are due to distributed capacitance of coupling transformers; "stray" capacitance between leads that connect the circuit makes up the rest. The net effect of the shunt capacitive reactances is to degrade the high-frequency response.

## Problem Set 4.5

4.5.1. Why is it impossible to produce an amplifier with unlimited frequency response?

4.5.2. Indicate the effect on the low-frequency response of the following components:

    a. Capacitance in series with signal path
    b. Inductance shunting the signal path
    c. Capacitance shunting the signal path
    d. Inductance in series with signal path

4.5.3. Repeat Prob. 4.5.2 for effects on the high-frequency response.

4.5.4. Can an inductor be used to perform the capacitors' coupling function in the circuit of Fig. 4.10a? Explain your answer.

### 4.5b  Bode Plots

Calculations to determine frequency response involve frequency-dependent quantities such as capacitive and inductive reactances. To adequately cover this topic, we will make use of a graphic procedure that simplifies the mathematics and allows for a greater understanding of the relationships involved. These graphs are referred to as asymptotic plots or Bode plots, after the scientist who made many contributions to the field of electrical engineering using such techniques.

Before proceeding to a study of Bode plots, you should review the principles of complex algebra (Appendix 6) so that you can manipulate the mathematical expressions to follow.

To begin, we must define two terms widely used to express frequency ratios. These are the *octave* and the *decade*. An octave represents a frequency ratio of two; for example, 20 Hz is an octave above 10 Hz. A decade represents a frequency ratio of 10; hence 1,000 Hz is a decade below 10,000 Hz.

If frequency $f_2$ is $n$ octaves above frequency $f_1$,

$$\frac{f_2}{f_1} = 2^n \tag{4.26}$$

or

$$n = \log_2 \frac{f_2}{f_1} \tag{4.26a}$$

As an example, to determine how many octaves 32 Hz is above 2 Hz,

$$n = \log_2 \frac{32}{2} = \log_2 16 = 4 \text{ octaves}$$

Similarly, if $f_2$ is $m$ decades above $f_1$,

$$\frac{f_2}{f_1} = 10^m \tag{4.27}$$

or

$$m = \log_{10} \frac{f_2}{f_1} \tag{4.27a}$$

If we wish to know what frequency is 3 decades above 60 Hz, we compute $f_2$ as follows:

$$f_2 = 10^m f_1 = 10^3 \times 60 = 60 \text{ kHz} \tag{4.27}$$

Now consider the expression for power gain as developed in Sec. 4.2:

$$G_{(\text{dB})} = 20 \log A_v + 10 \log \frac{R_i}{R_L} \tag{4.17}$$

Since $A_v$ is usually the only frequency-dependent term, we normally plot only the first term of Eq. (4.17). Such a term often takes the following general form:

$$A_v = \frac{K(1 + jf/f_1)}{1 + jf/f_2} \tag{4.28}$$

where $K$ is a constant, $f$ is the frequency in Hz, and $f_1$ and $f_2$ are specific constant values of frequency. This expression, because of the $j$ operator, may be resolved into a *magnitude* and a *phase* component. The magnitude component represents the actual amplification, while the phase component indicates phase shift between output and input voltages. Each of these components must be plotted separately versus $f$, yielding the following plots:

1. $20 \log |A_v|$ versus $f$
2. Phase of $A_v$ versus $f$

To plot Eq. (4.28), assume $K = 100$, $f_1 = 10$ kHz, and $f_2 = 100$ kHz. The magnitude plot is obtained as follows:

$$20 \log |A_v| = 20 \log \left| \frac{100(1 + f/10^4)}{1 + jf/10^5} \right| = 20 \log \frac{100|1 + jf/10^4|}{|1 + jf/10^5|}$$

$$20 \log |A_v| = 20 \log 100 + 20 \log \left|1 + \frac{jf}{10^4}\right| - 20 \log \left|1 + \frac{jf}{10^5}\right|$$

The first term, $20 \log 100$, is equal to 40 dB; this term is independent of frequency and is plotted in Fig. 4.11a. The second term, $20 \log |1 + jf/10^4|$, requires additional consideration. There are three distinct frequency ranges of interest:

a. $f \ll 10^4$ Hz: The quantity $f/10^4 \ll 1$. We can therefore approximate the total term as $20 \log |1| = 0$ dB. Thus, for frequencies much lower than $10^4$ Hz, the expression $20 \log |1 + jf/10^4|$ represents a constant 0 dB. The frequency $10^4$ Hz is referred to as the *break* or *corner* frequency.
b. $f = 10^4$ Hz: The quantity $f/10^4 = 1$. The expression $20 \log |1 + jf/10^4| = 20 \log |1 + j1| = 20 \log 1.41 = 3$ dB.
c. $f \gg 10^4$ Hz: The quantity $f/10^4 \gg 1$. We can neglect the real number 1, which yields

$$20 \log \left|1 + \frac{jf}{10^4}\right| = 20 \log \left|\frac{jf}{10^4}\right| = 20 \log \frac{f}{10^4}$$

As $f$ increases beyond $10^4$ Hz, the term $20 \log (f/10^4)$ also increases. Each time $f$ doubles (rises by one octave), the value of $20 \log (f/10^4)$ increases by 6 dB. To prove this, assume any $f$, say $f = 10^5$ Hz; this yields 20 log

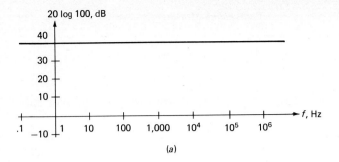
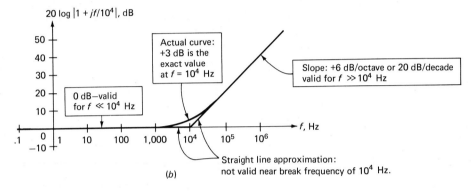
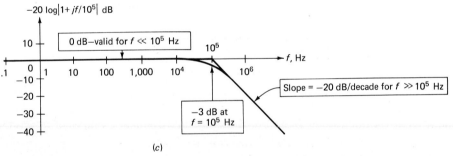
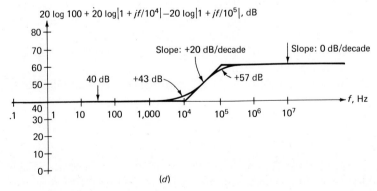

Fig. 4.11  *Development of magnitude plots using asymptotic approximation techniques.*

$(10^5/10^4) = 20 \log 10 = 20$ dB. Now if we double $f$, we get $20 \log 20 = 26$ dB. Hence the term $20 \log (f/10^4)$ can be plotted on semilog paper as a straight line whose slope is 6 dB per octave. The slope can also be defined in terms of decades, rather than octaves. Since a decade corresponds to a frequency ratio of 10, the term $20 \log (f/10^4)$ increases at the rate of 20 dB per decade. Again, this can be shown with an example. At $f = 10^5$ Hz the value of our term was 20 dB. A decade above this frequency, $f = 10^6$ Hz, and $20 \log (10^6/10^4) = 20 \log 100 = 40$ dB. Hence the magnitude of our term has increased from 20 to 40 dB, a net 20-dB rise, in one decade.

To summarize, the term $20 \log |1 + jf/10^4|$ may be approximated with a straight line along 0 dB for $f \ll 10^4$ Hz; it has an exact value of 3 dB at $f = 10^4$ Hz and rises beyond the break frequency at the rate of 6 dB per octave, or 20 dB per decade. The complete plot for this term is shown in Fig. 4.11$b$. Note that the straight-line approximation is quite accurate for frequencies far removed from the break frequency. At the break frequency, however, the asymptotic approximation is 3 dB in error.

We now come to the third term, $-20 \log |1 + jf/10^5|$. Again, there are three distinct regions of interest:

a. $f \ll 10^5$ Hz, in which case the term is a constant 0 dB.
b. $f = 10^5$ Hz, which yields $-20 \log |1 + j1| = -3$ dB.
c. $f \gg 10^5$ Hz, which results in a straight line whose slope is $-20$ dB per decade. The negative slope is due to the negative sign in front of the term. This last term is plotted in Fig. 4.11$c$.

The graph for the complete expression is obtained by combining the three plots from Fig. 4.11$a$, $b$, and $c$. For $f \ll 10^4$ Hz, the sum of all three plots is 40 dB. Beyond $f = 10^4$ Hz, the second term rises at the rate of 20 dB per decade, and since no other term is yet significant, the plot for the total expression does the same. Beyond $f = 10^5$ Hz, the third term contributes a negative slope of 20 dB per decade which cancels the positive slope due to the second term. The complete plot is given in Fig. 4.11$d$, including the 3-dB corrections at the two break frequencies.

We now come to the phase response. In general, the phase for a complex number $a + jb$ is

$$\phi(a + jb) = \tan^{-1} \frac{b}{a} \qquad (4.29)$$

as developed in Appendix 6, Eq. (A.12).

For the numerical example just discussed,

$$A_v = \frac{100(1 + jf/10^4)}{1 + jf/10^5}$$

The phase of $A_v$ is

$$\phi(A_v) = \tan^{-1}\frac{f}{10^4} - \tan^{-1}\frac{f}{10^5}$$

We will plot each term in the phase equation separately and then combine the two. The first term is plotted by considering three distinct possibilities:

a. $f \ll 10^4$ Hz: The quantity $f/10^4$ is very small, and the angle whose tangent is $f/10^4$ is also very small. Thus, as $f$ decreases, the phase angle approaches zero.
b. $f = 10^4$ Hz: The quantity $f/10^4 = 1$, and $\tan^{-1} 1 = 45°$.
c. $f \gg 10^4$ Hz: Since $f/10^4$ is large, $\tan^{-1}(f/10^4)$ approaches 90°. Hence, as $f$ increases beyond $10^4$ Hz, the phase angle approaches 90°.

In conclusion, the phase is zero for frequencies much below the break frequency; as the frequency is increased, the phase also increases until it is 45° at the break frequency. Beyond this point the phase continues to rise toward 90° for frequencies much higher than the break frequency. A complete graph for this term is shown in Fig. 4.12a. To obtain more accuracy when plotting such a graph, the phase can be computed at two additional points, one decade below and one decade above the break frequency. This yields

One decade below $10^4$ Hz: $\phi = \tan^{-1}\frac{10^3}{10^4} \simeq 6°$

One decade above $10^4$ Hz: $\phi = \tan^{-1}\frac{10^5}{10^4} \simeq 84°$

The second phase term, $-\tan^{-1}(f/10^5)$, follows the same pattern:

a. $f \ll 10^5$ Hz: $\phi \to 0°$
b. $f = 10^4$ Hz (one decade below break frequency): $\phi \simeq -6°$
c. $f = 10^5$ Hz (break frequency): $\phi = -45°$
d. $f = 10^6$ Hz (one decade above break frequency): $\phi \simeq -84°$
e. $f \gg 10^5$ Hz: $\phi \to -90°$

The graph for this phase term is given in Fig. 4.12b; the total phase shift is obtained by combining the two phase plots, yielding the composite curve of Fig. 4.12c.

Phase plots provide information regarding the phase shift through a particular circuit. In the example just completed, the output voltage is in phase with the input voltage for $f < 100$ Hz and $f > 10^7$ Hz. Between these extremes, the output leads the input by 6° at 1,000 Hz to 45° near 25,000 Hz and down to 6° again near $10^6$ Hz.

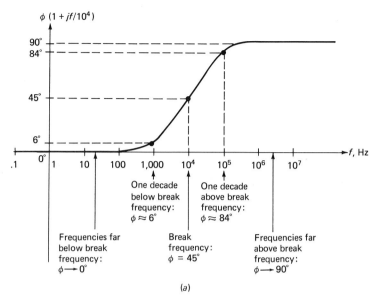

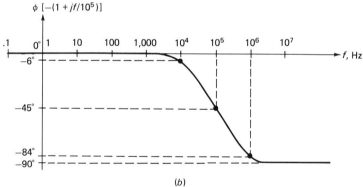

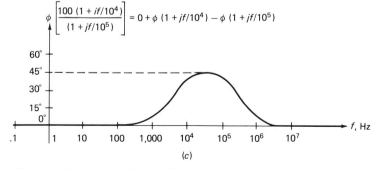

Fig. 4.12  Development of phase plots.

## Example 4.11

Obtain the magnitude and phase response for $A_v$ in the circuit of Fig. 4.13a. Determine the midfrequency amplification; also determine the low- and high-frequency cutoff, where applicable.

## Solution

The passive network in Fig. 4.13a does not amplify, because the output voltage $v_2$ is smaller than the input voltage $v_1$. The ratio $v_2/v_1$, however, can still be defined as $A_v$, although $|A_v|$ will, of course, be less than 1. Another name for $v_2/v_1$ is *transfer function*. Either term, $A_v$ or transfer function, may be used to denote the ratio of two voltages.

$$A_v = \frac{v_2}{v_1} = \frac{1/(j\omega C)}{R + 1/(j\omega C)} = \frac{1}{j2\pi f RC + 1} = \frac{1}{1 + jf/[1/(2\pi RC)]}$$

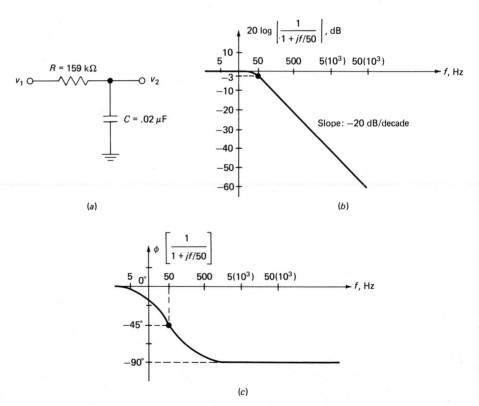

Fig. 4.13 (a) Circuit for Example 4.11; (b) magnitude plot for (a); (c) phase plot for (a).

which corresponds to the standard form previously developed. Substituting numerical values,

$$A_v = \frac{1}{1 + [jf/(1/0.02)]} = \frac{1}{1 + jf/50}$$

The magnitude response is

$$20 \log |A_v| = 20 \log 1 - 20 \log \left|1 + \frac{jf}{50}\right|$$

$$20 \log |A_v| = 0 - 20 \log \left|1 + \frac{jf}{50}\right|$$

The asymptotic approximation is 0 dB up to 50 Hz and a straight line with $-20$ dB per decade slope beyond that. After applying the 3-dB correction factor at the break frequency, the result is as shown in Fig. 4.13b. The midfrequency amplification is 0 dB; there is no low-frequency cutoff, and the high-frequency cutoff $f_H = 50$ Hz.

The phase response is

$$\phi(A_v) = \phi\left(\frac{1}{1 + jf/50}\right)$$

$$\phi(A_v) = 0° - \tan^{-1}\frac{f}{50}$$

This may be approximated as follows:
1. $f \ll 50$ Hz: $\phi \to 0°$
2. $f = 50$ Hz (break frequency): $\phi = -45°$
3. $f \gg 50$ Hz: $\phi \to -90°$

The graph for phase response is displayed in Fig. 4.13c. Note that the phase is $-45°$ at $f_H$.

**Example 4.12**

Obtain the magnitude and phase response for the circuit of Fig. 4.14. Determine $|A_{v(\text{mid})}|$, in decibels; also determine $f_H$ and $f_L$, if applicable.

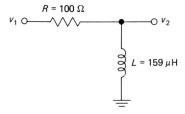

Fig. 4.14 Circuit for Example 4.12.

## Solution

$$A_v = \frac{v_2}{v_1} = \frac{j\omega L}{j\omega L + R} = \frac{j2\pi fL}{R[1 + j2\pi f/(R/L)]}$$

$$A_v = \frac{jf2\pi L/R}{1 + jf/[R/(2\pi L)]} = \frac{10^{-5}jf}{1 + jf/10^5}$$

$$20 \log |A_v| = 20 \log 10^{-5} + 20 \log |jf| - 20 \log \left|1 + \frac{jf}{10^5}\right|$$

We will plot each term in the above expression separately and then combine the results:

1. $20 \log 10^{-5}$: This represents a constant level of $20(-5) = -100$ dB, as indicated in the graph of Fig. 4.15a.
2. $20 \log |jf|$: This term is different from those previously discussed. It is actually very easy to graph, since $20 \log |jf| = 20 \log f$, which is the equation of a straight line whose slope is 20 dB per decade. Because there is no break frequency as such, the term is significant at all frequencies and may be plotted by evaluating it at some arbitrary frequency. If we choose $f = 1$ Hz, we get $20 \log 1 = 0$ dB. Hence the term may be plotted as a straight line that crosses the 0-dB axis at $f = 1$ Hz with a slope of 20 dB per decade. The plot is shown in Fig. 4.15b.
3. $-20 \log |1 + jf/10^5|$: This is 0 dB for $f \ll 10^5$ Hz, $-3$ dB at the break frequency ($10^5$ Hz), and asymptotically approaches a straight line whose slope is $-20$ dB per decade beyond $10^5$ Hz. Figure 4.15c displays the plot.

When the three plots of Fig. 4.15a, b, and c are combined, the graph of Fig. 4.15d results. From this we can read the midfrequency amplification as 0 dB and the low-frequency cutoff as $10^5$ Hz; there is no high-frequency cutoff.

The phase response is

$$\phi(A_v) = \phi\left(\frac{10^{-5}jf}{1 + jf/10^5}\right)$$

$$\phi(A_v) = \phi(jf) - \phi\left(1 + \frac{jf}{10^5}\right)$$

The term $jf$ has a constant phase of 90° at all frequencies, since no real component is involved; this is shown in Fig. 4.15e. The second term involves $-45°$ at the break frequency with the curve approaching 0° below and $-90°$ above the break frequency; the appropriate graph is shown in Fig. 4.15f. The composite curve of Fig. 4.15g provides the complete phase response. Note that there is a phase shift of 45° at the cutoff frequency $f_L$.

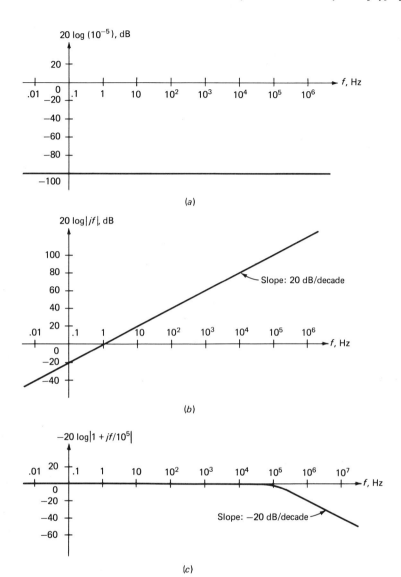

Fig. 4.15  Bode plots for the circuit of Fig. 4.14.

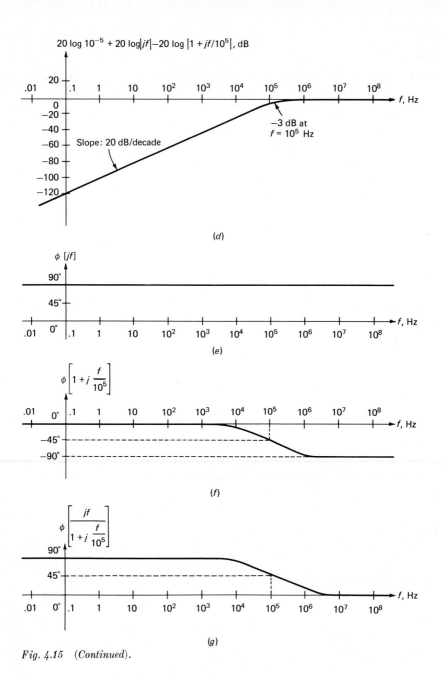

Fig. 4.15 (Continued).

## Example 4.13

Determine the magnitude and phase response for the circuit of Fig. 4.16a. What is $|A_{v(\text{mid})}|$ in decibels? What are the cutoff frequencies?

## Solution

This circuit could represent the input to an amplifier. $v_1$ might be the signal generator with an internal resistance $R_g$ of 600 Ω; $C_1$ could be the coupling capacitor. $R_1$ and $C_2$ could be the amplifier's input resistance and capacitance, respectively.

The midfrequency response is obtained by assuming $C_1$ to behave as a short circuit and $C_2$ as an open circuit. This is valid because at midfrequencies the reactance of $C_1$ is very small ($C_1$ determines the low-frequency response) while the reactance of $C_2$ is very high ($C_2$ controls the high-frequency cutoff). Using the midfrequency equivalent circuit of Fig. 4.16b, we can write

$$A_{v(\text{mid})} = \frac{v_2}{v_1} = \frac{R_1}{R_g + R_1} = \frac{1.4 \times 10^3}{2 \times 10^3} = 0.7 \qquad (a)$$

To obtain the low-frequency response, we can assume $C_2$ to be an open circuit, which simplifies the analysis. This assumption must be checked later

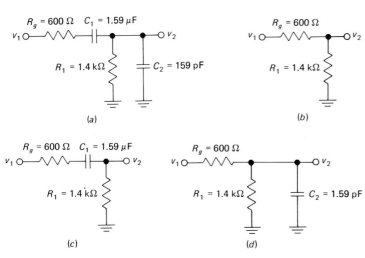

Fig. 4.16 (a) Circuit for Example 4.13; (b) midfrequency equivalent for (a); (c) low-frequency equivalent for (a); (d) high-frequency equivalent for (a).

on to determine whether it is actually valid. Using the equivalent circuit of Fig. 4.16c, we have

$$A_{v(\text{low})} = \frac{v_2}{v_1} = \frac{R_1}{R_1 + R_g + 1/(j\omega C_1)} = \frac{j2\pi f R_1 C_1}{1 + j2\pi f C_1 (R_1 + R_g)}$$

$$A_{v(\text{low})} = \frac{0.014 jf}{1 + jf/50} \qquad (b)$$

The above indicates a low-frequency break of 50 Hz. To check our earlier assumption that $C_2$ can be neglected, let us compute the reactance of $C_2$ at 50 Hz:

$$\frac{1}{j\omega C_2} = \frac{1}{(j6.28)(50)(1.59 \times 10^{-12})} = -j(2 \times 10^9) \; \Omega$$

The magnitude of $2 \times 10^9 \; \Omega$ is certainly large compared to the parallel resistance of 1.4 k$\Omega$, and hence no significant error in the low-frequency response results by neglecting $C_2$.

To obtain the high-frequency response we neglect $C_1$ ($C_1$ behaves as a short circuit at high frequencies) as shown in the equivalent circuit of Fig. 4.16d; this yields

$$A_{v(\text{high})} = \frac{v_2}{v_1} = \frac{Z_p}{R_g + Z_p}$$

where $Z_p$ is the impedance of the $R_1 - C_2$ combination:

$$Z_p = \frac{R_1 1/(j\omega C_2)}{R_1 + 1/(j\omega C_2)} = \frac{R_1}{1 + j2\pi f R_1 C_2}$$

$$A_{v(\text{high})} = \frac{R_1/(1 + j2\pi f R_1 C_2)}{R_g + [R_1/(1 + j2\pi f R_1 C_2)]} = \frac{R_1}{R_g + j2\pi f R_1 R_g C_2 + R_1}$$

$$A_{v(\text{high})} = \frac{1.4 \times 10^3}{2 \times 10^3 + jf(84 \times 10^{-7})} = \frac{0.7}{1 + jf/(23.8 \times 10^7)} \qquad (c)$$

The magnitude response curves corresponding to Eqs. (a), (b), and (c) are displayed in Fig. 4.17a, b, and c, respectively. The combined response is shown in Fig. 4.17d. From this curve, the following information is readily available:

   Midfrequency amplification:   $A_{v(\text{mid})} = -3.1$ dB
   Low-frequency cutoff:       $f_L = 50$ Hz
   High-frequency cutoff:      $f_H = 238$ MHz

The phase response curves corresponding to Eqs. (a), (b), and (c) are

Performance Measures for Amplifying Circuits    131

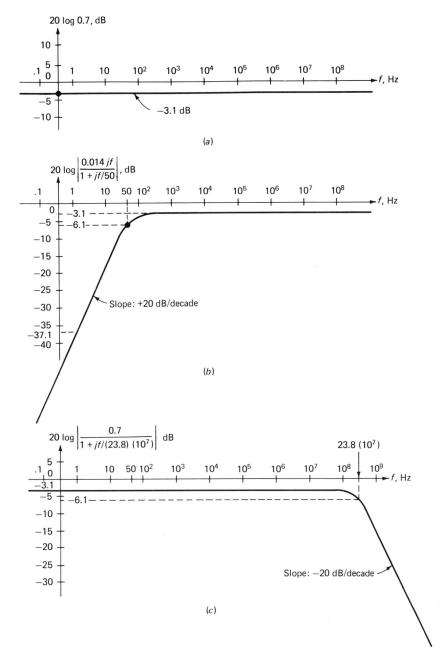

Fig. 4.17  Plots of $|A_v|$ for the circuit of Fig. 4.16a.

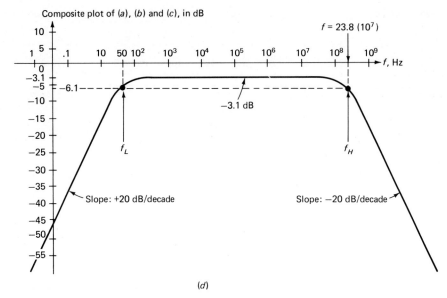

*Fig. 4.17* (*Continued*).

displayed in Fig. 4.18a, b, and c, respectively. The total phase shift is shown in Fig. 4.18d. Note the 45° phase at the break frequencies.

Although we solved the previous example by actually writing the appropriate equations, it is possible to sketch the frequency response curves by calculating the break frequencies and midfrequency amplification directly. The low-frequency response is affected only by $C_1$ (usually the reactance of $C_2$ is negligible at $f_L$); hence as $f$ drops below $f_L$, the response drops at the rate of 20 dB per decade. The break frequency is the frequency for which the reactance of $C_1$ is equal in magnitude to the Thevenin resistance seen by $C_1$:

$$\frac{1}{2\pi f_L C_1} = R_g + R_1$$

$$f_L = \frac{1}{2\pi C_1 (R_g + R_1)} = 50 \text{ Hz}$$

$C_2$ controls the high-frequency cutoff; again, $f_H$ is the frequency for which the magnitude of $C_2$'s reactance is equal to the Thevenin resistance it sees ($C_1$ is a short circuit at $f_H$):

$$\frac{1}{2\pi f_H C_2} = \frac{R_1 R_g}{R_1 + R_g}$$

$$f_H = \frac{R_1 + R_g}{2\pi C_2 R_1 R_g} = 238 \text{ MHz}$$

Performance Measures for Amplifying Circuits   133

(a)

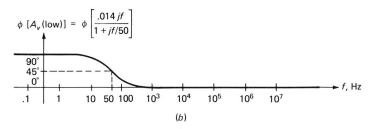

(b)

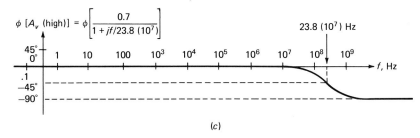

(c)

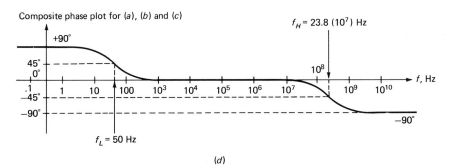

(d)

Fig. 4.18  Phase plots for the circuit of Fig. 4.16a.

Beyond $f_H$ the response rolls off at 20 dB per decade, which is the result obtained earlier.

## Problem Set 4.6

4.6.1. How many octaves is 10 Hz below 100 Hz?

4.6.2. How many decades is 50 Hz above 10 Hz?

4.6.3. $f_1$ is 2.4 octaves below 10 kHz. Determine $f_1$.

4.6.4. $f_2$ is 1.2 decades above 1,000 Hz. Determine $f_2$.

4.6.5. The range of frequencies audible by the human ear extends approximately from 30 to 15,000 Hz. Express this range in number of octaves and number of decades.

4.6.6. The frequency response for a particular amplifier indicates a mid-frequency voltage amplification of 60 dB, dropping beyond 100 kHz at the rate of 40 dB per decade. Determine $|A_v|$ at 2 MHz.

4.6.7. The straight-line approximation to an expression such as $1 + jf/f_1$ was shown to be in error by 3 dB at the break frequency. Determine the error at each of the following frequencies:

  a. One decade above $f_1$
  b. One octave above $f_1$
  c. One decade below $f_1$
  d. One octave below $f_1$

4.6.8. Determine the phase for an expression of the general form $1 + jf/f_1$ at each of the following frequencies:

  a. One octave above $f_1$
  b. One octave below $f_1$
  c. Two octaves above $f_1$
  d. Two octaves below $f_1$

4.6.9. Sketch on semilog graph paper, including the break-frequency corrections, the Bode plots (magnitude and phase) for the following expressions:

$$A_v = \frac{10^4 jf}{(1 + jf/10^6)(1 + jf/10)} \quad (a)$$

$$A_v = \frac{0.005 jf}{(1 + jf/400)(1 + jf/10^4)(1 + jf/10^5)} \quad (b)$$

$$A_v = \frac{(2{,}000)(1 + jf/500)(1 + jf/5{,}000)}{(1 + jf/2{,}000)(1 + jf/20{,}000)(1 + jf/10^5)^2} \quad (c)$$

4.6.10. Estimate the magnitude and phase for the expressions in Prob. 4.6.9 at the following frequencies:

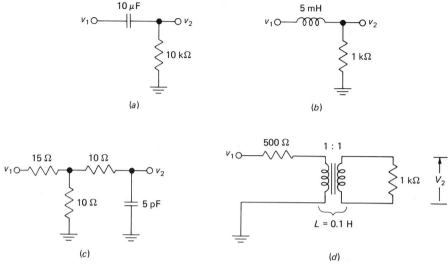

Fig. 4.19  Circuits for Prob. 4.6.11.

Equation (a):  $f = 10^{-3}, 1, 10, 40$ Hz; 500 kHz; $10^6, 10^7$ Hz.
Equation (b):  $f = 10^{-4}, 10^2, 400, 10^3, 10^5, 2 \times 10^6$ Hz.
Equation (c):  $f = 500, 2{,}000, 5{,}000, 10^4, 50 \times 10^3, 10^6$ Hz.

4.6.11. Sketch magnitude and phase plots, including break-frequency corrections, for each of the circuits in Fig. 4.19. Also determine the appropriate low or high cutoff frequencies.

4.6.12. Determine the transfer function $I_2/I_1$, and construct magnitude and phase plots for the circuit of Fig. 4.20.

4.6.13. Design an $RC$ circuit whose frequency response ($20 \log |A_v|$) is flat from DC to 10 kHz, down 3 dB at 10 kHz, and drops at 20 dB per decade as $f$ increases beyond 10 kHz. Use $R = 100$ k$\Omega$.

4.6.14. Design an $RC$ circuit whose frequency response ($20 \log |A_v|$) rises from DC up to 100 Hz at the rate of 20 dB per decade and remains flat at 0 dB beyond that (except for the 3-dB correction at 100 Hz). Use $R = 100$ k$\Omega$.

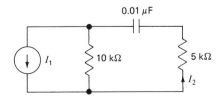

Fig. 4.20  Circuit for Prob. 4.6.12.

### 4.5c Cascaded Stages

It is often necessary to *cascade* two or more circuits in order to achieve greater amplification, better filtering, or some other desirable function. Two circuits are cascaded if the output of one serves as the input to the next.

Suppose, for example, that amplifier $A$ has a voltage amplification of 2,000. If the output of $A$ is fed to the input of $B$, the overall amplification would appear to be the product of 1,000 and 2,000, which is equal to $2 \times 10^6$. This is true as long as the input impedance to amplifier $B$ does not load amplifier $A$ to the point where its voltage amplification is altered. Expressed in decibels, the two gains are 60 and 63 dB, respectively (these are voltage, not power, gains). As long as the impedance looking into the input of $B$ is much greater than the output impedance of $A$, the overall gain is 123 dB.

Consider the cascaded circuits of Fig. 4.21a. If $C_1$ is not appreciably loaded by the input impedance to circuit $B$, we can write

$$\frac{v_2}{v_1} = \frac{1/(j\omega C_1)}{[1/(j\omega C_1)] + R_1} = \frac{1}{1 + j\omega R_1 C_1} = \frac{1}{1 + jf/f_1}$$

where

$$f_1 = \frac{1}{2\pi R_1 C_1}$$

The fact that $C_1$ is not significantly loaded by the input impedance of circuit $B$ can be verified by noting that $C_2$ is one-tenth of the value of $C_1$, and hence $X_{C2}$ is always 10 times greater than $X_{C1}$. Thus the load across $C_1$, which is the series combination of $R_2$ and $X_{C2}$, is always at least 10 times greater than $X_{C1}$; such loading can usually be neglected without introducing significant errors. Similarly, if no significant load is connected across $C_2$, the voltage transfer function for circuit $B$ may be written as

$$\frac{v_3}{v_2} = \frac{1}{1 + jf/f_2}$$

where

$$f_2 = \frac{1}{2\pi R_2 C_2}$$

With the substitution of numerical values, the transfer functions become

$$\frac{v_2}{v_1} = \frac{1}{1 + jf/10^5} \tag{a}$$

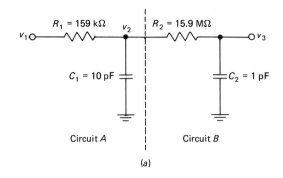

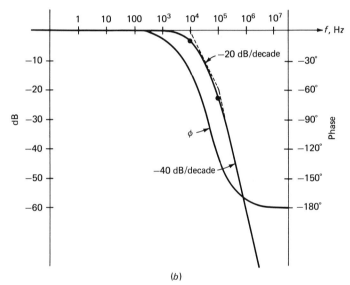

**Fig. 4.21** (a) Cascaded RC circuits; (b) magnitude and phase plots for (a).

$$\frac{v_3}{v_2} = \frac{1}{1 + jf/10^4} \tag{b}$$

The overall voltage transfer function is

$$\frac{v_3}{v_1} = \frac{v_3}{v_2}\frac{v_2}{v_1} = \frac{1}{1 + jf/10^4}\frac{1}{1 + jf/10^5} \tag{c}$$

Bode plots for the overall response of Eq. (c) are shown in Fig. 4.21b. Note that as long as the assumption of no loading is valid, the plots for the overall response are the sum of Bode plots for the response of each individual circuit; this applies to both magnitude and phase.

## Example 4.14

The block diagram of Fig. 4.22 represents a system in which all circuits are designed to be fed from 50-Ω sources and work into 50-Ω loads. Hence, the various gains and losses indicated apply only as long as the source and load impedances are both 50 Ω. Determine

a. The midfrequency value of each output level, expressed in dBm.
b. The high-frequency cutoff for each output, and the rate at which the outputs drop beyond the cutoff frequency.

## Solution

a. Whenever $R_L = R_i$, the expression for power gain is

$$G = 20 \log A_v \tag{4.19}$$

This applies in our example, since all circuits are fed from 50-Ω sources and see 50-Ω loads. We can now write

$$V_1 = \frac{E_g 50}{50 + 50} = \left(\frac{1}{2}\right) E_g = 224 \text{ mV}$$

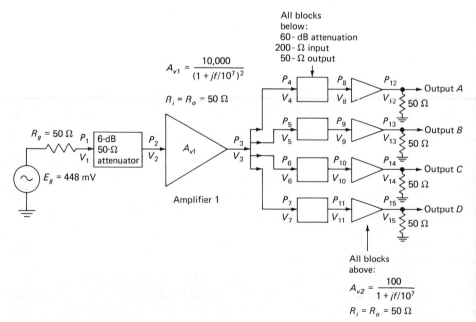

Fig. 4.22 Block diagram of system for Example 4.14.

$$P_1 = \frac{V_1^2}{50} = \frac{0.224^2}{50} = 1 \text{ mW} = 0 \text{ dBm}$$
$$P_2 = P_1 - 6 \text{ dB} = -6 \text{ dBm}$$
$$P_3 = P_2 + 20 \log |A_{v1}| = -6 \text{ dBm} + 20 \log 10{,}000$$
$$= 74 \text{ dBm} \quad \text{(at midfrequencies)}$$

The output of amplifier 1 is fed to four parallel attenuators. Since the input resistance to each of these attenuators is 200 Ω, the load seen by amplifier 1 is $200\|200\|200\|200 = {}^{200}\!/_4 = 50$ Ω. The total power $P_3$ is now split four ways. This means that

$$P_4 = P_5 = P_6 = P_7 = \tfrac{1}{4}P_3 \quad \text{(in watts)}$$

Since a power ratio of 1:4 corresponds to $-6$ dB, each attenuator receives $74 - 6 = 68$ dBm:

$$P_4 = P_5 = P_6 = P_7 = 68 \text{ dBm}$$

The 60-dB attenuators reduce each of the above power levels to 8 dBm. Hence the power input to each of the last amplifier stages is 8 dBm. The midfrequency power gain for each of these stages is $20 \log 100 = 40$ dB, and hence the outputs are

$$P_{12} = P_{13} = P_{14} = P_{15} = 8 + 40 = 48 \text{ dBm}$$

b. The high-frequency response falls off because of the $(1 + jf/10^7)^2$ term in the denominator of $A_{v1}$ and the $1 + jf/10^7$ term in the denominator of $A_{v2}$. Thus, the net response for any one output is subjected to a total high-frequency roll-off of 60 dB per decade (20 dB per decade for each $1 + jf/10^7$ term). At the corner frequency of $10^7$ Hz, the response is down from the midfrequency value by 3 dB for each term, or a total of 9 dB. This is indicated in the Bode plot of Fig. 4.23. By plotting a couple of points on each side of $10^7$ Hz, a reasonably accurate curve may be drawn which merges with the asymptotes in both directions away from the break frequency. The 3-dB high-frequency cutoff may be read from the resulting graph as

$$f_H = 5.1 \text{ MHz}$$

Note that if there had been only one $1 + jf/10^7$ term, the response would be falling off at 20 dB per decade, and $f_H$ would equal 10 MHz. This would be called a first-order response. In our case, however, there are three such terms involved, two from $A_{v1}$ and one from $A_{v2}$. This yields a third-order system in which the roll-off is 60 dB per decade, but $f_H$ is also affected. Although each term has the same break frequency of 10 MHz, the total response is down 3 dB at 5.1 MHz. Thus the bandwidth for the whole system is narrower than it is for either of the two amplifiers.

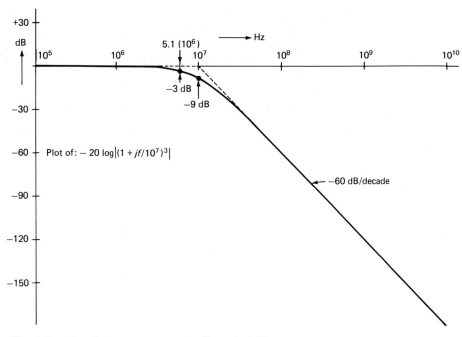

Fig. 4.23 Magnitude response curve for Example 4.14.

### Problem Set 4.7

4.7.1. Sketch overall magnitude and phase plots for the circuits in Fig. 4.24. Assume no loading between sections. Determine low and high 3-dB frequencies whenever they apply.

4.7.2. Two amplifier stages with the same low and high 3-dB frequencies ($f'_H$ and $f'_L$) are cascaded. The roll-off is 20 dB per decade on each side of $f'_H$ and $f'_L$. Prove that the 3-dB frequencies ($f''_H$ and $f''_L$) for the overall response are $f''_H = 0.64 f'_H$ and $f''_L = f'_L/0.64$.

4.7.3. Assuming $f_L \ll f_H$, what is the bandwidth reduction for a third-order system compared to a first-order system?

## 4.6 DISTORTION

Amplifiers are supposed to produce an output which differs from the input only in magnitude; that is, the output is expected to be larger than the input, but should be the same in all other respects. Typical input and output voltage

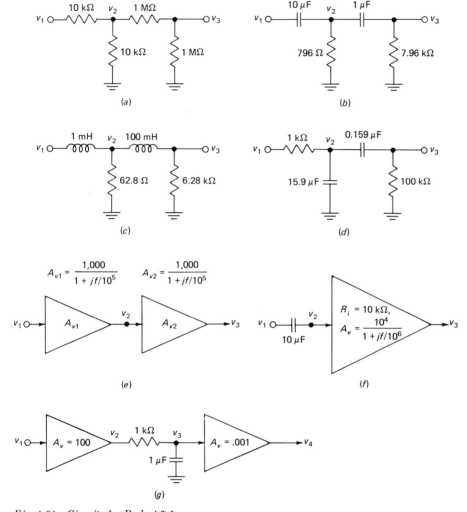

Fig. 4.24 Circuits for Prob. 4.7.1.

waveforms for a distortionless amplifier are shown in Fig. 4.25. Such an amplifier exhibits a *linear* transfer function, also shown in the figure. As implied by the graph, the instantaneous values of input and output voltage ($v_i$ and $v_o$) are *linearly* related: if $v_i$ doubles, $v_o$ doubles; if $v_i$ triples, $v_o$ triples, and so on.

Real amplifiers are not this good. For one reason or another, all amplifiers exhibit nonlinearities in their characteristics that become significant whenever large amplitude signals are involved. These nonlinearities have the effect of altering the amplification received by different parts of the same signal; for

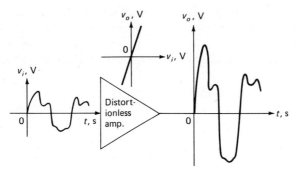

Fig. 4.25 Input and output waveforms in a distortionless amplifier.

example, the positive half-cycle of a signal may be amplified by a smaller factor than the negative half-cycle. Such deviations from linear amplification represent waveform distortion.

Before discussing distortion in greater depth, a few words on the frequency properties of signals might be in order. We often analyze circuits assuming a sinusoidal signal, not because the signal is necessarily sinusoidal, but because any signal may be expressed in terms of many different sinusoidal components, each with its own amplitude, frequency, and phase. As an example, a 1-V peak square wave may be described in terms of an infinite series of sinusoidal components as follows:[1]

$$e = 1.273 \sin (\omega t) + 0.425 \sin (3\omega t) + 0.253 \sin (5\omega t) + \cdots \quad (4.30)$$

where $\omega$ is the angular frequency of the square wave, in radians per second. If the square-wave period is 1 ms,

$$T = 10^{-3} \text{ s}$$

$$\omega = 2\pi f = \frac{2\pi}{T} = \frac{2\pi}{10^{-3}} = 6{,}280 \text{ rad/s}$$

Equation (4.30) states that a square wave whose peak value is 1 V is equivalent to an infinite series of sine waves; the first three of these sinusoidal components have the following parameters:

| | | |
|---|---|---|
| Fundamental: | $\omega = 6{,}280$ rad/s | peak value = 1.273 V |
| Third harmonic: | $3\omega = 18{,}840$ rad/s | peak value = 0.425 V |
| Fifth harmonic: | $5\omega = 31{,}400$ rad/s | peak value = 0.253 V |

For simplicity only the first three terms of the infinite series have been considered; the remaining components get progressively smaller, and hence their exclusion is justified. Figure 4.26 demonstrates how the first three components

---

[1] See pp. 414–415 of Reference 4 at the end of this chapter.

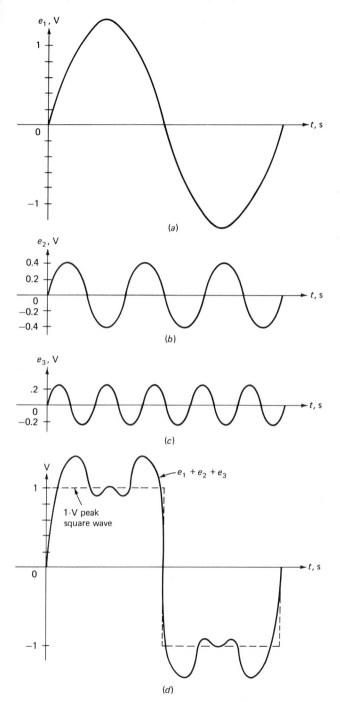

**Fig. 4.26** Square-wave representation in terms of first three components of infinite series: $e_1 = 1.273 \sin 6{,}280t$; $e_2 = 0.425 \sin 18{,}840t$; $e_3 = 0.253 \sin 31{,}400t$.

yield an approximate square wave; if additional components are used, the square-wave approximation improves.

An alternative method for indicating the relative amplitude of each frequency component is presented in Fig. 4.27. This is the frequency spectrum of our signal; it is much simpler to construct than sine waves, and yet it provides the same basic information.

A distortionless amplifier treats all frequency components alike. In practice, however, this is true only for components within the flat frequency range. If, for example, the signal contains frequency components outside the flat response region, they are amplified less than the other components, and the output waveform is not an exact replica of the input. This *frequency distortion* can be further complicated because of phase shift near and beyond the cutoff frequencies; a different phase shift for each frequency component tends to distort the signal. Such distortion is not always objectionable. For example, an audio amplifier whose high-frequency response is poor does not necessarily produce unpleasant sounds; the higher-frequency components are simply not as loud. Phase distortion, although objectionable in the transmission of picture signals, presents no problems with sound, because the human ear cannot distinguish the relative phase of different sound frequencies.

To illustrate some of the effects of frequency and phase distortion, consider the perfect square-wave voltage input of Fig. 4.28. Depending on the amplifier's frequency characteristics, various outputs may result. Generally, the flat portion of the square-wave output provides information about how the

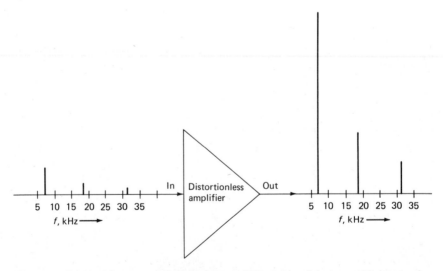

*Fig. 4.27 Input and output of distortionless amplifier expressed in terms of frequency spectrum for a square-wave signal (only the first three components are shown). Note that the relative magnitude of each component is not changed by the amplifier.*

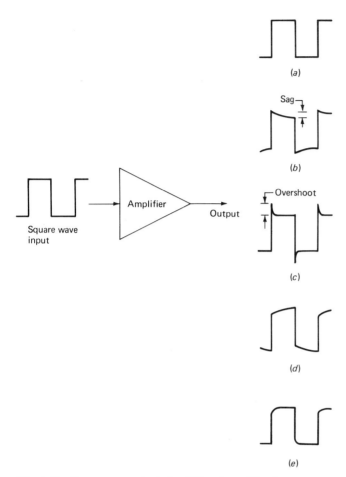

*Fig. 4.28 Square-wave outputs for different amplifier frequency responses;* (a) *good low- and high-frequency response;* (b) *amplification drops and phase shift increases at low frequencies;* (c) *amplification rises with frequency;* (d) *rising low-frequency response and increasing phase lag at low frequencies;* (e) *poor high-frequency response.*

amplifier treats slowly changing inputs (low-frequency response). The rising and falling portions, on the other hand, tell us how well the amplifier responds to fast-changing signals (high-frequency response). We will now analyze each output waveform in turn:

a. No distortion—good low- and high-frequency response
b. Poor low-frequency response as evidenced by the "sag" (also called "tilt")—positive phase shift at low frequencies (output leads input)
c. Rising high-frequency response—note overshoot

d. Rising low-frequency response—negative phase shift at low frequencies (output lags input)
e. Poor high-frequency response—note loss in steepness of square-wave rise and fall

Frequency and phase distortion of the type just discussed is referred to as *linear* distortion. Linear distortion results whenever the signal's frequency components are unequally amplified or phase-shifted. The output contains no frequencies other than those at the input; only their relative amplitude and phase are altered.

*Nonlinear* distortion causes frequency components to appear at the output that are not present in the input; this type of distortion is not due to unequal frequency or phase response but results from nonlinearities in the characteristics of amplifier circuits. These nonlinearities may cause clipping or clamping, in which case part of the waveform may be lost. Additionally, nonlinear distortion may be classified as follows:

a. Harmonic distortion
b. Intermodulation distortion

Harmonic distortion produces new frequencies that are harmonically related to (exact multiples of) those present in the signal. For example, a 1,000-Hz signal passing through an amplifier that produces harmonic distortion may yield an output containing, in addition to the desired 1,000-Hz signal, second harmonics (2,000 Hz), third harmonics (3,000 Hz), and so on. Usually, harmonically related frequency components are not by themselves objectionable in sound reproduction.

The percent distortion $D_n$ due to any harmonic $n$ may be defined as follows:

$$D_n = \frac{A_n}{A_1} 100 \qquad (4.31)$$

where $A_n$ is the amplitude of the $n$th harmonic and $A_1$ the amplitude of the fundamental frequency. The total distortion due to all harmonics is

$$D_{\text{tot}} = \sqrt{D_1^2 + D_2^2 + \cdots D_n^2} \qquad (4.32)$$

The amplitude of each harmonic for any given signal may be obtained experimentally using a distortion or spectrum analyzer; graphic techniques may also be used.

Amplifiers are generally linear only as long as the voltage, current, and power levels remain within certain limits. If these limits are exceeded, the amplifier is said to be overdriven. The characteristics of an overdriven amplifier become progressively more nonlinear as the signal level increases. The resulting distortion is undesirable for amplifiers, but there are many other applications for which maximum distortion is wanted. This is true, for example, in frequency-multiplier circuits. Here it is desired to generate a signal whose frequency is an integral multiple of some other, more readily available, signal. In this case an amplifier may be intentionally overdriven so that it no longer produces linear amplification but generates harmonic distortion of the proper frequency.

Intermodulation (IM) distortion is a type of nonlinear distortion that generates frequency components not harmonically related to the signal frequencies; instead, they involve the sum and difference of signal frequencies and possibly the sum and difference of harmonics. As an example, assume that the signal is made up of just two frequencies, 1,000 and 1,500 Hz; harmonic distortion may produce new frequencies such as 2,000, 3,000, 4,000 Hz and 3,000, 4,500, 6,000 Hz. Intermodulation distortion, however, may generate $1,500 + 1,000 = 2,500$ Hz and $1,500 - 1,000 = 500$ Hz; the harmonics may also interact to produce additional sum and difference frequencies. These by-products are quite undesirable in amplifiers, and great care is taken to minimize them. In audio amplifiers, for example, IM distortion is very objectionable because its components, not being harmonically related to the signal, are easily detected by the human ear.

High-fidelity audio amplifiers are generally specified as having no more than a certain percent of harmonic and IM distortion at a given output power level. The power level for which the distortion specification applies must be stated, since distortion is a function of how hard the amplifier is driven. Generally, keeping harmonic distortion down minimizes the generation of sum and difference of harmonic frequencies (which is IM distortion), therefore lowering IM distortion as well.

## Problem Set 4.8

4.8.1. What does linear amplification mean?
4.8.2. Is distortion always objectionable?
4.8.3. Why is distortion most common when large signals are involved?
4.8.4. The input to an amplifier contains the following frequencies and their relative magnitudes: 300 Hz: 1 mV; 500 Hz: 0.2 mV; 1,000 Hz: 3 mV. The output of the amplifier consists of 300 Hz: 20 mV; 500 Hz: 4 mV; 1,000 Hz: 45 mV. Sketch the input and output frequency spectrum. What type of distortion is involved here?
4.8.5. Phase shift has the effect of delaying each frequency component of a signal by a brief time interval. When phase shift is different for each frequency component, significant distortion of the total waveform may result. Determine in what manner phase shift should depend on frequency if it is to have no effect on the signal waveform.
4.8.6. What is the difference between linear and nonlinear distortion?

4.8.7. What is the difference between harmonic and intermodulation distortion?

4.8.8. Given input frequencies of $10^6$ and $1.2 \times 10^6$ Hz, identify the following output frequencies, indicating whether they are harmonics (state the fundamental in each case) or intermodulation frequencies (state which frequencies are intermodulating in each case): $3 \times 10^6$, $3.6 \times 10^6$, $2.2 \times 10^6$, $2 \times 10^5$, $2 \times 10^6$, $2.4 \times 10^6$, $6.6 \times 10^6$, $6 \times 10^5$, $4 \times 10^5$, $4.4 \times 10^6$ Hz.

## 4.7 NOISE[1]

Just as distortion is a limiting factor in the amplification of large signals, noise is the limiting factor when small signals are involved. Ideally any signal, no matter how small (but not zero), should be capable of being amplified as much as it takes to make it useful. Electric noise, however, makes it impossible to achieve such an ideal situation, because it tends to obscure small signals, thus making it difficult to distinguish noise from signal.

A figure of merit for the quality of any signal is the signal-to-noise power ratio $S/N$. For good signal intelligibility, $S/N$ should be high. Signals to be amplified may be noisy to start, as a result of transmission through a noisy medium (the atmosphere or a long cable); therefore there is an input $S/N$ that specifies the quality of signal before it is passed through the amplifier. All amplifiers contribute noise of their own to the signal being processed; therefore the output $S/N$ is bound to be lower than the input $S/N$. High-quality amplifiers are designed to minimize their noise contribution in order to achieve an output $S/N$ as close to the input $S/N$ as possible.

In general, the total input power consists of a signal and noise component:

$$P_i = S_i + N_i \qquad (4.33)$$

The output power can be similarly expressed:

$$P_o = S_o + N_o \qquad (4.34)$$

In the above expressions, $S_i$ and $S_o$ represent signal input and output power, while $N_i$ and $N_o$ are the noise input and output power, respectively.

An important measure of performance for amplifiers is the ratio of input $S/N$ to output $S/N$, called noise factor:

$$\text{Noise factor:} \quad F = \frac{S_i/N_i}{S_o/N_o} \qquad (4.35)$$

The noise factor is usually expressed in decibels, in which case it is called the noise figure:

$$\text{Noise figure:} \quad NF = 10 \log F \qquad (4.36)$$

[1] See chap. 13 of Reference 2 at the end of this chapter.

Basically, the noise figure tells us how "noisy" an amplifier is. If an amplifier could be built that generated no noise of its own, the input $S/N$ and the output $S/N$ would be the same; the noise factor would be 1 and the noise figure 0 dB. All real amplifiers have noise figures greater than 0 dB.

The output signal power $S_o$ is the input signal power $S_i$ multiplied by the power gain $G$:

$$S_o = S_i G \tag{4.37}$$

The output noise power $N_o$ consists of amplified input noise $N_i G$ plus noise $N_A$ generated by the amplifier itself:

$$N_o = N_i G + N_A \tag{4.38}$$

**Example 4.15**

The signal input to a given amplifier is made up of 100-$\mu$W signal power and 1-$\mu$W noise power. The amplifier contributes an additional 100 $\mu$W of noise and has a power gain of 20 dB. Determine the input $S/N$, output $S/N$, noise factor, and noise figure.

**Solution**

The input signal-to-noise ratio is

$$\frac{S_i}{N_i} = \frac{100 \times 10^{-6}}{1 \times 10^{-6}} = 100$$

Before proceeding further, we must convert $G$ to a ratio:

$10 \log G = 20$ dB
$\log G = 2$
$G = 100$

The output signal is

$$S_o = S_i G \tag{4.37}$$
$$S_o = (100 \times 10^{-6})(100) = 10^{-2} \text{ W}$$

The output noise is

$$N_o = N_i G + N_A \tag{4.38}$$
$$N_o = (10^{-6} \times 100) + (100 \times 10^{-6})$$
$$N_o = 2 \times 10^{-4} \text{ W}$$

The output $S/N$ ratio is

$$\frac{S_o}{N_o} = \frac{10^{-2}}{2 \times 10^{-4}} = 50$$

The noise factor is

$$F = \frac{S_i/N_i}{S_o/N_o} \qquad (4.35)$$

$$F = \frac{100}{50} = 2$$

The noise figure is

$$NF = 10 \log F \qquad (4.36)$$
$$NF = 10 \log 2 = 3 \text{ dB}$$

### Problem Set 4.9

4.9.1. Why does noise matter only when small signals are involved?
4.9.2. What is the absolute minimum value of noise factor? Noise figure?
4.9.3. An amplifier with a power gain of $10^4$ produces a noise output of 10 μW when the noise input is 500 pW. Determine the noise figure.
4.9.4. An amplifier has a noise figure of 6 dB, a power gain of 30 dB, and an input noise level of $-75$ dBm. Determine the output noise level in dBm.

## REFERENCES

1. Joyce, M. V., and K. K. Clarke: "Transistor Circuit Analysis," Addison-Wesley Publishing Company, Inc., Reading, Mass., 1961.
2. Ristenbatt, M. P., and R. L. Riddle: "Transistor Physics and Circuits," Prentice-Hall, Inc., Englewood Cliffs, N.J., 1966.
3. Angelo, E. J.: "Electronic Circuits," McGraw-Hill Book Company, New York, 1964.
4. Skilling, H. H.: "Electrical Engineering Circuits," John Wiley & Sons, Inc., New York, 1957.

# chapter 5

# introduction to junction transistors

## 5.1 THE JUNCTION TRANSISTOR

A transistor is formed by sandwiching a thin layer of doped semiconductor between two thicker layers of oppositely doped semiconductor. The resulting device is called a junction, or bipolar, transistor. For example, if the middle layer is N and the other two layers P type, the result is a PNP transistor as shown in Fig. 5.1a. Opposite doping yields an NPN transistor. Schematic symbols for a PNP and NPN transistor are shown in Fig. 5.1b and c.

The center region of a junction transistor structure, in addition to being physically thin, is lightly doped for reasons which will be discussed soon. The names assigned to each region are emitter, base, and collector; the base is the thin region in the middle. Transistors are fabricated using a single crystal; impurities are introduced during the manufacturing process using a variety of techniques, some of which are discussed later in this chapter.

When properly connected in a circuit, a transistor is capable of amplifying weak signals such as those picked up by a receiving antenna and making them powerful enough to be heard, as in a radio. In addition to its usefulness as an amplifying device, the transistor may be used as an extremely fast electronic "switch" with almost endless applications in computer and control circuits.

For purposes of illustrating transistor action, consider first an NPN transistor as in Fig. 5.2a. The DC supply $V_{CC}$ may be batteries (as in a portable

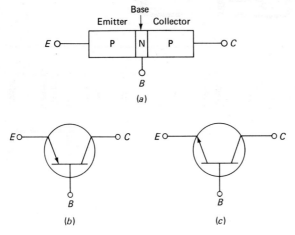

*Fig. 5.1* (a) *Structure of PNP transistor;* (b) *symbol for PNP transistor;* (c) *symbol for NPN transistor.*

transistor radio) or the filtered output of a half- or full-wave rectifier (for line-operated circuits). Note that there are two PN junctions, one between collector and base, the other between emitter and base. For linear amplification, one of these junctions must be reverse-biased while the other is forward-biased. By applying the DC voltage shown, the collector-base junction is reverse-biased (high resistance). $V_{CC}$ tends to forward-bias the base-emitter junction, but as long as the base is open, nothing of great significance can happen. The corresponding diagram for transistor and DC supply is shown in Fig. 5.2b.

We now apply an external voltage $V_{BB}$ from base to emitter whose polarity forward-biases this junction; the circuit is completed as shown in Fig. 5.3a. For an NPN transistor, the P-type base is connected to the positive terminal and the N-type emitter to the negative terminal of $V_{BB}$. We now have a forward-

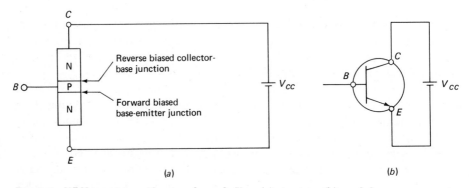

*Fig. 5.2* NPN *transistor with external supply* $V_{CC}$; (a) *structure;* (b) *symbol.*

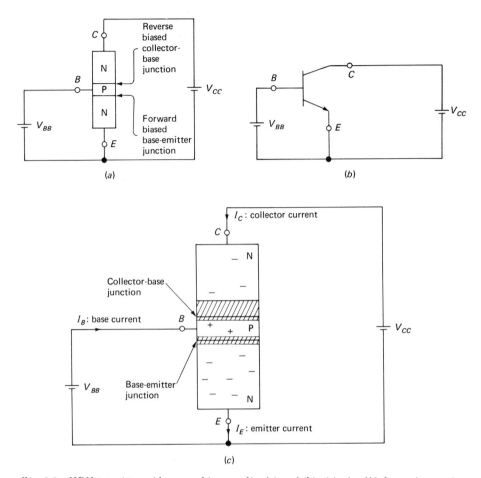

Fig. 5.3  NPN transistor with proper bias supplies (a) and (b); (c) simplified transistor action.

biased base-emitter junction and a reverse-biased collector-base junction; the diagram is shown in Fig. 5.3b.

To explain junction transistor action, consider the diagram of Fig. 5.3c. The $V_{BB}$ supply sees a forward-biased diode whose *anode* is the P-type base and whose *cathode* is the N-type emitter. As with any forward-biased diode, significant current flows once the threshold voltage is exceeded.

Electrons in the N-type emitter are the majority carriers. The forward bias causes these majority carriers to diffuse from emitter into base; this "injection" of electrons results in emitter current $I_E$ as indicated on the diagram.

Once emitter-injected electrons reach the base, they are no longer majority carriers because the base is P type. One would normally expect some fairly quick recombination of electrons and holes to take place, and to a small degree, this does occur. But the base is thin and lightly doped; thus relatively few holes are available to combine with electrons injected from the emitter. The net result is a long "lifetime" for these electrons; that is, they can remain in the base region for a relatively long time before a hole is available for recombination.

Note that there is a positive potential on the collector side of the collector-base junction due to $V_{CC}$. Electrons in the base region that have not recombined come under the influence of this potential. Once this happens, these electrons are swept out of the base, through the collector-base junction, and come out through the collector terminal. This flow of electrons results in collector current $I_C$ which completes its path through $V_{CC}$.

Thus we have electrons leaving the emitter ($I_E$), of which a large part reach the collector ($I_C$). What happens to the small percentage of electrons that do not reach the collector? These recombine with holes in the base to form covalent bonds. Meanwhile valence electrons from the base region near the external circuit break their bonds and leave the base, completing their external path through $V_{BB}$. Therefore, emitter-injected electrons ($I_E$) diffuse into the base where they either are "collected" by the collector ($I_C$) or cause an equivalent number of electrons to leave through the base terminal, resulting in base current ($I_B$). Applying Kirchhoff's current law to our circuit, we must conclude that total emitter current equals the sum of collector and base currents:

$$I_E = I_B + I_C \tag{5.1}$$

A factor used to relate that part of emitter current that becomes collector current is $\alpha$, called the DC alpha:

$$\alpha = \frac{I_C}{I_E}\text{[1]} \tag{5.2}$$

Typical values of $\alpha$ range between 0.9 and 0.99. Since $\alpha$ is the ratio of emitter carriers that actually reach the collector, it can approach but not exceed 1. To maximize $\alpha$, emitter injection should be maximized while recombinations in the base are minimized. To increase emitter injection, the emitter is heavily doped; to minimize recombinations, the base is thin and lightly doped.

**Example 5.1**

The transistor in the diagram of Fig. 5.3b has an $\alpha = 0.95$. If $I_E = 1$ mA, determine $I_C$ and $I_B$.

---

[1] This equation is not entirely correct because of an additional component of reverse current through the collector-base junction. We will modify the expression later on to include this factor; however, the equation is quite adequate for our present discussion. The parameter $h_{FB}$ may also be used instead of $\alpha$; $h_{FB}$ is defined as $-\alpha$, and so $I_C/I_E = -h_{FB}$.

## Solution

If we know $I_E$, we can find $I_C$ using Eq. (5.2):

$$I_C = \alpha I_E \tag{5.2}$$
$$I_C = 0.95 \times 10^{-3} = 0.95 \text{ mA}$$

Once any two currents are known, the third is determined using the relationship of Eq. (5.1):

$$I_B = I_E - I_C \tag{5.1}$$
$$I_B = (1 - 0.95)(10^{-3})$$
$$I_B = 50 \ \mu A$$

### Example 5.2

The following currents are measured in a transistor circuit of the type in Fig. 5.3b: $I_B = 20 \ \mu A$; $I_E = 2 \text{ mA}$. Determine $I_C$ and $\alpha$.

### Solution

$$I_C = I_E - I_B \tag{5.1}$$
$$I_C = (2 \times 10^{-3}) - (20 \times 10^{-6})$$
$$I_C = 1.98 \text{ mA}$$
$$\alpha = \frac{I_C}{I_E} \tag{5.2}$$
$$\alpha = \frac{1.98 \times 10^{-3}}{2 \times 10^{-3}} = 0.99$$

Further insight into transistor operation may be gained by considering the sequence of events from another point of view. For this purpose we will analyze a PNP transistor—that is, one in which the base is N type. In order that both junctions be properly biased, the polarity of external voltage sources must be opposite to those used for the NPN transistor. Except for this polarity change and the consequent reversal of all current directions, there are no differences in the external behavior, and both Eqs. (5.1) and (5.2) apply. The appropriate circuit diagram is shown in Fig. 5.4.

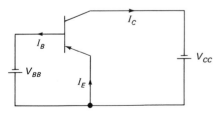

*Fig. 5.4* PNP transistor bias voltages and currents.

This is what physically happens. Electrons leave the negative terminal of $V_{BB}$ and travel toward the base. The flow of these electrons results in $I_B$. As soon as an electron enters, the base acquires a net negative charge. Holes are injected from the emitter $I_E$ into the base in an effort to neutralize this excess negative charge. The majority of injected holes $\alpha I_E$ are swept across the base to the collector. The probability that a hole actually succeeds in neutralizing the excess negative charge is relatively low. If, for example, $\alpha = 0.9$, then 90 percent of all emitter-injected holes reach the collector, and 10 percent recombine in the base. In other words, to neutralize each electron entering the base, 10 holes are injected by the emitter. Out of these 10 holes, 90 percent, which is 9, reach the collector, and 10 percent, which is 1, recombine in the base. Therefore, *one unit of charge entering the base results in nine units of charge leaving the collector*. This represents amplification.

Just as we defined the ratio of collector to emitter current as $\alpha$, we may define the ratio of collector to base current as $h_{FE}$; this is the common emitter DC forward current transfer ratio, also called the DC beta ($\beta$). The mathematical relationship is

$$h_{FE} = \frac{I_C}{I_B} \tag{5.3}$$

Obviously, there is a relationship between $\alpha$ and $h_{FE}$. It can be developed as follows:

$$I_C = \alpha I_E \tag{5.2}$$
$$I_E = I_B + I_C \tag{5.1}$$
$$I_C = \alpha(I_B + I_C)$$
$$I_C(1 - \alpha) = \alpha I_B$$
$$\frac{I_C}{I_B} = \frac{\alpha}{1 - \alpha}$$

but

$$\frac{I_C}{I_B} = h_{FE} \tag{5.3}$$

therefore,

$$h_{FE} = \frac{\alpha}{1 - \alpha} \tag{5.4}$$

Just as $h_{FE}$ can be expressed in terms of $\alpha$, $\alpha$ can be expressed in terms of $h_{FE}$.

$$h_{FE}(1 - \alpha) = \alpha \tag{5.4}$$
$$h_{FE} = h_{FE}\alpha + \alpha = \alpha(h_{FE} + 1)$$
$$\frac{h_{FE}}{h_{FE} + 1} = \alpha \tag{5.5}$$

The relationships of Eqs. (5.4) and (5.5) apply to either PNP or NPN transistors, provided, of course, that the polarities of external supplies are correct for the transistor type.

## Example 5.3

The transistor in the circuit of Fig. 5.4 has $h_{FE} = 100$ and $I_B = 40 \ \mu A$. Determine $I_C$, $\alpha$, and $I_E$.

### Solution

Since $h_{FE}$ and $I_B$ are known, $I_C$ is determined using Eq. (5.3):

$$I_C = h_{FE} I_B \tag{5.3}$$
$$I_C = 100(40 \times 10^{-6})$$
$$I_C = 4 \text{ mA}$$

The parameter $\alpha$ can be calculated using Eq. (5.5):

$$\alpha = \frac{h_{FE}}{h_{FE} + 1} \tag{5.5}$$

$$\alpha = \frac{100}{101} = 0.99$$

The emitter current is simply the sum of base and collector currents:

$$I_E = I_B + I_C \tag{5.1}$$
$$I_E = (40 \times 10^{-6}) + (4 \times 10^{-3})$$
$$I_E = 4.04 \text{ mA}$$

## Problem Set 5.1

5.1.1. Do you think it is possible to produce a junction transistor by connecting two diodes together? Explain.

5.1.2. What are the three regions of a bipolar transistor?

5.1.3. Draw the schematic symbol for each of the two types of transistor. What is the relationship between the directions of emitter current and the arrow on the schematic symbol for each type of transistor?

5.1.4. The following quantities are measured in a transistor circuit:

$I_B = 0.1$ mA
$I_C = 5$ mA

Determine $I_E$, the transistor's $\alpha$, and $h_{FE}$.

5.1.5. Which junction in a transistor is normally forward-biased? Which junction is normally reverse-biased?

**5.1.6.** Why must the polarity of all external DC voltages applied to a PNP transistor be opposite to those applied to an NPN transistor?

**5.1.7.** What is meant by injection?

**5.1.8.** Why do you think the emitter is usually heavily doped?

**5.1.9.** Why do you think the base is usually lightly doped and physically thin?

**5.1.10.** What is the significance of $\alpha$? Explain both qualitatively and in equation form.

**5.1.11.** The base current for a particular transistor is measured at 0.05 mA. If $h_{FE} = 70$, determine $I_C$, $I_E$, and $\alpha$.

**5.1.12.** If the base current in Prob. 5.1.11 increases by an amount $\Delta I_B$, what is the corresponding increase in collector current ($\Delta I_C$)?

## 5.2 STATIC CHARACTERISTICS

Transistor relationships are often expressed using graphic displays. One such type of display is a set of *static characteristics* that relate various transistor voltages and currents to each other.

Transistors may be operated in any one of three possible configurations. Usually one of the three terminals (emitter, base, or collector) is common to input and output; this yields common (or grounded) emitter, base, or collector configurations, as shown in Fig. 5.5. External DC voltage sources have been omitted for simplicity. For any of these configurations we may define four variables: input and output voltage, input and output current.

Consider the transistor circuit diagram of Fig. 5.6. Again, for simplicity, external DC sources have been omitted. Note that the emitter is common to both input and output circuits. The four transistor variables in the common emitter configuration are therefore

a. Input voltage:   $V_{BE}$
b. Input current:   $I_B$
c. Output voltage:  $V_{CE}$
d. Output current:  $I_C$

Graphs may be drawn relating the above variables to each other; however, not all such graphs are necessary. The most useful set of data are generally provided by

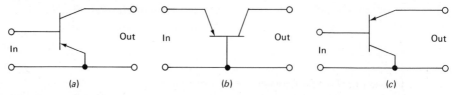

*Fig. 5.5* (a) *Common emitter*, (b) *common base*, and (c) *common collector transistor configurations. External voltage supplies have been left out for simplicity.*

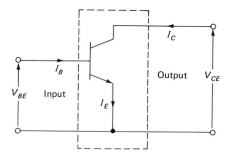

Fig. 5.6  Common emitter input and output voltages and currents.

the output characteristics, a plot of transistor output voltage $V_{CE}$ versus output current $I_C$.

Any plot of $V_{CE}$ versus $I_C$ depends on $V_{BE}$, the forward bias across the base-emitter junction. Before there is significant base current, for example, the threshold voltage of the base-emitter diode must be overcome.

The common emitter output characteristics are now developed starting with the graph of Fig. 5.7. Initially, assume that the base-emitter drive is not sufficient to cause base current flow, hence $I_B = 0$. According to Eq. (5.3), $I_C = h_{FE}I_B$; therefore, if $I_B = 0$, $I_C$ is also zero. In practice, however, even with zero $I_B$ there is some collector current, for reasons to be discussed later in this chapter. For now, curve 1 in the graph of Fig. 5.7 represents a plot of $I_C$ versus $V_{CE}$ for the special case when $I_B = 0$.

If we now increase $V_{BE}$ beyond the threshold voltage, base current flows, emitter injection takes place, and collector current results. This can be seen as curve 2 in Fig. 5.7. Note that until point $A$ is reached, $I_C$ increases almost directly with $V_{CE}$. Beyond $A$, however, further increases in $V_{CE}$ do not result in significant increases in $I_C$. This is because when the reverse voltage $V_{CE}$ is very small, any increase in $V_{CE}$ also increases the likelihood of injected carriers being collected. Once $V_{CE}$ reaches a certain level, though, almost all injected carriers

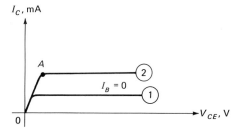

Fig. 5.7  Development of common emitter output characteristics.

that reach the collector-base region are swept across; hence increasing $V_{CE}$ does not cause appreciable increases in $I_C$.

Using a similar procedure, additional curves may be plotted to yield the complete output characteristics of Fig. 5.8. Each curve corresponds to a different value of $I_B$, which is controlled in turn by the forward voltage $V_{BE}$. If $V_{CE}$ is increased further, breakdown of the collector-base junction eventually results; this phenomenon is not shown on the characteristics of Fig. 5.8, but will be considered in due time.

What information do output characteristics provide? For one, note that in a common emitter configuration, $I_C$ and $I_B$ are the transistor's output and input currents; therefore, the ratio $I_C/I_B$ is the current amplification for a common emitter transistor. The parameter $h_{FE}$, which corresponds to $I_C/I_B$, is now more easily defined. The $h_F$ part denotes a forward current transfer, or more simply, current amplification; the $E$ part indicates a common emitter configuration. Capital subscripts are used to denote DC (static) parameters.

The characteristics of Fig. 5.8 indicate that $I_C$ increases as $I_B$ increases. This should be obvious, since whenever the forward base-emitter drive is increased, $I_B$ also increases, resulting in further emitter injection and hence more

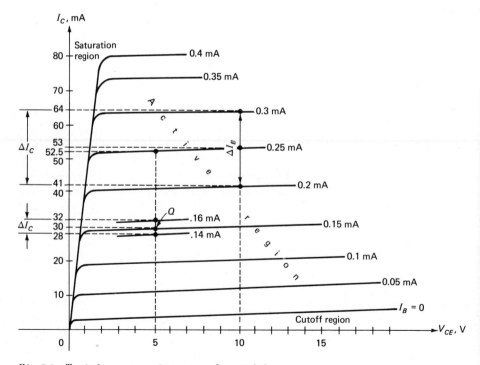

Fig. 5.8  Typical common emitter output characteristics.

collector current. The relationship between $I_B$ and $I_C$ can be used to determine $h_{FE}$. For example, if $V_{CE} = 5$ V and $I_B = 0.25$ mA, the corresponding value of $I_C$ is 52.5 mA. This yields

$$h_{FE} = \frac{I_C}{I_B} \tag{5.3}$$

$$h_{FE}\bigg|_{\substack{\text{at}\\I_C=50\text{ mA}\\V_{CE}=5\text{ V}}} = \frac{52.5 \times 10^{-3}}{25 \times 10^{-5}} = 210$$

The notation $I_C = 50$ mA and $V_{CE} = 5$ V is included above to indicate the $Q$ (operating) point; this value of $h_{FE}$ may not be valid for other $Q$ points.

When dealing with amplifier circuits, not only are we interested in static conditions—that is, voltages and currents that remain the same—but also we deal with voltages and currents that vary with time in accordance with the signal. As such, if the base current were to vary, the collector current should vary in a similar manner, but by a greater amount. The magnitude of these changes depends on the $Q$ point; for our purpose we select $V_{CE} = 10$ V and $I_C = 53$ mA. This corresponds to $I_B = 0.25$ mA, as indicated on the construction in Fig. 5.8. As $I_B$ changes from 0.2 to 0.3 mA, $I_C$ changes from approximately 41 to 64 mA. Now we may write

$\Delta I_B = 0.3 - 0.2 = 0.1$ mA
$\Delta I_C = 64 - 41 = 23$ mA

The common emitter dynamic (AC) forward current transfer ratio may be defined as follows:

$$h_{fe} = \frac{\Delta I_C}{\Delta I_B}\bigg|_{V_{CE},\, I_C} \tag{5.6}$$

For our example this yields

$$h_{fe} = \frac{23 \times 10^{-3}}{0.1 \times 10^{-3}}\bigg|_{\substack{\text{at}\\V_{CE}=10\text{ V}\\I_C=53\text{ mA}}} = 230$$

The parameter $h_{fe}$ is, therefore, the ratio of a small change in $I_C$ to the corresponding change in $I_B$. When the operating point is $V_{CE} = 10$ V and $I_C = 53$ mA, $h_{fe} = 230$.

The increments chosen to calculate $h_{fe}$ should be sufficiently small to remain within the linear region of the transistor. Also, note that the value of $h_{fe}$ calculated at one operating point is not necessarily the same as that calculated at a different point. More will be said on these aspects in Chap. 6.

## Example 5.4

Using the static characteristics of Fig. 5.8, determine $h_{FE}$ at an operating point of $V_{CE} = 5$ V and $I_C = 30$ mA. Also determine $h_{fe}$ at the same operating point.

### Solution

The operating point is located on the curves and labeled $Q$. At this point $I_B = 0.15$ mA, yielding

$$h_{FE} = \frac{I_C}{I_B} \tag{5.3}$$

$$h_{FE} = \frac{30 \times 10^{-3}}{0.15 \times 10^{-3}} = 200$$

To determine $h_{fe}$, small changes about the operating point are required. These are constructed on the graph, yielding

$$\Delta I_B = (0.16 - 0.14)(10^{-3}) = 0.02 \text{ mA}$$
$$\Delta I_C = (32 - 28)(10^{-3}) = 4 \text{ mA}$$

$$h_{fe} \Big|_{\substack{\text{at} \\ V_{CE}=5\text{ V} \\ I_C=30\text{ mA}}} = \frac{4 \times 10^{-3}}{0.02 \times 10^{-3}} = 200 \tag{5.6}$$

### Problem Set 5.2

5.2.1. Assuming a transistor is properly biased, what controls the magnitude of collector current?

5.2.2. Why is the effect of collector voltage on $I_C$ much greater in the region to the left of point $A$ on the output characteristics of Fig. 5.7?

5.2.3. Estimate $h_{fe}$ (using small increments) at a $Q$ point of $V_{CE} = 5$ V and $I_C = 77.5$ mA on the curves of Fig. 5.8. Compare this value of $h_{fe}$ with the value obtained at the operating point of Example 5.4. What conclusions can you draw regarding the dependence of $h_{fe}$ on $I_C$?

## 5.3 GRAPHIC ANALYSIS

### 5.3a Locating the Q Point

In order for a transistor to amplify, it has to be properly biased. This involves forward-biasing the base-emitter junction and reverse-biasing the collector-base junction. For linear amplification, this must result in transistor operation in the active region as opposed to the cutoff or saturation regions of Fig. 5.8; these other regions will be discussed later. There are many ways of accomplishing this; we will start with the common emitter circuit of Fig. 5.9a. The input circuit involves $V_{BB}$, $R_B$, and the base-emitter junction. Resistor $R_C$, supply $V_{CC}$, and the transistor itself make up the output circuit.

The $V_{CC}$ supply reverse-biases the collector-base junction. Normally 4 or 5 V of reverse bias is used, although there are applications for which less than 1 V or several hundred volts may be needed. The $V_{BB}$ supply forward-biases the base-emitter junction. Usually a few tenths of a volt of forward bias is required. $V_{BB}$ is not the actual base-emitter voltage because $I_B$ develops a voltage drop across $R_B$, as indicated in Fig. 5.9b. The actual value of $V_{BE}$ is

$$V_{BE} = V_{BB} - I_B R_B \tag{5.7}$$

In practice, both $V_{BB}$ and $R_B$ are known quantities. Here $V_{BB} = 1$ V and $R_B = 8$ kΩ. This yields

$$V_{BE} = 1 - 8{,}000 I_B \tag{a}$$

This equation involves two unknowns. It cannot be solved unless the relationship between $V_{BE}$ and $I_B$ is established. Figure 5.10 displays the transistor's input characteristics; this is a plot of input voltage $V_{BE}$ versus input current $I_B$. The curves will be recognized as being very similar to the forward characteristics of a

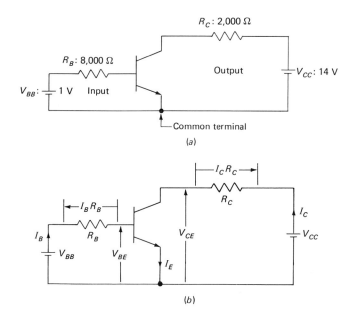

Fig. 5.9 (a) *Common emitter transistor amplifier (DC circuit only)*; (b) *circuit of (a) with currents and voltage drops shown.*

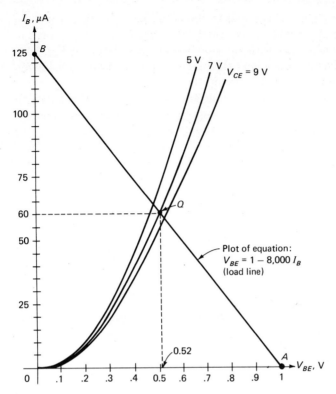

*Fig. 5.10 Common emitter input characteristics and load line construction.*

semiconductor diode. They are, in effect, a plot of voltage versus current for the forward-biased base-emitter diode. The reason why more than one curve is shown is that in a transistor there is some interaction between the collector-base and base-emitter junctions. As such, different values of $V_{CE}$ result in slightly different curves of $V_{BE}$ versus $I_B$.

We will assume that $V_{CE} = 7$ V; therefore, the characteristic labeled 7 V applies. Equation (a) can be solved graphically on the curves of Fig. 5.10 by constructing an appropriate load line. Setting $I_B = 0$ yields

$$V_{BE} = 1 \text{ V} \tag{b}$$

while if $V_{BE} = 0$, we get

$$0 = 1 - 8{,}000 I_B$$
$$1 = 8{,}000 I_B$$
$$I_B = \frac{1}{8{,}000} = 125 \text{ }\mu\text{A} \tag{c}$$

The condition $I_B = 0$ and $V_{BE} = 1$ V is point $A$ in Fig. 5.10. Similarly, the condition $I_B = 125$ μA and $V_{BE} = 0$ is point $B$ on the same curves. These points are the intercepts of Eq. (a); when they are joined, the load line results.

Our $Q$ point is located at the intersection of the load line and the characteristic for $V_{CE} = 7$ V; this is the solution to Eq. (a). Reading from the graph,

$$V_{BE} = 0.52 \text{ V} \quad \text{and} \quad I_B = 60 \text{ μA}$$

We now consider the output circuit in Fig. 5.9b. Again, the actual voltage $V_{CE}$ across the transistor is not the full $V_{CC}$. Instead we have

$$V_{CE} = V_{CC} - I_C R_C \tag{5.8}$$

Since $V_{CC} = 14$ V and $R_C = 2$ kΩ,

$$V_{CE} = 14 - 2,000 I_C \tag{d}$$

Equation (d) is also solved using load line techniques. To do this we need a set of output characteristics (plot of $V_{CE}$ versus $I_C$) as in Fig. 5.11. The horizontal and vertical intercepts are points $A$ and $B$.

*Horizontal: Point A*

$I_C = 0$
$V_{CE} = 14$ V

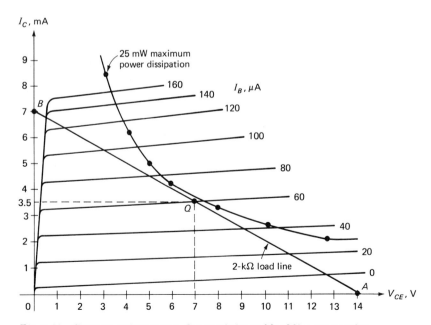

Fig. 5.11  Common emitter output characteristics and load line construction.

*Vertical: Point B*

$V_{CE} = 0$
$0 = 14 - 2{,}000 I_C$
$I_C = 7$ mA

Points $A$ and $B$ are joined to yield the 2,000-Ω load line. The solution to Eq. (d) is the intersection of this load line and the proper base current characteristic ($I_B = 60$ μA). The $Q$ point is therefore

$V_{CE} = 7$ V  $\quad I_C = 3.5$ mA

The DC levels for the complete circuit are given in Fig. 5.12, as follows:

## DC Operating Conditions

$I_B = 60$ μA
$I_C = 3.5$ mA
$V_{BE} = 0.52$ V
$V_{CE} = 7$ V

The selection of an operating point is generally based on a number of factors, such as linearity of characteristics and maximum power rating of the device.

To ensure operation within the transistor's power-dissipation capabilities, a curve may be constructed on the output characteristics that represents the locus of all points equal to the device's power rating. As an example, suppose the transistor of Fig. 5.12 is rated at 25 mW. Since $P = VI$, the transistor can operate with 10 V and 2.5 mA, 5 V and 5 mA, or any $VI$ combination whose product is 25 mW. By plotting a number of such points, the maximum power hyperbola shown in Fig. 5.11 results. For safe operation the $Q$ point must lie to the left and below the hyperbola, as it does in this case.

The circuit of Fig. 5.12 involves two different sources of DC voltage. For practical reasons, it is preferable to have only one such source. Since both $V_{BB}$

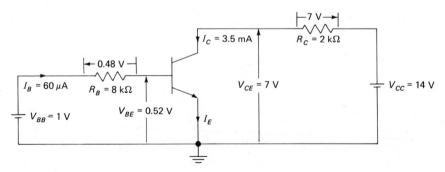

*Fig. 5.12  Common emitter circuit with voltage and current values shown.*

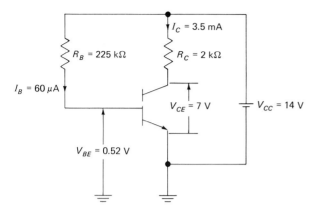

Fig. 5.13 Circuit of Fig. 5.12 modified to work with only one DC supply.

and $V_{CC}$ are positive with respect to ground, it should be possible to eliminate the $V_{BB}$ supply and connect $R_B$ to the positive terminal of $V_{CC}$, as shown in Fig. 5.13. Since $V_{CC} = 14$ V and $V_{BB}$ is only 1 V, $R_B$ must obviously be larger when connected to $V_{CC}$ to maintain the same DC operating conditions. The actual voltage across $R_B$ is

$$14 - 0.52 = 13.48 \text{ V}$$

which must cause 60 μA to flow through $R_B$. Hence

$$R_B = \frac{13.48}{60 \times 10^{-6}} = 225 \text{ k}\Omega$$

We now have a simpler circuit with the same operating point as before.

### Problem Set 5.3

5.3.1. The circuit of Fig. 5.9b is to be biased at $V_{BE} = 0.6$ V and $I_B = 90$ μA. Calculate $R_B$ if

    a. $V_{BB} = 1$ V
    b. $V_{BB} = 14$ V

5.3.2. For the circuit of Fig. 5.13, assume $V_{CC} = 10$ V and the input conditions established in Prob. 5.3.1. It is desired to set the operating point on the characteristics of Fig. 5.11 at $V_{CE} = 3$ V. Determine the corresponding value of $I_C$ and the value of $R_L$ required. Draw the load line and locate the $Q$ point. Also draw a 10-mW power-dissipation hyperbola. Does the $Q$ point lie within a safe power region?

## 5.3b  Determining Amplification

The circuit just discussed is now ready to be used as an amplifier. The operating point has been established; a signal may be introduced at the input and an amplified version of the same signal should appear at the output. For now, we will assume the signal voltage to be made up of a single frequency component. The signal is introduced into the amplifier of Fig. 5.14a (which is the same circuit of Fig. 5.13) using coupling capacitor $C_1$.

The capacitor prevents DC current from flowing between transistor and generator. If the generator were coupled directly to the base of the transistor, without $C_1$, $R_g$ would shunt the base to ground. If $R_g$ were low, a definite change in DC operating levels from those established in the circuit of Fig. 5.13 would result (the AC generator offers no opposition to DC). $C_1$ should be large enough to offer negligible reactance at the signal frequency.

Let $E_{gm}$ (the peak value of $e_g$) equal 25 mV and $R_g = 0\ \Omega$; the changing

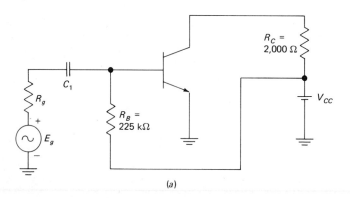

(a)

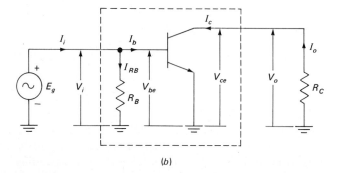

(b)

*Fig. 5.14* (a) *Common emitter amplifier circuit with signal generator;* (b) *AC equivalent circuit for* (a).

signal voltage is entirely coupled through $C_1$, to the transistor base. The DC base voltage has been set at 0.52 V; hence a 25-mV swing on each side of 0.52 V yields extreme values of $v_{BE}$ as follows:

$$\text{Maximum } v_{BE} = 0.52 + 0.025 = 0.545 \text{ V}$$
$$\text{Minimum } v_{BE} = 0.52 - 0.025 = 0.495 \text{ V}$$

A plot of instantaneous $v_{BE}$ versus time is given in Fig. 5.15a; the symbols used in this figure follow the standard conventions of Appendix 2 that apply to any alternating current or voltage. In our case we have

$v_{be}$: instantaneous value of AC component only
$v_{BE}$: instantaneous value of composite signal (DC and AC components)
$V_{bem}$: peak value of AC component
$V_{BE}$: DC component
$V_{be}$: RMS value of AC component

For example, in Fig. 5.15a, at some point in time $v_{BE}$ takes on a value of, say, 0.5 V. When this happens, we say that $v_{BE} = 0.5$ V, or $v_{be} = -0.02$ V (the AC component is measured from the DC reference level of 0.52 V, and hence 0.5 V is 0.02 V below this reference). Also $V_{bem} = 0.545 - 0.52 = 0.025$ V; $V_{BE} = 0.52$ V, and $V_{be} = 0.707 V_{bem} = 0.707(0.025) = 17.7$ mV.

The sinusoidal swing of $v_{BE}$ about the Q point and the extreme values of $v_{BE}$ (0.545 and 0.495 V) are shown on the construction of Fig. 5.16, which uses the characteristics of Fig. 5.10. The changing component of $v_{BE}$ causes corresponding base current variations. These can be determined by drawing vertical lines from each extreme of $v_{BE}$ on the $V_{BE}$ axis to the $V_{CE} = 7$ V characteristic and by extending horizontal lines to the base current axis as shown. This procedure is not entirely accurate, because $V_{CE}$ does not remain at 7 V while $v_{BE}$ and $i_B$ are

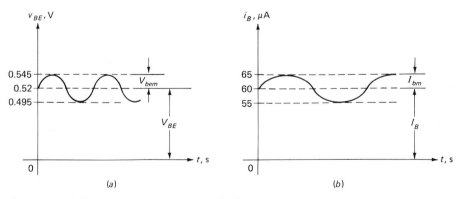

Fig. 5.15  (a) Base-emitter voltage notation; (b) base current notation.

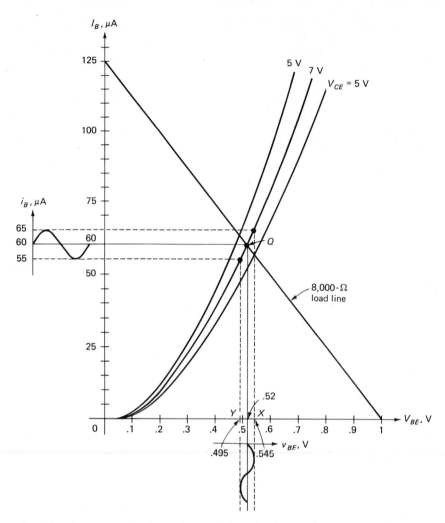

*Fig. 5.16  Common emitter input characteristics with voltage and current increments.*

changing; the error, however, is reasonably small. The extreme values of base current are

   Maximum $i_B = 65 \ \mu A$
   Minimum $i_B = 55 \ \mu A$

As a result of the applied signal, the base current alternates between 55 and 65 μA. Transistor action takes place, yielding collector current variations corresponding to those in the base; these variations can be determined by working with the output characteristics of Fig. 5.17, which are the same as in Fig. 5.11.

Introduction to Junction Transistors  171

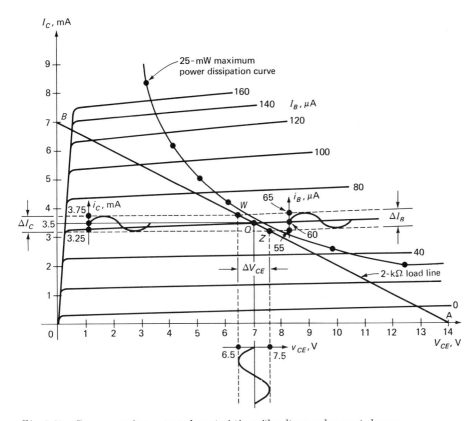

*Fig. 5.17*  Common emitter output characteristics with voltage and current changes.

Point $W$ on the load line corresponds to a base current of 65 μA. Similarly, point $Z$ corresponds to 55 μA. If we drop vertical lines to the $V_{CE}$ axis from these points, we get the values of $v_{CE}$ corresponding to our base current extremes. By extending horizontal lines from $Z$ and $W$ to the vertical axis, the collector current variations can be established. The results are

| $v_{CE}$ | $i_C$ |
|---|---|
| Maximum: 7.5 V | Maximum: 3.75 mA |
| Minimum: 6.5 V | Minimum: 3.25 mA |

Note that, as base current rises from 55 to 65 μA, collector current rises from 3.25 to 3.75 mA. Collector and base currents are therefore in phase. The peak base current $I_{bm} = 65 - 60 = 5$ μA. The peak collector current $I_{cm} =$

3.75 − 3.5 = 0.25 mA. Also note that as the base voltage rises, the collector voltage drops[1]; hence these voltages are 180° out of phase.

To compute $A_i$, $A_v$, and $G$, the AC midfrequency equivalent circuit of Fig. 5.14b is used. Note that $R_g = 0$; also the DC supply $V_{CC}$ behaves as an AC short circuit; therefore the bottom end of $R_B$ is at AC ground potential. The peak base-emitter voltage, from Fig. 5.15a, is 25 mV. The pertinent quantities are

## RMS Input Voltage

$$V_i = E_g = V_{be} = 0.707(25 \times 10^{-3}) = 17.7 \text{ mV}$$

## RMS Input Current

The generator must supply current both to the base and to the biasing resistor; hence

$$I_i = I_b + I_{R_B} = 0.707 I_{bm} + \frac{V_i}{R_B}$$

$$I_i = 0.707(5 \times 10^{-6}) + \frac{17.7 \times 10^{-3}}{225 \times 10^{-3}} = 3.61 \ \mu\text{A}$$

## RMS Output Voltage

$$V_o = V_{ce} = 0.707 V_{cem} = 0.707(0.5)$$
$$V_o = 0.353 \text{ V}$$

## RMS Output Current

$$I_o = I_c = 0.707 I_{cm} = 0.707(0.25 \times 10^{-3})$$
$$I_o = 0.177 \text{ mA}$$

## RMS Output Voltage

$$V_o = -I_o R_C$$

The negative sign is due to the assumed polarity of $V_o$, which is opposite to that given by the direction of $I_o$.

$$V_o = (-0.177 \times 10^{-3})(2{,}000) = -0.354 \text{ V}$$

## Voltage Amplification

$$A_v = \frac{V_o}{V_i} \tag{4.6}$$

$$A_v = \frac{-0.354}{0.0177} = -20$$

---

[1] By *base voltage* is meant the voltage between base and ground, that is, $v_{BE}$. Similarly *collector voltage* means $v_{CE}$.

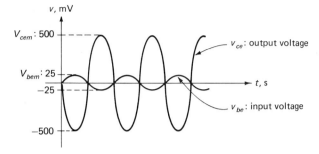

Fig. 5.18  Input and output voltage waveforms.

The negative sign indicates phase reversal. Time graphs for $v_{ce}$ and $v_{be}$ are shown in Fig. 5.18. Note the phase reversal.

## Current Amplification

$$A_i = \frac{I_o}{I_i} \tag{4.7}$$

$$A_i = \frac{0.177 \times 10^{-3}}{3.61 \times 10^{-6}} = 49$$

## Power Gain

$$G = |A_v A_i| \tag{4.12}$$
$$G = 20(49) = 980 = 29.9 \text{ dB}$$

### Problem Set 5.4

Assume the output circuit of Fig. 5.13 is biased at $V_{CE} = 3$ V and $I_C = 5$ mA. The DC supply $V_{CC} = 10$ V. Using the output characteristics of Fig. 5.11:

5.4.1.  Draw the load line; what is the value of $R_C$?
5.4.2.  If $I_B$ changes by 10 μA on each side of the $Q$ point, what are the extreme values of $I_B$, $I_C$, and $V_{CE}$?
5.4.3.  In Prob. 5.4.2, what are the values of $I_c$ and $V_{ce}$ (RMS AC components)? What is the AC output power?

### 5.3c  AC Loading

In the circuit of Fig. 5.14a, $i_C$ flows through $R_C$, $V_{CC}$, and the transistor; $i_C$ consists of an alternating component superimposed on a DC level. The DC level biases the transistor, while the AC component is due to base current changes produced by the signal.

In many cases, the output of a transistor amplifier must drive additional circuits. For example, when stages are cascaded to achieve greater signal amplification, the output of one stage becomes the input to the next. The load seen by any stage therefore includes whatever components make up the input to the next stage. In general, the input impedance to an amplifier may include reactive as well as resistive elements. For the moment we will consider only resistive inputs, a common situation at low and middle frequencies.

The circuit of Fig. 5.19 is a modified version of that in Fig. 5.14a. The only change is the addition of coupling capacitor $C_2$ and load $R_L$. $C_2$ is used to isolate the transistor's DC levels from $R_L$. $R_L$ therefore has no effect on the DC levels previously established. The 2,000-$\Omega$ load line and $Q$ point on the characteristics of Fig. 5.17 still apply; the term *DC load line* is used, however, to distinguish it from the AC load line, soon to be discussed. Both the $Q$ point ($V_{CE} = 7$ V, $I_C = 3.5$ mA) and DC load line are reproduced on the characteristics of Fig. 5.20.

In the example of Sec. 5.3b the load seen by the transistor was 2,000 $\Omega$; this load was common to both the DC and AC components of voltage and current. The 2,000-$\Omega$ load line could therefore be used to establish DC levels ($Q$ point) and AC voltage and current swings.

The situation in the circuit of Fig. 5.19 is different. Since $C_2$ behaves as an open circuit to DC, the DC load is due only to $R_C$, and it is equal to 2,000 $\Omega$. The dynamic or AC load, however, is the parallel combination of $R_C$ and $R_L$:

AC load: $R'_L = R_C \| R_L = 2,000 \| 2,000 = 1,000\ \Omega$

An AC equivalent circuit is shown in Fig. 5.21; both capacitors have been replaced with short circuits (a valid assumption at midfrequencies and above) as well as $V_{CC}$. The AC load line is constructed on the characteristics of Fig. 5.20 by noting that the voltage-to-current ratio anywhere along its length must equal 1,000 $\Omega$ and that it must pass through the $Q$ point. This last point can be established by noting that, in the absence of any signal, the voltages and currents in the transistor are those at the $Q$ point; therefore, both load lines must cross the $Q$ point. To determine where the AC load line intersects the horizontal axis we can

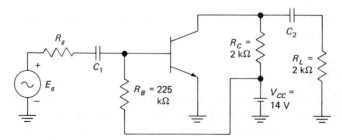

*Fig. 5.19* Common emitter circuit with capacitively coupled load.

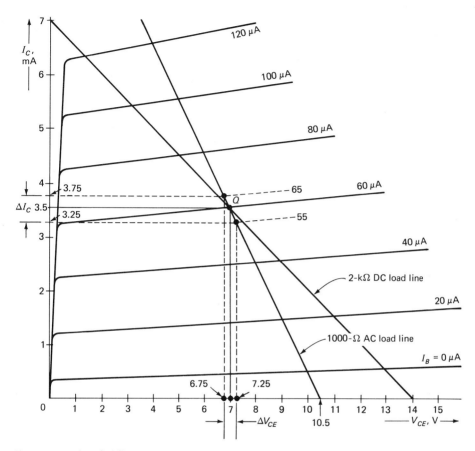

Fig. 5.20  DC and AC load lines on common emitter output characteristics.

use the fact that, if $I_C$ drops from its $Q$ point level to zero ($\Delta I_C = 3.5$ mA), $V_{CE}$ must rise from its $Q$ point value by an amount $\Delta I_C R_L' = (3.5 \times 10^{-3})(1,000) = 3.5$ V. The horizontal-axis intercept is therefore at $V_{CE} = 7 + 3.5 = 10.5$ V, as shown in Fig. 5.20.

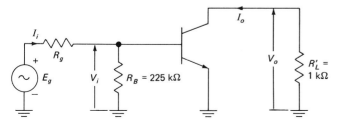

Fig. 5.21  AC equivalent for the circuit of Fig. 5.19.

From the graphic construction on the output characteristics of Fig. 5.20 we find that, as base current changes from 55 to 65 μA, the corresponding collector current change is from 3.25 to 3.75 mA, the same as in the example of Sec. 5.3b. The change in collector voltage, however, is now $\Delta V_{CE} = 7.25 - 6.75 = 0.5$ V, exactly one-half of the earlier value. This is because the same alternating current now flows through one-half the resistance, yielding one-half the voltage.

The current through $R_L$ is pure AC; $C_2$ removes any DC component; its peak value is 0.25 mA, on each side of 0. Similarly, the voltage across $R_L$ is pure AC; its peak value is 0.25 V, on each side of 0.

The current amplification appears unchanged since the total change in collector current is still 0.5 mA. The AC component of collector current, however, divides between $R_C$ and $R_L$; since $R_C = R_L$, it follows that $R_L$ actually receives only one-half. In terms of power gain, the unloaded circuit provided 29.9 dB. For the loaded circuit:

$$I_o = \tfrac{1}{2}I_c = \tfrac{1}{2}(0.707 I_{cm}) = (0.5)(0.707)(0.25 \times 10^{-3}) = 0.088 \text{ mA}$$
$$V_o = -I_o R'_L = (-0.088 \times 10^{-3})(2{,}000) = -0.176 \text{ V}$$
$$V_i = 17.7 \text{ mV} \quad \text{and} \quad I_i = 3.61 \text{ μA (from Sec. 5.3b)}$$

$$A_v = \frac{V_o}{V_i} = \frac{-0.176}{.0177} \simeq -10 \tag{4.6}$$

$$A_i = \frac{I_o}{I_i} = \frac{0.088 \times 10^{-3}}{3.61 \times 10^{-6}} = 24.4 \tag{4.7}$$

$$G = |A_v A_i| = 244 = 23.9 \text{ dB} \tag{4.12}$$

Note that the power gain is one-fourth of (6 dB less than) its previous value; this is due to the additional load $R_L$ placed across the transistor's output.

### Problem Set 5.5

5.5.1. Using the characteristics and $Q$ point of Fig. 5.20, draw the AC load line when $R_L$ in the circuit of Fig. 5.19 is 8 kΩ. What is the $V_{CE}$ axis intercept?

5.5.2. If the base current in Prob. 5.5.1 changes by 5 μA on each side of the $Q$ point, what are the resulting values of $\Delta I_C$ and $\Delta V_{CE}$? Also determine, in decibels,

    a. $A_v$          b. $A_i$          c. $G$

## 5.4 DC CIRCUIT ANALYSIS

Graphic methods of transistor circuit analysis are not always necessary, since a number of simplifying assumptions allow for the solution of most problems using nothing more than Ohm's and Kirchhoff's laws. It was shown earlier that

$$I_E = I_B + I_C \tag{5.1}$$
$$I_C = \alpha I_E \tag{5.2}$$

There is, however, an additional component of collector current that cannot always be neglected. This is due to the flow of minority carriers across the reverse-biased collector-base junction; it is called $I_{CBO}$ and depends heavily on temperature. To be complete, Eq. (5.2) must be modified as follows:

$$I_C = \alpha I_E + I_{CBO} \tag{5.2a}$$

$I_{CBO}$ is defined as the collector current with the collector-base junction suitably reverse-biased and the emitter open. $I_{CBO}$ is often called *cutoff current;* it is nothing more than the reverse current of the collector-base diode, because as long as the emitter is open ($I_E = 0$), no transistor action takes place. More will be said about $I_{CBO}$ in due time; however, it should be noted that, just like diodes, silicon devices have much lower reverse currents than germanium. By substituting Eq. (5.1) in Eq. (5.2a), another form of the same expression is obtained:

$$I_C = I_B \frac{\alpha}{1-\alpha} + I_{CBO} \frac{1}{1-\alpha} \tag{5.9}$$

and since $\alpha = h_{FE}/(h_{FE} + 1)$ \hfill (5.5)

$$I_C = h_{FE} I_B + I_{CBO}(h_{FE} + 1) \tag{5.10}$$

A number of circuits are now discussed to develop the skills needed to analyze transistor circuits in general.

Consider the circuit of Fig. 5.22. The $-10$ V indicates that an external DC source holds point $A$ at 10 V below ground potential. The transistor is controlled by the base current; hence it is convenient to calculate $I_B$ first. Once this is done, all other currents and voltages can be computed. Note that $I_B$ flows through the

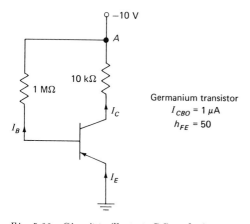

*Fig. 5.22  Circuit to illustrate DC analysis.*

1-MΩ resistor. We can therefore compute $I_B$ if we know the voltage across this resistor; the top end is at $-10$ V, while the lower end is tied to the base. Usually the voltage between base and emitter is around a few tenths of a volt (0.2 V for Ge and 0.6 V for Si). This is a germanium transistor, and hence the base may be assumed to be 0.2 V negative with respect to the emitter (ground). The net voltage across the 1-MΩ resistor is therefore $10 - 0.2 = 9.8$ V; however, it is doubtful that the accuracy achieved by taking the $V_{BE}$ drop into account is really significant. Unless extreme accuracy is necessary, we can say that the full 10 V is dropped across the 1-MΩ resistor, which yields

$$I_B = \frac{10}{10^6} = 10 \; \mu A$$

The collector current is found using Eq. (5.10):

$$I_C = h_{FE}I_B + I_{CBO}(h_{FE} + 1) = (50)(10)(10^{-6}) + (51 \times 10^{-6}) \tag{5.10}$$
$$I_C \simeq 550 \; \mu A$$

The voltage drop across the 10-kΩ resistor is

$$I_C \times 10^4 = (550 \times 10^{-6})(10^4) = 5.5 \; V$$

while the actual voltage across the transistor is

$$V_{CE} = -(10 - 5.5) = -4.5 \; V$$

This transistor, then, is operating at the following DC levels:

$$V_{CE} = -4.5 \; V \quad \text{and} \quad I_C = 0.55 \; mA$$

Now consider the circuit of Fig. 5.23. The emitter-base junction is forward-biased by $V_{EE}$. The collector-base junction is reverse-biased by $V_{CC}$. $V_{EE}$ determines the current in the input (emitter) circuit. The 10-V source sees the series combination of a 10-kΩ resistor and the emitter-base junction; since this is a

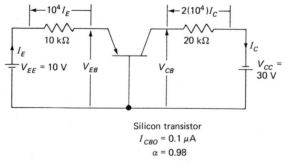

Fig. 5.23   Circuit to illustrate DC analysis.

silicon device, we can expect a $V_{EB}$ drop of about 0.6 V. The voltage across our 10-kΩ resistor is therefore $10 - 0.6 = 9.4$ V. The emitter current is

$$I_E = \frac{10 - 0.6}{10^4} = 0.94 \text{ mA}$$

Once we know $I_E$, we can compute $I_C$. With $\alpha = 0.98$ and $I_{CBO} = 0.1$ μA, we have

$$I_C = \alpha I_E + I_{CBO} \quad (5.2a)$$
$$I_C = 0.98(0.94 \times 10^{-3}) + (0.1 \times 10^{-6}) = 0.92 \text{ mA}$$

$I_C$ develops a voltage drop across the 20-kΩ resistor of $(0.92 \times 10^{-3})(20 \times 10^3) = 18.4$ V. Since $V_{CC} = -30$ V, $V_{CB} = -30 + 18.4 = -11.6$ V. Note that $V_{CC}$ has little effect on $I_C$, which is controlled by the input circuit as long as the collector-base junction is adequately reverse-biased.

The circuit of Fig. 5.24 is now discussed. The input consists of the 1.5-V DC source, 35-kΩ resistor, and base-emitter junction. Since the transistor is made of silicon, a typical $V_{BE} = -0.6$ V will be used. The $V_{BE}$ drop here is not negligible compared to the applied 1.5 V. The voltage across the 35-kΩ resistor is therefore $1.5 - 0.6 = 0.9$ V. This yields

$$I_B = \frac{0.9}{35 \times 10^3} = 25.7 \text{ μA}$$

Since $I_{CBO}$ is specified to be negligible,

$$I_C \simeq \frac{\alpha I_B}{1 - \alpha} \quad (5.9a)$$

$$I_C \simeq \frac{0.99 \, (25.7 \times 10^{-6})}{1 - 0.99} \simeq 2.57 \text{ mA}$$

$I_C$ develops a voltage drop across the 5-kΩ resistor of $(2.57 \times 10^{-3})(5 \times 10^3) = 12.9$ V. This leaves 12.1 V between collector and emitter ($V_{CE} = -12.1$ V), which provides a collector-base reverse bias of $V_{CB} = V_{CE} - V_{BE} = -12.1 - (-0.6) = -11.5$ V.

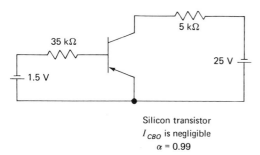

Silicon transistor
$I_{CBO}$ is negligible
$\alpha = 0.99$

Fig. 5.24  Circuit to illustrate DC analysis.

The circuit of Fig. 5.24 is now modified to include a resistor in the emitter lead. This resistor, as will be shown later, is desirable to prevent excessive changes in the operating point whenever the transistor's parameters change. For good temperature stabilization, the emitter resistor should be as large as possible, but, of course, it cannot be too large or it will require an excessive portion of the supply voltage. The modified circuit is shown in Fig. 5.25a.

Since $I_B$ is also the current through the 15-V source, $I_B$ can be calculated if the equivalent resistance seen by the 15-V source is known. This consists of the 50-k$\Omega$ resistor, the base-emitter junction (whose voltage drop is $V_{BE}$), and $R_E$. $R_E$, however, is in the emitter terminal, and it will be necessary to determine its effect when viewed from the base circuit before we can proceed. As long as $I_{CBO}$ is negligible, we can write

$$I_C = h_{FE} I_B \tag{5.3}$$

but

$$I_B = I_E - I_C \tag{5.1}$$

Therefore

$$I_B = I_E - h_{FE} I_B$$
$$I_E = I_B(h_{FE} + 1) \tag{5.11}$$

The total voltage from base to ground is

$$V_B = V_{BE} + I_E R_E$$
$$V_B = V_{BE} + I_B(h_{FE} + 1)R_E$$

The second term in the above expression can be considered to be the voltage drop due to a current $I_B$ flowing through a resistance $R_E(h_{FE} + 1)$. Thus emitter resistance appears multiplied by $h_{FE} + 1$ when seen from the base, as shown in Fig. 5.25b. The converse is true. Any resistance in the base circuit

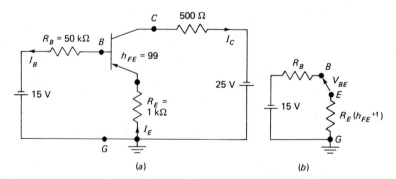

Fig. 5.25 Circuits to illustrate effective emitter resistance seen from base of transistor.

appears divided by $h_{FE} + 1$ when seen from the emitter. These relationships are quite useful for the simplification of transistor circuits; they are fully valid, however, only as long as $I_{CBO}$ is negligible.

Going back to the circuit of Fig. 5.25a, the 15-V DC source sees 50 k$\Omega$ in series with $R_E(h_{FE} + 1)$, for a total of 150 k$\Omega$. The $V_{BE}$ drop can be neglected, since it is small compared to 15 V. The base current is therefore $I_B = 15/(150 \times 10^3) = 0.1$ mA; when multiplied by $h_{FE}$, this yields $I_C$ (assuming negligible $I_{CBO}$). Thus, $I_C \simeq 10$ mA. The voltage drop across the 500-$\Omega$ resistor is 5 V, while the voltage drop across $R_E$ is roughly 10 V. The assumption has been made here that $I_E \simeq I_C$. Since 15 V is dropped across the two resistors in the output circuit, there is 10 V between collector and emitter.

The circuit of Fig. 5.25a may be simplified by eliminating one of the batteries. The 15-V supply is replaced with a resistive voltage divider that makes use of the 25-V battery as shown in Fig. 5.26. To prove that these two circuits are equivalent, the Thevenin circuit seen by the base will be compared in each case. For the circuit in Fig. 5.26, if the connection at $B$ is broken, the Thevenin voltage on the left side of $B$ is

$$V_B = -25 \frac{125 \times 10^3}{(125 \times 10^3) + (83.5 \times 10^3)} = -15 \text{ V}$$

This is equivalent to the $-15$ V of Fig. 5.25a. The Thevenin resistance seen by the base is obtained by shorting the 25-V battery, which places $R_1$ and $R_2$ in parallel:

$$R_1 \| R_2 = 83.5 \times 10^3 \| 125 \times 10^3 = 50 \text{ k}\Omega$$

This 50 k$\Omega$ is equivalent to $R_B$ in the circuit of Fig. 5.25a. These two circuits are therefore electrically equivalent. The advantage of the modified circuit is that only one DC supply is required.

It is now desired to determine the DC operating point ($V_{CE}$ and $I_C$) for

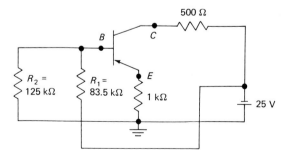

Fig. 5.26  Circuit of Fig. 5.25 modified to require only one DC source.

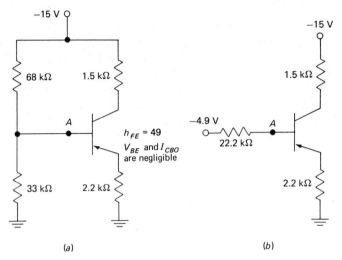

Fig. 5.27  Circuits to illustrate DC analysis.

the circuit of Fig. 5.27a. Our first task is to calculate $I_B$; when multiplied by $h_{FE}$, this yields the approximate value of $I_C$. To do this, we determine the Thevenin circuit that feeds the base. If the base connection is broken, the voltage at point $A$ is

$$V_A = -15\,\frac{33 \times 10^3}{(68 \times 10^3) + (33 \times 10^3)} = -4.9 \text{ V}$$

The Thevenin resistance is the parallel combination of 33 and 68 k$\Omega$, namely 22.2 k$\Omega$. The equivalent circuit is shown in Fig. 5.27b. $I_B$ is found by transforming the emitter resistance into its effective value seen from the base and applying Ohm's law. The 2.2 k$\Omega$ becomes $(2.2 \times 10^3)(h_{FE} + 1) = 110$ k$\Omega$; the total resistance seen by the 4.9-V battery is therefore $(22.2 \times 10^3) + (110 \times 10^3) = 132.2$ k$\Omega$. This yields

$$I_B = \frac{4.9}{132 \times 10^3} = 37.1 \ \mu\text{A}$$

from which

$$I_C \simeq h_{FE} I_B = 49(37.1 \times 10^{-6}) \simeq 1.82 \text{ mA}$$
$$V_{CE} \simeq -15 - (-I_C)(1.5 + 2.2)(10^3)$$
$$V_{CE} \simeq -15 + (1.82 \times 10^{-3})(3.7 \times 10^3) = -15 + 6.72 = -8.18 \text{ V}$$

The above results are not very accurate, but are acceptable for most applications. If greater precision is required, $I_{CBO}$ and $V_{BE}$ should be taken into account.

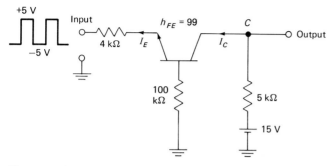

**Fig. 5.28**  *Circuit with square-wave input.*

The circuit of Fig. 5.28 is now discussed. The input to the transistor is a square wave which alternates between extremes of $+5$ V and $-5$ V. This input alternately forward- and reverse-biases the base-emitter junction. The circuit is therefore not operating as a linear amplifier, because to do so the base-emitter junction must always remain forward-biased.

When the input is 5 V, the emitter-base junction is reverse-biased, and there is no appreciable emitter current. Assuming an ideal transistor, no appreciable collector current flows either; there is no voltage across the 5-kΩ resistor, and the output, which is taken at point $C$, is simply equal to the 15-V supply.

The negative input, on the other hand, forward-biases the base-emitter junction. The emitter current is determined by transforming the 100-kΩ resistor in the base to its effective value seen from the emitter. Since base resistance appears divided by $h_{FE} + 1$ when seen from the emitter, the 100 kΩ is transformed to $10^5/100 = 1$ kΩ. Neglecting $V_{BE}$, the emitter current is the ratio of applied 5 V to the total resistance, which is 5 kΩ. Therefore $I_E \simeq 5/5{,}000 = 1$ mA. Since $h_{FE} = 99$, $\alpha = 0.99$ and $I_C = 0.99$ mA. $I_C$ develops a 4.95-V drop across the 5-kΩ resistor, leaving 10.05 V at point $C$, which is our output terminal.

The results of the analysis indicate that, when the input is 5 V, the output is 15 V; when the input is $-5$ V, the output is 10.05 V.

The circuit of Fig. 5.29 is now analyzed. It is desired to determine the value of $V_C$ (collector-to-ground voltage). Since the applied voltage is known (30 V), all we need to calculate is the voltage drop across the 10-kΩ resistor; the difference is $V_C$. We cannot, however, compute the voltage drop across the 10-kΩ resistor until we know the current. With this in mind, note that $I_B$ flows through the 50-kΩ resistor. Neglecting the $V_{BE}$ drop, the base can be considered at ground potential. The voltage across the 50-kΩ resistor, then, is simply $V_C$. This yields

$$I_B = \frac{V_C}{50 \times 10^3} \qquad (a)$$

The current through the 10-kΩ resistor is

# 184 Solid State Electronic Circuits

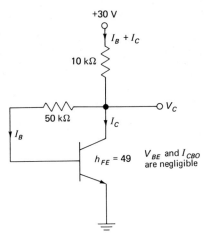

*Fig. 5.29 Circuit to illustrate DC analysis.*

$$I_B + I_C \simeq I_B + h_{FE}I_B = 50 I_B$$

Using Eq. (a) this is

$$50 I_B = 50 \frac{V_C}{50 \times 10^3} = 0.001 V_C \qquad (b)$$

The voltage across the 10-kΩ resistor is therefore $(10^4 \times 0.001) V_C = 10 V_C$, which, when subtracted from the applied 30 V, must yield $V_C$:

$V_C = 30 - 10 V_C$
$11 V_C = 30$
$V_C = 2.73$ V

## Problem Set 5.6

5.6.1. For each of the circuits in Fig. 5.30, calculate all three transistor currents $(I_B, I_C, I_E)$ and the voltage across the transistor ($V_{CB}$ or $V_{CE}$). Assume that both $V_{BE}$ and $I_{CBO}$ can be neglected.

5.6.2. Calculate the power dissipated by each transistor in the circuits of Fig. 5.30f, g, m, and q. *Hint:* Since $V_{BE} = 0$, only the collector-base junction dissipates power; this power can be calculated by taking the product of $V_{CB}$ and $I_C$.

5.6.3. Each of the transistor amplifier circuits in Fig. 5.31 must be biased at $V_{CB}$ or $V_{CE} = 5$ V and $I_C = 5$ mA. Determine the value of all unknown resistors. $V_{BE}$ and $I_{CBO}$ are to be neglected.

5.6.4. Determine the new operating point in each circuit of Prob. 5.6.3 if each transistor is replaced with another whose $h_{FE}$ is double the original value.

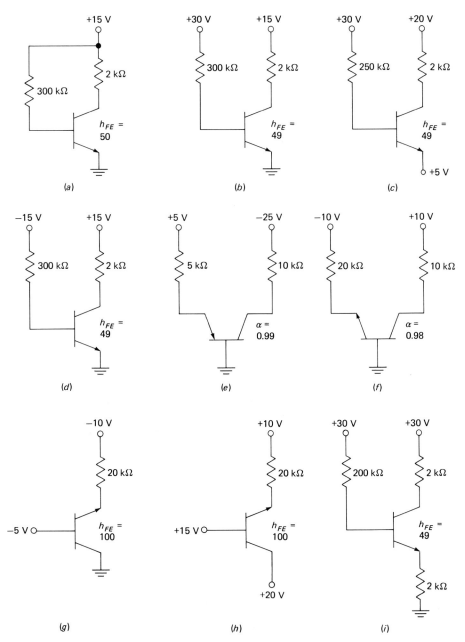

Fig. 5.30 Circuits for Prob. 5.6.1.

186   Solid State Electronic Circuits

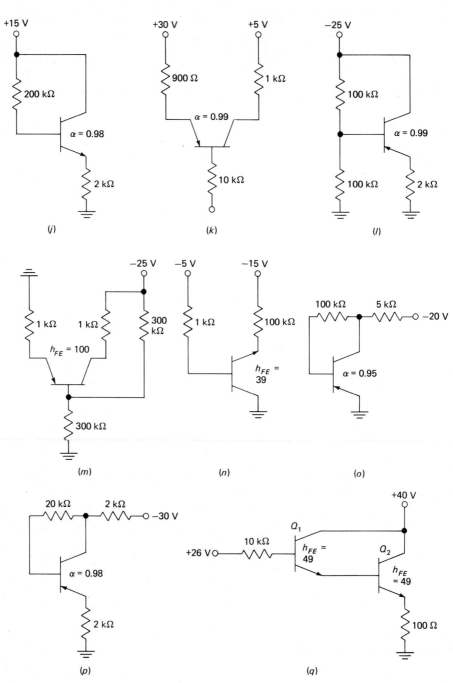

Fig. 5.30   (Continued).

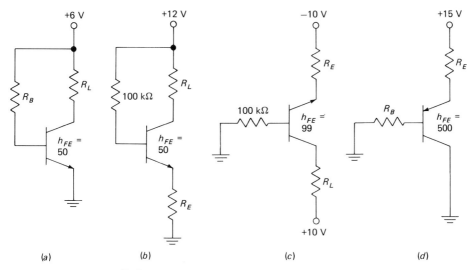

Fig. 5.31  Circuits for Prob. 5.6.3.

## 5.5 CUTOFF CURRENT

Cutoff current $I_{CBO}$ has already been introduced as an additional component of collector current. It is an undesirable component because of its temperature-dependence. $I_{CBO}$ is made up of a *bulk* component (originates in the body of the semiconductor) and a *surface* component (due to surface impurities). The bulk component is temperature-dependent, while the surface component depends on the applied voltage. The voltage-dependence is due to the semiconductor surface acting as an effective resistance (due to surface impurities), thus obeying Ohm's law.

$I_{CBO}$ is usually much more dependent on temperature than on voltage (as long as the voltage is below the reverse breakdown level), especially in Ge transistors, whose cutoff current may be 1,000 times greater than in Si. Most data sheets usually specify $I_{CBO}$ at 25°C; the temperature-dependence may also be given. If not, $I_{CBO}$ can be estimated at other temperatures using the rule of thumb that it doubles for every 10°C rise in Ge and for every 6°C rise in Si (this is approximate—if accuracy is required it should be measured at whatever temperature is necessary).

The notation $I_{CBO}$ is based on a convenient system for identifying cutoff currents. The first subscript indicates the terminal in which the current is measured; the second states the common terminal; the last specifies the connection between the third and common terminals. In the circuit of Fig. 5.32, the $C$ in $I_{CBO}$ stands for collector; the $B$ stands for common base configuration, and the $O$ specifies that the third terminal (emitter) is open. Other possible terminations

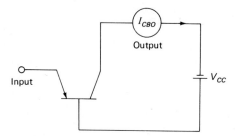

*Fig. 5.32  Circuit for measuring* $I_{CBO}$.

between the third terminal and common terminal are $S$ for shorted, $R$ for a resistor, and $X$ if a reverse bias is applied.

Cutoff current in a common emitter configuration may be analyzed using the circuit of Fig. 5.33. Reverse bias for the collector-base junction is provided by $V_{CC}$. Cutoff current is measured in the collector terminal; therefore the first subscript is $C$. The emitter is common; therefore $E$ is the second subscript. The last subscript depends on the termination between the third and common terminal.

With the base open, the collector current is $I_{CEO}$. Here the path for cutoff current is through the collector, the external supply, the emitter, and back to the collector. There is no base current, and $I_C = I_E = I_{CEO}$. The circuit for $I_{CEO}$ is shown in Fig. 5.34a.

Let us consider this further. The collector-base depletion region supports a reverse current which is mainly due to minority carriers. In an NPN transistor, electrons in the base and holes in the collector are minority carriers. A hole reaching the base from the collector represents excess positive charge that the emitter tries to neutralize by injecting electrons. Transistor action, however, takes place, so that the emitter has to inject a lot more than one electron to neutralize one hole. Those electrons injected into the base that do not recombine with the hole come out of the collector terminal, resulting in collector current.

For example, if $\alpha = 0.95$, 95 percent of all emitter-injected carriers reach

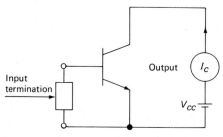

*Fig. 5.33  Circuit for measuring collector cutoff current in a common emitter configuration.*

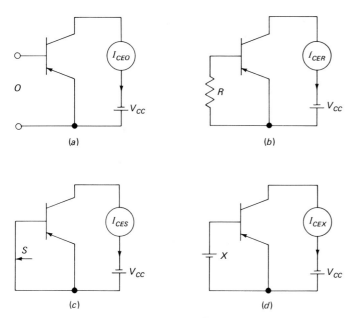

Fig. 5.34 Circuits for measuring: (a) $I_{CEO}$; (b) $I_{CER}$; (c) $I_{CES}$; (d) $I_{CEX}$.

the collector, and only 5 percent recombine in the base. To neutralize one hole, 20 electrons must be injected by the emitter; of these, 19 reach the collector, resulting in collector current.

The relationship between $I_{CBO}$ and $I_{CEO}$ may be established as follows:

$$I_C = \alpha I_E + I_{CBO} \tag{5.2a}$$

In this case $I_C = I_E = I_{CEO}$; therefore,

$$I_{CEO} = \alpha I_{CEO} + I_{CBO}$$
$$I_{CEO}(1 - \alpha) = I_{CBO}$$
$$I_{CEO} = \frac{I_{CBO}}{1 - \alpha} = I_{CBO}(h_{FE} + 1) \tag{5.12}$$

Equation (5.12) can be verified for the previous example. If $\alpha = 0.95$,

$$I_{CEO} = \frac{I_{CBO}}{1 - 0.95} = 20 I_{CBO}$$

That is, each unit of $I_{CBO}$ is amplified 20 times to produce $I_{CEO}$.

$I_{CEO}$ can be rather large, especially if the transistor's $\alpha$ is high. To reduce $I_{CEO}$ we must reduce accumulation of charge in the base, thus preventing emitter injection; this can be accomplished by providing a path for excess charge to leak out of the base. Resistor $R$ in the circuit of Fig. 5.34b does just this. The lower the value of $R$, the lower is the collector cutoff current $I_{CER}$. Ultimately we can make $R = 0$ as in the circuit of Fig. 5.34c. At this point the collector cutoff current $I_{CES}$ is very low, since there is practically no emitter injection required to neutralize excess base charge.

$I_{CES}$ and $I_{CEO}$ are really special cases of $I_{CER}$; the former when $R = 0$, and the latter when $R$ is infinite. $I_{CEO}$ is the largest, while $I_{CES}$ is the lowest value of $I_{CER}$.

A further improvement is possible by reverse-biasing the base-emitter junction as in Fig. 5.34d. This may be necessary in applications where cutoff current has to be minimized. Transistor data sheets specify this value as $I_{CEX}$, where $X$ is the base-emitter reverse bias. $I_{CEX}$ can be very low. Test voltages are given for any cutoff current specification, because all currents tend to rise as the magnitude of reverse voltage increases.

## Problem Set 5.7

5.7.1. For each position of the switch in the circuit of Fig. 5.35a, determine

    a. The correct symbol for the collector current
    b. The actual value of collector current at 55°C, using the manufacturer's graph provided in Fig. 5.35b

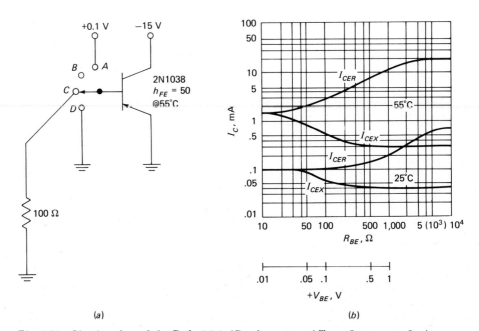

Fig. 5.35 Circuit and graph for Prob. 5.7.1. (Graph courtesy of Texas Instruments Inc.)

5.7.2. Draw a circuit diagram to measure $I_{EBO}$ in an NPN transistor.
5.7.3. Determine, using the data sheets in Appendix 10 for the 2N3199 silicon transistor, the following parameters (typical values):

   a. $I_{CEO}$ at $V_{CE} = \frac{1}{2} V_{CE(max)}$ and $T_A = 25°C$
   b. $I_{CEX}$ at $V_{CE} = \frac{1}{2} V_{CE(max)}$ and $T_A = 150°C$

5.7.4. Explain, from a physical point of view, why $I_{CEO} = I_{CBO}(h_{FE} + 1)$.

## 5.6 BREAKDOWN VOLTAGE

If the reverse bias on a PN junction is increased, a point known as voltage breakdown, characterized by a rise in current, is eventually reached. This current increase must be limited to avoid device destruction.

The symbol for breakdown voltage is $BV$ (sometimes $BR$) with subscripts as defined for cutoff current. All other things being equal, the greater the cutoff current, the lower the breakdown voltage. For example, the lowest voltage breakdown is $BV_{CEO}$ (because of $I_{CEO}$, the highest cutoff current). Similarly, with the emitter open, $I_{CBO}$ flows leading to $BV_{CBO}$, the highest voltage breakdown rating. There are other factors that influence breakdown, such as device geometry and manufacturing technique. In any case, as a general rule, $BV_{CBO}$ is the highest and $BV_{CEO}$ the lowest rating for any given unit.

Most modern transistors such as silicon devices manufactured by diffusion techniques break down by the avalanche mechanism which exhibits a positive temperature coefficient (PTC—breakdown voltage increases as the temperature rises). Transistor voltage ratings vary; in some cases they now exceed 1,000 V.

Typical transistor voltage breakdown characteristics are shown in Fig. 5.36. Since the initiation of breakdown is not always sharp, a sensing current

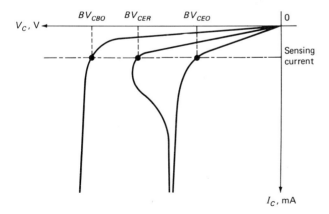

Fig. 5.36  Transistor breakdown voltages.

may be established to arbitrarily define a point where the transistor is said to be in breakdown.

## Problem Set 5.8

5.8.1. Using the appropriate data sheets in Appendix 10, determine the following absolute maximum ratings for the 2N2137:

    a. $BV_{CBO}$                                c. $BV_{CEO}$
    b. $BV_{CES}$                             d. $BV_{EBO}$

5.8.2. What is sensing current and why is it used?

5.8.3. Draw circuit diagrams that can be used to measure

    a. $BV_{CEX}$                b. $BV_{EBO}$                c. $BV_{CES}$

5.8.4. What happens to $BV_{CER}$ as $R$ varies from infinity to zero?

## 5.7 THE TRANSISTOR AS A SWITCH

Consider the circuit diagram of Fig. 5.37a. A 10-V DC source is connected through a 1-k$\Omega$ resistor to a switch $S$ (single pole, single throw); the switch may be open or closed. Let us plot the circuit current along the vertical axis of a graph and the voltage across the switch along the horizontal axis of the same graph, as shown in Fig. 5.37b. Since the switch can either be open or closed, there are only two possible states. If the switch is open, the current is zero and the voltage across the switch is 10 V. This gives us point $O$ on the graph. If the switch is closed, the circuit current is 10 mA, and the voltage across the switch is zero. This yields point $C$ on the graph. The function of the switch is simple; if it is closed, current flows; if it is open, no current flows. When the switch is closed, it has no voltage across its terminals; when it is open, it has no current.

    Here the switching is done manually. Electronic switching, on the other hand, is possible by applying appropriate control signals to switching devices, such as diodes, relays, or transistors. Although none of these devices are ideal,

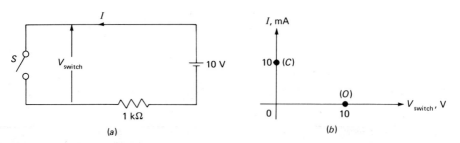

Fig. 5.37   Circuit to illustrate elementary principles of switching.

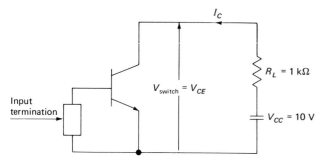

Fig. 5.38  Common emitter transistor as a switch.

they have advantages of speed over the manual switch; this is especially true for semiconductors.

Now consider the common emitter transistor circuit of Fig. 5.38. The collector and emitter terminals are analogous to the switch contacts of Fig. 5.37a. Again, there are two extreme possibilities. If the transistor were ideal, its OFF current would be zero, and the entire supply would be dropped across the transistor, yielding

$I_C = 0$
$V_{CE} = V_{CC}$

The above conditions represent point $D$ on the graph of Fig. 5.39. Similarly, the

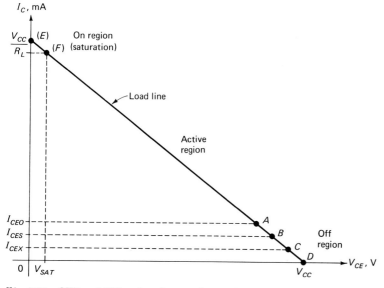

Fig. 5.39  OFF and ON regions for transistor switch.

ON voltage would be zero, and all $V_{CC}$ would be dropped across $R_L$. The current would then be

$$I_C = \frac{V_{CC}}{R_L}$$

and the voltage

$$V_{CE} = 0$$

yielding point $E$ on the graph of Fig. 5.39.

A real transistor does not have zero OFF current or zero ON voltage. The best we can hope for is minimization of these parameters. To turn the transistor OFF, as an example, we might consider leaving the base open, as in Fig. 5.34a. With the base open, $I_B = 0$, but $I_{CEO}$ flows through the collector; this may be in the milliampere range. To reduce the collector cutoff current, we could connect a resistor from base to emitter as in Fig. 5.34b. This yields $I_{CER}$, possibly a few hundred microamperes. Should this still not be satisfactory, the base may be shorted to the emitter, which reduces the collector current to $I_{CES}$ (Fig. 5.34c). For the numerical examples used so far, $I_{CES}$ might be around 100 μA. This can be further reduced by applying a reverse voltage from base to emitter (Fig. 5.34d). A reverse voltage of 0.1 V might give us an $I_{CEX}$ of 40 μA. This is the lowest cutoff current we can get in this transistor configuration. $I_{CBO}$ might be lower, but this requires a common base configuration.

It is important to note that when we reverse-bias the base-emitter junction, the breakdown voltage rating $BV_{EBO}$ must be observed; usually this rating is lower than for the collector-base junction of the same transistor.

How well our transistor can be turned OFF depends on the control signals that we apply at the input. The diagram of Fig. 5.39 shows point $A$ (base open: $I_{CEO}$ flows); $B$ (base shorted: $I_{CES}$ flows); $C$ (base is reverse-biased: $I_{CEX}$ flows); $D$ (ideal: $I = 0$, cannot be obtained). The region around which points $A$, $B$, $C$, and $D$ cluster is the OFF region. It corresponds to the region below the $I_B = 0$ characteristic on the output curves of Fig. 5.8. With the transistor in the OFF state, the current should be very small. For silicon transistors using a planar structure (see Sec. 5.8b) OFF currents are usually in the nanoampere range. Note that when a transistor is cut off, most of the external voltage is dropped from collector to emitter; therefore the external voltage should be less than the voltage breakdown rating. As a safety measure the lowest breakdown rating $BV_{CEO}$ should not be exceeded.

A transistor may be turned ON by applying base-emitter voltage whose polarity forward-biases the junction, causing base current to flow. For the NPN transistor of Fig. 5.40, a positive-going base voltage is required. Initially, as the magnitude of forward $V_{BE}$ increases, $I_B$ and $I_C$ also increase. The transistor is then in its active region, where linear amplification is possible. If we keep in-

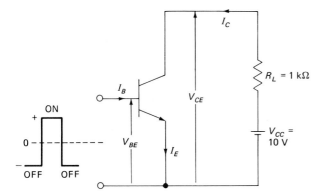

Fig. 5.40  Turning an NPN transistor switch ON (+ input); OFF (− input).

creasing the input drive, eventually $I_C$ reaches its maximum value, as determined by the supply voltage and total resistance in the collector circuit. For an ideal device this would be $V_{CC}/R_L$, but in practice there is a small resistance from collector to emitter $r_{(ON)}$ across which a saturation voltage $V_{CE(SAT)}$ or more simply $V_{SAT}$ is developed. On the graph of Fig. 5.39, this is shown as point $F$. When manufacturing transistors for switching applications, efforts are made to minimize $r_{(ON)}$. In general, saturation voltages of a few tenths of a volt are to be expected. When the voltage across a transistor is equal to $V_{SAT}$, the transistor is said to be in saturation; this is the high-current–low-voltage region in the output characteristics of Fig. 5.8. Because of silicon's higher resistivity, $r_{(ON)}$ and $V_{SAT}$ are greater in Si than in Ge transistors.

We will now consider the $V_{BE}$ drive needed to accomplish various functions. To cut off a Ge transistor, about 0.1-V base-emitter reverse bias is usually required. This causes $I_{CEX}$ to flow through the collector terminal. For Si transistors, with their lower cutoff current, $V_{BE} = 0$ is usually adequate to cut off the device.

The base-emitter forward bias for which a transistor begins conducting is called the cut-in voltage. Typical values of $V_{BE(cut-in)}$ are 0.1 V for Ge and 0.5 V for Si devices.

As $V_{BE}$ is increased beyond the cut-in value, operation of the transistor moves away from the cutoff into the active region. For Ge transistors a typical value of $V_{BE} = 0.2$ V; for Si, around 0.6 V. Further increasing $V_{BE}$ to about 0.3 V for Ge and 0.7 V for Si drives the transistor in saturation. Here $V_{CE}$ drops to $V_{CE(SAT)}$, with typical values of 0.1 V for Ge and 0.3 V for Si. Note that in saturation $V_{CE} < V_{BE}$; therefore the collector-base junction becomes forward-biased. Saturation, in fact, is characterized by both junctions being forward-biased.

## Example 5.5

The transistor in Fig. 5.40 is to be alternatively turned OFF and ON by applying the proper base-emitter drive. $V_{CC} = 10$ V and $R_L = 1$ kΩ.

a. Assuming an ideal device, determine voltage and current for both ON and OFF states.
b. Repeat (a) assuming a real Ge device. What base-emitter drive is required in each case?
c. Repeat (a) assuming a real Si device. What base-emitter drive is required in each case?

### Solution

a. *Ideal Transistor*

OFF current: $I_C = 0$
OFF voltage: $V_{CE} = V_{CC} = 10$ V
ON current: $I_C = V_{CC}/R_L = 10$ mA
ON voltage: $V_{CE} = 0$

b. *Real Germanium Transistor* The transistor may be turned OFF with $V_{BE} = -0.1$ V. The OFF current is then $I_{CEX}$. The voltage across the transistor is $V_{CE} = V_{CC} - I_{CEX}R_L$. The transistor may be turned ON using $V_{BE} = 0.3$ V. The ON voltage is typically $V_{CE(SAT)} = 0.1$ V, yielding an ON current $I_C = (10 - 0.1)/10^3 = 9.9$ mA.

c. *Real Silicon Transistor* The transistor may be turned OFF with $V_{BE} = 0$. The OFF current is then $I_{CES}$, which for Si is very low. The OFF voltage is $V_{CE} = V_{CC} - I_{CES}R_L$. To turn the transistor ON, $V_{BE} \simeq 0.7$ V is required. The ON voltage is $V_{CE(SAT)} = 0.3$ V, and the ON current $I_C = (10 - 0.3)/10^3 = 9.7$ mA.

## Example 5.6

The Si transistor in Fig. 5.41a has the following parameters: negligible reverse cutoff current; $h_{FE} = 30$. Estimate the transistor's $I_C$ and $V_{CE}$ for each position of the switch.

### Solution

*A:* The base-emitter junction is reverse-biased, hence the device is cut off. $I_C = 0$ and $V_{CE} = V_{CC} = -15$ V. Point A is shown on the graph of Fig. 5.41b.

*B:* Here $V_{BE} = -0.8$ V; this is sufficient forward bias to saturate the transistor, yielding a typical $V_{CE(SAT)} = -0.3$ V. Note that $V_{CB} = V_{CE} -$

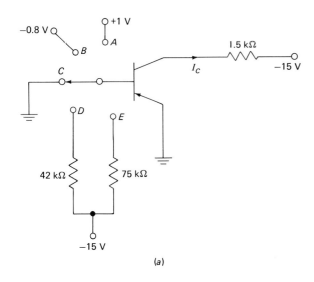

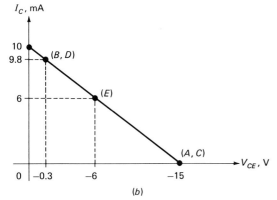

Fig. 5.41 Circuit and VI diagram for Example 5.6.

$V_{BE} = 0.5$ V, which represents a forward-biased collector-base junction. The current is

$$I_C = \frac{15 - 0.3}{1.5 \times 10^3} = 9.8 \text{ mA}$$

Point B is also shown on the graph.

C: Grounding the base forces $V_{BE} = 0$; the transistor is cut off and the same conditions as in A apply.

D: The $V_{BE}$ drop here can be neglected, since it is likely to be small compared with the $-15$-V supply. The base current is therefore

$$I_B \simeq \frac{15}{42 \times 10^3} = 0.357 \text{ mA}$$

If the transistor is in the active region (we cannot be sure about this until we have checked it), $I_C$ can be determined as

$$I_C = h_{FE} I_B$$

yielding

$$I_C = 30(0.357 \times 10^{-3}) = 11.71 \text{ mA}$$

But $V_{CC}/R_L = 15/(1.5 \times 10^3) = 10$ mA, and this is the maximum possible $I_C$ with zero $V_{CE(SAT)}$. The above answer is therefore not accurate; instead the transistor is in saturation, and the conditions in $B$ apply.

$E$: The same initial procedure as in $D$ can be followed here, and

$$I_B \simeq \frac{15}{75 \times 10^3} = 0.2 \text{ mA}$$

Assuming, for the moment, that the transistor is in the active region, we have

$$I_C = h_{FE} I_B$$
$$I_C = 30(0.2 \times 10^{-3}) = 6 \text{ mA}$$
$$\text{and } V_{CE} = -15 + (6 \times 10^{-3})(1.5 \times 10^3)$$
$$V_{CE} = -6 \text{ V}$$

The point ($I_C = 6$ mA and $V_{CE} = -6$ V) is shown at $E$ on the graph. This point is in the active region, and hence our analysis is valid.

In conclusion, the limiting factors in approaching an ideal switch are

OFF state:   cutoff current
ON state:    saturation voltage

These factors determine how well a transistor switch discriminates between one state and another. In computing and control applications, an additional factor, speed, is very important. The manual switch discussed earlier is obviously very slow; the electromagnetic relay is also very slow in that it requires a few milliseconds to be switched from one state to another. Transistors can be switched very quickly. The state of the art is such that it is now possible to switch a transistor from one state to another in a few nanoseconds.

### Problem Set 5.9

5.9.1. What polarity of voltage is required at the base of a PNP transistor with respect to the emitter to turn it ON? OFF?

5.9.2. Repeat Prob. 5.9.1 for an NPN transistor.

5.9.3. What transistor parameters would you consider important when analyzing the OFF behavior? The ON behavior?

5.9.4. Consider the common emitter output characteristics of Fig. 5.8. Compare the transistor power dissipation for each of the three regions shown (saturation, active, cutoff).

5.9.5. For each value of $V_{BE}$ given below, determine whether the NPN Ge transistor in Fig. 5.40 is cut off, cut in, saturated, or in the active region:

a. 0.3 V
b. 0.1 V
c. 0.4 V
d. 0.2 V
e. −0.1 V
f. −0.3 V

5.9.6. For each value of $V_{BE}$ given below, determine the region of operation of a PNP Si transistor suitably biased:

a. −0.8 V
b. 0.3 V
c. −0.5 V
d. 0 V
e. −0.7 V
f. −0.6 V

Problems 5.9.7 through 5.9.10 apply to the circuit of Fig. 5.42a.

5.9.7. What happens to $I_C$ as the potentiometer wiper is moved from A to B?

5.9.8. Repeat Prob. 5.9.7 for $V_C$.

5.9.9. Find the Thevenin equivalent circuit seen by the base when the potentiometer is three-quarters of the way between A and B (closer to B).

5.9.10. Using the results of Prob. 5.9.9, determine $I_C$ and $V_C$. In what region is the transistor operating?

5.9.11. Estimate $I_C$ for each position of the switch in the circuit of Fig. 5.42b.

## 5.8 MANUFACTURING TECHNIQUES

The fabrication of practical junction transistors begins with wafers of Si or Ge that have been purified to an extremely high degree (undesirable impurities less than one part in $10^9$). The process ends when leads are attached, and the transistor is hermetically sealed and electrically tested to verify that it meets all performance specifications.

The quantity and manner in which desired impurities (dopants) are introduced into the semiconductor determine, along with the physical geometry, the electrical performance of the device.

Considerations involved in manufacturing transistors depend on the ultimate application. For small-signal amplification we may want high gain and good frequency response; for power amplifiers large $BV_{CEO}$ and power-dissipation ratings are usually the main criteria. For switching applications $I_{CBO}$, $V_{SAT}$, and switching time must all be minimized. The transistor engineer must make

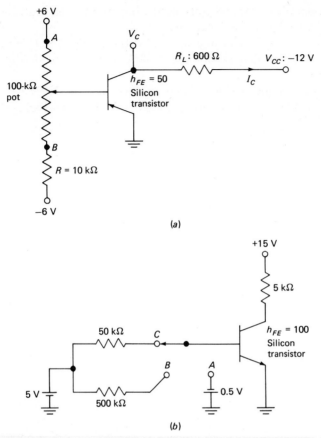

Fig. 5.42 (a) Circuit for Probs. 5.9.7 through 5.9.10; (b) Circuit for Prob. 5.9.11.

many compromises and trade-offs, since improvement in one parameter is often achieved at the expense of another.

### 5.8a  Junction-forming Processes

To produce transistors, appropriate PN junctions must be formed on a semiconductor structure. Junctions may be grown, alloyed, diffused, or etched. Each of these junction-forming processes is briefly described below.

*GROWN JUNCTIONS.* A semiconductor in the molten state is "pulled" from a crucible to which impurities are added in carefully controlled amounts

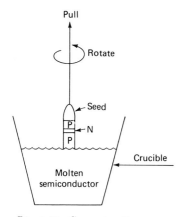

Fig. 5.43  *Grown junction process.*

(Fig. 5.43). As the semiconductor crystal (the *seed*) is pulled out, it is also rotated and the liquid melt *grows* on it. By periodically changing the impurity concentrations of the melt, alternate P- and N-type regions are produced.

ALLOYED JUNCTIONS.  Pellets of the appropriate impurity are *alloyed* on a semiconductor wafer by heating the combination to a temperature greater than the impurity's but lower than the semiconductor's melting point (Fig. 5.44). When the material cools, the impurity and semiconductor form a continuous crystal whose doping varies to produce PNP or NPN transistors.

The collector-base junction area is large to permit high power-dissipation ratings. The base, however, is relatively thick, because the alloying process does not lend itself to extremely fine control; this means that the high-frequency response is limited. Alloy transistors therefore find their greatest application in audio power circuits.

ETCHED JUNCTIONS.  A very thin base can result (with a consequent improvement in high-frequency response) if a semiconductor wafer is etched

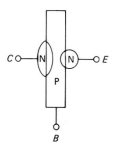

Fig. 5.44  *Alloyed junction.*

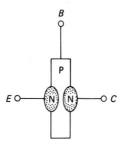

Fig. 5.45 Etched junction.

on each side using electrochemical techniques. Suitable impurities are then electroplated in the depressed regions (Fig. 5.45).

*DIFFUSED JUNCTIONS.* Most modern transistors are formed by this process. Impurities in a gaseous form are diffused into the crystal either through the surface of the semiconductor or from inside the crystal itself. Diffusion is temperature-sensitive; this permits very fine control of both area and depth of penetration, allowing for fabrication of regions with nonuniform doping, which is useful in improving certain performance characteristics.

Silicon lends itself very well to diffusion techniques because a *passivating* oxide ($SiO_2$) layer can be grown on its surface, which is then selectively etched, allowing diffusion to take place only in desired regions.

### 5.8b  Transistor Structures

Junction-forming processes are intimately tied with the geometric structure of the device. There are many variations possible, each with its own specific advantages; the most common structures are treated here.

*MESA.* The mesa transistor derives its name from the flat-top appearance (Fig. 5.46). The device may be manufactured as follows:

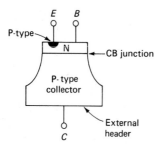

Fig. 5.46  Mesa structure.

a. A P-type semiconductor serves as the collector. Its area of contact with the external header on which it is mounted is relatively large, thus providing good thermal conductivity (high power dissipation).
b. An N-type base is diffused on the P-type collector. The P-type emitter is then formed using an alloy or vacuum evaporation process.
c. The sides of the structure are etched off to reduce the collector-base area of contact, thus minimizing collector-base capacitance while retaining the large common area between collector and header. The structure is now narrower at the top than it is at the bottom.

*PLANAR.* This is a variation of the mesa in which all junctions are terminated on a *plane* surface at the top. Its development involves several diffusion steps leading to *planar-diffused transistors* which exhibit improved surface-sensitive characteristics. For example, $I_{CBO}$ is greatly reduced because surface passivation (the process of growing the oxide layer) prevents the accumulation of impurities on the semiconductor surface.

*EPITAXIAL.* One of the problems in producing switching transistors is the need for both a low $V_{SAT}$ and a high breakdown voltage. A low $V_{SAT}$ can be achieved with heavy collector doping (low resistance, therefore low voltage drop when ON), which results in a narrow depletion region (see Sec. 1.5) which breaks down at lower voltages. Hence there is a conflict between these two requirements. The epitaxial structure gets around this conflict by providing both a high-resistivity (epitaxial) layer for large $BV$ ratings and a low-resistivity substrate for low $V_{SAT}$. Since the epitaxial layer is very thin, it does not substantially add to the total collector resistance.

Basically the structure is similar to the mesa (Fig. 5.47). A large, heavily doped (N⁺) substrate acts as the collector; the large area of contact with the header provides good thermal conductivity. A thin film of lightly doped (N) semiconductor is then *grown* on the substrate. This is the epitaxial layer, and it represents an extension of the substrate crystal. The P-type base and N-type emitter are then diffused into the semiconductor and contacts added. The

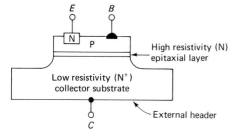

*Fig. 5.47  Epitaxial transistor structure.*

resulting device has the low capacitance of the mesa with high $BV$ rating and low $V_{SAT}$.

### 5.8c  Integrated Circuits

The planar and epitaxial processes developed in conjunction with bipolar transistors have been extended to fabricate entire circuits on silicon wafers as small as 0.05 in. square; these are called monolithic IC (integrated circuits).

The process starts with a Si slice, typically 2 in. in diameter and 0.005 in. thick. On each slice there are up to 1,000 wafers, each capable of containing a circuit involving as many as 50 components (transistors, diodes, resistors, capacitors) or more. Thus one of the advantages of ICs is their very small size.

IC resistors can be formed by varying the silicon's doping level and geometry; capacitors by reverse-biased diodes. The fabrication process is similar to that for planar transistors. Depending on circuit complexity, several diffusion steps may be required. The procedure is carried out on each wafer simultaneously. After each IC wafer has been electrically tested (using high-precision pointed probes), the slice is scribed and each wafer or chip is broken off. Imperfect chips are rejected; good ones are packaged, sealed, and subjected to final performance tests.

There are practical limitations in fabricating IC resistors (they are temperature-dependent and limited in value to less than 20 kΩ) and IC capacitors (they are a function of reverse voltage and limited to less than about 100 pF). Most ICs are therefore designed to minimize the number of resistors and capacitors while relying mainly on diodes and transistors. In some cases, passive components that cannot be easily integrated are connected to ICs using a thin conducting film deposited on a ceramic substrate; the result is called a hybrid circuit.

The economies of scale inherent in large-volume IC manufacturing have made ICs very attractive to use in consumer products of all types. Also, since the interconnections between components in an IC are made in the manufacturing process on hundreds of wafers at a time, wiring errors are virtually eliminated.

### Problem Set 5.10

5.10.1. How heavily doped is the base of a junction transistor compared to the emitter? Explain your answer.

5.10.2. How can the ability of a transistor to dissipate power be increased? What other characteristics suffer as a consequence?

5.10.3. How does doping level affect the breakdown voltage of PN junctions?

5.10.4. What conflicts exist between obtaining a low $V_{CE(SAT)}$ and a high $BV_{CBO}$?

5.10.5. How does the doping obtained by the diffusion process differ from that obtained in the alloy process?

5.10.6. What is surface passivation? Why is it used?
5.10.7. What is an epitaxial transistor?
5.10.8. What advantages does the mesa transistor have over the alloy transistor?
5.10.9. What advantages do ICs have over conventional circuits that use discrete components?

# REFERENCES

1. Phillips, A. B.: "Transistor Engineering and Introduction to Integrated Semiconductor Circuits," McGraw-Hill Book Company, New York, 1962.
2. Basic Theory and Application of Transistors, TM 11-690, Headquarters, Department of the Army, 1959.
3. Kiver, M. S.: "Transistors," McGraw-Hill Book Company, New York, 1962.
4. Malvino, A. P.: "Transistor Circuit Approximations," McGraw-Hill Book Company, New York, 1968.
5. Amos, S. W.: "Principles of Transistor Circuits," Hayden Book Company, Inc., New York, 1965.
6. Kahn, M., and J. M. Doyle: "The Synthesis of Transistor Amplifiers," Holt, Rinehart and Winston, Inc., New York, 1970.
7. Cowles, L. G.: "Transistor Circuits and Applications," Prentice-Hall, Inc., Englewood Cliffs, N.J., 1968.
8. Joyce, M. V., and K. K. Clarke: "Transistor Circuit Analysis," Addison-Wesley Publishing Company, Inc., Reading, Mass., 1961.
9. Cowles, L. G.: "Analysis and Design of Transistor Circuits," D. Van Nostrand Company, Inc., Princeton, N.J., 1966.
10. Millman, J., and C. C. Halkias: "Electronic Devices and Circuits," McGraw-Hill Book Company, New York, 1967.
11. Cutler, P.: "Semiconductor Circuit Analysis," McGraw-Hill Book Company, New York, 1964.
12. Texas Instruments: "Transistor Circuit Design," McGraw-Hill Book Company, New York, 1963.
13. Ristenbatt, M. P., and R. L. Riddle: "Transistor Physics and Circuits," Prentice-Hall, Inc., Englewood Cliffs, N.J., 1966.
14. Gibbons, J. F.: "Semiconductor Electronics," McGraw-Hill Book Company, New York, 1966.
15. Fitchen, F. C.: "Transistor Circuit Analysis and Design," 2d ed., D. Van Nostrand Company, Inc., Princeton, N.J.
16. "General Electric Transistor Manual," 7th ed., Syracuse, 1964.
17. Kendall, E. J. M.: "Transistors," Pergamon Press, New York, 1969.
18. Roddy, D.: "Introduction to Microelectronics," Pergamon Press, New York, 1970.

# chapter 6

# junction transistor small-signal models

Graphic techniques for the AC analysis of active devices were introduced in Chap. 5. The graphic approach involves locating a load line and determining changes occurring in one parameter as a result of variations in another. This is time-consuming and not necessary unless the signals are very large.

This chapter discusses *models* suitable for the analysis of small-signal junction transistor amplifiers. The parameters needed for each model are obtained from manufacturers' data; techniques for experimental determination of parameters are also discussed.

DC analysis is excluded from this chapter. Techniques for selecting and maintaining proper DC levels (operating point) in small-signal amplifiers are treated in Chap. 8. The inherent assumption here is that the operating point has been established in the linear (active) region of the operating characteristics. We must now determine how a "small signal" is affected by passing through such a circuit.

## 6.1 JUNCTION TRANSISTOR MODELS

Many different models can be constructed that predict transistor performance under a variety of operating conditions. All these models utilize combinations of passive components ($R$, $L$, $C$) and controlled sources (voltage or current) to simulate transistor behavior. Each component or source has a *parameter* associated with it that specifies its value.

Some models utilize impedance ($z$) or admittance ($y$) parameters; others use a combination of both: hybrid parameters, or $h$ parameters, for short. This chapter treats two basic models: the $h$ and hybrid pi; both are developed in the sections to follow.

## 6.2  THE HYBRID (*h*) PARAMETER MODEL

Consider the diagram of Fig. 6.1a. The electric behavior of the amplifying device can be described using a model involving *hybrid* parameters. The term *hybrid* refers to the use of both Thevenin and Norton equivalent circuits, as will be seen shortly. The $h$-parameter model is particularly useful for the analysis of small-signal, low-frequency junction transistor amplifiers. None of the components in the model can be identified exclusively with a specific region of an actual transistor. Only the *external* behavior, as determined by terminal (input and output) voltage and current measurements, is simulated.

There are four quantities required to describe the external behavior of an amplifying device. These are $V_1$, $I_1$, $V_2$, and $I_2$, as shown on the diagram of Fig. 6.1b. Either RMS or peak values may be used, provided we are consistent. We will use RMS values unless otherwise indicated. Both input and output currents ($I_1$ and $I_2$) are assumed to flow into the black box; input and output voltages ($V_1$ and $V_2$) are assumed positive with respect to ground. These are standard conventions and do not necessarily correspond to actual directions or polarities.

Let us now consider each component in the $h$-parameter model. The input circuit appears as a resistance ($h_i$) in series with a voltage generator ($h_r V_2$). The voltage generator takes into account the transistor's nonunilateral behavior. An ideal amplifying device should respond only to signals applied at the input; it should not do the reverse—that is, reproduce at the input any portion of a signal applied at the output. Such an ideal *one-way* device would be called *unilateral*. A real transistor cannot be unilateral because of interaction between output and input circuits (the base-emitter and collector-base junctions are part of the same crystal). Thus not only does the output respond to the input, but to a lesser degree, the input also responds to the output. The generator $h_r V_2$ is referred to as a voltage-controlled generator because its value is determined by the voltage $V_2$. The parameter $h_r$ is a dimensionless constant referred to as the reverse voltage transfer ratio; its value depends on the transistor type and configuration as well as the operating point. Note that the resistance $h_i$ and voltage generator $h_r V_2$ in the input represent a Thevenin equivalent circuit.

The output also involves two components, a current generator and a shunt conductance. These make up a Norton equivalent circuit. The $h_o$ com-

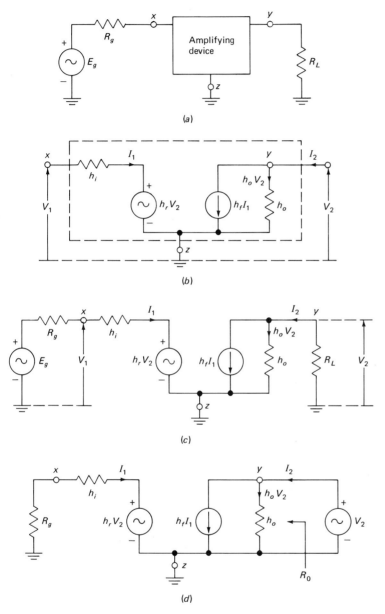

**Fig. 6.1** The amplifying device in (a) may be replaced by the model in (b), yielding the circuit in (c); the output resistance is the ratio $V_2/I_2$ in the circuit of (d).

ponent takes into account the fact that an effective conductance is seen when looking into the transistor from the output terminals. The units of the output conductance $h_o$ are mhos; the parameter $h_o$, of course, corresponds to a resistance $1/h_o$ whose units are ohms ($\Omega$). The $h_f I_1$ generator is current-controlled; its magnitude is determined by the value of $I_1$. It is this current generator that simulates the transistor's ability to amplify. The parameter $h_f$ is the forward current transfer ratio; it is a dimensionless constant that tells us by what factor the input current ($I_1$) is amplified.

The $h$-parameter model of Fig. 6.1b can be described mathematically using two equations:

## Input Circuit

$V_1$ = the sum of voltage drops from $x$ to $z$

$$V_1 = h_i I_1 + h_r V_2 \tag{6.1}$$

## Output Circuit

$I_2$ = the sum of all currents leaving junction $y$

$$I_2 = h_f I_1 + h_o V_2 \tag{6.2}$$

Note that $h_o V_2$ is the current through $h_o$. Equations (6.1) and (6.2) involve four variables ($V_1$, $I_1$, $V_2$, $I_2$) and four constants ($h_i$, $h_r$, $h_f$, $h_o$). The variables are the voltage and current at each set of terminals; they depend on the applied signal. The $h$ parameters are constant for any given circuit as long as the operating conditions do not change.

Consider now the circuit of Fig. 6.1c. Both a signal generator and load have been added to the transistor model. Performance measures for this circuit can be defined as follows:

$$R_i = \frac{V_1}{I_1} \tag{6.3}$$

The output resistance is obtained by reducing $E_g$ to 0, removing $R_L$, and applying a signal generator $V_2$ at the output, as shown in Fig. 6.1d. The ratio $V_2/I_2$ taken under these conditions is $R_o$:

$$R_o = \frac{V_2}{I_2}\bigg|_{E_g=0} \tag{6.4}$$

The voltage amplification is

$$A_v = \frac{V_2}{V_1} \tag{6.5}$$

The current amplification is

$$A_i = \frac{I_2}{I_1} \tag{6.6}$$

Equation (6.2) can be rewritten using the information that $V_2 = -I_2 R_L$. The negative sign is due to the polarity of $V_2$, which is opposite to that produced by the assumed direction of $I_2$.

$$I_2 = h_f I_1 - h_o I_2 R_L \tag{6.2a}$$

Equations (6.1) and (6.2a) can be used to obtain mathematical expressions for all four performance measures. Only one such derivation (current amplification) is carried out here; the remaining ones are left for a student exercise.

$$A_i = \frac{I_2}{I_1} \tag{6.6}$$

$$I_2 + I_2 h_o R_L = h_f I_1 \tag{6.2a}$$

$$A_i = \frac{I_2}{I_1} = \frac{h_f}{1 + h_o R_L} \tag{6.6a}$$

If $h_o R_L \ll 1$    then $A_i \simeq h_f$ \hfill (6.6b)

The simplification of Eq. (6.6b) is often useful. To say that $h_o R_L \ll 1$ is the same as saying that $R_L \ll 1/h_o$. This occurs when $R_L$ is much smaller than the output resistance ($1/h_o$), shunting the $h_f I_1$ current generator; most of the generator current then bypasses the transistor's output resistance in favor of $R_L$. This means that $I_2 \simeq h_f I_1$; hence $I_2/I_1 \simeq h_f$, as given by Eq. (6.6b).

Using a similar procedure, an expression for input resistance can be derived, yielding

$$R_i = h_i - \frac{h_f h_r}{(1/R_L) + h_o} \tag{6.3a}$$

Note that $R_i$ depends on $R_L$. If either $h_r$ or $R_L$ are very small, though, the second term drops and

$$R_i \simeq h_i \tag{6.3b}$$

This last simplification depends on the magnitude of reverse voltage developed by generator $h_r V_2$, compared to the voltage drop across $h_i$; $h_r V_2$ is small if either $h_r$ or $V_2$ is small. Since $V_2 = -I_2 R_L$, a small load resistance also means a low value for $V_2$, in which case the simplification of Eq. (6.3b) is valid.

The expression for output resistance is obtained using Eq. (6.4):

$$R_o = \frac{V_2}{I_2} \bigg|_{E_g = 0} \tag{6.4}$$

this yields

$$R_o = \frac{1}{h_o - [h_r h_f/(h_i + R_g)]} \tag{6.4a}$$

Note that $R_o$, in addition to the circuit's $h$ parameters, also depends on $R_g$. In cases where $R_g$ is very large (the circuit is driven from a current source), or $h_r$ is negligible, Eq. (6.4a) simplifies to

$$R_o \simeq \frac{1}{h_o} \tag{6.4b}$$

The expression for voltage amplification is

$$A_v = \frac{-h_f R_L}{h_i + R_L(h_i h_o - h_f h_r)} \tag{6.5a}$$

If $h_i \gg R_L(h_i h_o - h_f h_r)$, then $A_v$ simplifies to

$$A_v \simeq \frac{-h_f R_L}{h_i} \tag{6.5b}$$

### Example 6.1

The following data apply to the circuit of Fig. 6.1c: $R_g = 600\ \Omega$; $R_L = 600\ \Omega$; $h_i = 2{,}000\ \Omega$; $h_o = 10^{-4}$ mho; $h_r = 10^{-3}$; $h_f = 50$. Determine $R_i$, $R_o$, $A_i$, and $A_v$ using exact equations first and then making reasonable approximations.

### Solution

a. $$R_i = h_i - \frac{h_f h_r}{(1/R_L) + h_o} \tag{6.3a}$$

$$R_i = 2{,}000 - \frac{50(0.001)}{(1/600) + 10^{-4}}$$

$$R_i = 1{,}972\ \Omega$$

Since the second term in Eq. (6.3a) is relatively small, the approximate expression of Eq. (6.3b) may be used:

$$R_i \simeq h_i = 2{,}000\ \Omega$$

b. $$R_o = \frac{1}{h_o - [h_r h_f/(h_i + R_g)]} \tag{6.4a}$$

$$R_o = \frac{1}{10^{-4} - [50 \times 10^{-3}/(2{,}000 + 600)]}$$

$$R_o = 12{,}400\ \Omega$$

If the second term in the denominator is neglected,

$$R_o \simeq \frac{1}{h_o} = 10{,}000 \; \Omega$$

c.  $\quad A_i = \dfrac{h_f}{1 + h_o R_L}$  \hfill (6.6a)

$$A_i = \frac{50}{1 + (600 \times 10^{-4})}$$

$$A_i = 47$$

Because $h_o R_L \ll 1$, Eq. (6.6b) may be used, which yields

$$A_i \simeq 50$$

d.  $\quad A_v = \dfrac{-h_f R_L}{h_i + R_L(h_i h_o - h_f h_r)}$ \hfill (6.5a)

$$A_v = \frac{-50(600)}{2{,}000 + 600(0.2 - 0.05)}$$

$$A_v = -14.9$$

The second term in the denominator is small compared with the first; therefore

$$A_v \simeq -15 \hspace{4em} (6.5b)$$

## Problem Set 6.1

6.1.1. Develop Eq. (6.4a). Modify this expression for the case when the circuit is driven from a constant voltage source.
6.1.2. Develop Eq. (6.5a).
6.1.3. Develop Eq. (6.3a).
6.1.4. Using the following data, determine, for the circuit of Fig. 6.1c, $R_i$, $R_o$, $A_i$, and $A_v$. Make reasonable approximations. $R_g = 0 \; \Omega$; $R_L = 1 \; k\Omega$; $h_i = 20 \; \Omega$; $h_o = 10^{-5}$ mho; $h_r = 10^{-4}$; $h_f = -0.95$.
6.1.5. What is the power gain in Prob. 6.1.4? Express your answer in decibels.
6.1.6. Under what conditions is the input resistance to the circuit of Fig. 6.1c equal to $h_i$?
6.1.7. Under what conditions is the output resistance in the circuit of Fig. 6.1c equal to $1/h_o$?

### 6.2a General Properties of Hybrid Parameters

Hybrid parameters may be obtained by any of the following methods:

a. Experimentally.
b. From manufacturer's data sheets.

c. Using characteristic curves supplied by the manufacturer or determined experimentally. This involves a graphic procedure.

The parameters thus obtained depend on a number of factors, as follows:

a. Transistor type
b. Configuration (common emitter, common base, or common collector)
c. Operating point
d. Temperature
e. Frequency

It is therefore necessary to specify the conditions under which a given set of parameters apply. Unless otherwise indicated, a standard temperature of 25°C and a standard frequency of 1 kHz are used. The configuration is specified by adding an $e$, $b$, or $c$ subscript to each parameter for common emitter, common base, or common collector, respectively. A few examples are

$h_{ie}$: common emitter input resistance
$h_{fb}$: common base forward current transfer ratio
$h_{rc}$: common collector reverse transfer ratio
$h_{oe}$: common emitter output conductance

The $h$ parameters for each of the three configurations are related; knowing one set, it is therefore possible to calculate the corresponding values for a different configuration. It is also possible to convert $h$ parameters to other types, such as $z$ or $y$ parameters, and vice versa.[1]

## 6.2b Experimental Determination of h Parameters

Consider the circuit of Fig. 6.1b. Only terminals $x$, $y$, and $z$ are externally accessible. All four $h$ parameters can be determined using voltage and current measurements made at these terminals.

Looking into terminal $x$, we see $h_i$ in series with $h_r V_2$. If the output is short-circuited, $V_2 = 0$ and the $h_r V_2$ generator is reduced to zero as shown in Fig. 6.2a. We can now write

$$V_1 = h_i I_1$$

$$h_i = \left.\frac{V_1}{I_1}\right|_{V_2=0} \tag{6.7}$$

The parameter $h_i$ is therefore defined as the input resistance with the output short-circuited ($V_2 = 0$). In a transistor amplifier, the output cannot always be shorted without affecting the $Q$ point. Instead a capacitor with negligible reactance at the test frequency is connected across the output. The capacitor does

---

[1] See p. 56 of Reference 11 at the end of this chapter.

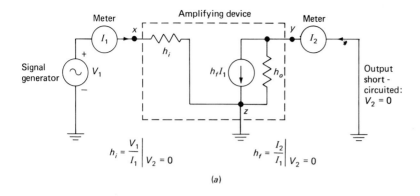

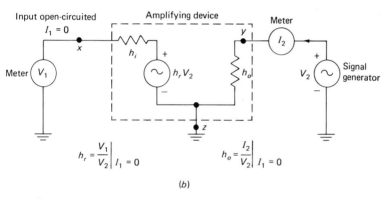

**Fig. 6.2** Connections for determining h parameters; (a) output short-circuited ($V_2 = 0$) to determine $h_i$ and $h_f$; (b) input open-circuited ($I_1 = 0$) to determine $h_r$ and $h_o$.

not change the DC levels but forces the AC component of output voltage ($V_2$) to equal zero.

Looking into terminal $y$ of Fig. 6.1b, we see a conductance $h_o$ shunted by the controlled-current source $h_f I_1$. By opening the input, $I_1$ is forced to zero; hence the $h_f I_1$ source disappears as shown in Fig. 6.2b. This allows us to write

$$I_2 = h_o V_2$$

$$h_o = \left. \frac{I_2}{V_2} \right|_{I_1 = 0} \tag{6.8}$$

The parameter $h_o$ is therefore defined as the output conductance with the input open ($I_1 = 0$).

The parameter $h_f$ is determined by AC short circuiting the output, as in Fig. 6.2a. Again, this may be accomplished by connecting a large capacitor across the output. Then all the current $h_f I_1$ is shunted around $h_o$ and flows through the short circuit, resulting in $I_2 = h_f I_1$. This yields

$$h_f = \left. \frac{I_2}{I_1} \right|_{V_2=0} \tag{6.9}$$

To determine $h_r$, the input is opened as in Fig. 6.2b. A generator $V_2$ is connected across the output, and the resulting voltage $V_1$ is measured. Since $I_1 = 0$, there is no voltage across $h_i$, and $V_1 = h_r V_2$. This results in

$$h_r = \left. \frac{V_1}{V_2} \right|_{I_1=0} \tag{6.10}$$

The foregoing procedure may be used to experimentally determine the $h$ parameters for any type of circuit, active or passive. To illustrate how such a procedure might be carried out for a common emitter transistor amplifier, consider the circuit of Fig. 6.3a. It is desired to determine the $h_f$ parameter, which, in a common emitter configuration, is referred to as $h_{fe}$. By shorting the output with $C_2$ as in Fig. 6.3b, we force $V_{ce} = 0$. In this circuit $I_b$ and $I_c$ correspond to $I_1$ and $I_2$, respectively. Similarly $V_{be}$ and $V_{ce}$ correspond to $V_1$ and $V_2$. We therefore have

$$h_{fe} = \left. \frac{I_c}{I_b} \right|_{V_{ce}=0} \tag{6.9a}$$

$I_c$ and $I_b$ are the AC RMS collector and base currents, respectively; the parameter $h_{fe}$ is also called $\beta$. Note that setting $V_{ce} = 0$ does not mean that $V_{CE}$ (the DC collector-emitter voltage) is zero. The changing component of collector current flows through $C_2$ instead of $R_L$ because of $C_2$'s extremely low reactance. The AC voltage developed across $C_2$ is therefore zero ($V_{ce} = 0$).

The same circuit configuration may be used to determine $h_{ie}$, since this also requires a shorted output. The general expression

$$h_i = \left. \frac{V_1}{I_1} \right|_{V_2=0} \tag{6.7}$$

becomes

$$h_{ie} = \left. \frac{V_{be}}{I_b} \right|_{V_{ce}=0} \tag{6.7a}$$

The remaining two parameters ($h_{re}$ and $h_{oe}$) require the input to be AC open-circuited. Figure 6.3c shows one way of doing this. An inductor $L$ is connected in series with $R_B$; the inductor's DC resistance is low, and therefore it does not disturb the operating point. The inductor's reactance, however, is large at the test frequency, so that little AC current flows through $R_B$. The

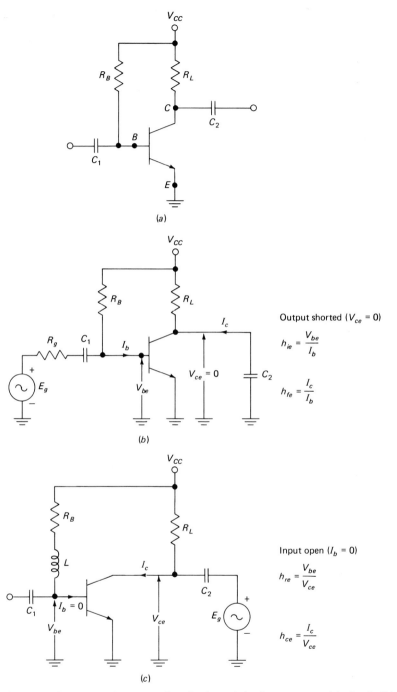

Fig. 6.3 Common emitter connections for determining h parameters; (a) circuit; (b) output shorted; (c) input open.

voltmeter used to measure $V_{be}$ also has a high-input impedance. Therefore there are no paths connected to the base with any appreciable AC current; this means that the base is effectively AC open-circuited and $I_b = 0$. How good an open circuit is achieved depends on the reactance (the reactance of $L$ increases as $L$ is made larger) and the actual impedance of the voltmeter used to measure $V_{be}$.

To determine $h_{re}$ a signal generator is applied across the output; the resulting voltage across the input ($V_{be}$) is then measured. Note that $V_{be}$ here corresponds to $V_1$ and $V_{ce}$ to $V_2$; therefore, the general expression

$$h_r = \left.\frac{V_1}{V_2}\right|_{I_1=0} \qquad (6.10)$$

may be written as

$$h_{re} = \left.\frac{V_{be}}{V_{ce}}\right|_{I_b=0} \qquad (6.10a)$$

The output admittance parameter requires only $V_2$ and $I_2$, which correspond to $V_{ce}$ and $I_c$, respectively. Equation (6.8) becomes

$$h_{oe} = \left.\frac{I_c}{V_{ce}}\right|_{I_b=0} \qquad (6.8a)$$

### Example 6.2

The following quantities are measured in a common emitter amplifier circuit:

a. *Output Short-circuited* ($V_{ce} = 0$)

$I_b = 10\ \mu\text{A}$
$I_c = 1\ \text{mA}$
$V_{be} = 10\ \text{mV}$

b. *Input Open-circuited* ($I_b = 0$)

$V_{be} = 0.65\ \text{mV}$
$I_c = 60\ \mu\text{A}$
$V_{ce} = 1\ \text{V}$

Determine all four $h$ parameters.

### Solution

1.  $h_{ie} = \left.\dfrac{V_{be}}{I_b}\right|_{V_{ce}=0} \qquad (6.7a)$

    $h_{ie} = \dfrac{10^{-2}}{10^{-5}} = 1{,}000\ \Omega$

2. $h_{fe} = \dfrac{I_c}{I_b}\bigg|_{V_{ce}=0}$ (6.9a)

$h_{fe} = \dfrac{10^{-3}}{10^{-5}} = 100$

3. $h_{re} = \dfrac{V_{be}}{V_{ce}}\bigg|_{I_b=0}$ (6.10a)

$h_{re} = \dfrac{0.65 \times 10^{-3}}{1} = 0.65 \times 10^{-3}$

4. $h_{oe} = \dfrac{I_c}{V_{ce}}\bigg|_{I_b=0}$ (6.8a)

$h_{oe} = \dfrac{60 \times 10^{-6}}{1} = 60\ \mu\text{mho}$

Circuit connections for obtaining $h$ parameters in the common base configuration are shown in Fig. 6.4. The operating point is established by $R_E$, $R_B$, and $R_L$ in conjunction with $V_{CC}$. Capacitors $C_1$ and $C_2$ AC-couple the input and output signals, while $C_B$ bypasses the base to ground.

With the output shorted, $V_{cb} = 0$. Here $V_1$, $I_1$, $V_2$, and $I_2$ correspond to $V_{eb}$, $I_e$, $V_{cb}$, and $I_c$, respectively. Note that both $I_e$ and $I_c$ are assumed by convention to flow into the transistor. In practice, however, if $I_e$ enters the emitter, $I_c$ leaves the collector. Therefore, the actual direction of $I_c$ is opposite to that assumed, and a negative sign must be associated with it. This means that $h_{fb}$ is negative. We therefore have

$h_{ib} = \dfrac{V_{eb}}{I_e}\bigg|_{V_{cb}=0}$ (6.7b)

$h_{fb} = \dfrac{I_c}{I_e}\bigg|_{V_{cb}=0}$ (6.9b)

The input is AC open-circuited by connecting a large coil in series with $R_E$ as in Fig. 6.4c. This causes $I_e$ to approach zero, allowing us to measure the quantities needed to calculate the remaining parameters:

$h_{ob} = \dfrac{I_c}{V_{cb}}\bigg|_{I_e=0}$ (6.8b)

$h_{rb} = \dfrac{V_{eb}}{V_{cb}}\bigg|_{I_e=0}$ (6.10b)

A common collector circuit and the connections required to determine its $h$ parameters are shown in Fig. 6.5. The operating point is set by $V_{CC}$, $R_B$, and $R_E$. $R_E$ is also the load resistor across which the output is developed. The collector is directly tied to the $V_{CC}$ supply; hence it is at AC ground potential. Here $V_1$, $I_1$, $V_2$, and $I_2$ are replaced with $V_{bc}$, $I_b$, $V_{ec}$, and $I_e$, respectively.

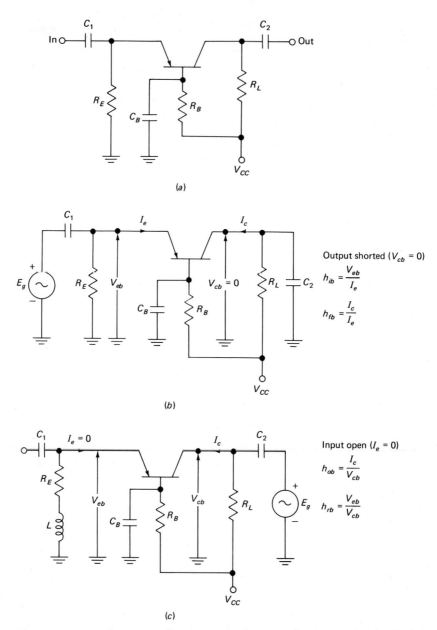

Fig. 6.4  Common base connections for determining h parameters; (a) circuit; (b) output shorted; (c) input open.

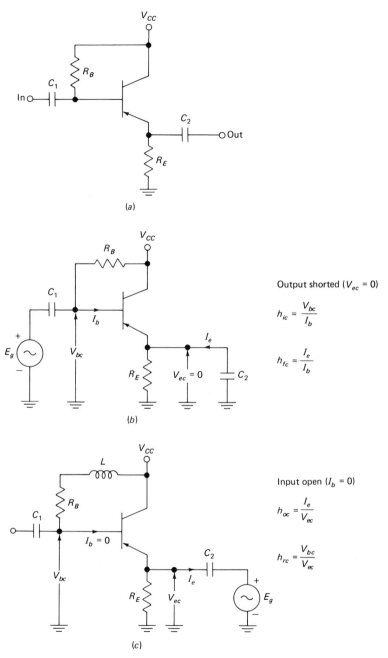

Fig. 6.5 *Common collector connections for determining* h *parameters;* (a) *circuit;* (b) *output shorted;* (c) *input open.*

With the output shorted, $V_{ec} = 0$. Note once more that, for the assumed direction of $I_b$, $I_e$ is in a direction opposite to that shown in Fig. 6.5b. This means that a negative sign must be associated with the measured value of $I_e$. The parameter $h_{fc}$ will therefore be negative. For the circuit of Fig. 6.5b, we have

$$h_{ic} = \left.\frac{V_{bc}}{I_b}\right|_{V_{ec}=0} \tag{6.7c}$$

$$h_{fc} = \left.\frac{I_e}{I_b}\right|_{V_{ec}=0} \tag{6.9c}$$

With the input AC open-circuited, there is no appreciable AC base current ($I_b = 0$), as indicated in Fig. 6.5c. The remaining parameters are

$$h_{oc} = \left.\frac{I_e}{V_{ec}}\right|_{I_b=0} \tag{6.8c}$$

$$h_{rc} = \left.\frac{V_{bc}}{V_{ec}}\right|_{I_b=0} \tag{6.10c}$$

### Example 6.3

The following quantities are measured in the circuits indicated:

| Fig. 6.4b | Fig. 6.4c |
|---|---|
| $V_{eb} = 10$ mV | $I_c = 1$ μA |
| $I_e = 0.2$ mA | $V_{cb} = 1$ V |
| $I_c = -0.198$ mA | $V_{eb} = 0.5$ mV |

Determine the common base $h$ parameters.

### Solution

Parameters $h_{ib}$ and $h_{fb}$ require short-circuited output data, as in the circuit of Fig. 6.4b:

$$h_{ib} = \left.\frac{V_{eb}}{I_e}\right|_{V_{cb}=0} \tag{6.7b}$$

$$h_{ib} = \frac{10^{-2}}{2 \times 10^{-4}} = 50\ \Omega$$

$$h_{fb} = \left.\frac{I_c}{I_e}\right|_{V_{cb}=0} \tag{6.9b}$$

$$h_{fb} = \frac{-1.98 \times 10^{-4}}{2 \times 10^{-4}} = -0.99$$

Parameters $h_{ob}$ and $h_{rb}$ require open-circuited input data, as in the circuit of Fig. 6.4c:

$$h_{ob} = \left.\frac{I_c}{V_{cb}}\right|_{I_e=0} \quad (6.8b)$$

$$h_{ob} = \frac{10^{-6}}{1} = 1 \ \mu\text{mho}$$

$$h_{rb} = \left.\frac{V_{eb}}{V_{cb}}\right|_{I_e=0} \quad (6.10b)$$

$$h_{rb} = \frac{5 \times 10^{-4}}{1} = 5 \times 10^{-4}$$

## Example 6.4

The following quantities are measured in the circuits indicated:

| Fig. 6.5b | Fig. 6.5c |
|---|---|
| $V_{bc} = 50$ mV | $I_e = 60\ \mu\text{A}$ |
| $I_b = 50\ \mu\text{A}$ | $V_{ec} = 1$ V |
| $I_e = -5$ mA | $V_{bc} = 1$ V |

## Solution

Parameters $h_{ic}$ and $h_{fc}$ may be calculated using data obtained from the circuit of Fig. 6.5b:

$$h_{ic} = \left.\frac{V_{bc}}{I_b}\right|_{V_{ec}=0} \quad (6.7c)$$

$$h_{ic} = \frac{5 \times 10^{-2}}{5 \times 10^{-5}} = 1{,}000 \ \Omega$$

$$h_{fc} = \left.\frac{I_e}{I_b}\right|_{V_{ec}=0} \quad (6.9c)$$

$$h_{fc} = \frac{-5 \times 10^{-3}}{5 \times 10^{-5}} = -100$$

The remaining parameters are calculated using the data obtained from the circuit of Fig. 6.5c:

$$h_{oc} = \left.\frac{I_e}{V_{ec}}\right|_{I_b=0} \quad (6.8c)$$

$$h_{oc} = \frac{60 \times 10^{-6}}{1} = 60 \; \mu\text{mho}$$

$$h_{rc} = \left.\frac{V_{bc}}{V_{ec}}\right|_{I_b=0} \tag{6.10c}$$

$$h_{rc} = \frac{1}{1} = 1$$

## Problem Set 6.2

6.2.1. The following two sets of data are obtained for the circuit of Fig. 6.1b:

    a.   $V_1 = 1$ V    $I_1 = 0.1$ mA    $V_2 = 0$    $I_2 = 7$ mA
    b.   $V_1 = 1 \; \mu$V    $I_1 = 0$    $V_2 = 1$ V    $I_2 = 2 \; \mu$A

    Determine $h_i$, $h_f$, $h_r$, and $h_o$.

6.2.2. How can we obtain an effective AC short circuit across the output of an amplifier? Does this affect the DC operating conditions?

6.2.3. When $h$ parameters are specified for a particular transistor, the operating point is usually given. Why is this necessary?

6.2.4. How can we obtain an effective AC open circuit at the input to an amplifier? Does this affect the DC operating conditions?

6.2.5. Why are the parameters $h_{fb}$ and $h_{fc}$ negative?

### 6.2c Graphic Evaluation of h Parameters

It is possible to obtain $h$ parameters, as well as other useful information, from a set of transistor characteristics. The curves provide a graphic display of the relationships among transistor voltages and currents.

There are several techniques that can be used to obtain characteristics. In some cases, the manufacturer may provide them on data sheets. They can also be plotted point by point using experimental data; this, of course, is very time-consuming and seldom done in practice. The most convenient method is to use an electronic curve tracer that displays the curves on an oscilloscope screen. The image can then be photographed, yielding an accurate set of curves very quickly.

A typical set of input and output characteristics for a transistor in the common emitter configuration is shown in Fig. 6.6. Note that the variables $V_{BE}$, $V_{CE}$, $I_B$, and $I_C$ correspond to $V_1$, $V_2$, $I_1$, and $I_2$, respectively.

Consider first the input characteristics in either Fig. 6.6b or Fig. 6.6d. The voltage-current relationship is similar to that of any forward-biased PN junction with the exception that there are different curves for different values of $V_{CE}$. $V_{CE}$ is referred to as the running parameter.

# Junction Transistor Small-signal Models 225

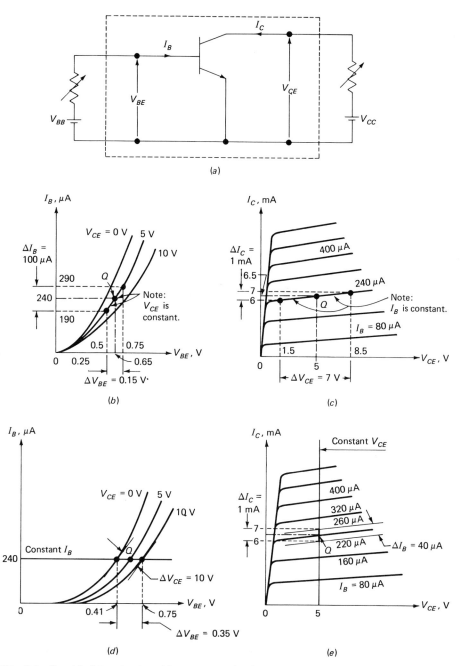

Fig. 6.6 Graphic determination of h parameters for the common emitter transistor in (a); input characteristics are shown in (b) and (d); output characteristics are shown in (c) and (e).

226  Solid State Electronic Circuits

The parameter $h_{ie}$ was defined earlier as

$$h_{ie} = \left. \frac{V_{be}}{I_b} \right|_{V_{ce}=0} \tag{6.7a}$$

where $V_{be}$ and $I_b$ are the AC components of base-emitter voltage and base current, respectively. The notation $V_{ce} = 0$ means that the AC component of collector-emitter voltage is zero.

The AC component of any voltage or current may be considered as a change in DC level. The characteristics of Fig. 6.6 are static (DC), but it is possible to obtain AC parameters by assuming small changes in DC levels. For example, the AC component of base-emitter voltage ($V_{be}$) may be represented as a small change in DC base-emitter voltage ($\Delta V_{BE}$). The changes must be small because the curves are linear only over a small portion of their length; if large changes are used, the resulting parameters may not be useful for predicting "small-signal" performance. As long as only the linear portion of a curve is used, however, the results are valid.

The parameter $h_{ie}$ can now be defined as follows:

$$h_{ie} = \left. \frac{\Delta V_{BE}}{\Delta I_B} \right|_{\Delta V_{CE}=0} \tag{6.7d}$$

where $\Delta V_{BE}$ and $\Delta I_B$ are *small changes* in $V_{BE}$ and $I_B$, respectively. The notation $\Delta V_{CE} = 0$ simply means that $V_{CE}$ remains constant.

The graphic constructions on the characteristics of Fig. 6.6 are used to illustrate how $h$ parameters are obtained; a $Q$ point at $V_{CE} = 5$ V, $I_C = 6.5$ mA, and $I_B = 240$ µA has been selected. To determine $h_{ie}$, the following procedure is followed:

a. Locate the $Q$ point on the characteristics of Fig. 6.6b.
b. Assume a small change on each side of the $Q$ point along the $V_{CE}$ curve for which the $Q$ point has been established. This satisfies the condition that $V_{CE}$ remain constant ($\Delta V_{CE} = 0$) while changes in $V_{BE}$ and $I_B$ are taking place.
c. Extend perpendiculars to both axes and read the resulting values of $\Delta V_{BE}$ and $\Delta I_B$. For this example, we have $\Delta V_{BE} = 0.65 - 0.5 = 0.15$ V and $\Delta I_B = 290 - 190 = 100$ µA, yielding

$$h_{ie} = \frac{0.15}{100 \times 10^{-6}} = 1.5 \text{ k}\Omega$$

A similar procedure is used to determine the remaining parameters. The appropriate equations are given below; the numerical values are obtained from graphic constructions on the appropriate characteristics.

$$h_{fe} = \left.\frac{\Delta I_C}{\Delta I_B}\right|_{\Delta V_{CE}=0} \tag{6.9d}$$

From the curves of Fig. 6.6e, we have

$$h_{fe} = \frac{10^{-3}}{40 \times 10^{-6}} = 25$$

$$h_{re} = \left.\frac{\Delta V_{BE}}{\Delta V_{CE}}\right|_{\Delta I_B=0} \tag{6.10d}$$

From the curves of Fig. 6.6d, we have

$$h_{re} = \frac{0.35}{10} = 0.035$$

$$h_{oe} = \left.\frac{\Delta I_C}{\Delta V_{CE}}\right|_{\Delta I_B=0} \tag{6.8d}$$

From the curves of Fig. 6.6c, we have

$$h_{oe} = \frac{10^{-3}}{7} = 143 \; \mu\text{mho}$$

Although common emitter parameters have been illustrated until now, the process for determining common base and common collector parameters is the same.

The common base input and output characteristics of Fig. 6.7 may be used to determine all four $h$ parameters. The appropriate expressions are

$$h_{ib} = \left.\frac{\Delta V_{EB}}{\Delta I_E}\right|_{\Delta V_{CB}=0} \tag{6.7e}$$

$$h_{ob} = \left.\frac{\Delta I_C}{\Delta V_{CB}}\right|_{\Delta I_E=0} \tag{6.8e}$$

$$h_{fb} = \left.\frac{\Delta I_C}{\Delta I_E}\right|_{\Delta V_{CB}=0} \tag{6.9e}$$

$$h_{rb} = \left.\frac{\Delta V_{EB}}{\Delta V_{CB}}\right|_{\Delta I_E=0} \tag{6.10e}$$

Similarly, the common collector input and output characteristics of Fig. 6.8 may be used to determine the common collector parameters, as follows:

$$h_{ic} = \left.\frac{\Delta V_{BC}}{\Delta I_B}\right|_{\Delta V_{EC}=0} \tag{6.7f}$$

$$h_{oc} = \left.\frac{\Delta I_E}{\Delta V_{EC}}\right|_{\Delta I_B=0} \tag{6.8f}$$

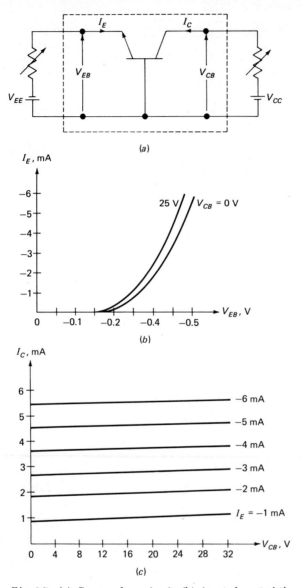

**Fig. 6.7** (a) *Common base circuit;* (b) *input characteristics for the circuit in* (a); (c) *output characteristics for the circuit in* (a). *Note that $I_E$ and $V_{EB}$ are shown on the characteristics as negative quantities. This is due to the fact that, for an NPN transistor, the actual polarities of $V_{EB}$ and $I_E$ are opposite to the conventional ones of the circuit in* (a).

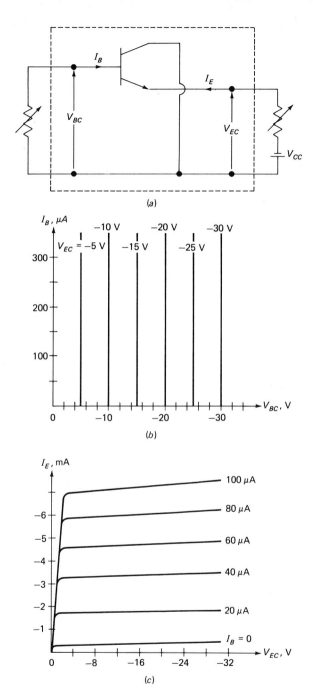

Fig. 6.8 (a) Common collector circuit; (b) input characteristics for the circuit in (a); (c) output characteristics for the circuit in (a). Note that $V_{BC}$, $V_{EC}$, and $I_E$ are shown on the characteristics as negative quantities. This is due to the fact that, for an NPN transistor, the actual polarities of $V_{BC}$, $V_{EC}$, and $I_E$ are opposite to the conventional ones of the circuit in (a).

$$h_{fc} = \frac{\Delta I_E}{\Delta I_B}\bigg|_{\Delta V_{EC}=0} \tag{6.9f}$$

$$h_{rc} = \frac{\Delta V_{BC}}{\Delta V_{EC}}\bigg|_{\Delta I_B=0} \tag{6.10f}$$

## Problem Set 6.3

6.3.1. A common base transistor amplifier whose characteristics are given in Fig. 6.7 is biased at $V_{CB} = 10$ V and $I_E = -4$ mA. Determine all four $h$ parameters.

6.3.2. A common collector transistor amplifier whose characteristics are given in Fig. 6.8 is biased at $V_{EC} = -15$ V and $I_E = -3$ mA. Determine all four $h$ parameters.

6.3.3. On the static characteristics of Fig. 6.7, $V_{EB}$ and $I_E$ are shown as negative quantities. If the transistor were changed to a PNP type, which variables would be positive and which would be negative?

6.3.4. If the transistor of Fig. 6.8a were PNP, what variables would have a negative sign?

6.3.5. What $h$ parameters are negative in the common emitter configuration? In the common base configuration? In the common collector configuration?

6.3.6. Why do you think there are different curves for different values of $V_{CE}$ in the static characteristics of Fig. 6.6b?

6.3.7. Using the characteristics of Fig. 6.6b, in what manner do you think $h_{ie}$ depends on the Q point? Specifically, compare $h_{ie}$ at low values of $V_{BE}$ with $h_{ie}$ at high values of $V_{BE}$.

6.3.8. Why must the changes about an operating point be small when computing small-signal parameters? Is this always necessary?

6.3.9. There are times when it is useful to consider *reverse transfer* characteristics. These are a plot of input voltage versus output voltage using the input current as the running parameter. The basic information required is contained on a set of input characteristics where the input current is plotted against the input voltage using the output voltage as the running parameter. It is therefore possible, through a graphic procedure, to obtain the reverse transfer characteristics. Carry out the above procedure for the characteristics of Fig. 6.6b.

6.3.10. The information contained in output characteristics may also be displayed in a different way, this time yielding a set of *forward transfer characteristics*. These are a plot of output current versus input current using the output voltage as the running parameter. Carry out the above procedure for the characteristics of Fig. 6.7c.

## 6.2d  Use of Data Sheets to Obtain h Parameters

Manufacturers' data sheets generally provide a wealth of information, including $h$ parameters when the transistor is designed for small-signal applications.

A set of data sheets for the 2N3392/3/4 transistors is reproduced in Appendix 10. This is a small-signal silicon NPN transistor. The common emitter $h$ parameters for each of the three transistor types are given as follows:

|  |  | 2N3392 | 2N3393 | 2N3394 | Units |
|---|---|---|---|---|---|
| Forward current transfer ratio | $h_{fe}$ | 208 | 150 | 100 |  |
| Input impedance | $h_{ie}$ | 6,000 | 3,400 | 2,750 | $\Omega$ |
| Output admittance | $h_{oe}$ | 14 | 10 | 7.7 | $\mu$mho |
| Voltage feedback ratio (reverse voltage transfer ratio) | $h_{re}$ | $0.33 \times 10^{-3}$ | $0.225 \times 10^{-3}$ | $0.175 \times 10^{-3}$ |  |

These parameters were obtained by the manufacturer, using experimental techniques. By having this information, the transistor user is saved the labor of experimentally determining each parameter. It should be pointed out, however, that these values are *typical* and that some variation for transistors of the same type, under the same operating conditions, should be expected. The test conditions for the above parameters are

Frequency:    1 kHz
Temperature:   25°C
Q point $\begin{cases} V_{CE} = 10 \text{ V} \\ I_C = 1 \text{ mA} \end{cases}$

If the operating conditions change, adjustments to the $h$ parameters must be made. The manufacturer provides a graph of "$h$ parameters versus temperature" which can be used to estimate any given $h$ parameter at a temperature other than the standard 25°C. As an example, at 80°C, $h_{ie}$ is twice the 25°C value; therefore, for the 2N3392, when $V_{CE} = 10$ V, $I_C = 1$ mA, $f = 1$ kHz, and $T = 80°C$, $h_{ie} = 2(6,000) = 12,000\ \Omega$.

Adjustments for different $Q$ points may be made using appropriate graphs also provided by the manufacturer. As an example, let us determine $h_{oe}$ for the 2N3393 at 25°C, 1 kHz, and $Q$ point of 1 V and 1 mA. Using the $h$ parameters versus voltage graph, we find that $h_{oe}$ at this $Q$ point is twice the value given earlier, namely $2(10) = 20\ \mu$mho.

Similarly, if $h_{fe}$ for the 2N3394 at 25°C, 1 kHz, and $Q$ point of 10 V and 0.15 mA is required, we use the $h$ parameters versus $I_C$ graph to yield an adjusted $h_{fe}$ equal to $0.8(100) = 80$.

Additional information as to how $h_{fe}$ varies with $I_C$ and frequency is provided by the graph on the third page of data sheets for the 2N3392/3/4, top righthand corner. This graph points out that $h_{fe}$[1] generally rises with increasing $I_C$ and drops with increasing frequency. The dependence of $h_{fe}$ on frequency is fairly important and will be discussed later.

Although this particular set of data sheets provides common emitter parameters almost exclusively, it is possible, using standard relationships, to convert to common base or common collector, as required. These relationships are given in the next section.

To summarize, $h$ parameters are normally provided by the manufacturer at a given $Q$ point, temperature, and frequency, and for a specific configuration. If the $h$ parameters required are for a set of conditions other than those specified by the manufacturer, appropriate adjustments must be made.

### Problem Set 6.4

*Note:* For all problems in this set, use the data sheets for the 2N3392/3/4 in Appendix 10.

6.4.1. Determine, for the 2N3392,
    a. $h_{ie}$ at $V_{CE} = 10$ V, $I_C = 1$ mA, $f = 1$ kHz, $T = 55°C$
    b. $h_{fe}$ at $V_{CE} = 1$ V, $I_C = 1$ mA, $f = 1$ kHz, $T = 25°C$
    c. $h_{oe}$ at $V_{CE} = 10$ V, $I_C = 10$ mA, $f = 1$ kHz, $T = 25°C$
    d. $h_{re}$ at $V_{CE} = 1$ V, $I_C = 1$ mA, $f = 1$ kHz, $T = 25°C$

6.4.2. What parameter(s) increase with temperature?
6.4.3. What parameter(s) increase with voltage?
6.4.4. What parameter(s) decrease with current?
6.4.5. How does $h_{fe}$ depend on $I_C$?
6.4.6. How does $h_{ie}$ depend on $I_C$?
6.4.7. How does $h_{fe}$ depend on frequency?
6.4.8. Determine, for the 2N3393, $h_{fe}$ at $V_{CE} = 5$ V, $I_C = 10$ mA, $T = 25°C$, and $f = 1$ kHz.
6.4.9. Determine, for the 2N3394, $h_{fe}$ at $25°C$, $V_{CE} = 10$ V, $I_C = 8$ mA, $f = 1$ MHz.

### 6.2e Conversion of h Parameters from One Configuration to Another

Consider the common emitter circuit of Fig. 6.9a. The same circuit is redrawn in a common base configuration in Fig. 6.9b. The parameter $h_{ib}$ was defined earlier as

---

[1] When using data sheets be careful not to confuse $h_{fe}$ with $h_{FE}$; uppercase subscripts are used for DC parameters, while lowercase subscripts are used for AC parameters.

Junction Transistor Small-signal Models   233

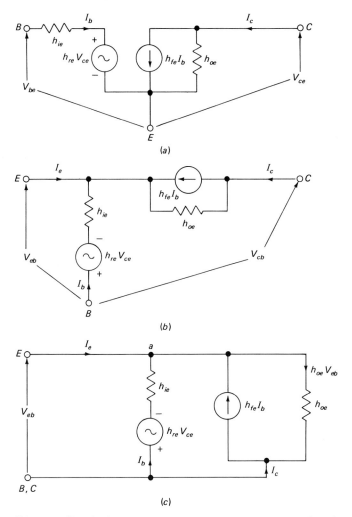

Fig. 6.9  Circuits for converting common emitter to common base h parameters.

$$h_{ib} = \left.\frac{V_{eb}}{I_e}\right|_{V_{cb}=0} \tag{6.7b}$$

The requirement that $V_{cb} = 0$ is satisfied by shorting the collector-base terminals, which yields the circuit of Fig. 6.9c. For this circuit the sum of all currents entering junction $a$ must equal the sum of all currents leaving:

$$h_{fe}I_b + I_b + I_e = h_{oe}V_{eb} \tag{a}$$

The algebraic sum of $h_{re}V_{ce}$ and the voltage across $h_{ie}$ must equal $V_{eb}$, as follows:

$$V_{eb} = -h_{re}V_{ce} - h_{ie}I_b$$

But, since base and collector terminals are tied together, $V_{eb} = -V_{ce}$; therefore

$$V_{eb} = h_{re}V_{eb} - h_{ie}I_b$$
$$V_{eb}(1 - h_{re}) = -h_{ie}I_b$$
$$V_{eb} = \frac{-h_{ie}I_b}{1 - h_{re}} \tag{b}$$

Substituting Eq. (b) into Eq. (a), we get

$$I_e = -h_{fe}I_b - I_b + h_{oe}\frac{-h_{ie}I_b}{1 - h_{re}} \tag{c}$$

and

$$h_{ib} = \frac{V_{eb}}{I_e} = \frac{-h_{ie}I_b/(1 - h_{re})}{I_b[-h_{fe} - 1 - (h_{ie}h_{oe})/(1 - h_{re})]}$$

$$h_{ib} = \frac{h_{ie}}{h_{fe} - h_{fe}h_{re} + 1 - h_{re} + h_{ie}h_{oe}}$$

$$h_{ib} = \frac{h_{ie}}{(h_{fe} + 1)(1 - h_{re}) + h_{ie}h_{oe}} \tag{6.11}$$

Usually $h_{re}$ is small compared to 1, and $h_{ie}h_{oe}$ is also negligible; therefore

$$h_{ib} \simeq \frac{h_{ie}}{h_{fe} + 1} \tag{6.11a}$$

Note that $h_{ie}$ and $h_{ib}$ represent approximately the same physical property. The fact that they do not have the same numerical value can be explained as follows: in the common emitter configuration the input resistance is obtained by looking into the base; this is $h_{ie}$, as indicated in the equivalent circuit of Fig. 6.10a:

$$h_{ie} = \frac{V}{I_b} \tag{d}$$

In the common base configuration the input resistance is obtained by looking into the emitter (see Fig. 6.10b). The result is

$$h_{ib} = \frac{V}{I_e} \tag{e}$$

For any given base-emitter voltage drive $V$, the emitter current is

$$I_e = I_b(h_{fe} + 1) \tag{f}$$

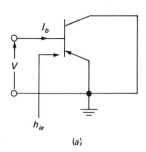 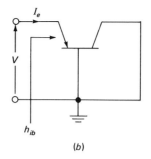

Fig. 6.10 Circuits to compare $h_{ie}$ and $h_{ib}$.

Therefore

$$h_{ib} = \frac{V}{I_b(h_{fe}+1)} = \frac{h_{ie}}{h_{fe}+1} \qquad (6.11a)$$

A simple way to illustrate the above is as follows:

## Common Emitter

Input terminal: base
Input current ($I_b$): small
Input resistance ($h_{ie}$): large

## Common Base

Input terminal: emitter
Input current $I_e = (h_{fe}+1)I_b$: large
Input resistance $h_{ib} = h_{ie}/(h_{fe}+1)$: small

The relationship between $h_{fb}$ and $h_{fe}$ may be developed using the circuit of Fig. 6.9c, as follows:

$$h_{fb} = \frac{I_c}{I_e}\bigg|_{V_{cb}=0} \qquad (6.9b)$$

But

$$I_c = h_{fe}I_b - h_{oe}V_{eb} \qquad (g)$$

which, when we substitute the relationship of Eq. (b), becomes

$$I_c = h_{fe}I_b + \frac{h_{oe}h_{ie}I_b}{1 - h_{re}} \qquad (h)$$

$I_e$ was found earlier to be

$$I_e = -I_b \left( h_{fe} + 1 + \frac{h_{oe} h_{ie}}{1 - h_{re}} \right) \qquad (c)$$

Therefore

$$h_{fb} = \frac{I_c}{I_e} = \frac{h_{fe} + [h_{oe} h_{ie}/(1 - h_{re})]}{-[h_{fe} + 1 + (h_{oe} h_{ie})/(1 - h_{re})]} \qquad (6.12)$$

The $h_{oe} h_{ie}/(1 - h_{re})$ term in the numerator is usually small compared to $h_{fe}$; if we neglect it, Eq. (6.12) becomes

$$h_{fb} = \frac{-h_{fe}}{h_{fe} + 1} \qquad (6.12a)$$

The relationship between $h_{ob}$ and $h_{oe}$ may be developed using Fig. 6.11. In the common base configuration of Fig. 6.11a the signal generator $V$ sees a conductance $h_{ob}$. The current supplied by the signal generator $V$ may be called $I_1$; thus $h_{ob} = I_1/V$. In a common emitter configuration (Fig. 6.11b), $V$ is con-

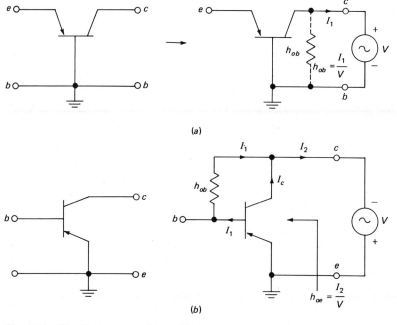

Fig. 6.11 Circuits to compare $h_{ob}$ and $h_{oe}$.

nected from collector to emitter. The resulting current may be called $I_2$, and the parameter $h_{oe} = I_2/V$. There is, however, the conductance $h_{ob}$ from collector to base, as shown. The voltage across $h_{ob}$ is approximately $V$, since the voltage drop from base to emitter is usually small in comparison. This means that the current through $h_{ob}$ is the same here as in the common base circuit, that is, $I_1 = (V)h_{ob}$. Note, however, that $I_1$ must now also flow out of the base; hence $I_1 = I_b$. This leads to the following:

$$I_2 = I_c + I_1$$
$$I_2 = h_{fe}I_1 + I_1 = I_1(h_{fe} + 1)$$
$$h_{oe} = \frac{I_2}{V} = \frac{I_1(h_{fe} + 1)}{V}$$
$$h_{oe} = \frac{I_1}{V}(h_{fe} + 1) = h_{ob}(h_{fe} + 1)$$
$$h_{oe} = h_{ob}(h_{fe} + 1) = \frac{h_{ob}}{1 + h_{fb}} \tag{6.13}$$

The derivations just carried out are very useful for the analysis of small-signal amplifiers. The remaining conversion formulas are not derived here because the process is quite lengthy; they are, however, listed in Table 6.1 with normal approximations already made.

### Problem Set 6.5

6.5.1. Convert the common emitter $h$ parameters for the 2N3393 to common base parameters.
6.5.2. Perform the conversion of Prob. 6.5.1 to common collector parameters.
6.5.3. What transistor configuration has the lowest value of $h_i$? The highest?
6.5.4. What transistor configuration has the lowest value of $h_f$? The highest?
6.5.5. What transistor configuration has the lowest value of $h_r$? The highest?
6.5.6. What transistor configuration has the lowest value of $h_o$? The highest?

## 6.3 THE HYBRID-PI MODEL

A junction transistor model useful to predict high-frequency performance is the hybrid-pi model. We will develop such a model here and apply it to determine the high-frequency response of simple amplifiers.

The $h$-parameter model is primarily useful in low-frequency applications. At higher frequencies we must consider the effects of transistor capacitances, and

Table 6.1. h Parameters for Each Transistor Configuration in Terms of the Other Two.

|  |  | Common base | Common collector |
|---|---|---|---|
| Common emitter | $h_{ie}$ | $\dfrac{h_{ib}}{h_{fb}+1}$ | $h_{ic}$ |
|  | $h_{fe}$ | $-\dfrac{h_{fb}}{h_{fb}+1}$ | $-(h_{fc}+1)$ |
|  | $h_{oe}$ | $\dfrac{h_{ob}}{h_{fb}+1}$ | $h_{oc}$ |
|  | $h_{re}$ | $\dfrac{h_{ib}h_{ob}}{1+h_{fb}} - h_{rb}$ | $1 - h_{rc}$ |

|  |  | Common emitter | Common collector |
|---|---|---|---|
| Common base | $h_{ib}$ | $\dfrac{h_{ie}}{h_{fe}+1}$ | $-\dfrac{h_{ic}}{h_{fc}}$ |
|  | $h_{fb}$ | $\dfrac{-h_{fe}}{h_{fe}+1}$ | $-\dfrac{h_{fc}+1}{h_{fc}}$ |
|  | $h_{ob}$ | $\dfrac{h_{oe}}{h_{fe}+1}$ | $-\dfrac{h_{oc}}{h_{fc}}$ |
|  | $h_{rb}$ | $\dfrac{h_{ie}h_{oe}}{h_{fe}+1} - h_{re}$ | $\dfrac{-h_{ic}h_{oc}}{h_{fc}} + h_{rc} - 1$ |

|  |  | Common emitter | Common base |
|---|---|---|---|
| Common collector | $h_{ic}$ | $h_{ie}$ | $\dfrac{h_{ib}}{1+h_{fb}}$ |
|  | $h_{fc}$ | $-(1+h_{fe})$ | $\dfrac{-1}{1+h_{fb}}$ |
|  | $h_{oc}$ | $h_{oe}$ | $\dfrac{h_{ob}}{1+h_{fb}}$ |
|  | $h_{rc}$ | $1 - h_{re} \simeq 1$ (since $h_{re} \ll 1$) | $\dfrac{1 - h_{ib}h_{ob}}{1+h_{fb}}$ |

although the $h$-parameter model could be adapted to do this, it is simpler to use the hybrid-pi model shown in Fig. 6.12.

To describe the hybrid-pi model we will consider each component in turn. The points labeled $b$, $e$, and $c$ correspond to the external base, emitter, and collector terminals, respectively. The point $b'$ corresponds to an area inside the base where the transistor's active region begins; there is a transverse resistance $r_{bb'}$ between $b$ and $b'$ through which base current flows. The thinner the

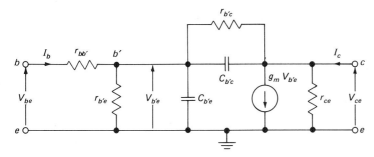

Fig. 6.12 Hybrid-pi model.

base region, the higher $r_{bb'}$; generally it ranges from 50 to 500 Ω and is constant for any given transistor.

The dynamic resistance of the forward-biased base-emitter junction is $r_{b'e}$. This parameter depends on the Q point; typical values range from about 1,000 to 8,000 Ω. At low frequencies capacitors $C_{b'e}$ and $C_{b'c}$ behave as open circuits.

$C_{b'e}$ represents an equivalent capacitance designed to take into account the transistor's reduction in gain and increase in phase shift at higher frequencies. There are two components to $C_{b'e}$. The first and usually much smaller component is the depletion layer (also called transition region) capacitance present across any PN junction. This capacitance is voltage-dependent and is described in Sec. 1.8. The larger component of $C_{b'e}$, usually referred to as a diffusion capacitance, provides a convenient way of taking into account the presence of excess charge in the base region whenever a forward-biasing voltage is applied. This charge is made up of minority carriers injected by the emitter and diffusing into the base. The relationship between voltage and charge can be described by an equivalent capacitance ($C = Q/V$), and this is why $C_{b'e}$ is used. The value of $C_{b'e}$ is not constant but proportional to the current, since the larger the current, the more charge there is at any given time.

$C_{b'c}$ is the depletion-region capacitance of the collector-base junction. It depends on voltage as indicated in Sec. 1.8. $C_{b'c}$ may be referred to as $C_{ob}$, since it is also the capacitance seen when looking into the output of a common base configuration.

Resistance $r_{b'c}$ is due partly to junction leakage. The main component, however, is caused by another phenomenon called the Early effect. To describe this effect, assume a constant emitter current $I_E$. The collector current $I_C = \alpha I_E$; therefore, as long as both $I_E$ and $\alpha$ are constant, $I_C$ is also constant. Suppose now that the reverse collector-base voltage increases; the depletion region becomes wider and extends further into the base. This produces an increase in $\alpha$ because of the narrower base region. Hence, even though $I_E$ remains constant, $I_C$ increases because of the rise in $\alpha$. The increase in collector-base voltage has therefore resulted in a corresponding increase of collector current. This effect can be simulated with a resistance, in our case $r_{b'c}$. Note that $r_{b'c}$ connects the base region to the collector; it is a physical link that produces interaction between output and input circuits and makes it necessary to have the parameter $h_{re}$.

We now come to the current generator. In place of the familiar $h_{fe}I_b$, a *voltage-controlled* current source is used. The voltage that controls how much current this generator produces is $V_{b'e}$. The parameter $g_m$ is called the forward transconductance and is constant for the given circuit and $Q$ point; the units of $g_m$ are mhos, so that when $V_{b'e}$ is multiplied by $g_m$, the result has the units of amperes. The parameter $g_m$ is related to the operating point by[1]

$$g_m \simeq 40 I_C \tag{6.14}$$

The parameter $r_{ce}$ takes into account the output resistance of the transistor; it may be neglected if $R_L \ll r_{ce}$.

All the resistive parameters described above may be related to $h$ parameters; we will do so as the need arises. Meanwhile an additional simplification may be made by neglecting $r_{b'c}$; this is equivalent to neglecting $h_{re}$ in the $h$-parameter model.

A simplified equivalent circuit valid at low frequencies (both capacitors appear as open circuits) is shown in Fig. 6.13a; a source and load have also been added. Note that, if the output were short-circuited, the input resistance would equal $r_{bb'} + r_{b'e}$; this is equivalent to $h_{ie}$. For this circuit we can write

$$I_c = g_m V_{b'e} = g_m (I_b r_{b'e})$$
$$\frac{I_c}{I_b} = g_m r_{b'e}$$
$$h_{fe} = g_m r_{b'e} \tag{6.15}$$

The ratio of $V_o$ to $V_{b'e}$ is

$$\frac{V_o}{V_{b'e}} = \frac{-I_c R_L}{V_{b'e}} = \frac{-g_m V_{b'e} R_L}{V_{b'e}}$$

$$\frac{V_o}{V_{b'e}} = -g_m R_L \tag{6.16}$$

This ratio is equivalent to $A_v$ if the input voltage is close to $V_{b'e}$ ($r_{bb'} \ll r_{b'e}$); we will use this expression later on.

A high-frequency measure of performance often given for transistors is the beta cutoff frequency $f_\beta$. At this frequency the magnitude of $h_{fe}$ is down

---

[1] The derivation required to prove this expression is beyond the scope of this text. (See p. 224 of Reference 12 at the end of this chapter.)

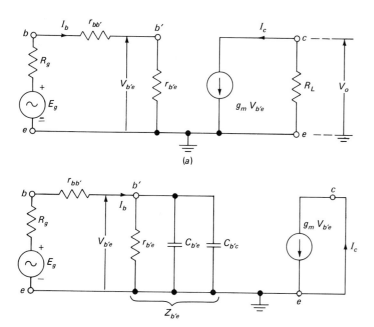

Fig. 6.13 (a) Low-frequency hybrid-pi equivalent circuit; (b) circuit for determining $f_\beta$.

3 dB from its midfrequency value. To determine $f_\beta$ we will use the circuit of Fig. 6.13b, in which the output has been short-circuited; the parallel combination of $r_{b'e}$, $C_{b'e}$, and $C_{b'c}$ is referred to as $Z_{b'e}$:

$$V_{b'e} = I_b Z_{b'e} = I_b \frac{r_{b'e}}{1 + j\omega r_{b'e}(C_{b'e} + C_{b'c})}$$

$$I_c = g_m V_{b'e} = \frac{g_m I_b r_{b'e}}{1 + j\omega r_{b'e}(C_{b'e} + C_{b'c})}$$

$$h_{fe} = \frac{I_c}{I_b} = \frac{g_m r_{b'e}}{1 + j\omega/\omega_\beta} \qquad (6.17)$$

where

$$\omega_\beta = \frac{1}{r_{b'e}(C_{b'e} + C_{b'c})} \qquad (6.17a)$$

$C_{b'e}$ is usually much larger than $C_{b'c}$; therefore the latter can be neglected, yielding

$$\omega_\beta = \frac{1}{r_{b'e} C_{b'e}}$$

$$f_\beta = \frac{1}{2\pi r_{b'e} C_{b'e}} \qquad (6.17b)$$

An examination of Eq. (6.17) indicates that at low frequencies ($\omega \ll \omega_\beta$), $h_{fe} = g_m r_{b'e}$, as given earlier by Eq. (6.15). Beyond $\omega_\beta$, however, $|h_{fe}|$ drops at the rate of 20 dB per decade; there is also an appropriate phase shift.

An additional measure of performance is $f_T$, the frequency for which $|h_{fe}| = 1$. This may be obtained by setting the magnitude of the expression in Eq. (6.17) equal to 1, as follows:

$$|h_{fe}| = \frac{g_m r_{b'e}}{\sqrt{1 + (\omega_T/\omega_\beta)^2}} = 1$$

$$g_m r_{b'e} = \sqrt{1 + (f_T/f_\beta)^2}$$

Since the second term under the radical is usually much larger than 1,

$$g_m r_{b'e} \simeq \frac{f_T}{f_\beta}$$

$$f_T = f_\beta g_m r_{b'e} = f_\beta h_{fe} \tag{6.18}$$

Using Eq. (6.17b), we can also write

$$f_T = \frac{g_m r_{b'e}}{2\pi r_{b'e} C_{b'e}} = \frac{g_m}{2\pi C_{b'e}} \tag{6.19}$$

Note that $f_\beta$, being the frequency at which $|h_{fe}|$ is down 3 dB, corresponds approximately to the transistor's current-amplification bandwidth. Now $f_T$ is equal to $h_{fe} f_\beta$, the product of gain $h_{fe}$ and bandwidth $f_\beta$, and therefore $f_T$ is called the gain bandwidth product. This measure of performance is usually given by manufacturers in data sheets; it is a function of operating point, since the parameter $g_m$, itself dependent on $I_C$, is involved.

### Example 6.5

A transistor is biased at $I_C = 2.5$ mA. If $f_T = 500$ MHz and $r_{b'e} = 1{,}000\ \Omega$, determine the following parameters: $g_m$, $C_{b'e}$, $h_{fe}$, and $f_\beta$.

### Solution

a. $g_m = 40 I_C$     (6.14)
$g_m = 40(2.5 \times 10^{-3}) = 100$ mmho

b. $f_T = \dfrac{g_m}{2\pi C_{b'e}}$     (6.19)

$C_{b'e} = \dfrac{g_m}{2\pi f_T} = \dfrac{10^{-1}}{6.28(500 \times 10^6)}$

$C_{b'e} \simeq 32$ pF

c. $h_{fe} = g_m r_{b'e}$ (6.15)
   $h_{fe} = 0.1(1{,}000) = 100$

d. $f_T = f_\beta h_{fe}$ (6.18)
   $f_\beta = \dfrac{f_T}{h_{fe}} = \dfrac{500 \times 10^6}{100} = 5 \text{ MHz}$

## Problem Set 6.6

6.6.1. Why does $r_{bb'}$ increase as the base is made thinner? Why do we want a thin base?

6.6.2. In what way does $r_{b'e}$ vary with current?

6.6.3. Is $C_{b'e}$ mainly a depletion-region capacitance? Explain your answer.

6.6.4. Why is $C_{b'e}$ proportional to current while $C_{b'c}$ is proportional to voltage?

6.6.5. What is the Early effect?

6.6.6. Prove that the low-frequency value of $h_{re} = r_{b'e}/(r_{b'e} + r_{b'c}) \simeq r_{b'e}/r_{b'c}$.

6.6.7. Prove that the low-frequency value of $h_{oe} = 1/r_{ce} + [(1 + g_m r_{b'e})/(r_{b'c} + r_{b'e})]$.

6.6.8. A given transistor is biased at $I_C = 5$ mA. Determine $g_m$.

6.6.9. Prove that $r_{b'c} \simeq 1/h_{ob}$.

6.6.10. If the midfrequency value of $h_{fe}$ for the transistor of Prob. 6.6.8 is 150, determine $r_{b'e}$.

6.6.11. $f_\alpha$ is defined as the frequency for which $|h_{fb}|$ is down 3 dB from its low-frequency value. Using the relationship $h_{fb} = -h_{fe}/(h_{fe} + 1)$ and Eq. (6.17), show that $f_\alpha = f_\beta(h_{fe} + 1)$.

## REFERENCES

1. Mulvey, J.: "Semiconductor Device Measurements," Tektronix, Inc., Beaverton, Oreg., 1968.
2. Phillips, A. B.: "Transistor Engineering and Introduction to Integrated Semiconductor Circuits," McGraw-Hill Book Company, New York, 1962.
3. Basic Theory and Application of Transistors, TM 11-690, Headquarters, Department of the Army, 1959.
4. Malvino, A. P.: "Transistor Circuit Approximations," McGraw-Hill Book Company, New York, 1968.
5. Joyce, M. V., and K. K. Clarke: "Transistor Circuit Analysis," Addison-Wesley Publishing Company, Inc., Reading, Mass., 1961.
6. Millman, J., and C. C. Halkias: "Electronic Devices and Circuits," McGraw-Hill Book Company, New York, 1967.
7. Cutler, P.: "Semiconductor Circuit Analysis," McGraw-Hill Book Company, New York, 1964.

8. Texas Instruments Incorporated: "Transistor Circuit Design," McGraw-Hill Book Company, New York, 1963.
9. Ristenbatt, M. P., and R. L. Riddle: "Transistor Physics and Circuits," Prentice-Hall, Inc., Englewood Cliffs, N.J., 1966.
10. Fitchen, F. C.: "Transistor Circuit Analysis and Design," 2d ed., D. Van Nostrand Company, Inc., Princeton, N.J., 1966.
11. "General Electric Transistor Manual," 7th ed., Syracuse, 1964.
12. Angelo, E. J.: "Electronic Circuits," McGraw-Hill Book Company, New York, 1964.

# chapter 7

## junction transistor small-signal analysis

Basic performance measures for junction transistor amplifiers in each of the three configurations are discussed. The dependence of $R_i$, $R_o$, $A_v$, $A_i$, and $G$ on circuit parameters and frequency is considered.

### 7.1 THE COMMON EMITTER AMPLIFIER

A common emitter circuit is shown in Fig. 7.1a. The DC supply, biasing resistors, and coupling capacitors are not shown, since we are performing an AC analysis. The transistor is replaced by its $h$-parameter model in Fig. 7.1b. We now analyze the circuit using a set of common emitter $h$ parameters for a 2N3394 silicon transistor:

$h_{ie} = 2{,}750 \ \Omega$
$h_{fe} = 100$
$h_{re} = 0.175 \times 10^{-3}$
$h_{oe} = 7.7 \ \mu\text{mho}$

The input resistance is considered first. This was derived in Chap. 6 for the general case and is rewritten here in terms of common emitter parameters:

$$R_i = h_{ie} - \frac{h_{fe}h_{re}}{h_{oe} + (1/R_L)} \qquad (6.3a)$$

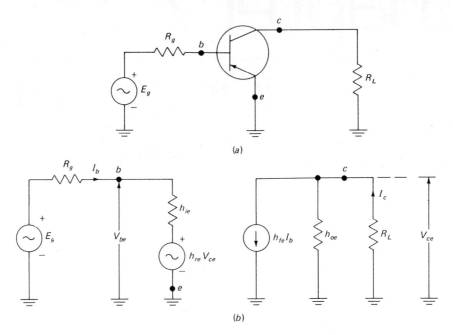

*Fig. 7.1 Common emitter amplifier and h-parameter model.*

The dependence of $R_i$ on $R_L$ is a matter of interest. If $R_L$ is very small, the second term in Eq. (6.3a) drops and $R_i \simeq h_{ie} = 2{,}750\ \Omega$. If $R_L$ is increased, however, the second term is no longer negligible, and it subtracts from $h_{ie}$; therefore $R_i$ drops as $R_L$ increases. As $R_L$ is made larger still, its effect disappears as soon as $R_L \gg 1/h_{oe}$, in which case the $1/R_L$ term in the denominator drops and $R_i \simeq h_{ie} - (h_{fe}h_{re}/h_{oe}) = 2{,}750 - 2{,}270 = 480\ \Omega$. The dependence of $R_i$ on $R_L$ is illustrated by the graph of Fig. 7.2a. Note that $R_i$ is not affected by values of $R_L$ smaller than 1 k$\Omega$ or larger than 10 M$\Omega$.

The output resistance for the common emitter amplifier is

$$R_o = \frac{1}{h_{oe} - [h_{re}h_{fe}/(h_{ie} + R_g)]} \qquad (6.4a)$$

$R_o$ is a function of $R_g$, as shown in Fig. 7.2b. As long as $R_g \ll h_{ie}$, $R_o$ is independent of $R_g$ and approximately equal to $1/[h_{oe} - (h_{re}h_{fe}/h_{ie})] = 755$ k$\Omega$. Larger values of $R_g$ cause $R_o$ to decrease with increasing $R_g$. Eventually, as $R_g$ increases beyond the range of 100 k$\Omega$, $R_o$ approaches $1/h_{oe} = 130$ k$\Omega$, in which case $R_o$ is again independent of $R_g$.

The basic expression for current amplification is

$$A_i = \frac{h_{fe}}{1 + h_{oe}R_L} \qquad (6.6a)$$

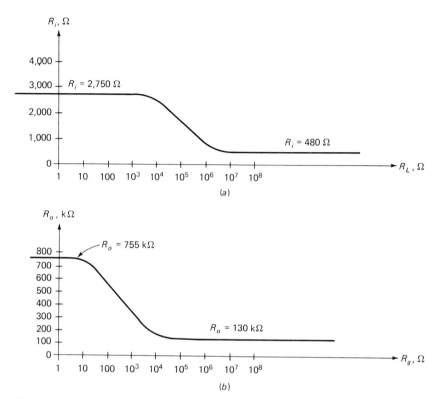

*Fig. 7.2 Performance measures for common emitter amplifier; (a) input resistance $R_i$ as a function of $R_L$; (b) output resistance $R_o$ as a function of $R_g$.*

Thus $A_i$ depends on $R_L$, as shown in Fig. 7.3a. If $R_L$ is very low, $h_{oe}R_L \ll 1$, $A_i \simeq h_{fe} = 100$. As $R_L$ increases, $A_i$ drops until it is equal to zero, which occurs when $R_L$ is an open circuit. Now all $h_{fe}I_b$ produced by the generator in Fig. 7.1b flows through $h_{oe}$, and none through $R_L$.

The expression for $A_v$ is

$$A_v = \frac{-h_{fe}R_L}{h_{ie} + R_L(h_{ie}h_{oe} - h_{fe}h_{re})}$$

$A_v$ is also a function of $R_L$, as shown in Fig. 7.3b. For low values of $R_L$, most of the $h_{fe}I_b$ generator current flows through $R_L$; therefore as $R_L$ increases, so does the output voltage and $A_v$. Once $R_L$ becomes $\gg 1/h_{oe}$, however, current is diverted from $R_L$ to $h_{oe}$. With $R_L$ very large, all the current flows through $h_{oe}$, and none through $R_L$. The output voltage is then determined exclusively by the product of $h_{fe}I_b$ and $1/h_{oe}$. Further increases in $R_L$ do not result in greater output voltage, and the $A_v$-versus-$R_L$ curve levels off to a constant value, as indicated by

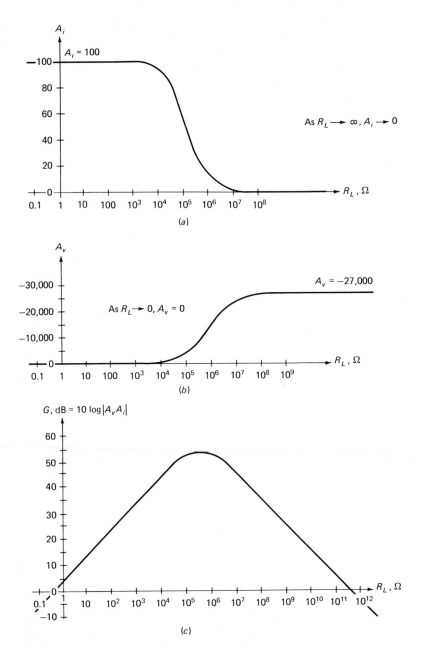

Fig. 7.3 Performance measures for common emitter amplifier as a function of $R_L$; (a) current amplification $A_i$; (b) voltage amplification $A_v$; (c) power gain G in decibels.

the plot. The negative sign for $A_v$ accounts for the 180° phase shift between input voltage $V_{be}$ and output voltage $V_{ce}$.

Having obtained $A_i$ and $A_v$, it is usually a simple matter to determine the power gain. This is done in Fig. 7.3c, where $G = 10 \log |A_v A_i|$ is plotted as a function of $R_L$. Note that $G$ is highest for values of $R_L$ in the 100-kΩ to 1-MΩ range; this corresponds to an approximate match between load resistance and the transistor's output resistance. The power gain falls off for lower values of $R_L$ because $A_v$ decreases; at higher values of $R_L$ the drop in $A_i$ causes $G$ to fall off.

We can now summarize the foregoing observations for our common emitter amplifier:

1. Low to moderate input resistance
2. Moderately high output resistance
3. Large current amplification
4. Large voltage amplification
5. Large power gain
6. 180° phase shift between input and output voltages

The graphs of Figs. 7.2 and 7.3 are intended to illustrate the dependence of amplifier performance on $R_L$ or $R_g$. The amplification and resistance extremes indicated on these graphs may not be possible in practice because of limitations imposed by biasing resistors and supply voltage.

## 7.2 THE COMMON BASE AMPLIFIER

The common base circuit of Fig. 7.4a is now considered. The $h$ parameters used for the common emitter amplifier just discussed are converted to their common base counterparts using the relationships of Table 6.1. The results are

$h_{ib} = 27\ \Omega$
$h_{fb} = -0.99$
$h_{ob} = 7.7 \times 10^{-8}$ mho
$h_{rb} = 37 \times 10^{-6}$

The plot of $R_i$ as a function of $R_L$ is shown in Fig. 7.5a. The appropriate equation is

$$R_i = h_{ib} - \frac{h_{fb} h_{rb}}{h_{ob} + (1/R_L)} \tag{6.3a}$$

When $R_L$ is less than about 100 kΩ, the second term disappears and $R_i \simeq h_{ib} = 27\ \Omega$. For very large values of $R_L$, the $1/R_L$ term becomes negligible compared to $h_{ob}$; $R_i$ is now independent of $R_L$ and approximately equal to $h_{ib} - (h_{fb} h_{rb} / h_{ob}) = 503\ \Omega$. Note that the range of $R_i$ is from 27 to 503 Ω; this points out that,

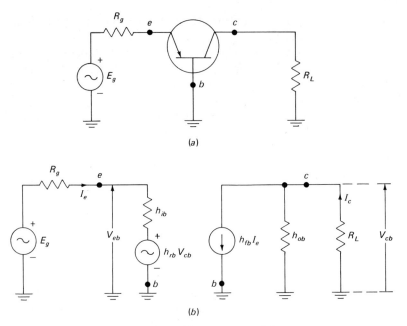

Fig. 7.4  Common base amplifier and h-parameter model.

for a common base amplifier, $R_i$ is relatively low. Normally, values of $R_L > 1/h_{ob}$ are impractical; therefore the 503-Ω extreme for $R_i$ is not feasible.

The output resistance

$$R_o = \frac{1}{h_{ob} - [h_{rb}h_{fb}/(h_{ib} + R_g)]} \tag{6.4a}$$

is plotted as a function of $R_g$ in Fig. 7.5b. Normally, $R_o$ is lower than the range predicted by the graph, because of the shunting effect of biasing resistors.

The current amplification

$$A_i = \frac{h_{fb}}{1 + h_{ob}R_L} \tag{6.6a}$$

is plotted as a function of $R_L$ in Fig. 7.6a. $A_i$ is always less than 1 and negative. The negative sign is due to $h_{fb}$; the less-than-unity magnitude is to be expected, since the ratio of output current $I_c$ to input current $I_e$ can never exceed 1. As $R_L$ becomes larger, more current from the $h_{fb}I_e$ generator is diverted from $R_L$ in favor of $h_{ob}$, and $A_i$ drops. As indicated by the graph, as $R_L$ approaches infinity, $A_i$ approaches zero.

The expression for voltage amplification is

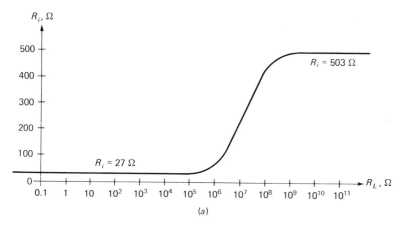

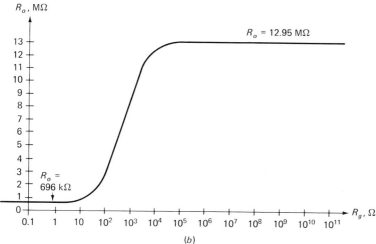

*Fig. 7.5  Performance measures for common base amplifier;* (a) *input resistance* $R_i$ *as a function of* $R_L$; (b) *output resistance* $R_o$ *as a function of* $R_g$.

$$A_v = \frac{-h_{fb}R_L}{h_{ib} + R_L(h_{ib}h_{ob} - h_{fb}h_{rb})} \tag{6.5a}$$

This is plotted in Fig. 7.6b as a function of $R_L$. Note that $A_v$ rises with $R_L$ until a maximum value of about 25,600 is reached.

The power gain curve is shown as a function of $R_L$ in Fig. 7.6c. $G$ is maximum for $R_L$ in the range of 1 M$\Omega$; this corresponds to an approximate match between $R_L$ and $R_o$. Note that the maximum power gain is less than that available in a common emitter amplifier.

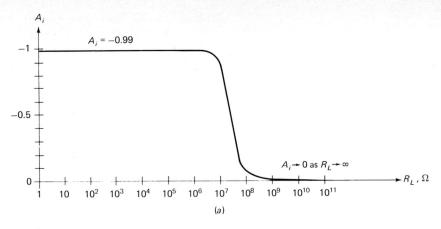

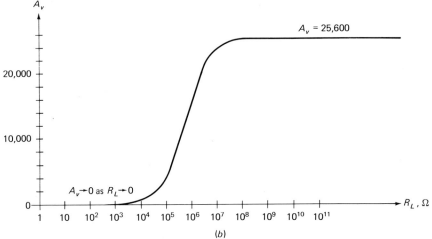

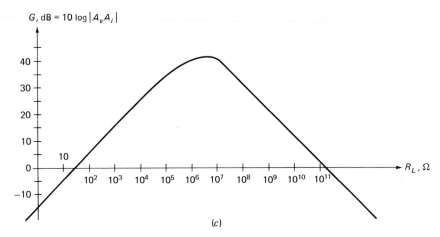

*Fig. 7.6* Performance measures for common base amplifier as a function of $R_L$; (a) current amplification $A_i$; (b) voltage amplification $A_v$; (c) power gain $G$ in decibels.

The performance of our common base amplifier may be summarized as follows:

1. Low input resistance
2. High output resistance
3. Current amplification less than 1
4. High voltage amplification and no phase inversion
5. Moderate power gain

Similar observations as in the common emitter amplifier apply here regarding the possibility of achieving some of the extreme resistance or amplification values indicated by the graphs.

## 7.3 THE COMMON COLLECTOR AMPLIFIER

A common collector amplifier and its $h$-parameter model are shown in Fig. 7.7. With this configuration the output current is $I_e$ and the output voltage is taken between emitter and the grounded collector.

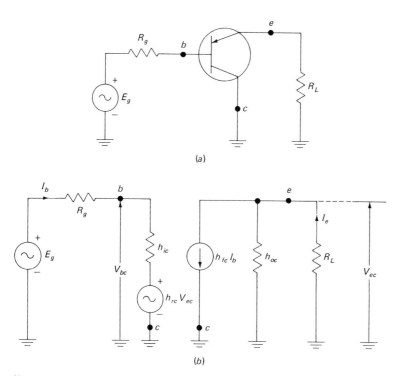

*Fig. 7.7* Common collector amplifier and h parameter model.

The appropriate $h$ parameters are obtained by converting those for the common emitter configuration using the relationships of Table 6.1. This yields

$h_{ic} = 2{,}750\ \Omega$
$h_{fc} = -101$
$h_{rc} = 1$
$h_{oc} = 7.7 \times 10^{-6}$ mho

Consider the plot of $R_i$ versus $R_L$, given in Fig. 7.8a. Except for very low or very high values of $R_L$, $R_i$ depends heavily on $R_L$, much more so than in common emitter or common base circuits. In fact, $R_i$ in the common collector circuit is

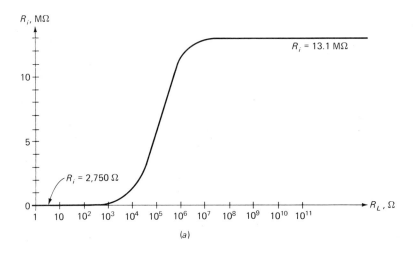

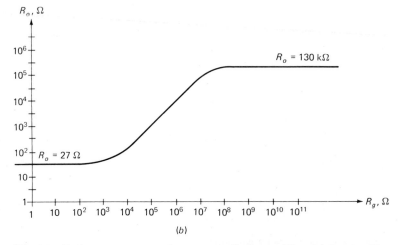

*Fig. 7.8  Performance measures for common collector amplifier; (a) input resistance $R_i$ as a function of $R_L$; (b) output resistance $R_o$ as a function of $R_g$.*

highest and most heavily dependent on $R_L$. The dependence of $R_i$ on $R_L$ may be traced to the reverse voltage transfer ratio $h_{rc}$ which is 1, the highest of any configuration. Such a large $h_r$ implies that there is poor isolation between output and input; in other words, conditions in the output circuit are reflected very strongly in the input circuit.

The above may be illustrated by developing a general expression for $R_i$ in a common collector amplifier. The input circuit to such an amplifier is shown in Fig. 7.9a. Note that $R_L$ is simultaneously in the output and input circuit; no wonder there is such poor isolation! For this circuit, the total applied voltage $V$ is equal to the sum of the base-emitter voltage and the voltage across $R_L$:

$$V = I_b h_{ie} + R_L I_e$$
$$V = I_b h_{ie} + R_L I_b (h_{fe} + 1)$$
$$V = I_b [h_{ie} + R_L (h_{fe} + 1)]$$
$$R_i = \frac{V}{I_b} = h_{ie} + R_L(h_{fe} + 1) \tag{7.1}$$

The expression for $R_i$ appears as the sum of two series resistances, $h_{ie}$ and $R_L(h_{fe} + 1)$, as shown in the figure. This implies that any resistance in the

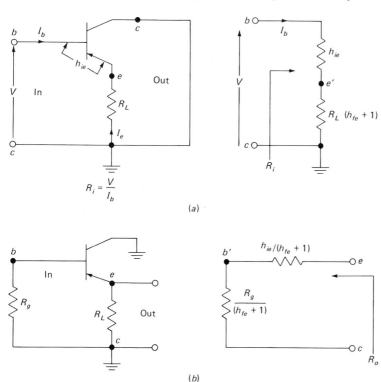

Fig. 7.9  Circuits to study $R_i$ and $R_o$ in common collector amplifiers.

emitter circuit is multiplied by the factor $h_{fe} + 1$ when seen from the base. The notation $e'$ on the circuit diagram represents the emitter terminal as seen from the base.

The technique for determining $R_i$ outlined above is simpler than the general equation developed earlier and allows us to work with an equivalent circuit in which the $h_r V_2$ generator may be omitted.

The output resistance is determined using the circuits of Fig. 7.9b. The input generator is reduced to zero, which places $R_g$ between base and ground; the load $R_L$ is removed, and the resistance looking into the emitter is calculated. Just as any emitter resistance is multiplied by $h_{fe} + 1$ when seen from the base, so is any base resistance divided by $h_{fe} + 1$ when seen from the emitter. We therefore have

$$R_o = \frac{R_g}{h_{fe} + 1} + \frac{h_{ie}}{h_{fe} + 1} = \frac{R_g}{h_{fe} + 1} + h_{ib} \tag{7.2}$$

Note that unless $R_g$ is unusually high, $R_o$ is fairly small. The graph of Fig. 7.8b illustrates how $R_o$ varies with $R_g$, from a low of 27 Ω to a high of 130 kΩ. The common collector circuit's low output resistance, in fact, is one of its advantages. It can be used to drive low-impedance, high-current loads. The combination of high $R_i$ and low $R_o$ also makes this circuit useful as an impedance-matching device, to couple high-impedance sources to low-impedance loads.

Plots of $A_i$, $A_v$, and $G$ are given in Fig. 7.10. $A_i$, which is the ratio of $I_e$ to $I_b$, is good as long as $R_L$ is relatively low; here $A_i = -101$.

$A_v$ is always less than 1. $R_L$ must be at least 1,000 Ω here for $A_v$ to even approach 1; because of this, our power gain is low, in fact the lowest of all three configurations. Because $A_v$ is usually close to 1, the output (emitter) voltage is said to follow the input, and the common collector amplifier is also called an emitter follower.

Properties of the common collector amplifier may be summarized as follows:

1. High input resistance
2. Low output resistance
3. Good current amplification
4. Voltage amplification less than 1
5. Lowest power gain of all configurations
6. Poorest isolation of all configurations

**Example 7.1**

The transistor parameters for the circuit of Fig. 7.11a are $h_{fe} = 100$, $h_{ie} = 2,000$ Ω, $h_{re}$ is negligible, $h_{oe} = 10^{-5}$ mho. Estimate, using reasonable approximations, $A_v$, $A_i$, $R_i$, and $R_o$.

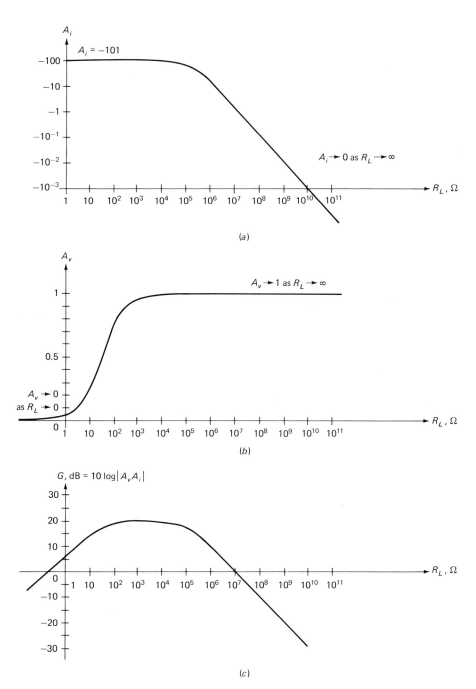

Fig. 7.10 Performance measures for common collector amplifier as a function of $R_L$; (a) current amplification $A_i$; (b) voltage amplification $A_v$; (c) power gain G in decibels.

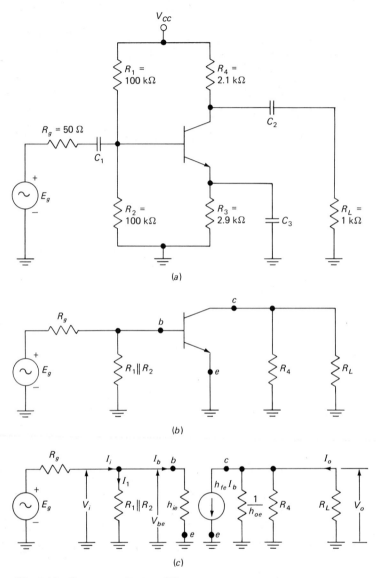

*Fig. 7.11*  *Common emitter amplifier.*

### Solution

The AC equivalent circuit is shown in Fig. 7.11b, and the h-parameter model is added in Fig. 7.11c. Since $h_{re}$ is negligible, the resistance looking into the base is $h_{ie} = 2{,}000\ \Omega$. The total resistance looking into the amplifier is

$R_i = R_1 \| R_2 \| h_{ie}$
$R_i = (50 \times 10^3) \| (2 \times 10^3) = 1{,}925 \; \Omega$

The output resistance is obtained by removing $R_L$ and looking into the transistor's output terminals (collector to ground) with the input terminated with $R_g$ (signal generator shorted to ground). With $E_g = 0$, $I_b = 0$ and the $h_{fe}I_b$ generator is also zero. The output resistance, including the shunting effect of $R_4$, is

$R_o = R_4 \| (1/h_{oe})$
$R_o = (2.1 \times 10^3) \| (100 \times 10^3) \simeq 2 \text{ k}\Omega$

To simplify the calculation of $A_i$, it sometimes helps, although it is not necessary, to assume a specific value of current, say $I_b = 0.1$ mA, and work the rest as follows:

$I_b = 0.1 \text{ mA}$   (assumption)
$V_{be} = I_b h_{ie} = 10^{-4}(2{,}000) = 0.2 \text{ V}$
$I_1 = \dfrac{V_{be}}{50 \times 10^3} = \dfrac{0.2}{50 \times 10^3} = 4 \; \mu\text{A}$

Total input current:  $I_i = I_b + I_1 = 104 \; \mu\text{A}$

$I_o$ is the current through the 1-k$\Omega$ load. Since the $1/h_{oe}$ resistor (100 k$\Omega$) is large compared with $R_4$ and $R_L$, all $h_{fe}I_b$ bypasses the former in favor of the parallel combination of $R_4$ and $R_L$; the load current is then

$I_o \simeq h_{fe}I_b \dfrac{R_4}{R_4 + R_L}$

$I_o \simeq 100 \times 10^{-4} \dfrac{2.1 \times 10^3}{(2.1 \times 10^3) + 10^3} = 6.78 \text{ mA}$

The current amplification is then

$A_i = \dfrac{I_o}{I_i} = \dfrac{6.78 \times 10^{-3}}{104 \times 10^{-6}} = 65$

To compute $A_v$ we can continue using the arbitrary value of 0.1 mA for $I_b$; hence $I_o = 6.78$ mA. The output voltage is

$V_o = -I_o R_L = (-6.78 \times 10^{-3})(10^3) = -6.78 \text{ V}$

The input voltage $V_i = V_{be}$; this was calculated above to be 0.2 V. The voltage amplification is therefore

$A_v = \dfrac{V_o}{V_i} = \dfrac{-6.78}{0.2} = -33.9$

260  Solid State Electronic Circuits

The negative sign for $A_v$ confirms the 180° phase shift between collector and base voltages (as base goes more positive, collector goes more negative).

**Example 7.2**

The circuit of Example 7.1 is to be operated in the common base configuration. Determine $R_i$, $R_o$, $A_i$, and $A_v$.

**Solution**

In a common base amplifier, the transistor is driven from the emitter, and the output is taken at the collector, as shown in Fig. 7.12a. $C_3$ has a low reactance at the frequency of operation, so that both $R_1$ and $R_2$ are effectively bypassed to ground.

The circuit is further simplified in Fig. 7.12b and c. Since $h_{re}$ is negligible, $h_{rb}$ is even less significant; hence the $h_{rb}V_{cb}$ generator is omitted. Common base amplifiers, in fact, suffer least from interaction between output and input circuits.

The input resistance seen from the emitter is simply $h_{ib}$, which can be calculated using the relationship given earlier:

$$h_{ib} = \frac{h_{ie}}{h_{fe} + 1} = \frac{2{,}000}{101} \simeq 20 \ \Omega \qquad (6.11a)$$

The input resistance is $R_3 \| h_{ib}$:

$R_i \simeq 20 \ \Omega$

The output conductance $h_{ob}$ is given in Table 6.1 as

$$h_{ob} = \frac{h_{oo}}{h_{fe} + 1} = \frac{10^{-5}}{101} \simeq 10^{-7} \ \text{mho}$$

This corresponds to a 10-MΩ resistance. The output resistance is obtained by shorting $E_g$, removing $R_L$, and looking into the collector terminals, in which case

$$R_o = R_4 \| \frac{1}{h_{ob}}$$

$R_o = (2.1 \times 10^3) \| 10^7 \simeq 2.1 \ \text{k}\Omega$

To determine $A_v$ and $A_i$, it may be helpful to assume an arbitrary value for $I_e$, say 10 μA. Since 20 Ω ≪ 2,900 Ω, the current through $R_3$ is much smaller than $I_e$, and the total input current $I_i \simeq I_e = 10$ μA. We can now write

$V_i \simeq I_i h_{ib} = 0.2$ mV
$I_c \simeq h_{fb} I_e$

$$h_{fb} = \frac{-h_{fe}}{h_{fe} + 1} = -0.99$$

$I_c = (-0.99 \times 10)(10^{-6}) = -9.9$ μA

Junction Transistor Small-signal Analysis  261

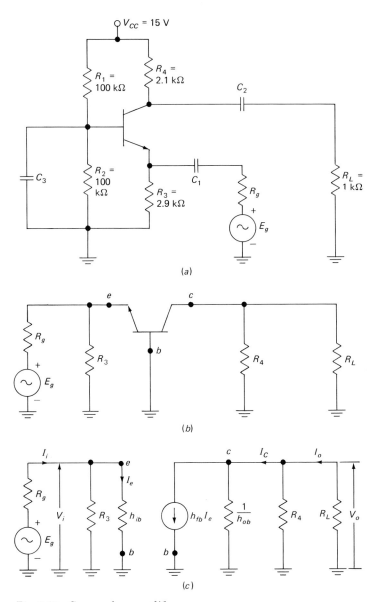

Fig. 7.12  Common base amplifier.

$$I_o = I_c \frac{R_4}{R_4 + R_L}$$

$$I_o = \frac{(-9.9 \times 10^{-6})(2.1 \times 10^3)}{3.1 \times 10^3} = -6.71 \ \mu A$$

$$V_o = -I_o R_L = -(-6.71 \times 10^{-6})(10^3) = 6.71 \text{ mV}$$
$$A_v = \frac{V_o}{V_i} = \frac{6.71}{0.2} = 33.5$$

Note that there is no phase shift between emitter voltage (input) and collector voltage (output); therefore $A_v$ is a positive quantity. The current amplification is

$$A_i = \frac{I_o}{I_i} = \frac{-6.71 \times 10^{-6}}{10 \times 10^{-6}} = -0.671$$

**Example 7.3**

The circuit of Example 7.1 is to be operated in the common collector configuration. Determine $R_i$, $R_o$, $A_i$, and $A_v$.

**Solution**

In a common collector stage, the input signal is coupled to the base, while the output is taken at the emitter. In each case, capacitors may be used to couple the signal generator and the load, as shown in the circuit of Fig. 7.13a.
  The collector may be AC-grounded by connecting a capacitor from collector to ground ($C_3$ in the figure) or by eliminating $R_4$ and tying the collector directly to the DC supply. In the latter case, $R_3$ would have to be increased to maintain the same operating point. No capacitor would be necessary, however, since the DC supply offers a path of negligible impedance to ground. The notation $e'$ in Fig. 7.13c identifies the emitter terminal as seen from the base. Using this approach, it is possible to determine $R_i$ quite readily; first we calculate the resistance looking into base:

$$R_b = h_{ie} + (R_3 \| R_L)(h_{fe} + 1) \simeq 2{,}000 + 75{,}000 = 77 \text{ k}\Omega$$

The input resistance is then

$$R_i = R_1 \| R_2 \| R_b$$
$$R_i = (50 \times 10^3) \| (77 \times 10^3)$$
$$R_i \simeq 30.3 \text{ k}\Omega$$

  $A_v$ is the ratio of $V_o$ to $V_i$. $V_i$ is the voltage from base to ground, while $V_o$ is the voltage between point $e'$ and ground. The ratio $V_o/V_i$ is therefore equal to the voltage divider action of the 75- and 2-k$\Omega$ resistances:

$$A_v = \frac{V_o}{V_i} = \frac{V_{e'}}{V_b} = \frac{75 \times 10^3}{77 \times 10^3} = 0.975$$

A value of 0.975 for $A_v$ is typical of common collector stages. $A_v$ is always less than 1 and positive (base and emitter voltages are in phase).

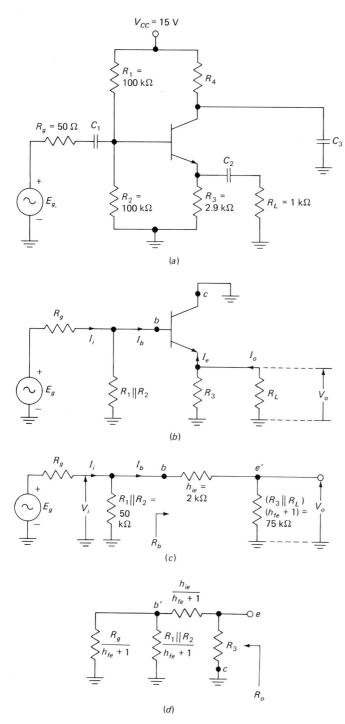

Fig. 7.13  Common collector amplifier.

To determine $A_i$, we will assume $I_i = 10~\mu A$. $I_i$ divides between the two available parallel paths (50- and 77-k$\Omega$ resistances in Fig. 7.13c), resulting in

$$I_b = I_i \frac{50 \times 10^3}{(50 \times 10^3) + (77 \times 10^3)} = 3.94~\mu A$$

For the assumed current directions of Fig. 7.13b, the emitter current is

$$I_e = -I_b(h_{fe} + 1) = -(3.94 \times 10^{-6})(101)$$
$$I_e = -398~\mu A$$

$I_o$ is that portion of $I_e$ that flows through $R_L$, as shown in Fig. 7.13b:

$$I_o = I_e \frac{R_3}{R_3 + R_L} = -398 \times 10^{-6} \frac{2.9 \times 10^3}{3.9 \times 10^3} = -296~\mu A$$

$$A_i = \frac{I_o}{I_i} = \frac{-296 \times 10^{-6}}{10 \times 10^{-6}} = -29.6$$

$R_o$ is estimated using the equivalent circuit of Fig. 7.13d. By removing $R_L$ and shorting the signal generator, the circuit seen from the emitter terminals is as shown in this figure. Note that all resistances in the base circuit appear divided by $h_{fe} + 1$. This yields

$$R_o = R_3 \left\| \left[ \frac{h_{ie}}{h_{fe} + 1} + \left( \frac{R_g}{h_{fe} + 1} \middle\| \frac{R_1 \| R_2}{h_{fe} + 1} \right) \right] \right.$$
$$R_o = (2.9 \times 10^3) \| 20.5 \simeq 20~\Omega$$

## Problem Set 7.1

7.1.1. Indicate the transistor configuration that exhibits the following properties:

    a. Highest input resistance
    b. Lowest input resistance
    c. Highest output resistance
    d. Lowest output resistance
    e. Highest voltage amplification
    f. Lowest voltage amplification
    g. Highest current amplification
    h. Lowest current amplification
    i. Highest power gain
    j. Lowest power gain
    k. Poorest output-input isolation
    l. 180° phase shift between input and output voltage

7.1.2. For which transistor configuration is $R_i$
   a. Least dependent on $R_L$?
   b. Most dependent on $R_L$?

7.1.3. Which transistor configuration has the best input-output isolation?

For all circuits in Fig. 7.14, assume transistor parameters as follows: $h_{fe} = 49$, $h_{oe} = 6 \times 10^{-6}$ mho, $h_{ib} = 25\ \Omega$, $h_{re}$ is negligible; all capacitors behave as short circuits at the frequency of operation. Estimate $A_i$, $A_v$, $R_i$, and $R_o$ for the circuit of

7.1.4. Fig. 7.14a
7.1.5. Fig. 7.14b
7.1.6. Fig. 7.14c
7.1.7. Fig. 7.14d
7.1.8. Fig. 7.14e
7.1.9. Fig. 7.14f
7.1.10. Fig. 7.14g
7.1.11. Fig. 7.14h

## 7.4 TRANSISTOR AMPLIFIER WITH UNBYPASSED EMITTER RESISTANCE

It is sometimes desirable to leave all or part of the emitter resistance in a transistor amplifier unbypassed, as indicated in the circuit diagram of Fig. 7.15a. This has no effect on the DC operating conditions but may produce certain improvements in amplifier performance. These improvements are due to negative feedback, a topic to be discussed more thoroughly in Chap. 10. For the circuit of Fig. 7.15a, the frequency response may be extended and the input resistance increased. The price we pay is a reduction in voltage amplification. In this section we discuss the manner in which $R_i$ and other performance measures are affected.

A first approximation to the input equivalent circuit of our amplifier is shown in Fig. 7.15b. The effective value of $R_E$, as seen from the base, is $R_E(h_{fe} + 1)$. This resistance is in series with $h_{ie}$. If $h_{ie} = 2,000\ \Omega$ and $h_{fe} = 50$, $R_i$, for any value of $R_E$, is

$$R_i = (50 \times 10^3) \| (2{,}000 + 51 R_E)$$

To see the dependence of $R_i$ on $R_E$, $R_i$ is given below for different values of $R_E$:

a.   $R_E = 10\ \Omega$
     $R_i = (50 \times 10^3) \| [2{,}000 + 10(51)]$
     $R_i = 2.4\ \text{k}\Omega$

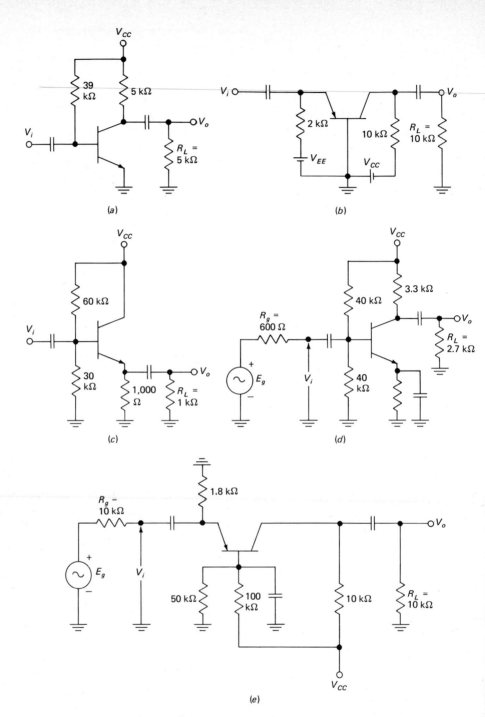

Fig. 7.14 Circuits for Problem Set 7.1.

Junction Transistor Small-signal Analysis 267

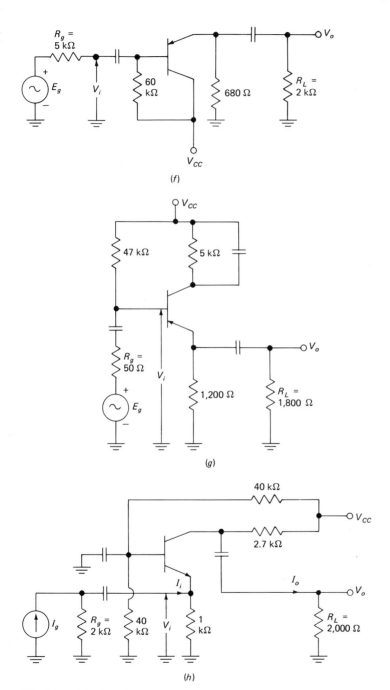

Fig. 7.14 (Continued)

268  Solid State Electronic Circuits

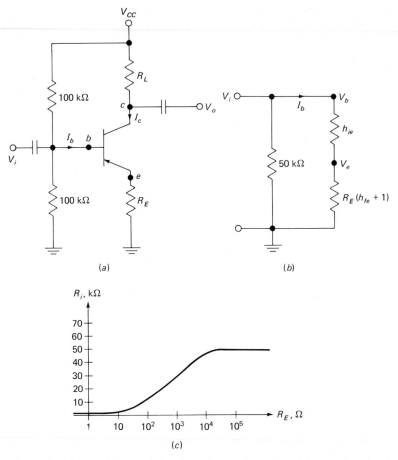

Fig. 7.15 (a) Transistor amplifier with unbypassed emitter resistance; (b) equivalent input circuit for (a); (c) $R_i$ for the circuit of (a) as a function of $R_E$.

b.   $R_E = 100\ \Omega$
    $R_i = (50 \times 10^3) \| [2{,}000 + 100(51)]$
    $R_i = 6.2\ \text{k}\Omega$

c.   $R_E = 1{,}000\ \Omega$
    $R_i = (50 \times 10^3) \| [2{,}000 + 1{,}000(51)]$
    $R_i = 25.7\ \text{k}\Omega$

d.   $R_E = 10{,}000\ \Omega$
    $R_i = (50 \times 10^3) \| [2{,}000 + 10{,}000(51)]$
    $R_i = 45.5\ \text{k}\Omega$

e.   $R_E = 100{,}000\ \Omega$
    $R_i = (50 \times 10^3) \| [2{,}000 + 100{,}000(51)]$
    $R_i = 49.5\ \text{k}\Omega$

The above data are used to construct the graph of Fig. 7.15c. Note that there is a limit of around 50 kΩ to the maximum input resistance that can be achieved. This limit is imposed by the biasing resistors that shunt the input circuit. In practice, it is unlikely that this limit could be approached, because large values of $R_E$ require large DC voltage drops. For example, if the transistor is biased with an emitter current of 1 mA and $R_E = 10$ kΩ, the DC voltage drop across $R_E$ is 10 V. The DC source must supply this in addition to voltage for $R_L$ and $V_{CE}$. Obviously, there is no point in making $R_E$ very large, first because of the large DC voltage required and then because the biasing resistors limit $R_i$ anyway.

The value of $A_v$ for this circuit can be simply obtained. The input voltage is

$$V_i = I_b[h_{ie} + R_E(h_{fe} + 1)]$$

The output voltage, if $R_L \gg 1/h_{oe}$, is

$$V_o \simeq -I_c R_L$$

The voltage amplification is

$$A_v = \frac{V_o}{V_i} = \frac{-I_c R_L}{I_b[h_{ie} + R_E(h_{fe} + 1)]}$$

Usually $R_E(h_{fe} + 1) \gg h_{ie}$, $h_{fe} \gg 1$, and $I_c = h_{fe} I_b$; therefore we can write

$$A_v \simeq \frac{-h_{fe} I_b R_L}{I_b R_E h_{fe}}$$

$$A_v \simeq -\frac{R_L}{R_E} \tag{7.3}$$

Note that $A_v$ here is independent of transistor parameters; it is also less than could be obtained if $R_E$ were bypassed. The reduction in $A_v$, as indicated earlier, is the price we pay for the higher input resistance and improved frequency response. On the other hand, the fact that $A_v$ is independent of transistor parameters such as $h_{fe}$ and $h_{ie}$ may be a definite advantage. It means that, as long as the criteria on which Eq. (7.3) is based are satisfied, $A_v$ is determined entirely by external resistors $R_E$ and $R_L$. If the Q point should change, or the transistor is replaced, $A_v$ remains the same.

**Example 7.4**

Estimate $A_v$ and $R_i$ in the circuit of Fig. 7.16 for each value of $R_5$ given below. $1/h_{oe}$ is large compared with the load seen by the transistor. All capacitors have negligible reactance at the test frequency.

a. $R_5 = 0$
b. $R_5 = 1{,}000$ Ω

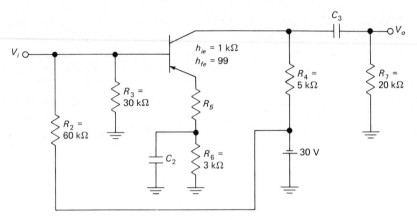

Fig. 7.16  Circuit for Example 7.4.

## Solution

a. The AC equivalent circuit is shown in Fig. 7.17a. The parallel combination of $R_4$ and $R_7$ yields $R_L = 4$ k$\Omega$; the parallel combination of $R_2$ and $R_3$ is 20 k$\Omega$. The voltage across $h_{ie}$ is $V_i$, yielding

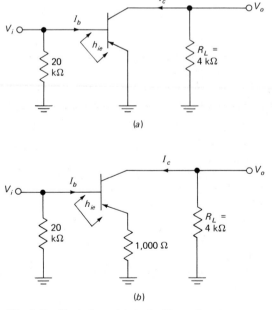

Fig. 7.17  Equivalent circuits for Example 7.4.

$$I_b = \frac{V_i}{h_{ie}}$$

$$I_c = h_{fe}I_b = \frac{h_{fe}V_i}{h_{ie}}$$

$$V_o = -I_cR_L = \frac{-h_{fe}V_iR_L}{h_{ie}}$$

$$\frac{V_o}{V_i} = \frac{-h_{fe}R_L}{h_{ie}} = \frac{-99(4 \times 10^3)}{10^3}$$

$$A_v \simeq -400$$

The input resistance is the parallel combination of 20 k$\Omega$ and $h_{ie}$:

$$R_i = (20 \times 10^3) \| 10^3$$
$$R_i \simeq 950 \; \Omega$$

b. The AC equivalent circuit is shown in Fig. 7.17b. The reflected value of emitter resistance is 1,000(100) = 100 k$\Omega$; this is in series with $h_{ie}$, yielding

$$R_i = (20 \times 10^3) \| (101 \times 10^3) = 16.7 \text{ k}\Omega$$

$$A_v \simeq \frac{-R_L}{R_E} = \frac{-4,000}{1,000} = -4$$

Note that, with an unbypassed resistance in the emitter, $A_v$ is reduced by a factor of 100. At the same time, the resistance looking into the base increases by about 100. $R_i$, however, increases only by a factor of about 17 because of the shunting effect of $R_2$ and $R_3$.

The first approximation of Fig. 7.15b neglected to take into account the effect of output resistance on $R_i$. For an accurate analysis, the output resistance must be considered, especially when the desired value of $R_i$ is very high.

The simplest approach is to assume an effective resistance $r_c$ between base and collector, as in the circuit of Fig. 7.18a. The value of $r_c$ is equivalent to $1/h_{ob}$, so if we know any of the output resistance h parameters, $r_c$ can be quickly calculated.

In common base circuits, $r_c$ appears only in the output (collector to base) and does not influence $R_i$. The common base configuration therefore has good isolation. In common emitter circuits, however, $r_c$ appears between input (base) and output (collector) terminals, therefore affecting both $R_i$ and $R_o$. Although this effect is indirectly taken into account in the various equations developed earlier, we will now treat it from a slightly different point of view.

Consider the diagram of Fig. 7.18b. The amplifier symbol represents voltage amplification from base to collector terminals. $I_1$ is the total current supplied by the generator $V_1$; it divides between the two available paths, $I_2$ through $r_c$ and $I_3$ to the amplifier input. The generator sees two equivalent

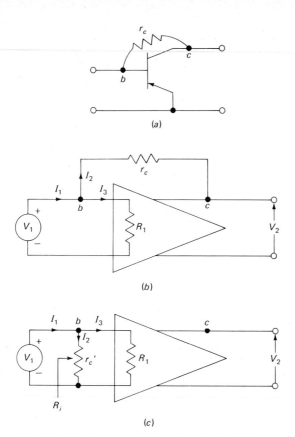

Fig. 7.18  Development of Miller effect relationship for $r_c$.

resistances: the input resistance ($R_1$) to the amplifier proper which draws $I_3$ and a resistance due to $r_c$ which draws $I_2$. This last component is $r'_c$, as shown in Fig. 7.18c. Its voltage and current are $V_1$ and $I_2$, respectively; therefore

$$r'_c = \frac{V_1}{I_2}$$

From the circuit of Fig. 7.18b we can write

$$I_2 = \frac{V_1 - V_2}{r_c}$$

Therefore

$$r'_c = \frac{V_1}{(V_1 - V_2)/r_c} = \frac{r_c}{1 - (V_2/V_1)}$$

The ratio $V_2/V_1$ is the voltage amplification from base to collector and can be replaced with $A_v$:

$$r'_c = \frac{r_c}{1 - A_v} \tag{7.4}$$

The loading effect on the input due to an impedance between input and output terminals is called the Miller effect. As indicated in Fig. 7.18c, the total resistance $R_i$ seen by $V_1$ is the parallel combination of $r'_c$ and $R_1$.

**Example 7.5**

Estimate $R_i$ for the circuit of Fig. 7.19a. The appropriate transistor parameters are $h_{fe} = 49$, $h_{ie} = 1$ k$\Omega$, $h_{oe} = 50$ $\mu$mho.

**Solution**

The input resistance is made up of three components as shown in Fig. 7.19b. The 20 k$\Omega$ is the parallel combination of the 30- and 60-k$\Omega$ biasing resistors. The 61 k$\Omega$ represents $h_{ie} + R_E(h_{fe} + 1)$, while $r'_c$ is the Miller effect equivalent of $r_c$. Since $h_{oe}$ is given, $r_c$ may be calculated as follows:

$$r_c = \frac{1}{h_{ob}}$$

$$h_{ob} = \frac{h_{oe}}{h_{fe} + 1} = \frac{50 \times 10^{-6}}{50} = 1 \ \mu\text{mho}$$

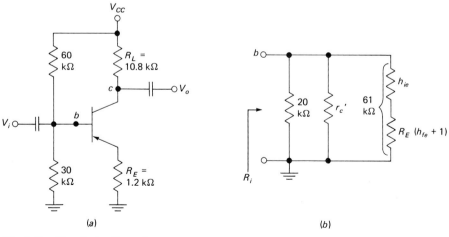

Fig. 7.19  Circuits for Example 7.5.

$$r_c = \frac{1}{10^{-6}} = 1 \text{ M}\Omega$$

The voltage amplification is

$$A_v \simeq -\frac{R_L}{R_E} = -\frac{10.8 \times 10^3}{1.2 \times 10^3} = -9$$

Therefore

$$r'_c = \frac{r_c}{1 - A_v} = \frac{10^6}{1 + 9} = 100 \text{ k}\Omega$$

The total input resistance is

$$R_i = (20 \times 10^3) \| (100 \times 10^3) \| (61 \times 10^3)$$
$$R_i \simeq 13 \text{ k}\Omega$$

The input loading due to $r_c$ can be minimized by reducing $A_v$. For example, the collector load resistor may be AC-bypassed or eliminated completely, as in Fig. 7.20a. With the collector at AC ground, $r'_c = r_c$. Since no AC voltage can be developed at the collector, the amplifier's output is taken at the emitter. This, of course, changes the circuit to a common collector configuration. The signal voltage amplification is slightly less than 1:

$$\frac{V_o}{V_i} = \frac{V_e}{V_b} \simeq 1$$

as it is for any common collector circuit. The voltage amplification between the two points connecting $r_c$ (base and collector) is

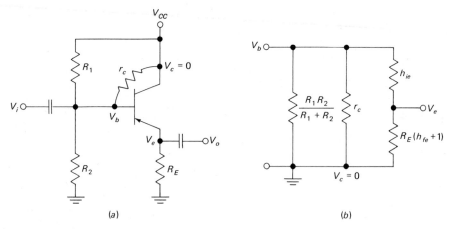

Fig. 7.20  Common collector circuit to illustrate effect of $r_c$ on $R_i$.

$$\frac{V_c}{V_b} = 0$$

making

$$r'_c = \frac{r_c}{1-0} = r_c$$

as stated above.

**Example 7.6**

Estimate $R_i$ and $A_v$ for the circuit of Fig. 7.21a. The following parameters apply to both transistors: $h_{ie} = 1$ k$\Omega$, $h_{ob} = 10^{-5}$ mho, $h_{fb} = -0.98$.

**Solution**

First we determine $r_c$ and $h_{fe}$:

$$r_c = \frac{1}{h_{ob}} = 100 \text{ k}\Omega$$

$$h_{fe} = \frac{-h_{fb}}{1+h_{fb}} = \frac{0.98}{0.02} = 49$$

$R_{L1}$ is the AC load seen by the collector of $Q_1$ as shown in Fig. 7.21b; here we can write

$$R_{L1} = R_3 \| R_4 \| R_5 \| r_c \| [h_{ie} + (h_{fe} + 1)(R_{E2})]$$
$$R_{L1} = (5 \times 10^3) \| (200 \times 10^3) \| (200 \times 10^3) \| (100 \times 10^3) \| (101 \times 10^3)$$
$$R_{L1} \simeq 4.4 \text{ k}\Omega$$

The voltage amplification for the first stage is

$$A_{v1} \simeq \frac{-R_{L1}}{R_{E1}} = -\frac{4.4 \times 10^3}{1.5 \times 10^3}$$
$$A_{v1} \simeq -2.9$$

The second stage is an emitter follower; therefore $A_{v2} \simeq 1$ and the overall $A_v$ is $-2.9$.

The equivalent circuit of Fig. 7.21c is used to estimate $R_i$. In addition to $R_1$ and $R_2$, the Miller effect equivalent of $r_c$ also shunts the input:

$$r'_c = \frac{r_c}{1 - A_{v1}} = \frac{100 \times 10^3}{1 + 2.9}$$
$$r'_c = 25.6 \text{ k}\Omega$$

The series combination of $h_{ie}$ and $R_{E1}(h_{fe} + 1)$ is 76 k$\Omega$. The total value of $R_i$ is therefore

276  Solid State Electronic Circuits

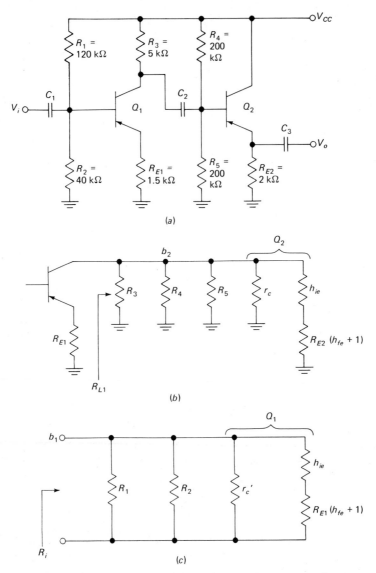

Fig. 7.21  (a) Circuit for Example 7.6; (b) equivalent circuit to determine $R_{L1}$; (c) equivalent circuit to determine $R_i$.

$R_i = R_1 \| R_2 \| r'_c \| [h_{ie} + R_{E1}(h_{fe} + 1)]$
$R_i = (40 \times 10^3) \| (120 \times 10^3) \| (25.6 \times 10^3) \| (76 \times 10^3)$
$R_i \simeq 11.7 \text{ k}\Omega$

The high-input impedance techniques discussed so far have obvious limitations. Although an unbypassed emitter resistance can raise the resistance looking into the base by an amount equal to $R_E(h_{fe} + 1)$, we must emphasize that

a. Very large values of $R_E$ are impractical because of the correspondingly large DC voltage drops required.
b. Biasing resistors shunt the input circuit, thereby limiting $R_i$.
c. The Miller effect resistance $r'_c$ also shunts the input circuit, again reducing $R_i$.

To minimize the shunting effects of (b), inductances in series with the biasing resistors could be used. If ideal, the coils would have no DC resistance and high AC reactance, thus increasing $R_i$. In practice, however, the values of inductance required would be impractically large, especially at low frequencies. Other techniques are therefore used to minimize the shunting effects of biasing resistors.

One such technique, called bootstrapping, is explained using the circuit of Fig. 7.22. In Fig. 7.22a, generator $V$ is directly across $R$; the resistance seen by the generator is

$$R_i = \frac{V}{I}$$
$$R_i = R \tag{a}$$

Let us now assume that the bottom end of $R$ is lifted above ground by $kV$ volts, where $k$ is a constant, as indicated in Fig. 7.22b. The generator $V$ now sees a different $R_i$ than in Fig. 7.22a, as follows:

$$R'_i = \frac{V}{I'} \tag{b}$$

where $I' = (V - kV)/R$. Therefore

$$R'_i = \frac{V}{(V - kV)/R}$$
$$R'_i = \frac{R}{1 - k} \tag{7.5}$$

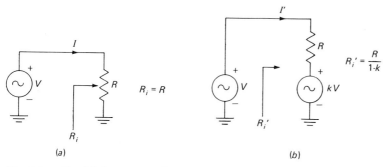

(a)    (b)

Fig. 7.22 Simplified circuits to illustrate bootstrapping action.

By tying the bottom end of $R$ to the voltage source $kV$, the voltage across $R$ is reduced from $V$ to $V - kV$; the current through $R$ is also reduced from $V/R$ to $(V - kV)/R$. The current through $R$ is also the current through the generator $V$; therefore $V$ sees $R'_i$, which is larger than $R_i$. Equation (7.5) gives $R'_i$ as a function of $k$. If $k = 1$, we have perfect bootstrapping, and $R'_i$ is infinite. This is so because when $k = 1$, both the top and bottom of $R$ are at the same potential, and the current is zero. Since the generator $V$ supplies no current, it sees an infinite resistance.

A circuit that employs bootstrapping to raise input resistance is shown in Fig. 7.23a. The transistor's $h_{fe} = 49$, $h_{ie} = 2,000$ $\Omega$, and $r_c = 100$ k$\Omega$. The DC equivalent circuit is shown in Fig. 7.23b; it is a standard biasing circuit except for $R_5$. This resistance develops a small DC voltage drop, easily compensated by an appropriate selection of the other biasing resistors.

The AC operation is explained using Fig. 7.23c. Capacitor $C$ in Fig. 7.23a behaves as a short circuit at the test frequency; therefore $R_1$, $R_2$, and $R_3$ are in parallel, yielding an equivalent $R_x = (60 \times 10^3) \| (30 \times 10^3) \| (5 \times 10^3) = 4$ k$\Omega$. Note that $R_5$ is tied to the emitter, whose voltage $V_e$, relative to ground, may be simulated with a generator, as shown in Fig. 7.23d. $V_e$ is in phase and approximately equal to $V_b$; therefore the voltage across $R_5$ is about zero and so is the current. The less current flows through $R_5$, the less the biasing resistors affect $R_i$.

To determine the shunting effect of $R_5$, we use the fact that current through $R_5$ is usually small compared to emitter current; therefore $h_{ie}$ and $R_x(h_{fe} + 1)$ are approximately in series, as shown in Fig. 7.23e. This allows us to write

$$V_e = V_b \frac{R_x(h_{fe} + 1)}{R_x(h_{fe} + 1) + h_{ie}}$$

$$V_e = V_b \frac{(4 \times 10^3)(50)}{(4 \times 10^3)(50) + 2,000} = 0.99 V_b$$

$$R'_5 = \frac{V_b}{I}$$

Where $I$ is the current through $R_5$ in Fig. 7.23d:

$$I = \frac{V_b - V_e}{R_5}$$

$$R'_5 = \frac{V_b}{(V_b - V_e)/R_5}$$

$$R'_5 = \frac{V_b R_5}{V_b - 0.99 V_b} = \frac{R_5}{1 - 0.99}$$

$$R'_5 = \frac{10^4}{0.01} = 1 \text{ M}\Omega$$

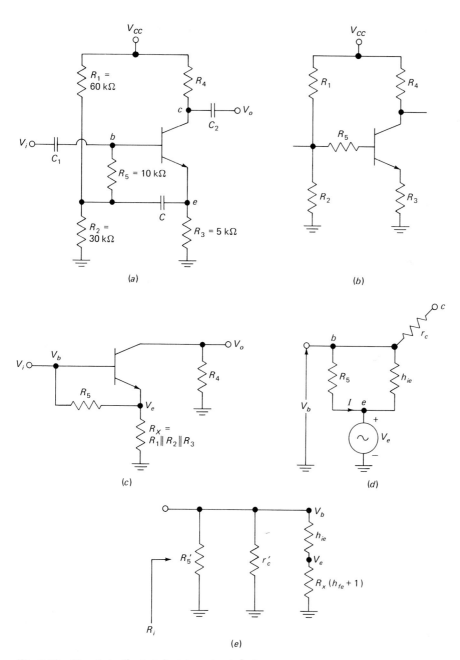

Fig. 7.23  Circuits to illustrate bootstrapping technique.

Through bootstrapping we have succeeded in making 10 kΩ appear as 1 MΩ. For a complete picture, however, the effects of $r_c$ must be taken into account. In Fig. 7.23e, $r'_c$ (the Miller effect equivalent of $r_c$) also shunts the input; $r'_c$ can be sufficiently low to prevent us from achieving a high input resistance.

The circuit of Fig. 7.24a involves bootstrapping the biasing resistors as well as the collector. This way it is possible to reduce the shunting effects of $r_c$ and the biasing resistors, yielding a relatively high input resistance. Bootstrapping of $r_{c1}$ is accomplished by tying the collector of $Q_1$, through capacitor $C_2$, to the emitter of $Q_2$. The voltage at the emitter of $Q_2$ is almost the same as the voltage at the base of $Q_1$; therefore the current through $r_{c1}$ is quite small.

A more complete analysis is possible with the equivalent circuit of Fig. 7.24b. Here $V_{e2}$ can be calculated in terms of $V_{e1}$. Note that because of $C_1$ in Fig. 7.24a, both $R_1$ and $R_2$ are in parallel with $R_5$; also, because of $C_2$, $R_4$ is in parallel with $R_5$. The net resistance between the emitter of $Q_2$ and ground is therefore

$$R_{e2} = R_2 \| R_1 \| R_5 \| R_4$$
$$R_{e2} = (60 \times 10^3) \| (120 \times 10^3) \| (20 \times 10^3) \| (20 \times 10^3) = 8 \text{ k}\Omega$$

When seen from the base of $Q_2$ (also the emitter of $Q_1$), $R_{e2}$ becomes

$$R'_{e2} = R_{e2}(h_{fe2} + 1) = (8 \times 10^3)(100) = 800 \text{ k}\Omega$$
$$V_{e2} = V_{e1} \frac{R'_{e2}}{h_{ie2} + R'_{e2}} = V_{e1} \frac{800 \times 10^3}{801 \times 10^3} \simeq 0.999 V_{e1}$$

We now calculate $R_{e1}$, the total resistance from the emitter of $Q_1$ to ground:

$$R_{e1} = r_{c2} \| (h_{ie2} + R'_{e2})$$
$$R_{e1} = 10^5 \| 801 \times 10^3 \simeq 90 \text{ k}\Omega$$

The voltage at the emitter of $Q_1$ is determined in terms of $V_{b1}$ using the circuit of Fig. 7.24c:

$$V_{e1} = V_{b1} \frac{R_{e1}(h_{fe1} + 1)}{R_{e1}(h_{fe1} + 1) + h_{ie1}}$$
$$V_{e1} = V_{b1} \frac{(90 \times 10^3)(100)}{(90 \times 10^3)(100) + 1{,}000}$$
$$V_{e1} \simeq 0.999 V_{b1}$$

Since $V_{e2} \simeq 0.999 V_{e1} = (0.999)(0.999)(V_{b1})$, it follows that $V_{e2} \simeq 0.998 V_{b1}$. This is used in the circuit of Fig. 7.24d to determine $R_i$:

$$R_i = R'_3 \| r'_{c1} \| [h_{ie1} + R_{e1}(h_{fe1} + 1)]$$

where $R'_3$ and $r'_{c1}$ refer to the equivalent value of $R_3$ and $r_{c1}$, respectively, as seen from the input:

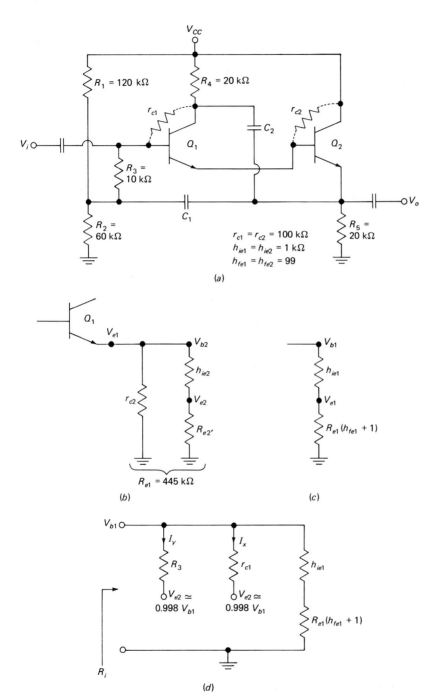

Fig. 7.24 Collector bootstrapped circuit.

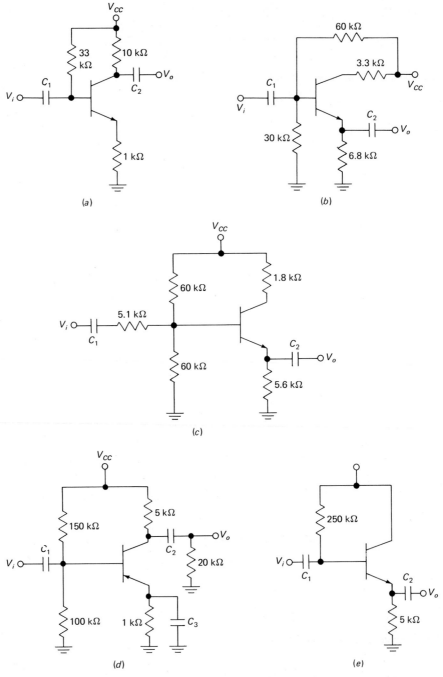

Fig. 7.25 Circuits for Problem Set 7.2. All $r_c = 400$ k$\Omega$, all $h_{ie} = 1500$ $\Omega$, and all $h_{fe} = 49$.

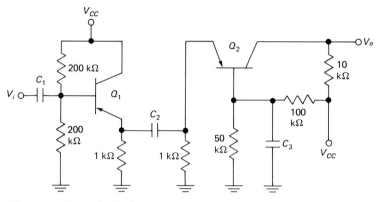

Fig. 7.26  Circuit for Prob. 7.2.5.

$$R_3' = \frac{V_{b1}}{I_y} = \frac{V_{b1}}{(V_{b1} - 0.998 V_{b1})/R_3}$$

$$R_3' = \frac{R_3}{1 - 0.998} = \frac{10^4}{0.002} = 5 \text{ M}\Omega$$

$$r_{c1}' = \frac{V_{b1}}{I_x} = \frac{V_{b1}}{(V_{b1} - 0.998 V_{b1})/r_{c1}}$$

$$r_{c1}' = \frac{r_{c1}}{1 - 0.998} = \frac{100 \times 10^3}{0.002} = 50 \text{ M}\Omega$$

The net input resistance is

$$R_i = (5 \times 10^6) \| (50 \times 10^6) \| [1{,}000 + (90 \times 10^3)(100)]$$
$$R_i \simeq 3 \text{ M}\Omega$$

## Problem Set 7.2

7.2.1. Estimate $A_v$ in the circuits of Fig. 7.25.
7.2.2. Estimate $R_i$ for each circuit in Fig. 7.25.
7.2.3. Estimate $R_o$ for the circuit of Fig. 7.25c.

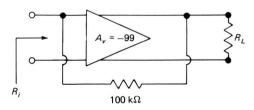

Fig. 7.27  Diagram for Prob. 7.2.6.

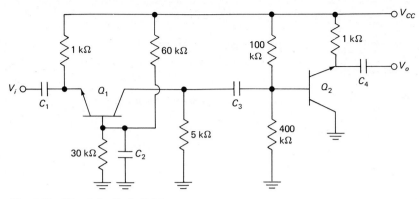

Fig. 7.28 Circuit for Prob. 7.2.7.

7.2.4. For an amplifier with an unbypassed emitter resistor, what is the approximate magnitude and phase relationship between base and emitter voltage?

7.2.5. Estimate $A_v$ for the circuit of Fig. 7.26. All $h_{ie} = 2$ k$\Omega$, all $h_{fe} = 49$.

7.2.6. Determine $R_i$ for the amplifier of Fig. 7.27.

7.2.7. Estimate $A_v$ and $R_i$ for the circuit of Fig. 7.28. All $h_{ie} = 2$ k$\Omega$, all $h_{fe} = 49$.

7.2.8. Estimate $R_i$ for the circuit of Fig. 7.29.

7.2.9. What does bootstrapping refer to?

7.2.10. Estimate $R_i$ for the circuit of Fig. 7.30.

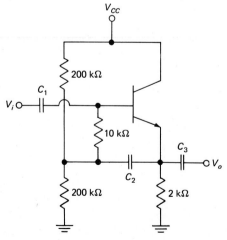

Fig. 7.29 Circuit for Prob. 7.2.8; $h_{ie} = 2$ k$\Omega$, $h_{fe} = 99$, $r_c = 200$ k$\Omega$.

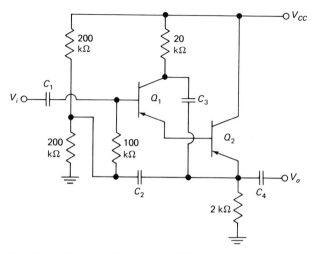

**Fig. 7.30** *Circuit for Prob. 7.2.10; all* $h_{ie} = 2\ k\Omega$, *all* $h_{fe} = 99$, *all* $r_c = 200\ k\Omega$.

**7.2.11.** Determine $A_v$ and $R_i$ for the circuit of Fig. 7.31a.
**7.2.12.** Estimate $A_v$ in the circuit of Fig. 7.31b.

## 7.5 LOW-FREQUENCY RESPONSE

A study of the low-frequency response of amplifiers is closely related to a study of coupling techniques. The components that couple an amplifying stage to its source and load determine, to a large extent, the low-frequency response.[1]

Consider the capacitor-coupled amplifier of Fig. 7.32a. We will determine $A_v$ at mid and low frequencies, using the hybrid-parameter model shown in Fig. 7.32b. To simplify the analysis, $h_{re}$ is assumed to be negligible; this allows us to treat input and output circuits independently. The remaining parameters are $h_{ie} = 4\ k\Omega$, $h_{fe} = 50$, $h_{oe} = 10^{-5}$ mho.

At midfrequencies, all capacitors behave as AC short circuits. If we let $R'_L = R_C \| R_L$, then $R'_L = 4\ k\Omega$. This is much smaller than $1/h_{oe}(= 10^5\ k\Omega)$; therefore the generator current $h_{fe}I_b$ flows almost entirely through $R'_L$, yielding

$$V_o \simeq -h_{fe}I_b R'_L = -(150 I_b)(4{,}000) = (-6 \times 10^5)(I_b)$$
$$V_i = I_b h_{ie} = 4{,}000 I_b$$
$$A_v = \frac{V_o}{V_i} = -\frac{(6 \times 10^5)(I_b)}{(4 \times 10^3)(I_b)} = -150$$

[1] See Sec. 4.5a for a discussion of the effects of coupling components on low-frequency response.

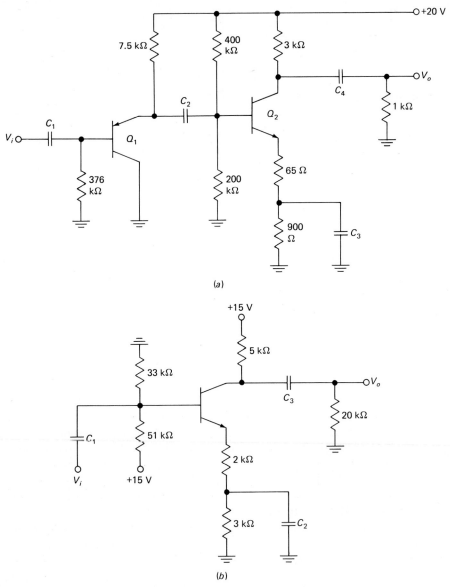

Fig. 7.31 All $h_{fe} = 100$, all $h_{ie} = 1\ k\Omega$, all $h_{re} = h_{oe} = 0$; circuits for (a) Prob. 7.2.11; (b) Prob. 7.2.12.

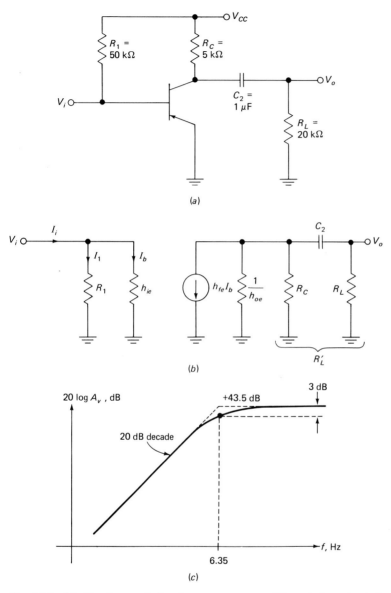

Fig. 7.32 (a) Circuit to study low-frequency response; (b) equivalent circuit for (a); (c) low-frequency response.

and $20 \log |A_v| = 43.5$ dB. The low-frequency cutoff is defined as the frequency for which $|A_v|$ is down 3 dB from its midfrequency value. The break frequency introduced by $C_2$ is computed as follows:[1]

$$f_L = \frac{1}{2\pi C_2 R_{T2}}$$

where $R_{T2}$ is the Thevenin resistance seen by $C_2$:

$$R_{T2} = R_L + R_C = (20 \times 10^3) + (5 \times 10^3) = 25 \text{ k}\Omega$$

$$f_L = \frac{1}{2\pi (1 \times 10^{-6})(25 \times 10^3)} = 6.35 \text{ Hz}$$

The low-frequency response graph is shown in Fig. 7.32c. $A_v$ is 43.5 dB at midfrequencies, then falls off at 20 dB per decade below 6.35 Hz. The response is down 3 dB from the midfrequency value at $f_L = 6.35$ Hz.

**Example 7.7**

For the circuit in Fig. 7.33a, determine

a. $20 \log |A_v|$ at midfrequencies
b. The low-frequency response of $A_v$

**Solution**

a. At midfrequencies, the circuit of Fig. 7.33b can be used to estimate $A_{v1}$. The second-stage amplification is close to unity ($A_{v2} \simeq 1$); hence the overall $A_v \simeq A_{v1}$. The load seen by the first stage is

$$R_{L1} = R_4 \| R_5 \| R_6 \| [h_{ie2} + R_7(h_{fe} + 1)]$$
$$R_{L1} = (10 \times 10^3) \| (40 \times 10^3) \| (40 \times 10^3) \| (20 \times 10^3) = 5 \text{ k}\Omega$$

We can now write

$$V_i = h_{ie} I_b$$
$$V_i = (2.5 \times 10^3)(I_b)$$
$$V_o = -h_{fe} I_b R_{L1}$$
$$V_o = (-34 I_b)(5 \times 10^3) = (-170 \times 10^3)(I_b)$$
$$A_v = \frac{V_o}{V_i} = -\frac{(170 \times 10^3)(I_b)}{(2.5 \times 10^3)(I_b)} = -68$$
$$20 \log |A_v| = 20 \log 68 = 36.6 \text{ dB}$$

[1] See Sec. 4.5b for a discussion of break frequencies.

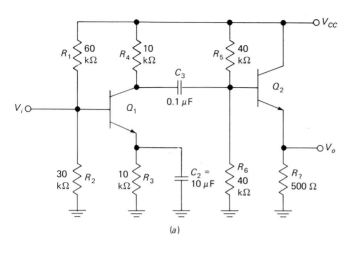

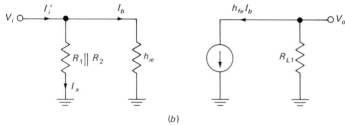

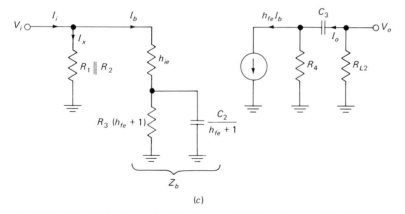

Fig. 7.33 Circuits for Example 7.7; all $h_{ie} = 2{,}500\ \Omega$, all $h_{fe} = 34$, all $h_{oe} = h_{re} = 0$.

b. The low-frequency response is determined using the circuit in Fig. 7.33c, as follows:

$$V_i = I_b Z_b \qquad (a)$$

where

$$Z_b = h_{ie} + \left[ R_2(h_{fe} + 1) \left\| \frac{1}{j\omega C_2/(h_{fe} + 1)} \right. \right] \qquad (b)$$

Substituting Eq. (b) into Eq. (a), including numerical values, yields

$$V_i = \frac{(1{,}000 I_b)(352.5 + 0.25 j\omega)}{1 + (j\omega/10)} \qquad (c)$$

The output voltage is

$$V_o = -I_o R_{L2}$$

where

$$R_{L2} = R_5 \| R_6 \| [h_{ie} + R_7(h_{fe} + 1)]$$
$$R_{L2} = (40 \times 10^3) \| (40 \times 10^3) \| [2{,}500 + 500(35)]$$
$$R_{L2} = 10 \text{ k}\Omega$$

and

$$I_o = \frac{h_{fe} I_b R_4}{R_4 + 1/(j\omega C_3) + R_{L2}}$$

$$I_o = \frac{(34 I_b)(10^4)}{10^4 + 1/[j\omega(10^{-7})] + 10^4}$$

$$I_o = \frac{I_b(0.034 j\omega)}{1 + (j\omega/500)}$$

$$V_o = -\frac{I_b(340 j\omega)}{1 + (j\omega/500)} \qquad (d)$$

The complete expression for $A_v$ is obtained by taking the ratio of Eq. (d) to Eq. (c):

$$A_v = \frac{V_o}{V_i}$$

$$A_v = -\frac{(I_b)(340 j\omega)(1 + j\omega/10)}{(1 + j\omega/500)(1{,}000 I_b)(352.5 + 0.25 j\omega)}$$

which simplifies to

$$A_v = \frac{(6.05 \times 10^{-3})(jf)(1 + jf/1.59)}{(1 + jf/79.6)(1 + jf/222)}$$

The decibel plot for $20 \log |A_v|$ is shown in Fig. 7.34. The low-frequency 3-dB point is $f_L = 222$ Hz.

The effects of transformer coupling are now discussed. The circuit of Fig. 7.35a shows a voltage source with internal resistance $R_g$ driving a load $R_L$ through a transformer. A model for the transformer is shown in Fig. 7.35b. All components are referred to the primary side. $R_p$ and $R_s$ account for the copper losses in the primary and secondary winding, respectively. Because perfect magnetic coupling is impossible to achieve, there is leakage inductance in the primary ($L_p$) and secondary ($L_s$) circuits. $R$ simulates the resistance due to core losses; $L$ is the magnetizing inductance. There is also shunt capacitance, but at low frequencies its reactance is high and can be neglected. The load, as seen from the primary side, is $(n_1/n_2)^2 R_L$.

A simplified low-frequency transformer model is shown in Fig. 7.35c. Here core losses, copper losses, and leakage inductance have been neglected. The reactance due to leakage inductance is negligible at low frequencies; core losses can also usually be neglected at low frequencies, or they can be lumped with $R_L$. To account for the copper losses, $R_p$ could be lumped with $R_g$ (they are in series) on the primary side, while $R_s$ and $(n_1/n_2)^2 R_L$ could be combined on the output side. The circuit of Fig. 7.35c can therefore be used in any case; what we decide to lump with $R_g$ and $(n_1/n_2)^2 R_L$ depends on the specific element values and the accuracy required.

The low-frequency response of $|V_o/E_g|$ in the circuit of Fig. 7.35c is down 3 dB when the inductor's reactance is equal to its Thevenin resistance:

$$2\pi f_L L = R_g \left\| \left[ R_L \left(\frac{n_1}{n_2}\right)^2 \right] \right.$$

$$f_L = \frac{R_g \| [R_L (n_1/n_2)^2]}{2\pi L} \tag{7.6}$$

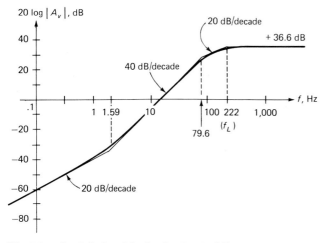

Fig. 7.34  Decibel plot of $A_V$ for the circuit of Fig. 7.33a.

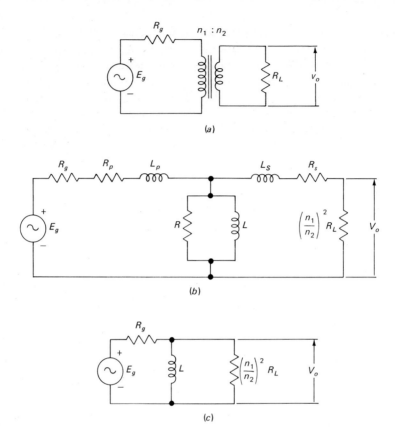

Fig. 7.35 Transformer equivalent circuit.

### Example 7.8

The following data apply to the transformer-coupled amplifier of Fig. 7.36a:

### Transistors

$h_{fe} = 75$, $h_{ie} = 1{,}500\ \Omega$, $h_{oe} = 20\ \mu$mho, $h_{re} \simeq 0$

### Transformers

$T_1$: $\dfrac{n_1}{n_2} = 2$, magnetizing inductance seen from primary: $L_1 = 850$ mH

$T_2$: $\dfrac{n_3}{n_4} = 15$, magnetizing inductance seen from primary: $L_2 = 300$ mH

Determine $f_L$, the frequency where $|V_o/E_g|$ is down 3 dB.

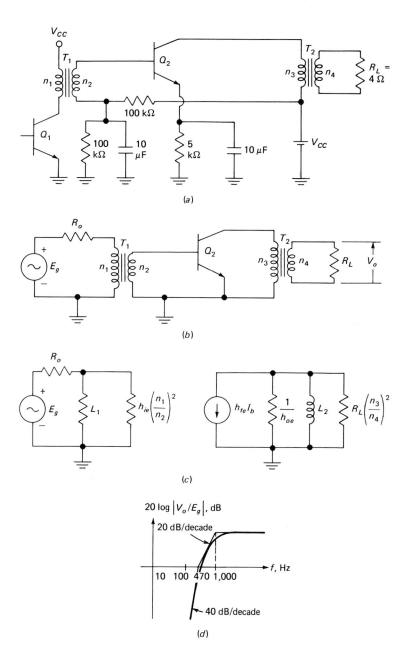

Fig. 7.36 Circuits for Example 7.8.

## Solution

The first step is to draw an AC equivalent circuit, as in Fig. 7.36b. The generator $E_g$ and $R_o$ replace the driving transistor $Q_1$. $R_o$ is $Q_1$'s output resistance; since $h_{re} \simeq 0$, $R_o = 1/h_{oe} = 50$ k$\Omega$. All capacitors are assumed to behave as short circuits near $f_L$; therefore it is the break frequencies produced by the transformers that determine the low-frequency cutoff. This assumption can always be verified later on.

The final model is shown in Fig. 7.36c. The input resistance to $Q_2$ is $h_{ie}$; because of $T_1$, this appears as $h_{ie}(n_1/n_2)^2 = 1{,}500(2^2) = 6$ k$\Omega$ on the primary side of $T_1$.[1] The output equivalent circuit is similarly developed; the load seen by $Q_2$ is $R_L(n_3/n_4)^2 = 4(15^2) = 900$ $\Omega$ shunted by $L_2$. Since $h_{re}$ is negligible, the input and output circuits can be treated independently.

The break frequency due to $T_1$ occurs when the reactance of $L_1$ is equivalent to the Thevenin resistance it sees:

$$2\pi f_1 L_1 = R_o \left\| \left[ h_{ie} \left(\frac{n_1}{n_2}\right)^2 \right] \right.$$

$(2\pi f_1)(0.85) = (50 \times 10^3) \| (6 \times 10^3) = 5.35$ k$\Omega$

$$f_1 = \frac{5{,}350}{5.35} = 100 \text{ Hz}$$

The break frequency due to $T_2$ is similarly determined:

$$2\pi f_2 L_2 = 1/h_{oe} \left\| \left[ R_L \left(\frac{n_3}{n_4}\right)^2 \right] \right.$$

$(2\pi f_2)(0.3) = (50 \times 10^3) \| 900 = 885$ $\Omega$

$$f_2 = \frac{885}{1.88} = 470 \text{ Hz}$$

The low-frequency response for $|V_o/E_g|$ is graphed in Fig. 7.36d; it is down 3 dB at 1,000 Hz and drops from there (as the frequency decreases) at 20 dB per decade until the next break frequency of 470 Hz, where the slope becomes 40 dB per decade. Therefore $f_L = 1{,}000$ Hz. We can now test our earlier assumption that each capacitor's reactance is negligible at $f_L$. At 1,000 Hz, the reactance of either 10-$\mu$F capacitor is $1/[(2\pi)(1{,}000 \times 10^{-5})] = 15.9$ $\Omega$. Since $15.9 \ll 5 \times 10^3$, the 5-k$\Omega$ resistor in the emitter of $Q_2$ is adequately bypassed; the same applies to the 100-k$\Omega$ resistor.

---

[1] We could show everything on the secondary side of $T_1$, where $h_{ie}$ would appear as 1.5 k$\Omega$ and the other components would be divided by $2^2$; this would be equally valid and yield the same results.

## Problem Set 7.3

7.3.1. Plot the mid- and low-frequency value of $20 \log |V_o/E_g|$ for the circuit in Fig. 7.37a. Make reasonable approximations. What is $f_L$?

7.3.2. Estimate $f_L$ in the circuit of Fig. 7.37b.

7.3.3. Sketch the low-frequency phase response of $V_o/E_g$ in the circuit of Fig. 7.37a.

7.3.4. Sketch $20 \log |V_o/E_g|$ for the transformer-coupled circuit of Fig. 7.38. What is $f_L$? Make reasonable approximations.

7.3.5. The interstage coupling transformer in Fig. 7.39 is to be chosen on the basis of maximum power transfer. Determine the required turns ratio and magnetizing inductance for a low-frequency cutoff of 200 Hz.

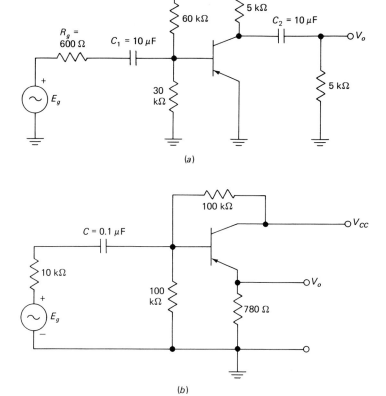

Fig. 7.37 (a) *Circuit for Prob. 7.3.1*; $h_{fe} = 150$, $h_{ie} = 5,000 \ \Omega$, $h_{re} = h_{oe} = 0$; (b) *circuit for Prob. 7.3.2*; $h_{fe} = 49$, $h_{ie} = 1 \ k\Omega$, $h_{oe} = h_{re} = 0$.

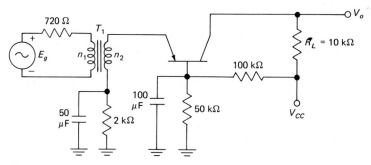

*Fig. 7.38  Circuit for Prob. 7.3.4;* $h_{ie} = 2{,}700\ \Omega$, $h_{fb} = -0.97$, $h_{ob} = h_{rb} = 0$; $n_1/n_2 = 5$; $L = 100\ mH$ *when seen from primary of* $T_1$.

## 7.6  HIGH-FREQUENCY RESPONSE

The hybrid-pi model introduced in Sec. 6.3 (see Fig. 6.12) is used to estimate high-frequency response; before doing so, however, a simplification involving $C_{b'c}$ must be made.

You will recall that, due to the Miller effect, any resistance connected between an amplifier's input and output terminals appears as an equivalent resistance divided by $1 - A_v$ across the input. The Miller effect also applies to reactances or impedances. In the case of a capacitor, dividing its reactance by $1 - A_v$ is equivalent to multiplying the capacitance by $1 - A_v$ (reactance is inversely proportional to capacitance). In the hybrid-pi model of Fig. 7.40, $C_{b'c}$ is therefore multiplied by $1 - A_v$ when seen from the input. $A_v$ here is not necessarily the circuit voltage amplification but the amplification between the two points across which $C_{b'c}$ appears, that is, $b'$ and $c$. This was computed in Chap. 6 as the ratio of $V_o$ to $V_{b'e}$:

$$\frac{V_o}{V_{b'e}} = -g_m R_L \qquad (6.16)$$

The Miller equivalent capacitance is therefore

$$C_m = C_{b'c}(1 + g_m R_L) \qquad (7.7)$$

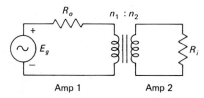

*Fig. 7.39  Circuit for Prob. 7.3.5;* $R_o = 100\ k\Omega$, $R_i = 1\ k\Omega$.

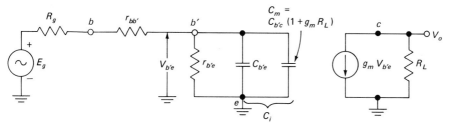

Fig. 7.40 Hybrid-pi model with Miller effect equivalent input capacitance.

The total input capacitance is

$$C_i = C_{b'e} + C_m \tag{7.8}$$

The quantity $|V_o/E_g|$ is down 3 dB at $f_H$, the frequency for which the magnitude of capacitive reactance is equal to the Thevenin resistance:

$$\frac{1}{2\pi f_H C_i} = r_{b'e} \| (r_{bb'} + R_g)$$

$$f_H = \frac{r_{b'e} + r_{bb'} + R_g}{2\pi [C_{b'e} + C_{b'c}(1 + g_m R_L)][r_{b'e}(r_{bb'} + R_g)]} \tag{7.9}$$

For the special case when the circuit is driven from a current source ($R_g = \infty$), $f_H$ simplifies to

$$f_H = \frac{1}{2\pi r_{b'e}[C_{b'e} + C_{b'c}(1 + g_m R_L)]} \tag{7.9a}$$

## Example 7.9

Determine $f_H$ for the voltage transfer ratio $V_o/E_g$ in the circuit of Fig. 7.41a. The following data are available for the transistor, which is biased at $I_C = 5$ mA: $f_T = 750$ MHz; $h_{fe} = 150$; $r_{bb'} = 200\ \Omega$; $C_{b'c} = 2$ pF.

## Solution

The AC equivalent circuit is shown in Fig. 7.41b. The pertinent parameters are as follows:

$$g_m = 40 I_C \tag{6.14}$$
$$g_m = 40(5 \times 10^{-3}) = 0.2 \text{ mho}$$
$$r_{b'e} = \frac{h_{fe}}{g_m} \tag{6.15}$$
$$r_{b'e} = \frac{150}{0.2} = 750\ \Omega$$

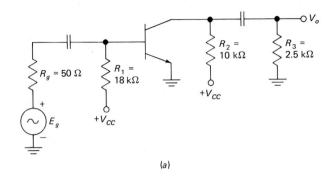

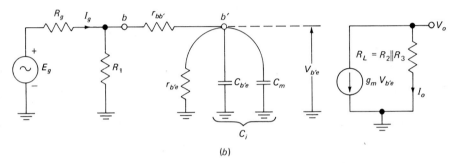

Fig. 7.41  Circuit to study high-frequency response (Example 7.9).

The Miller input capacitance is

$$C_m = C_{b'c}(1 + g_m R_L) \tag{7.7}$$

where

$R_L = R_2 \| R_3 = 10,000 \| 2,500 = 2,000 \ \Omega$
$C_m = (2 \times 10^{-12})[1 + 0.2(2,000)]$
$C_m = 800 \text{ pF}$

$$C_{b'e} = \frac{g_m}{2\pi f_T} \tag{6.19}$$

$$C_{b'e} = \frac{0.2}{6.28(750 \times 10^6)} = 43 \text{ pF}$$

The total input capacitance is

$C_i = C_m + C_{b'e} = 843 \text{ pF}$

The Thevenin resistance seen by $C_i$ is

$R_T = r_{b'e} \| (r_{bb'} + R_1 \| R_g)$
$R_T = 750 \| (200 + 18,000 \| 50)$
$R_T = 750 \| 250 = 187 \ \Omega$

The upper 3-dB frequency occurs when the reactance of $C_i = 187\ \Omega$:

$$\frac{1}{(2\pi)(f_H)(843)(10^{-12})} = 187$$

$$f_H = \frac{10^{12}}{(6.28)(843)(187)} = 1\ \text{MHz}$$

### Problem Set 7.4

7.4.1. Determine the upper 3-dB frequency for $I_o/I_g$ in the circuit of Example 7.9.
7.4.2. Determine $f_H$ in Example 7.9 if $R_g = 10\ \text{k}\Omega$.
7.4.3. Express Eq. (7.9a) in terms of $f_\beta$.
7.4.4. Modify Eq. (7.9) for the special case when the circuit is driven from a constant voltage source.

## REFERENCES

1. Phillips, A. B.: "Transistor Engineering and Introduction to Integrated Semiconductor Circuits," McGraw-Hill Book Company, New York, 1962.
2. Basic Theory and Application of Transistors, TM 11-690, Headquarters, Department of the Army, 1959.
3. Malvino, A. P.: "Transistor Circuit Approximations," McGraw-Hill Book Company, New York, 1968.
4. Amos, S. W.: "Principles of Transistor Circuits," Hayden Book Company, Inc., New York, 1965.
5. Kahn, M., and J. M. Doyle: "The Synthesis of Transistor Amplifiers," Holt, Rinehart and Winston, Inc., New York, 1970.
6. Cowles, L. G.: "Transistor Circuits and Applications," Prentice-Hall, Inc., Englewood Cliffs, N.J., 1968.
7. Joyce, M. V., and K. K. Clarke: "Transistor Circuit Analysis," Addison-Wesley Publishing Company, Inc., Reading, Mass., 1961.
8. Cowles, L. G.: "Analysis and Design of Transistor Circuits," D. Van Nostrand Company, Inc., Princeton, N.J., 1966.
9. Millman, J., and C. C. Halkias: "Electronic Devices and Circuits," McGraw-Hill Book Company, New York, 1967.
10. Cutler, P.: "Semiconductor Circuit Analysis," McGraw-Hill Book Company, New York, 1964.
11. Texas Instruments Incorporated: "Transistor Circuit Design," McGraw-Hill Book Company, New York, 1963.
12. Ristenbatt, M. P., and R. L. Riddle: "Transistor Physics and Circuits," Prentice-Hall, Inc., Englewood Cliffs, N.J., 1966.

13. Angelo, E. J.: "Electronic Circuits," McGraw-Hill Book Company, New York, 1964.
14. Gibbons, J. F.: "Semiconductor Electronics," McGraw-Hill Book Company, New York, 1966.
15. Fitchen, F. C.: "Transistor Circuit Analysis and Design," 2d ed., D. Van Nostrand Company, Inc., Princeton, N.J., 1966.
16. "General Electric Transistor Manual," 7th ed., Syracuse, 1964.

# chapter 8

# junction transistor biasing

## 8.1 GENERAL CONSIDERATIONS

This chapter discusses criteria for setting the DC operating point in small-signal transistor amplifiers and techniques for maintaining the $Q$ point near its desired value.

Consider the common emitter output characteristics and circuit of Fig. 8.1. A load line is drawn on the characteristics between points $A$ and $B$. At point $A$ the transistor is cut off, $I_C = 0$, and $V_{CE} = V_{CC}$. At point $B$ the transistor is on, $V_{CE} = 0$, and $I_C = V_{CC}/R_L$. For linear amplification the $Q$ point must be on the load line in the active region, away from the cutoff and saturation regions. If the $Q$ point is selected on this basis and the signal is not too large, there should be no appreciable distortion.

For large-signal amplification, the $Q$ point is usually chosen at about the midpoint of the usable range to allow maximum swing on each alternate half-cycle. Large-signal considerations are treated in more detail in Chap. 12; therefore, we will limit our discussion here to small-signal amplifiers. In general, however, the load line should not

a. Be too close to the breakdown region of the transistor
b. Cross the maximum power-dissipation curve

Typical small-signal performance measures that one may want to optimize

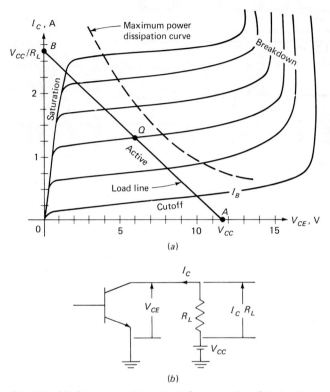

Fig. 8.1 (a) *Common emitter output characteristics;* (b) *circuit.*

through appropriate selection of the $Q$ point are power gain, frequency response, and noise figure. Power gain is related to $h_{fe}$, so that if power gain is the main criterion, a common emitter configuration with a high $h_{fe}$ transistor is used. As an example, data sheets for the 2N3392/3/4 series of silicon transistors in Appendix 10 indicate a typical $h_{fe}$ of 208 for the 2N3392 at $f = 1$ kHz, $V_{CE} = 10$ V, and $I_C = 1$ mA. The dependence of $h_{fe}$ on $Q$ point is given on a graph in the same data sheets; $h_{fe}$ increases with $I_C$ up to about $I_C = 8$ mA, at which point it starts to drop. Therefore, if maximum $h_{fe}$ were desired and other considerations would allow it (maximum power dissipation, etc.), a $Q$ point near $I_C = 8$ mA would be selected.

Noise-figure considerations are important in preamplifiers (first stages of amplification). Noise figure is a function of operating point and signal source resistance. For the 2N3392/3/4 series, noise voltage $E_n$ and noise current $I_n$ are given at different frequencies as a function of $Q$ point; note that both $E_n$ and $I_n$ generally increase with $I_E$, indicating that for low-noise applications the quiescent current should be low.

Frequency response is also affected by $Q$ point selection, because $f_T$ is a function of quiescent current. In addition to $Q$ point considerations, frequency response may indicate that a common base configuration is preferred, because $f_\alpha > f_\beta$ (see Prob. 6.6.11); hence a higher $f_H$ is possible with a common base amplifier.

Once a $Q$ point is chosen, the problems of keeping it constant must be considered. The most serious cause of $Q$ point shifting is changes in transistor parameters; these may be due to temperature or device replacement. Changes in $Q$ point are undesirable because they affect the transistor's AC parameters, causing variations in gain, impedance levels, noise figure, and frequency response. If the shift is small, changes in amplifier performance may be tolerable, but if the $Q$ point substantially moves toward the cutoff or saturation regions, severe distortion results.

The sections to follow discuss a number of circuit configurations with varying degrees of operating point stability. The basic objective is to maintain the $Q$ point as close to its originally established value as possible.

## 8.2 CONSTANT BASE CURRENT BIAS

The circuit of Fig. 8.2 employs *constant base current* biasing. The parameters that affect the $Q$ point are now considered. The base current is

$$I_B = \frac{V_{CC} - V_{BE}}{R_B} \tag{8.1}$$

$V_{BE}$ is usually small compared with $V_{CC}$; therefore $I_B$ is almost entirely dependent on $V_{CC}$ and $R_B$. In practice neither $R_B$ nor $V_{CC}$ vary a great deal; hence

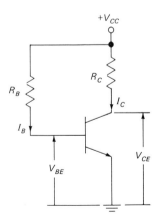

Fig. 8.2 *Constant base current bias circuit.*

$I_B$ remains constant. This circuit would then appear to be quite stable. Actually, a constant base current produces the worst possible type of biasing, as will be shown in the following discussion.

Remember that we want the $Q$ point shown on the output characteristics of Fig. 8.1 to be stable. This $Q$ point represents a specific value of $I_C$ and $V_{CE}$. Hence it is basically $I_C$ that must be stabilized. It was shown earlier that

$$I_C = h_{FE}I_B + I_{CBO}(h_{FE} + 1) \tag{5.10}$$

If $I_B$ is constant, only variations in $h_{FE}$ and $I_{CBO}$ can affect $I_C$. This is indeed the problem, since $h_{FE}$ and $I_{CBO}$ vary with temperature and with transistor replacement. If, in the circuit of Fig. 8.2, $V_{CC} = 10$ V and $R_B = 100$ k$\Omega$,

$$I_B \simeq \frac{V_{CC}}{R_B} = \frac{10}{10^5} = 100 \ \mu A$$

Assuming a nominal $h_{FE} = 50$, $I_{CBO} = 10 \ \mu A$, and $R_C = 1$ k$\Omega$, the collector current is

$$I_C = h_{FE}I_B + I_{CBO}(h_{FE} + 1) \tag{5.10}$$
$$I_C = 50(100 \times 10^{-6}) + (10 \times 10^{-6})(51)$$
$$I_C \simeq 5.5 \text{ mA}$$

The collector-emitter voltage is

$$V_{CE} = V_{CC} - I_C R_C = 10 - 5.5 = 4.5 \text{ V}$$

If the transistor is now replaced with one whose $h_{FE} = 75$, $I_B$ is unchanged, but

$$I_C = 75(100 \times 10^{-6}) + (10 \times 10^{-6})(76)$$
$$I_C \simeq 8.26 \text{ mA}$$

and

$$V_{CE} = 10 - 8.26 = 1.74 \text{ V}$$

Note the drastic change in operating point; it was caused by a variation in $h_{FE}$ (50 to 75) that is quite common.

Another factor to consider is $I_{CBO}$. This can be quite a problem, especially in germanium transistors, because of its temperature-dependence. $I_{CBO}$ doubles roughly for every 10°C rise in Ge and for every 6°C rise in Si units. Except for large-power or high-temperature applications, $I_{CBO}$ is usually negligible in Si transistors. If $I_{CBO}$ is 10 $\mu$A at 25°C in a Ge transistor and the temperature increases to 55°C, the new $I_{CBO}$ is 80 $\mu$A (doubles three times). Using the previous data, we have

$$I_C = 50(100 \times 10^{-6}) + (80 \times 10^{-6})(51)$$
$$I_C \simeq 9 \text{ mA}$$

and

$$V_{CE} \simeq 10 - 9 = 1 \text{ V}$$

again, drastically different from the original values.

Such large $Q$ point variations are unacceptable. We can show mathematically how poor the circuit is by determining the change in collector current $\Delta I_C$ caused by a change in cutoff current $\Delta I_{CBO}$. To determine the effects of a change in $I_{CBO}$, we hold $h_{FE}$ constant in Eq. (5.10) and write

$$I_C = h_{FE}I_B + I_{CBO}(h_{FE} + 1) \qquad (5.10)$$
$$\Delta I_C = h_{FE}\Delta I_B + \Delta I_{CBO}(h_{FE} + 1)$$

But $I_B$ is constant; thus $\Delta I_B = 0$, and

$$\Delta I_C = \Delta I_{CBO}(h_{FE} + 1) \qquad (8.2)$$

Equation (8.2) shows that a change in $I_{CBO}$ causes a corresponding change in $I_C$ that is $h_{FE} + 1$ times as large. In effect, the transistor amplifies its own cutoff current. This process can lead not only to large $Q$ point changes, but also to self-destruction, since the rise in $I_C$ may increase the power, which in turn heats the transistor even more, producing further increases in $I_{CBO}$.

### Problem Set 8.1

8.1.1. Construct a 5-W power-dissipation curve on the characteristics of Fig. 8.1a.

8.1.2. For the circuit of Fig. 8.2, find the value of $R_B$ and $R_C$ to yield a bias of 10 V and 1 mA. Assume the germanium transistor has negligible $V_{BE}$ but $I_{CBO} = 5$ μA at 25°C. The supply $V_{CC} = 20$ V, and $h_{FE} = 100$ at 25°C.

8.1.3. Using the results and data from Prob. 8.1.2, determine the operating point at a temperature of 75°C; assume $h_{FE}$ doubles every 50°C.

## 8.3 SERIES (EMITTER) FEEDBACK BIAS

To overcome some of the problems associated with constant base current biasing, an emitter resistor may be added, as shown in Fig. 8.3a. Recall that our basic objective is to keep $I_C$ constant. $I_C$ is mainly determined by the input drive, namely $I_B$. Keeping $I_B$ constant, however, is no guarantee that $I_C$ stays the same, as shown in the last example. If $I_C$ rises in the circuit of Fig. 8.3a, the voltage across $R_E$ also rises; this reduces the voltage across $R_B$ and $V_{BE}$, thus lowering $I_B$. The decrease in $I_B$ also causes $I_C$ to decrease. Hence, the rise in $I_C$ is limited, although not completely.

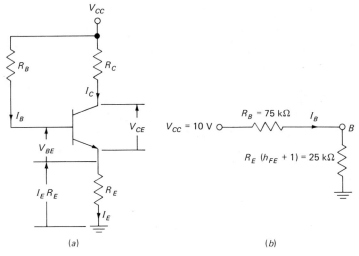

Fig. 8.3  (a) *Emitter feedback bias circuit;* (b) *Thevenin circuit seen by base.*

The action just described involves *negative feedback* in that an *increase* in $I_C$ produces *feedback* which tends to *reduce* $I_C$.

A numerical example can best show circuit performance. Let us assume $V_{CC} = 10$ V, $R_B = 75$ kΩ, $R_E = 500$ Ω, and $R_C = 500$ Ω. The transistor's $V_{BE}$ can be neglected, $h_{FE} = 50$, and $I_{CBO} = 10$ µA. The $Q$ point is determined using the Thevenin[1] equivalent circuit seen by the base, as shown in Fig. 8.3b. The base current is

$$I_B \simeq \frac{10}{10^5} = 0.1 \text{ mA}$$

$$I_C = h_{FE}I_B + I_{CBO}(h_{FE} + 1) \quad (5.10)$$
$$I_C \simeq 50(0.1 \times 10^{-3}) + (10 \times 10^{-6})(51)$$
$$I_C \simeq 5.5 \text{ mA}$$
$$V_{CE} = V_{CC} - I_C R_C - I_E R_E \simeq V_{CC} - I_C(R_C + R_E)$$
$$V_{CE} \simeq 10 - (5.5 \times 10^{-3})(1{,}000) = 4.5 \text{ V}$$

The $Q$ point of 5.5 mA and 4.5 V is identical to that established for the constant $I_B$ bias circuit discussed earlier.

Suppose now a new transistor is used whose $h_{FE}$ is 75 but with the same $I_{CBO}$. The equivalent value of $R_E$ is

$$R_E(h_{FE} + 1) = 500(76) = 38 \text{ kΩ}$$

$R_B$ is still 75 kΩ; hence $I_B \simeq 10/[(75 + 38)(10^3)] \simeq 0.0885$ mA.

---

[1] The transformation leading to $R_E(h_{FE} + 1)$ as seen from the base is valid only as long as $I_{CBO}$ is not significant.

$I_C \simeq 75(0.0885 \times 10^{-3}) + (10 \times 10^{-6})(76)$
$I_C \simeq 7.4$ mA

Hence

$V_{CE} \simeq 10 - (7.4 \times 10^{-3})(10^3) = 2.6$ V

Note the improvement in $Q$ point stability over the circuit of Fig. 8.2; in that circuit $I_C$ changed from 5.5 to 9 mA, while here the change is from 5.5 to 7.4 mA.

Let us now determine the change in $Q$ point due to an increase in temperature to 55°C. We will assume that $h_{FE}$ remains constant (in practice $h_{FE}$ also increases with temperature). $I_{CBO}$ doubles every 10°C; therefore it rises to 80 μA. Since $I_{CBO}$ is no longer negligible, the approximations involving $R_E(h_{FE} + 1)$ cannot be used; instead $I_C$ is calculated as follows:

$$10 \simeq I_B R_B + I_E R_E \quad \text{(neglect } V_{BE}) \tag{a}$$

But

$$I_C = \alpha I_E + I_{CBO} \tag{5.2a}$$

therefore

$$I_E = \frac{I_C - I_{CBO}}{\alpha}$$

$$I_B = I_E - I_C \tag{5.1}$$

$$I_B = \frac{I_C - I_{CBO}}{\alpha} - I_C \tag{8.3}$$

Equation (a) can now be written in terms of $I_C$:

$$10 \simeq \left(\frac{I_C - I_{CBO}}{\alpha} - I_C\right)(R_B) + \frac{I_C - I_{CBO}}{\alpha} R_E$$

$$10 \simeq \frac{I_C R_B - I_{CBO} R_B - \alpha I_C R_B + I_C R_E - I_{CBO} R_E}{\alpha}$$

$$10\alpha + I_{CBO}(R_B + R_E) \simeq I_C R_B (1 - \alpha) + R_E$$

$$I_C \simeq \frac{10\alpha + I_{CBO}(R_B + R_E)}{R_B(1 - \alpha) + R_E} \tag{b}$$

Equation (b) is accurate except for the absence of $V_{BE}$. Substituting numerical values, we have

$$I_C \simeq \frac{10(0.98) + (80 \times 10^{-6})(75{,}500)}{(75 \times 10^3)(0.02) + 500}$$

$I_C \simeq 7.92$ mA

## Problem Set 8.2

**8.2.1.** In the circuit of Fig. 8.3a $V_{CC} = 30$ V, $R_C = 500$ Ω, and at room temperature $I_{CBO} = 10$ μA, $h_{FE} = 50$, and $V_{BE}$ can be neglected.

    a. Select $R_B$ and $R_E$ to yield a Q point of $V_{CE} = 4.5$ V and $I_C = 5.5$ mA.
    b. Determine the change in $I_C$ when $h_{FE} = 75$.
    c. Determine the change in $V_{CE}$ when $h_{FE} = 75$.

## 8.4 STABILITY FACTORS

We now develop a more complete derivation that allows us to identify conditions affecting Q point stability. This is done for the circuit of Fig. 8.4a, which is extensively used. The base equivalent circuit is shown in Fig. 8.4b; the Thevenin voltage and resistance are

$$V_{BB} = V_{CC} \frac{R_2}{R_1 + R_2} \tag{c}$$

$$R_B = \frac{R_1 R_2}{R_1 + R_2} \tag{d}$$

Substitution of Eq. (d) into Eq. (c) yields

$$V_{BB} = \frac{V_{CC} R_B}{R_1} \tag{e}$$

Writing Kirchhoff's voltage law around the loop involving $V_{BB}$, $R_{BB}$, and $R_E$, we get

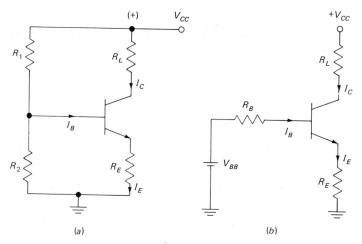

Fig. 8.4 (a) General bias circuit; (b) Thevenin equivalent circuit for (a).

$$V_{BB} = I_B R_B + V_{BE} + I_E R_E \tag{f}$$

But

$$I_E = \frac{I_C - I_{CBO}}{\alpha} \tag{5.2a}$$

and

$$I_B = \frac{I_C - I_{CBO}}{\alpha} - I_C \tag{8.3}$$

Substituting Eqs. (e), (5.2a), and (8.3) into Eq. (f) yields

$$\frac{V_{CC} R_B}{R_1} = \frac{I_C - I_{CBO}}{\alpha}(R_E + R_B) - I_C R_B + V_{BE}$$

from which

$$I_C = \alpha \left[ \frac{V_{CC} R_B / R_1 + I_{CBO} R_B - V_{BE}}{R_E + R_B(1 - \alpha)} \right] + I_{CBO} \tag{8.4}$$

Equation (8.4) indicates that $I_C$ is a function of many quantities, such as $I_{CBO}$, $\alpha$, $V_{CC}$, and $V_{BE}$. The resistors are usually considered constant; however, there are cases when their tolerances and temperature dependence may have to be considered.

$I_{CBO}$ depends mainly on temperature, approximately doubling every 10°C in Ge and every 6°C in Si transistors. Usually $I_{CBO}$ is much greater in Ge transistors; therefore it causes more problems than it does in Si devices.

$V_{BE}$ decreases with temperature at the rate of about 2.5 mV/°C for either Si or Ge transistors. The transistor's $h_{FE}$, and hence $\alpha$, generally rises with temperature, the exact amount depending on the particular device. $V_{CC}$ may also change, usually because of poor power supply regulation or ripple. The temperature dependences of $h_{FE}$ and $V_{BE}$ predominate in Si transistors.

We can determine to what extent $I_C$ depends on changes in any of the quantities of Eq. (8.4) by defining appropriate stability factors; these factors are now calculated to determine how $I_C$ can be made insensitive to parameter variations. The first and most common stability factor is

$$S_I = \frac{\Delta I_C}{\Delta I_{CBO}} \tag{8.5}$$

$S_I$ is calculated by holding everything on the righthand side of Eq. (8.4) constant except for $I_{CBO}$. In this case we have

$$\Delta I_C = \alpha \left[ \frac{\Delta I_{CBO} R_B}{R_E + R_B(1-\alpha)} \right] + \Delta I_{CBO}$$

$$\Delta I_C = \Delta I_{CBO} \left[ 1 + \frac{\alpha R_B}{R_E + R_B(1-\alpha)} \right] = \Delta I_{CBO} \left[ 1 + \frac{h_{FE} R_B}{R_E(h_{FE}+1) + R_B} \right]$$

$$S_I = \frac{\Delta I_C}{\Delta I_{CBO}} = 1 + \frac{h_{FE} R_B}{R_E(h_{FE}+1) + R_B} \tag{8.6}$$

Equation (8.6) indicates that the minimum value of $S_I$ is 1; that is, the change in $I_C$ is always as great as or larger than the change in $I_{CBO}$ that causes it. In order to reduce the second term of the equation to as low a value as possible, $R_B$ should be low and $R_E$ should be high. In practice, values of $S_I$ as low as 5 can usually be achieved, and this is satisfactory for most applications. With $R_B$ relatively low, $I_{CBO}$ is prevented from being amplified through emitter injection (see Sec. 5.5). If we make $R_B$ too small, however, the amplifier's input impedance may be substantially reduced. We must therefore reach a compromise based on how much loading of the input signal source we can accept and how much protection against the temperature dependence of $I_{CBO}$ we require.

A large emitter resistor, often called a swamping resistor, also leads to difficulties. If $R_E$ is too large, it may require an excessively large DC voltage drop. The limiting factor here is $V_{CC}$, which must supply voltage for $R_E$ as well as the transistor and $R_C$.

A second stability factor that relates the dependence of $I_C$ to $V_{BE}$ is

$$S_V = \frac{\Delta I_C}{\Delta V_{BE}} \tag{8.7}$$

Again, by holding everything except $V_{BE}$ constant in Eq. (8.4), we have

$$I_C = \alpha \frac{-\Delta V_{BE}}{R_E + R_B(1-\alpha)} = \frac{-h_{FE} \Delta V_{BE}}{R_E(h_{FE}+1) + R_B}$$

$$S_V = \frac{\Delta I_C}{\Delta V_{BE}} = \frac{-h_{FE}}{R_B + R_E(h_{FE}+1)} \tag{8.8}$$

Equation (8.8) indicates that, to reduce the dependence of $I_C$ on $V_{BE}$, both $R_B$ and $R_E$ should be large. Note that the voltages across $R_B$ and $R_E$ are in series with $V_{BE}$. Making these resistors large increases their DC voltage drop relative to $V_{BE}$. This means that if $V_{BE}$ varies, its effect on $I_C$ is insignificant because of the more stable DC voltages across $R_B$ and $R_E$. But a large $R_B$ degrades $S_I$. $R_B$, however, is not as significant as $R_E$ in Eq. (8.8) ($R_E$ is multiplied by $h_{FE}+1$, while $R_B$ is not). The practical solution, then, is to make $R_E$ as large and $R_B$ as low as circumstances allow.

The dependence of $I_C$ on the transistor's $h_{FE}$ is more complex to derive. For this reason it is not developed here, but the results are given below:

$$S_H = \frac{\Delta I_C}{\Delta h_{FE}} \tag{8.9}$$

$$S_H = \frac{\Delta I_C}{\Delta h_{FE}} = \frac{I_{C1} - I_{CBO1}}{h_{FE1}} \frac{R_B + R_E}{R_B + R_E(1 + h_{FE1})} \tag{8.10}$$

The subscript 1 refers to nominal values, that is, before $h_{FE}$ changes. Equation (8.10) indicates that the larger $R_E$, the less dependent $I_C$ is on $h_{FE}$. A numerical example to illustrate stability factors follows.

### Example 8.1

Compute the percent change in $I_C$ when $h_{FE}$ varies from 50 to 100 in the circuit of Fig. 8.4a. Neglect $V_{BE}$ and $I_{CBO}$. The relevant information is $R_1 = 150$ k$\Omega$, $R_2 = 33$ k$\Omega$, $R_L = 1$ k$\Omega$, $V_{CC} = 10$ V, $R_E = 0$.

### Solution

We can use Eq. (8.4) to calculate $I_C$. Since $\alpha = h_{FE}/(h_{FE} + 1)$, we get

$$\alpha_1 = \frac{50}{51} = 0.98 \qquad \alpha_2 = \frac{100}{101} = 0.99$$

$R_B$ can be calculated as

$$R_B = R_1 \| R_2 = (150 \times 10^3) \| (33 \times 10^3) = 27 \text{ k}\Omega$$

$$I_{C1} = 0.98 \left[ \frac{10(27 \times 10^3)/(150 \times 10^3)}{(27 \times 10^3)(0.02)} \right] = 3.26 \text{ mA} \tag{8.4}$$

$$I_{C2} = 0.99 \left[ \frac{10(27 \times 10^3)/(150 \times 10^3)}{(27 \times 10^3)(0.01)} \right] = 6.6 \text{ mA} \tag{8.4}$$

$$\Delta I_C = I_{C2} - I_{C1} = 3.34 \text{ mA}$$

The percent change is

$$\frac{3.34}{3.26} 100 = 102 \text{ percent}$$

### Example 8.2

Repeat Example 8.1 with $R_E = 1{,}000$ $\Omega$.

### Solution

$$I_{C1} = 0.98 \left[ \frac{10(27 \times 10^3)/(150 \times 10^3)}{10^3 + (27 \times 10^3)(0.02)} \right] = 1.145 \text{ mA} \tag{8.4}$$

$$I_{C2} = 0.99 \left[ \frac{10(27 \times 10^3)/(150 \times 10^3)}{10^3 + (27 \times 10^3)(0.01)} \right] = 1.39 \text{ mA} \tag{8.4}$$

$$\Delta I_C = I_{C2} - I_{C1} = 0.245 \text{ mA}$$

The percent change is

$$\frac{0.245}{1.145} 100 = 21.4 \text{ percent}$$

Note that, as $h_{FE}$ doubles, $I_C$ also doubles when $R_E = 0$. With $R_E = 1$ k$\Omega$, however, the change in $I_C$ is 21.4 percent.

We will now determine the dependence of $I_C$ on $V_{CC}$. A stability factor can be defined for this purpose:

$$S_E = \frac{\Delta I_C}{\Delta V_{CC}} \tag{8.11}$$

Holding everything in Eq. (8.4) constant except $V_{CC}$, we have

$$\Delta I_C = \alpha \left[ \frac{\Delta V_{CC} R_B / R_1}{R_E + R_B(1 - \alpha)} \right]$$

$$S_E = \frac{\Delta I_C}{\Delta V_{CC}} = \frac{\alpha R_B / R_1}{R_E + R_B(1 - \alpha)} \tag{8.12}$$

Equation (8.12) indicates that a large $R_E$ tends to improve this situation too. In fact, it can be concluded that making the ratio $R_E/R_B$ as large as other considerations allow results in total improved stabilization of the $Q$ point. In general,

$$I_{C2} = I_{C1} + \Delta I_{C(\text{tot})} \tag{8.13}$$

where

$$\Delta I_{C(\text{tot})} = S_I \Delta I_{CBO} + S_V \Delta V_{BE} + S_H \Delta h_{FE} + S_E \Delta V_{CC} \tag{8.14}$$

**Example 8.3**

Calculate $I_C$ in the circuit of Fig. 8.4 at 25 and 50°C. The following data apply at 25°C: $h_{FE} = 50$, $V_{BE} = 0.3$ V, $I_{CBO} = 10$ µA. $I_{CBO}$ doubles every 10°C, $h_{FE} = 70$ at 50°C, and $V_{BE}$ drops at the rate of 2.5 mV/°C. $V_{CC} = 10$ V, $R_1 = 200$ k$\Omega$, $R_2 = 200$ k$\Omega$, $R_E = 0$.

**Solution**

At 25°C, $I_C$ is

$$I_{C1} = 0.98 \left[ \frac{(10 \times 10^5)/(2 \times 10^5) + 10^{-5}(10^5) - 0.3}{10^5(0.02)} \right] + 10^{-5} \qquad (8.4)$$

$$I_{C1} = 2.8 \text{ mA}$$

The stability factors are

$$S_I = \frac{\Delta I_C}{\Delta I_{CBO}} = 1 + \frac{70 \times 10^5}{10^5} = 71 \qquad (8.6)$$

$$S_V = \frac{\Delta I_C}{\Delta V_{BE}} = \frac{-50 \text{ A}}{100 \times 10^3 \text{ V}} = \frac{-0.5 \times 10^{-3}}{\Omega} \qquad (8.8)$$

$$S_H = \frac{\Delta I_C}{\Delta h_{FE}} = \frac{2.8 \times 10^{-3} - 10^{-5} \, 10^5}{50 \quad 10^5} \qquad (8.10)$$

$$S_H = \frac{2.79 \times 10^{-3}}{50} = 55.7 \text{ μA}$$

$$S_E = \frac{\Delta I_C}{\Delta V_{CC}} = \frac{0.98(100 \times 10^3)/(200 \times 10^3)}{(100 \times 10^3)(0.02)} = \frac{0.245 \times 10^{-3}}{\Omega} \qquad (8.12)$$

$I_{CBO}$ doubles every 10°C; therefore $I_{CBO} \simeq 60$ μA at 50°C. Thus $\Delta I_{CBO} = 50$ μA, and

$$\Delta I_C \text{ (due to } \Delta I_{CBO}) = S_I \, \Delta I_{CBO} = 71(50 \times 10^{-6}) = 3.55 \text{ mA} \qquad (8.5)$$

$V_{BE}$ drops at the rate of 2.5 mV/°C; hence the change from 25 to 50°C is $(-2.5 \times 10^{-3})(25) = -62$ mV, and

$$\Delta I_C \text{ (due to } \Delta V_{BE}) = S_V \, \Delta V_{BE} = (-0.5 \times 10^{-3})(-62 \times 10^{-3}) = 31 \text{ μA} \qquad (8.7)$$

The transistor's $h_{FE}$ changes from 50 to 70. Thus $\Delta h_{FE} = 20$, and

$$\Delta I_C \text{ (due to } \Delta h_{FE}) = S_H \, \Delta h_{FE} = (55.7 \times 10^{-6})(20) = 1.114 \text{ mA} \qquad (8.9)$$

Assuming $V_{CC}$ remains constant, the total change in $I_C$ is

$$\Delta I_{C(\text{tot})} = (3.55 \times 10^{-3}) + (31 \times 10^{-6}) + (1.114 \times 10^{-3}) \simeq 4.7 \text{ mA}$$

which, when added to the original $I_{C1}$, yields

$$I_{C2} = (4.7 + 2.8)(10^{-3}) = 7.5 \text{ mA}$$

Naturally, $I_{C2}$ could have been calculated directly using Eq. (8.4):

$$I_{C2} = \frac{70}{71} \left[ \frac{(10 \times 10^5)/(200 \times 10^3) + (60 \times 10^{-6})(10^5) - 0.238}{10^5(0.0141)} \right] + 60 \times 10^{-6}$$

$$I_{C2} = 0.985 \left[ \frac{5 + 6 - 0.238}{1.41 \times 10^3} \right] + (60 \times 10^{-6})$$

$$I_{C2} \simeq 7.5 \text{ mA}$$

The large increase in $I_C$ is due to the very poor bias stability created by the lack of an emitter resistor.

## Problem Set 8.3

8.3.1. The following data apply to the circuit of Fig. 8.4a: $R_1 = R_2 = 50$ k$\Omega$; $V_{BE} = 0.6$ V at 25°C; $R_L = 100$ $\Omega$; $R_E = 2$ k$\Omega$; $V_{CC} = 40$ V; $I_{CBO} = 20$ nA at 25°C; $h_{FE} = 75$ at 25°C. Determine

    a. The transistor material (Si or Ge).
    b. The $Q$ point at 25°C.
    c. The following stability factors: $S_I$, $S_H$, and $S_V$.
    d. $V_{BE}$, $I_{CBO}$, and $h_{FE}$ at 100°C, assuming that $V_{BE}$ drops at the rate of 2.5 mV/°C, $I_{CBO}$ doubles every 6°C, and $h_{FE}$ doubles for a 75°C rise.
    e. Using the results of (c) and (d), the change in $I_C$ due to each parameter change at 100°C. Also determine the total change in $I_C$ due to the combined effects of these changes.
    f. The $Q$ point at 100°C.

8.3.2. Explain qualitatively (rather than through the use of mathematical equations) why $Q$ point stability improves as $R_E$ is made larger.

8.3.3. Do you think that $R_L$ has any effect on the operating point stability in the circuit of Fig. 8.4? Explain your answer.

8.3.4. In the circuit of Fig. 8.4, $I_C$ tends to increase with temperature. What is the limit to the maximum value of $I_C$? In what state is the transistor when this limit has been reached?

8.3.5. Operating point stability can be quite good if $R_E$ is increased in value. What limits the maximum usable value of $R_E$?

## 8.5 SHUNT FEEDBACK BIASING

Another method of stabilizing the operating point is to use shunt feedback. This is accomplished by connecting a resistor from collector to base as shown in Fig. 8.5.

Note that the current through $R_C$ is the sum of $I_B$ and $I_C$. Suppose now that $I_C$ were to increase; the voltage across $R_C$ and $R_E$ would also increase. Because $V_{CC}$ is constant, $V_{CB}$ would have to decrease. But $V_{CB}$ is the voltage across $R_B$ whose current is $I_B$; therefore as $V_{CB}$ decreases, so does $I_B$. The decrease in $I_B$ tends to reduce $I_C$ toward its original value. This type of biasing is most effective when $R_C$ is relatively high, since even a small increase in $I_C$ develops a relatively large increase in voltage across $R_C$ and a corresponding reduction in voltage across $R_B$.

An expression for $I_C$ can be developed by writing equations based on Kirchhoff's laws; the algebraic sum of all voltage drops from supply to ground must equal $V_{CC}$:

$$V_{CC} = (I_B + I_C)(R_C) + I_B R_B + V_{BE} + I_E R_E \qquad (a)$$

But

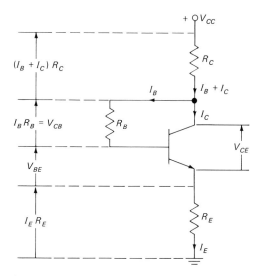

Fig. 8.5  Shunt feedback bias circuit.

$$I_B = \frac{I_C(1-\alpha) - I_{CBO}}{\alpha} \tag{8.3}$$

and

$$I_E = \frac{I_C - I_{CBO}}{\alpha} \tag{5.2a}$$

Therefore

$$V_{CC} = \left[\frac{I_C(1-\alpha) - I_{CBO}}{\alpha}\right](R_B + R_C) + I_C R_C + V_{BE} + \left[\frac{I_C - I_{CBO}}{\alpha}\right] R_E$$

which yields

$$I_C = \frac{\alpha V_{CC} + I_{CBO}(R_E + R_B + R_C) - \alpha V_{BE}}{R_E + R_C + R_B(1-\alpha)} \tag{8.15}$$

The above equation is exact; no simplifying assumptions have been made. To determine how $I_C$ depends on $I_{CBO}$, let us assume a change in $I_{CBO}$ while holding the other parameters constant. This yields

$$\Delta I_C = \frac{\Delta I_{CBO}(R_E + R_B + R_C)}{R_E + R_C + R_B(1-\alpha)}$$

$$S_I = \frac{\Delta I_C}{\Delta I_{CBO}} = \frac{1 + [R_B/(R_E + R_C)]}{1 + [R_B(1-\alpha)/(R_E + R_C)]} \tag{8.16}$$

To minimize $S_I$, both $R_E$ and $R_C$ should be large. In fact, if $R_E + R_C \gg R_B$, $S_I$ approaches 1, which is ideal. This agrees with our earlier statement that $R_C$ must be relatively large for this type of compensation to be effective.

Circuits that use $R_B$ only ($R_E \simeq 0$) are referred to as *shunt feedback*, because corrective action is obtained through a *shunt* resistor ($R_B$) connected between output and input. If both $R_E$ and $R_B$ are used, the combined benefits of shunt and series feedback result.

### Problem Set 8.4

8.4.1. The $Q$ point stability of the shunt feedback circuit of Fig. 8.5 depends on the value of collector resistor, while $R_L$ has no effect on the stability of the circuit in Fig. 8.4. Why do you think this is so?

8.4.2. Explain qualitatively the dependence of $Q$ point stability on $R_B$ and $R_C$ for the circuit of Fig. 8.5.

8.4.3. The following data apply to the circuit of Fig. 8.5: $V_{CC} = 40$ V, $R_C = 40$ kΩ, $R_B = 8$ kΩ, $R_E = 0$, $V_{BE} = 0.2$ V at 25°C, $I_{CBO} = 10$ μA at 25°C, $h_{FE} = 50$. Determine

    a. The $Q$ point at 25°C
    b. The stability factor $S_I$
    c. The change in $I_C$ due to $I_{CBO}$ for a 30°C rise (this is a Ge transistor)
    d. The new $Q$ point at 55°C, assuming that changes in $V_{BE}$ and $h_{FE}$ with temperature are relatively insignificant

## 8.6 NONLINEAR BIAS STABILIZATION

Stabilization of the operating point using standard linear resistors can never be perfect, nor can it, in most cases, even approach the ideal. There are nonlinear components, however, whose temperature characteristics can be utilized to produce a very high degree of $Q$ point stability. A variety of components is available, some of which are described here.

    Thermistors are resistors whose resistance depends on temperature. They are available with either a positive temperature coefficient (PTC) or negative temperature coefficient (NTC); the resistance of a PTC thermistor increases, while that of an NTC thermistor decreases, with temperature. By the proper series and/or parallel combination of thermistors and linear resistors, it is possible to achieve virtually any desired temperature coefficient.

    Semiconductor diodes are also temperature-dependent. When forward-biased, the threshold voltage drops with temperature (usually at the rate of 2.5 mV/°C). When reverse-biased but not in breakdown, the reverse current increases with temperature, thus producing a decrease in the effective resistance. Hence a diode exhibits a negative resistance temperature coefficient whether it is forward- or reverse-biased. The dependence of transistor characteristics can also be used to cancel temperature effects in other transistor circuits.

Breakdown diodes can maintain a relatively constant voltage. These diodes, however, are temperature-dependent, PTC for avalanche breakdown, NTC in Zener devices. Breakdown diodes can be temperature-compensated by adding forward-biased diodes in series; this way the breakdown diode's PTC is canceled by the forward-biased diodes' NTC.

In order for these schemes to work, the compensating device must be in close thermal contact with the rest of the circuit so that temperature changes occur throughout the circuit more or less simultaneously. In some applications, a transistor and its compensating elements may share the same housing or heat sink to maintain a uniform temperature.

Consider the circuit of Fig. 8.6. Connected between the DC supply and the emitter there is a PTC thermistor. As the temperature increases, $I_C$ tends to increase. The resistance of the thermistor also increases, thereby lowering the voltage across $R_E$. This reduces $I_E$, thus canceling the tendency of $I_C$ to rise. It is therefore possible to maintain $I_C$ relatively constant over a wide temperature range.

As an example of thermistor compensation, consider the bias circuit of Fig. 8.7. The circuit is placed in an oven whose temperature can be accurately measured and controlled. Leads are brought out from the oven to make the connections to $V_{CC}$ and an ammeter to measure $I_C$. As the oven temperature is varied, $R_{1V}$ is also varied until $I_C$ returns to its room-temperature value. This is done at several temperatures, yielding data which show how $R_{1V}$ must vary with temperature to keep $I_C$ constant. $R_1$ is then replaced with a thermistor whose resistance varies with temperature in the proper manner, yielding almost perfect stabilization of $I_C$.

## Problem Set 8.5

8.5.1. Why can temperature-sensitive components provide better $Q$ point stability than linear resistors alone?

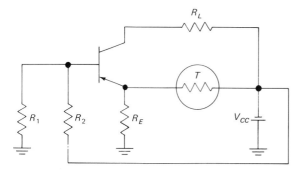

*Fig. 8.6* Thermistor compensated bias circuit.

318  Solid State Electronic Circuits

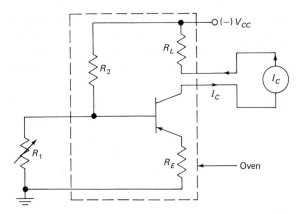

*Fig. 8.7  Experimental set-up to determine temperature dependence of resistance required to keep $I_C$ constant.*

8.5.2. What temperature-sensitive devices can be used for stabilizing a transistor's $Q$ point?

8.5.3. If $R_E$ in the circuit of Fig. 8.7 is replaced with a thermistor, should a PTC or NTC unit be used? Explain your choice.

8.5.4. Do you think $R_L$ in the circuit of Fig. 8.7 could be replaced with a thermistor for purposes of temperature stabilization? Explain your answer.

# REFERENCES

1. Basic Theory and Application of Transistors, TM 11-690, Headquarters, Department of the Army, 1959.
2. Malvino, A. P.: "Transistor Circuit Approximations," McGraw-Hill Book Company, New York, 1968.
3. Amos, S. W.: "Principles of Transistor Circuits," Hayden Book Company, Inc., New York, 1965.
4. Joyce, M. V., and K. K. Clarke: "Transistor Circuit Analysis," Addison-Wesley Publishing Company, Inc., Reading, Mass., 1961.
5. Millman, J., and C. C. Halkias: "Electronic Devices and Circuits," McGraw-Hill Book Company, New York, 1967.
6. Cutler, P.: "Semiconductor Circuit Analysis," McGraw-Hill Book Company, New York, 1964.
7. Texas Instruments Incorporated: "Transistor Circuit Design," McGraw-Hill Book Company, New York, 1963.
8. Ristenbatt, M. P., and R. L. Riddle: "Transistor Physics and Circuits," Prentice-Hall, Inc., Englewood Cliffs, N.J., 1966.
9. "General Electric Transistor Manual," 7th ed., Syracuse, 1964.

# chapter 9

# field effect transistors and circuits

Although the theoretical knowledge necessary to produce field effect transistors has been available for quite some time, practical difficulties prevented the entry of this promising device into the commercial semiconductor family until recently.

Field effect transistors (FETs) generally have a higher input impedance than junction transistors. They are also less sensitive to nuclear radiation and can exhibit low noise figures.

The input to junction transistor amplifiers involves a forward-biased PN junction with its inherently low dynamic impedance. The input to FETs, on the other hand, involves a reverse-biased PN junction; hence the high input impedance. This can be a definite advantage in many applications where the current drawn from a signal source must be minimized. A disadvantage of the FET, however, is the relatively large spread in parameters for devices of the same type. This will, no doubt, be improved in due time.

Amplification is accomplished in FETs by establishing current through a *channel;* the resistance of the channel is varied by an electric potential at right angles to the flow of current. Thus changes in the transverse voltage cause corresponding variations of the channel resistance, which in turn controls the magnitude of current flow. This basic principle underlies all FET devices; however, there is more than one way of accomplishing the operation, which leads to two main classifications, the junction field effect transistor (JFET) and the metal oxide semiconductor field effect transistor (MOSFET), also called insulated gate field effect transistor (IGFET). We will study the structure and characteristics of the JFET and MOSFET separately.

## 9.1 THE JUNCTION FIELD EFFECT TRANSISTOR (JFET)

The operation of JFETs can be demonstrated using Fig. 9.1. We start with a sample of N-type semiconductor, as shown in Fig. 9.1a. This is the *channel*, and it is electrically equivalent to a resistance, as shown in Fig. 9.1b. Ohmic contacts are then added on each side of the channel to permit external connections. Thus, if voltage is applied, current flows. The terminal where majority carriers enter the channel is the *source*, and the terminal through which majority carriers leave is the *drain*. For an N-channel device, electrons are the majority carriers. Hence, the circuit in Fig. 9.1a and b involves nothing more than a DC voltage $V_{DS}$ applied across a resistance $R_{DS}$. The resulting current is the drain current $I_D$. In this simple circuit Ohm's law applies; that is, if $V_{DS}$ increases, $I_D$ increases proportionately.

We now add P-type regions on each side of the N-type channel, as in Fig. 9.1c. Note that the polarity of the drain-source voltage is the same as in Fig. 9.1a. Ohmic contacts, referred to as gate 1 and gate 2, are also added to each P-type region. For now, both gates are grounded, yielding zero gate-source voltage ($V_{GS} = 0$). The word *gate* is used because the potential applied between this terminal and the source controls the channel width and hence the current.

As with all PN junctions, a depletion region is formed as shown in Fig. 9.1c. The depletion-region width increases with the magnitude of reverse bias. Here the potential at any point along the channel depends on the distance of that point from the drain; points close to the drain are at a higher positive potential, relative to ground, than points close to the source. Both depletion regions are therefore subjected to greater reverse voltage near the drain; this explains why the depletion regions extend further into the channel at points near the drain.

The flow of electrons from source to drain is now restricted to the narrow channel between the nonconducting depletion regions. The width of this channel determines the resistance between drain and source.

Consider now the behavior of drain current $I_D$ versus drain-source voltage $V_{DS}$, as shown in Fig. 9.1d. The gate-source voltage $V_{GS}$ is still zero; the resulting curve is therefore valid only for $V_{GS} = 0$, as indicated. Suppose that $V_{DS}$ is gradually increased from 0 V; $I_D$ also increases, since the channel behaves as a resistance, therefore following Ohm's law. This rise of current with voltage is evident in the initial portion of the curve, which is referred to as the ohmic region. As $V_{DS}$ is further increased, $I_D$ begins to level off until a specific value of $V_{DS}$ is reached, called the *pinch-off voltage* $V_P$. At this point further increases in $V_{DS}$ do not produce corresponding increases in $I_D$. Instead, as $V_{DS}$ increases,

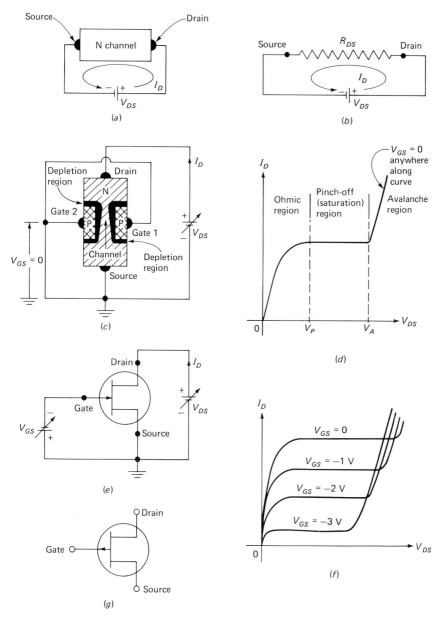

*Fig. 9.1* (a) N channel; (b) effective resistance of N channel; (c) physical structure of junction field-effect transistor (JFET); (d) plot of drain current $I_D$ versus drain-source voltage $V_{DS}$ for gate-source voltage $V_{GS} = 0$; (e) schematic symbol of N channel JFET with $V_{GS}$ and $V_{DS}$ voltages shown; (f) family of curves of $I_D$ versus $V_{DS}$ for different values of $V_{GS}$; (g) schematic symbol for P channel JFET.

both depletion regions extend further into the channel, resulting in a narrower cross section, and hence a higher channel resistance. Thus, even though there is more voltage, the resistance is also greater and the current remains relatively constant. We are now in the pinch-off, or saturation, region.

As with all PN junctions, breakdown occurs when the reverse voltage exceeds a certain level; FETs are no exception. When $V_{DS}$ exceeds $V_A$, the avalanche breakdown voltage, $I_D$ rises very rapidly, as shown on the characteristic of Fig. 9.1d.

Consider now an N-channel JFET with a reverse gate-source voltage, as shown in Fig. 9.1e. For any specific value of $V_{GS}$, a characteristic curve of the type in Fig. 9.1d results. A family of such curves is displayed in Fig. 9.1f; note that pinch-off occurs at a lower drain current when the magnitude of reverse gate-source voltage is increased ($V_{GS}$ more negative). This can be verified using an example. Suppose that $V_{GS} = 0$ and that due to $V_{DS}$, a specific point along the channel is at $+5$ V with respect to ground. The reverse voltage across either PN junction is now 5 V. If $V_{GS}$ is decreased from 0 to $-1$ V, the net reverse bias near the same point is $5 - (-1) = 6$ V. Thus, for any fixed $V_{DS}$, the channel is narrower as $V_{GS}$ is made more negative.

Note that changes in $V_{GS}$ produce changes in channel width, which in turn cause $I_D$ variations corresponding to the original changes in $V_{GS}$. If, in fact, an AC voltage corresponding to a signal is superimposed on the DC voltage $V_{GS}$, we should expect $I_D$ to vary according to the AC signal. Since the gate-source circuit consists of a reverse-biased PN junction with its inherently high resistance, the AC signal generator does not have to supply appreciable current to the gate of the FET.

Although the theory of the JFET has been developed using an N-type channel, it is also possible to use a P channel with N-type gates. The resulting P-channel JFET, whose symbol is given in Fig. 9.1g, would require opposite polarity DC levels for proper operation.

Note that only one type of carrier makes up the channel current, electrons in N channels, and holes in P channels; this leads to the *unipolar transistor* designation for FETs as opposed to bipolar transistor in which both types of carriers are involved.

## 9.2 THE METAL OXIDE SEMICONDUCTOR FIELD EFFECT TRANSISTOR (MOSFET)

The MOSFET can be explained by following the sequence starting in Fig. 9.2a. A P-type *substrate* serves as the basic structure upon which N-type regions are created by the diffusion of appropriate impurities. An oxide layer, which acts as an insulator, covers the entire substrate and N regions. By etching suitable openings through the oxide, metal contacts for source and drain connections are

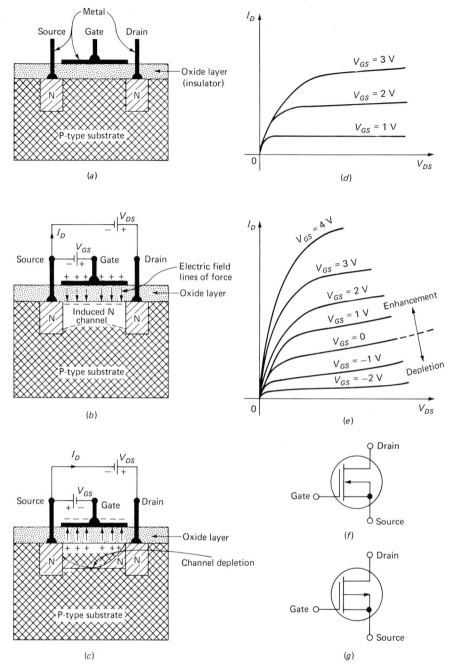

*Fig. 9.2* (a) Basic metal oxide semiconductor field-effect transistor (MOSFET) structure; (b) MOSFET structure with applied potentials and induced channel; (c) MOSFET structure with channel depletion; (d) characteristics for enhancement mode MOSFET; (e) characteristics for both enhancement and depletion mode MOSFET; (f) schematic symbol for N channel MOSFET; (g) schematic symbol for P channel MOSFET. (Structure diagrams courtesy of Motorola Inc.)

made to the N-type regions. The gate contact is then formed on the surface of the oxide layer; hence, the gate is electrically insulated from both N-type regions and substrate. As things are now, no current can flow from source to drain, because the N-type source, P-type substrate, and N-type drain behave as two diodes back to back; thus, regardless of the polarity of applied voltage, one of the PN junctions is always reverse-biased.

Suppose now that a positive potential is applied between gate and source; this is shown in Fig. 9.2b. Because the oxide layer is an insulator sandwiched between conductive regions, an equivalent capacitor is formed; the metal surface which is part of the gate and the conducting substrate below act as the capacitor *plates*. Whenever a positive charge is applied to one plate of a capacitor, a corresponding negative charge is *induced* on the opposite plate by the action of the electric field within the dielectric. In our case, the positive potential on the gate induces a negative charge in the P substrate; this charge results from minority carriers (electrons) attracted toward the area below the gate. As the number of electrons reaching this region increases, the relative density of majority carriers (holes) decreases until there are more free electrons than holes. This, of course, is true only in the relatively small region of the substrate directly below the gate. An N-type region now extends continuously from source to drain. The N channel below the gate is said to be "induced" because it is produced by the process of electric induction; if the positive gate potential is removed, the induced channel disappears and we revert to the previous situation.

As the gate voltage is further increased, a greater number of negative charge carriers are attracted toward the induced channel; this results in a decrease of the effective channel resistance (because of the presence of more charge carriers). Thus, the resistance seen by the drain-source supply $V_{DS}$ depends on the voltage applied to the gate: the greater the gate potential, the lower the channel resistance and the higher the drain current $I_D$. This process is referred to as *enhancement*, and the resulting device is an enhancement-type MOSFET. The resistance looking into the gate is extremely high, since the oxide layer is an insulator. If the oxide layer were perfect, the resistance looking into the gate would be infinite. In practice, resistances as high as $10^{14}$ Ω can be obtained. There is, of course, an effective capacitance as well.

It is also possible to produce MOSFETs of the *depletion* type. In this case, as the gate voltage increases, the channel is depleted of carriers, thus increasing the effective channel resistance. Such a structure is shown in Fig. 9.2c, where the region below the gate is doped N type. There is, therefore, a continuous N-type channel from source to drain through which current can flow. If a negative voltage is applied from gate to source, however, the negative charge on the gate induces an equal but positive charge (holes) on the other side of the oxide layer. The recombination of holes with electrons in the N-type channel reduces the number of free carriers available; thus the channel is "depleted" of charge carriers, which increases its effective resistance. As $V_{GS}$ is further decreased, the channel resistance becomes even higher, resulting in less drain current for any given $V_{DS}$. If a positive gate potential is applied instead, a situation similar to the enhancement device previously discussed occurs; that is, as $V_{GS}$ increases, negative charges are induced in the N channel. This enhances the channel's ability to conduct; hence $I_D$ increases. Thus the structure in Fig. 9.2c can be operated either in the enhancement or depletion mode. Characteristics for both devices are given in Fig. 9.2d and e.

MOSFETs lend themselves to direct coupling; this feature, along with the extremely high input impedance, indicates a large number of potential applications such as multistage amplifiers, logic circuits, and so on.

## 9.3 GRAPHIC ANALYSIS OF FET AMPLIFIER

An example will now be used to illustrate amplification in the FET circuit of Fig. 9.3b; we will assume a Q point at $V_{GS} = -1.5$ V and $V_{DS} = 9.5$ V.

The drain characteristics given in Fig. 9.3a are for the common source configuration. By locating the Q point, we can read $I_D$ as 1.65 mA. Since a 16-V supply is used, the DC load line must extend from the 16-V point on the $V_{DS}$ axis through the Q point, yielding a 4-mA intercept on the $I_D$ axis. The DC resistance is therefore

$$R_{DC} = \frac{16}{4 \times 10^{-3}} = 4 \text{ k}\Omega$$

This is made up of $R_L$ and $R_S$. Note that the gate is at zero DC potential; therefore

$$V_{GS} + R_S I_D = 0 \tag{9.1}$$
$$R_S = -\frac{V_{GS}}{I_D}$$
$$R_S = \frac{1.5}{1.65 \times 10^{-3}} \simeq 900 \text{ }\Omega$$
$$R_L = 4{,}000 - 900 = 3{,}100 \text{ }\Omega$$

$R_S$ is bypassed with a capacitor which AC grounds the source; therefore all $E_g$ is developed across the gate-source junction. The purpose of $R_S$ is for Q point stabilization. If $I_D$ should rise, for example, the larger voltage drop across $R_S$ would make $V_{GS}$ more negative, thus reducing $I_D$. Since $R_S$ is bypassed, the AC load line consists only of $R_L$; it is constructed on the characteristics using the fact that it must cross the Q point and that the voltage-current ratio along its length has to equal the AC load, namely $R_L = 3{,}100$ $\Omega$. Hence, as we move along the AC load line from the Q point toward the $V_{DS}$ axis, the current increment of

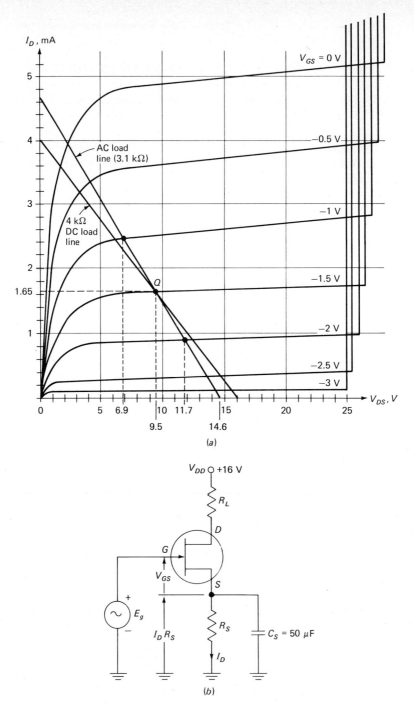

Fig. 9.3 (a) Family of drain characteristics for common source JFET configuration; (b) JFET amplifier circuit.

1.65 mA must cause a voltage change (from 9.5 V) of $(1.65 \times 10^{-3})(3{,}100) = 5.1$ V, yielding the 14.6-V intercept.

Having constructed the AC load line, it is now possible to determine the output voltage and current swings resulting from the AC input voltage. Assuming a 1-V peak-to-peak input signal, $V_{GS}$ swings by 0.5 V on each side of its quiescent level of $-1.5$ V, yielding $V_{GS}$ extremes of $-1$ and $-2$ V. The corresponding variations in $V_{DS}$ are read from the graph as 6.9 and 14.6 V, respectively. The voltage amplification for the circuit is

$$A_v = \frac{V_o}{V_i} = \frac{\Delta V_{DS}}{\Delta V_{GS}}$$

$$A_v = \frac{6.9 - 14.6}{-1 - (-2)} = -7.7$$

The negative sign indicates phase reversal; as $V_i$ rises, $V_o$ drops.

## Problem Set 9.1

9.1.1. What basic advantages do FETs have over bipolar transistors?

9.1.2. The FET can be characterized as a voltage-dependent resistor. What properties of the FET give rise to this description?

9.1.3. Describe briefly the following terms:

    a. Channel      c. Drain
    b. Source      d. Gate

9.1.4. What are the two main classifications of field effect transistors?

9.1.5. Draw the schematic symbol for the N- and P-channel JFET.

9.1.6. Indicate the normal polarity of gate-source voltage for the N- and P-channel JFET.

9.1.7. Why is the depletion-region width not the same at all points along the channel-gate junction of the JFET?

9.1.8. The characteristics in Fig. 9.1f indicate that avalanche breakdown is lower as $V_{GS}$ becomes more negative. Why do you think this is so?

9.1.9. Why does pinch-off occur at lower values of $I_D$ when the magnitude of $V_{GS}$ increases?

9.1.10. What physical characteristics of the JFET structure are responsible for the high input impedance? How does this differ from bipolar transistors?

9.1.11. Briefly define the following terms:

    a. Substrate      c. Depletion mode
    b. Induced channel      d. Enhancement mode

9.1.12. What is the difference between an *induced* channel and one which is diffused into the structure during the manufacturing process?

9.1.13. What gives rise to the high input resistance of MOSFETs?

9.1.14. Using the circuit and characteristics of Fig. 9.3 and the same $Q$ point, but $V_{DD} = 20$ V, determine

    a. The DC load line
    b. The value of $R_L$ and $R_S$
    c. The AC load line
    d. The output voltage swing for a 1-V peak-to-peak input voltage
    e. The voltage amplification

## 9.4 STATIC CHARACTERISTICS AND RATINGS

As with junction transistors, FETs have maximum ratings that cannot be exceeded without damaging them. These ratings generally include maximum power-dissipation, maximum current, maximum voltage, and storage temperature range. For the 2N2386 silicon JFET whose data sheet is reproduced in Appendix 10, the following ratings apply:

    Gate current: 10 mA
    Dissipation at or below 25°C air temperature: 0.5 W
    Storage temperature range: $-65$ to $+300°C$

In addition to maximum ratings, the electrical characteristics of FETs must also be considered. In this section we discuss static, as opposed to dynamic, characteristics; these are useful for biasing and general DC analysis.

### 9.4a Leakage Current

An inportant specification for any FET is the gate leakage current. Ideally there should not be any gate current since, looking into the gate of a JFET, we see a reverse-biased PN junction, and looking into the gate of a MOSFET, we see an insulator. In practice, however, reverse-biased PN junctions have leakage currents; insulators also conduct, although to a lesser degree.

The circuit of Fig. 9.4a may be used to obtain $I_{GSS}$, a standard value of gate leakage current. The subscripts here are interpreted in a manner similar to bipolar transistors, so that $I_{GSS}$, commonly known as the gate cutoff current, represents gate current ($G$) in a common source configuration ($S$) with the drain (third terminal) shorted ($S$) to the common terminal. The gate-source junction is reverse-biased by $V_{GS}$; since this is a P-channel JFET, $V_{GS}$ is positive. When specifying $I_{GSS}$, the test conditions, namely the value of $V_{GS}$ and temperature, must be indicated. As an example, the 2N2386 (see Appendix 10) has a gate cutoff current $I_{GSS}$ of 0.01 $\mu$A for $V_{GS} = 10$ V at a free air temperature of 25°C. At 100°C, $I_{GSS}$ is 1 $\mu$A. These specifications for $I_{GSS}$ represent maximum values,

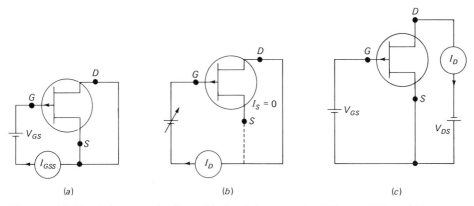

Fig. 9.4  (a) *Circuit for measuring* $I_{GSS}$; (b) *circuit for measuring* $BV_{DGO}$ *or* $BV_{GSS}$; (c) *general circuit with* $V_{GS}$ *and* $V_{DS}$ *supplies.*

which are needed for worst-case analysis. The temperature dependence of $I_{GSS}$ is demonstrated by the graph of "Gate cutoff current versus Free air temperature" provided in the manufacturer's data sheet. Note that $I_{GSS}$ rises by a factor of 10 for every 45°C (approximately) rise in the free air temperature.

The same basic technique may be used to determine gate cutoff current for MOSFETs; in this case the temperature dependence of gate cutoff current is much lower.

There are other ways of specifying gate leakage current (by opening the drain or source), but $I_{GSS}$ is the most common since it represents worst-case conditions.

### 9.4b  Voltage Breakdown in JFETs

The avalanche breakdown voltage may be expressed in a number of ways. In JFETs the gate junction breaks down whenever the combination of voltages applied to the other terminals exceeds a specific level. A common specification is $BV_{DGO}$; this is the value of $V_{DG}$ that will cause junction breakdown when the source is left open ($I_S = 0$), as shown in the circuit of Fig. 9.4b.

Recall that breakdown is characterized by a rise in current; therefore specifications of breakdown voltage generally include an arbitrary sensing current for which the measurement is made. For the 2N2386, the manufacturer specifies a minimum $BV_{DGO}$ of $-20$ V using a drain current of 10 μA as the criterion for sensing breakdown.

If the source is tied to the drain in the circuit of Fig. 9.4b, as indicated by the dashed lines, there is no appreciable change in the drain-gate voltage required to initiate gate junction breakdown. The parameter given by this connection is $BV_{GSS}$ (gate-source voltage with drain shorted). Note that, except for polarity

($BV_{DGO}$ is negative while $BV_{GSS}$ is positive in a P-channel JFET), the parameters $BV_{DGO}$ and $BV_{GSS}$ have the same value:

$$BV_{GSS} = -BV_{DGO} \tag{9.2}$$

Gate junction breakdown may also result from the application of excessive voltage between drain and source. The general case is to have both a drain-source voltage $V_{DS}$ and a gate-source voltage $V_{GS}$, as indicated in the circuit of Fig. 9.4c. Any combination of these two voltages that exceeds $BV_{DGO}$ results in breakdown. In general,

$$V_{DS} = V_{DG} + V_{GS} \tag{9.3}$$

The maximum allowable value of $V_{DS}$ is the maximum allowable value of $V_{DG} + V_{GS}$, from which it follows that

$$BV_{DSX} = BV_{DGO} + V_{GSX} \tag{9.3a}$$

where $X$ refers to a specific value of $V_{GS}$.

As an example, consider the 2N2386's common source drain characteristics in Appendix 10. When $V_{GS} = 0$, breakdown is initiated for $V_{DS} \simeq -27.5$ V. Using Eq. (9.3a), we have

$$-27.5 = BV_{DGO} + 0 \tag{9.3a}$$

Therefore $BV_{DGO} = -27.5$ V. Note that the earlier specification of $-20$ V represented a guaranteed *minimum* value for $BV_{DGO}$. The $-27.5$ V obtained from the characteristics is a *typical* value, but it is not guaranteed by the manufacturer. As $V_{GS}$ increases, the magnitude of $V_{DS}$ needed to initiate breakdown decreases; for the typical data given above, when $V_{GS} = 2.5$ V,

$$BV_{DSX} = -27.5 + 2.5 = -25 \text{ V}$$

which agrees with the characteristics.

### 9.4c  Voltage Breakdown in MOSFETs

There are two possibilities here. If excessive voltage is applied between gate and source or gate and drain, the insulating oxide layer may break down. This applies to any insulator; the voltage required for breakdown depends on how thick the oxide layer is and the concentration of impurities. Manufacturers will generally specify a maximum allowable value for $V_{GS}$ and $V_{DS}$. Some care in handling MOSFETs is also necessary, as the accumulation of static charge on the gate may result in rupture of the oxide layer. To overcome this problem the manufacturer may build into the MOSFET package an integrated breakdown diode that clamps the gate to the body of the device; this way the gate static charge is prevented from building up to the point where damage can occur (see

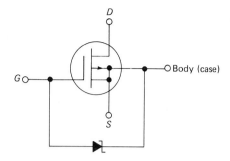

*Fig. 9.5 P channel MOSFET with built-in breakdown diode for preventing static charge build-up.*

Fig. 9.5). Another precaution is to short leads together when units are shipped or stored.

The drain-source breakdown voltage is specified as with JFETs. In cases where the MOSFET operates only in the enhancement mode, a $BV_{DSS}$ rating is given. Here the gate is shorted to the drain and minimum drain current flows (device is cut off). For MOSFETs that operate in either the enhancement or depletion mode, the device is cut off by applying the appropriate voltage to the gate. The rating usually specified here is $BV_{DSX}$, where $X$ refers to the specific value of $V_{GS}$. As an example, the $BV_{DSX}$ rating for the 2N3797 MOSFET is given in the data sheets (see Appendix 10) as 20 V minimum for $V_{GS} = -7$ V. The 5-$\mu$A drain current condition is given to indicate that breakdown voltage was defined as the voltage for which $I_D$ exceeds 5 $\mu$A.

### Example 9.1

Using the circuit of Fig. 9.4c and the typical data obtained from the drain characteristics of the 2N2386, determine the following:

a. $BV_{GSS}$
b. $BV_{DSX}$    $V_{GS} = 1$ V
c. $BV_{GSX}$    $V_{DS} = -20$ V

### Solution

a. $BV_{GSS}$ is the gate-source voltage required to initiate breakdown when the third terminal (drain) is shorted ($V_{DS} = 0$):

$$BV_{GSS} = -BV_{DGO} = 27.5 \text{ V} \tag{9.2}$$

b. $BV_{DSX} = BV_{DGO} + V_{GSX} \tag{9.3a}$
$BV_{DSX}$ (at $V_{GS} = 1$ V) $= -27.5 + 1 = -26.5$ V

c.  Here $BV_{GSX}$ represents the maximum allowable gate-source voltage when $V_{DS} = -20$ V. Equation (9.3a) may be rewritten as

$$V_{DS} = BV_{DGO} + BV_{GSX} \tag{9.3b}$$
$$BV_{GSX} = -20 - (-27.5) = 7.5 \text{ V}$$

## 9.4d  Drain Current Specifications

$I_{DSS}$ is the drain current with zero gate-source voltage for a specific value of $V_{DS}$. The circuit for measuring $I_{DSS}$ is shown in Fig. 9.6. Generally, $V_{DS}$ is chosen somewhere in the pinch-off (saturation) region.

Using the typical characteristics for the 2N2386 JFET, at $V_{DS} = -20$ V, $I_{DSS} = -4$ mA. Note that $I_{DSS}$ is not heavily dependent on $V_{DS}$ as long as $V_{DS}$ remains in the saturation region.

An additional parameter, the ON drain current, is also often specified. For depletion-mode devices, such as JFETs, $I_{D(ON)}$ is measured at a specific value of $V_{DS}$ and $V_{GS} = 0$. In this case $I_{DSS}$ and $I_{D(ON)}$ are the same.

For enhancement-mode devices, such as MOSFETs, $I_{D(ON)}$ is measured with a gate-source voltage larger than zero. In this case $I_{DSS}$ is not the maximum drain current. For example, the data sheets for the 2N3796 (Appendix 10) give the following typical values of drain current:

Zero gate-voltage drain current ($V_{DS} = 10$ V, $V_{GS} = 0$):  $I_{DSS} = 1.5$ mA
ON drain current ($V_{DS} = 10$ V, $V_{GS} = 3.5$ V):  $I_{D(ON)} = 8.3$ mA

For MOSFETs that operate *only* in the enhancement mode, $I_{DSS}$ represents a minimum value of drain current, since minimum conduction occurs with $V_{GS} = 0$. In general, both $I_{DSS}$ and $I_{D(ON)}$ are useful data for switching applications in which the FET is being turned ON. The variation in these parameters from one unit to another, however, can be rather severe (as high as a 3:1 factor), and this is a problem that must be coped with when designing FET circuits.

A measure of how well a FET can be turned OFF, also useful in switching applications, is the pinch-off drain current $I_{D(OFF)}$. A manufacturer will usually guarantee an $I_{D(OFF)}$ not to exceed some maximum value for a given $V_{GS}$ and

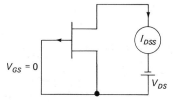

Fig. 9.6  Circuit for measuring $I_{DSS}$.

$V_{DS}$. For the 2N2386, the maximum $I_{D(OFF)}$ is 10 µA for a gate-source reverse bias of 8 V and $V_{DS} = -12$ V. For the 2N3796 MOSFET, similar information is provided by specifying $V_{GS(OFF)}$—that is, the value of gate-source reverse bias needed to reduce the drain current to some specific low value. In this case, not more than 4 V (typically 3 V) is necessary to reduce $I_D$ to 0.5 µA when $V_{DS} = 10$ V. It should be pointed out that the main factor controlling drain current is $V_{GS}$ and not $V_{DS}$. $V_{DS}$ has only a slight effect on $I_D$ beyond the pinch-off voltage.

## Problem Set 9.2

9.2.1. Why do you think leakage current is both lower and less temperature-dependent in MOSFETs than in JFETs?

9.2.2. Determine, using appropriate data sheets from Appendix 10, the typical 50°C value of $I_{GSS}$ for the 2N2386 JFET. The gate-source voltage is 10 V.

9.2.3. What is the maximum value of $I_{GSS}$ for the 2N3796/7 MOSFETs at 25°C and $V_{GS} = -10$ V?

9.2.4. Using appropriate data sheets, determine the minimum $BV_{DSX}$ for the 2N3796 MOSFET at 25°C and $V_{GS} = -4$ V. What is the corresponding value of $BV_{DGO}$?

9.2.5. Using the common source transfer characteristics for the 2N3797 MOSFET, determine $V_{GS(OFF)}$ at $V_{DS} = 10$ V, $T_A = 25°C$, for an $I_{D(OFF)} = 30$ µA.

## 9.5 BIASING THE FET

As with junction transistors, before a FET can amplify, it must be properly biased. Biasing generally involves three considerations:

a. Selection of $Q$ point
b. DC circuit design to yield desired $Q$ point
c. Circuit optimization to minimize shifts in $Q$ point due to device replacement or temperature variations

Selection of an operating point is partly based on AC performance specifications. These may include gain, frequency response, noise figure, impedance levels, distortion, and so on. Information is usually provided in manufacturer's data sheets to indicate the dependence of certain AC parameters on operating point. The stage configuration (common source, drain or gate) to be used is also usually determined on the basis of AC performance. Other considerations that must be taken into account in $Q$ point selection include the device's power, voltage, and current ratings.

Consider the common source drain characteristics of a JFET shown on the righthand side of Fig. 9.7. For linear amplification a $Q$ point should be selected

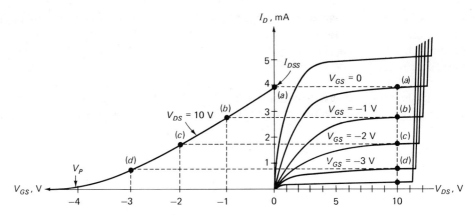

Fig. 9.7  Graphic development of transfer characteristics.

somewhere in the saturation region. On the lefthand side of Fig. 9.7 a transfer characteristic for the same device is given. This graph is obtained by transferring the information contained in the plots of $I_D$ versus $V_{DS}$ to a different form, namely a plot of $I_D$ versus $V_{GS}$ for a constant value of $V_{DS}$. The transfer characteristic contains no information other than what is available from the drain characteristics; it is simply a rearranged plot of the same data.

To construct the transfer characteristic, a constant value of $V_{DS}$ is selected. Since the drain characteristics are fairly flat in the saturation region, it does not really matter what value of $V_{DS}$ is chosen within this region. We have selected $V_{DS} = 10$ V here, as indicated by the vertical line along $V_{DS} = 10$ V. The intersection of this line with each characteristic yields a specific value of $V_{GS}$ and $I_D$. These values are then used to develop the transfer characteristic, which involves two endpoints: $I_{DSS}$, which is the value of $I_D$ for $V_{GS} = 0$, and the pinch-off voltage $V_P$, which represents the value of $V_{GS}$ when $I_D = 0$.

The transfer characteristic for the JFET can be approximated with an equation of the general form[1]

$$I_D = (I_{DSS} + I_{GSS})\left(1 - \frac{V_{GS}}{V_P}\right)^2 - I_{GSS} \tag{9.4}$$

which, for negligible $I_{GSS}$, simplifies to

$$I_D = I_{DSS}\left(1 - \frac{V_{GS}}{V_P}\right)^2 \tag{9.4a}$$

Thus if the end points $V_P$ and $I_{DSS}$ are known, a transfer characteristic for the device can be constructed. It is also possible to solve for the operating point of a JFET stage using the relationship of Eq. (9.4a) and a DC analysis of the circuit

---

[1] See p. 21 of Reference 9 at the end of this chapter.

involved. To do this, both $I_{DSS}$ and $V_P$ must be known, and the analysis remains valid only as long as there are no changes in these parameters.

## Example 9.2

Determine the quiescent values of $V_{GS}$, $I_D$, and $V_{DS}$ for the circuit of Fig. 9.8 using the following data: $I_{DSS} = 5$ mA, $V_P = -5$ V, $R_S = 5$ k$\Omega$, $R_L = 2$ k$\Omega$, $V_{DD} = 10$ V.

## Solution

The gate is at DC ground potential. Since $I_S \simeq I_D$, Eq. (9.1) applies, as follows:

$$0 = V_{GS} + I_D R_S \qquad (9.1)$$
$$V_{GS} = -I_D R_S$$
$$V_{GS} = -5{,}000 I_D \qquad (a)$$

Equation (a) cannot be solved because there are two unknowns, but the relationship of Eq. (9.4a) may be used to eliminate one of these unknowns:

$$I_D = I_{DSS}\left(1 - \frac{V_{GS}}{V_P}\right)^2 \qquad (9.4a)$$

$$I_D = (5 \times 10^{-3})\left(1 - \frac{V_{GS}}{-5}\right)^2 \qquad (b)$$

Substitution of Eq. (b) into Eq. (a) yields

$$V_{GS} = (-5{,}000)(5 \times 10^{-3})(1 + 0.2 V_{GS})^2$$
$$V_{GS}^2 + 11 V_{GS} + 25 = 0$$

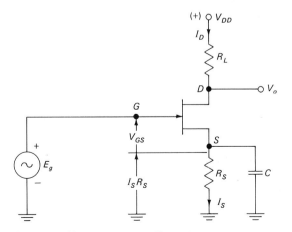

Fig. 9.8 Self-bias circuit for Example 9.2.

which has two solutions: $V_{GS} = -3.2$ V and $V_{GS} = -7.8$ V. The correct answer is $V_{GS} = -3.2$ V, since $-7.8$ V exceeds $V_P$, the voltage for which $I_D = 0$. Having determined $V_{GS}$, we can now solve for $I_D$ using Eq. (b):

$$I_D = (5 \times 10^{-3})(1 - 0.64)^2 = 0.65 \text{ mA}$$

The DC voltage drop across $R_L$ is

$$I_D R_L = 2{,}000 I_D = 1.3 \text{ V}$$

The DC voltage drop across $R_S$ is

$$I_D R_S \simeq 5{,}000 I_D = 3.25 \text{ V}$$

Therefore

$$V_{DS} = 10 - 1.3 - 3.25 = 5.45 \text{ V}$$

Now consider the self-bias circuit of Fig. 9.9. The term *self-bias* is used because the $V_{GS}$ bias is obtained from the FETs own source current flowing through $R_S$. This bias is the same as that in Example 9.2, except for the addition of $C_1$ and $R_G$.

If the signal generator is directly coupled to the gate (by replacing $C_1$ with a short circuit), the gate is at DC ground potential. In this case $R_G$ is not needed and the DC voltage drop across $R_S$ is also the source-gate voltage. In many applications, however, the FET may be driven from a previous circuit whose DC levels are incompatible with the FET's zero gate voltage. In this case, a coupling capacitor $C_1$ is used to keep each stage's DC levels independent. But unless a resistor $R_G$ is connected as indicated, the gate cannot be maintained at DC ground potential because of $C_1$. The function of $R_G$ is therefore to provide a DC path for gate leakage current, thus maintaining the gate at some DC voltage close to ground.

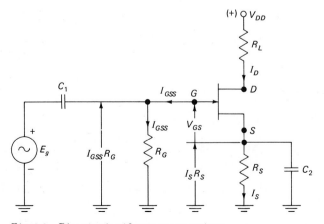

Fig. 9.9  Bias circuit with gate return resistor.

The magnitude of $R_G$ is dictated by two constraints. If $R_G$ is relatively low, its DC drop is also low, since the gate leakage current is usually in the nanoampere range. This arrangement keeps the gate pretty close to DC ground, and the relationship of Eq. (9.1) applies. But a low value of $R_G$ also reduces the input impedance to the FET, thus destroying one of the FET's main advantages. A lower limit is thus placed on $R_G$ based on the maximum loading that the driving source can tolerate.

If in an effort to increase the input impedance we make $R_G$ very large, other undesirable features result. With a gate leakage current of 10 nA, for example, and $R_G = 100$ M$\Omega$, the gate is 1 V above ground. To ensure a gate-source reverse bias of, say, 3 V, the source must be held at 4 V above ground. This is done by selecting an $I_D$ and $R_S$ whose product is 4 V. But $I_{GSS}$ is a function of temperature, and at elevated temperatures it could easily reach 100 nA. This would place the gate at $(100 \times 10^{-9})(R_G) = 10$ V above ground, thus altering the operating point substantially (the gate junction is now forward-biased). Based on these considerations there are usually lower and upper limits on the value of $R_G$.

In the circuit of Fig. 9.9, $I_D$ is the sum of $I_S$ and $I_{GSS}$:

$$I_D = I_S + I_{GSS} \tag{9.5}$$

The voltage across $R_G$ is also the voltage from gate to ground, as follows:

$$I_{GSS}R_G = V_{GS} + I_S R_S$$
$$I_{GSS}R_G = V_{GS} + (I_D - I_{GSS})(R_S)$$
$$V_{GS} = -I_D R_S + I_{GSS}(R_S + R_G) \tag{9.6}$$

When $I_{GSS}$ is negligible, Eq. (9.6) simplifies to Eq. (9.1).

In practice, both $I_{DSS}$ and $V_P$ can vary widely for devices of the same type, operating under identical conditions. Thus it becomes necessary for worst-case analysis to construct two transfer curves for $I_D$ versus $V_{GS}$, one for each extreme of $I_{DSS}$ and $V_P$. There are also temperature variations to consider, not only for $I_{GSS}$ but for $I_{DSS}$ as well (the resistance of silicon channels has a positive temperature coefficient, which means that $I_{DSS}$ drops with increasing temperature).

**Example 9.3**

The bias circuit of Fig. 9.9 must operate between temperature extremes of 25 and 75°C. $I_{GSS} = 10$ nA at 25°C and increases by a factor of 10 for every 50°C rise. The extreme values of $I_{DSS}$ and $V_P$, also corrected for temperature variations between 25 and 75°C, are

$\underline{V_P = 4.5 \text{ V}} \qquad \overline{V_P = 7.5 \text{ V}}$

$\underline{I_{DSS} = 5 \text{ mA}} \qquad \overline{I_{DSS} = 9 \text{ mA}}$

Construct the transfer characteristics and identify the range of $Q$ point variations for $R_G = 10$ MΩ, (a) with $R_S = 500$ Ω; (b) with $R_S = 5$ kΩ.

**Solution**

The transfer characteristics are constructed in Fig. 9.10 using the data given above and the relationship of Eq. (9.4). Each value of $R_S$ is now considered.

a.  $R_S = 500$ Ω. At 25°C Eq. (9.6) becomes

$$V_{GS} = -I_D(500) + (10 \times 10^{-9})(500 + 10^7) \quad (9.6)$$
$$V_{GS} \simeq -500 I_D + 0.1 \quad (a)$$

At 75°C, $I_{GSS} = 100$ nA, and our equation is modified to

$$V_{GS} \simeq -500 I_D + (100 \times 10^{-9})(10^7)$$
$$V_{GS} \simeq -500 I_D + 1 \quad (b)$$

Equations (a) and (b) are plotted on the transfer characteristics; the cross-hatched area between the two lines represents the range of $Q$ point variations for $R_S = 500$ Ω.

b.  $R_S = 5$ kΩ. At 25°C

$$V_{GS} = -I_D(5{,}000) + (10 \times 10^{-9})(5{,}000 + 10^7)$$
$$V_{GS} \simeq -5{,}000 I_D + 0.1 \quad (c)$$

At 75°C

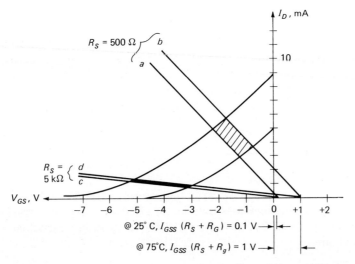

Fig. 9.10 Comparison of $Q$ point variation for $R_S = 500$ Ω and $R_S = 5$ kΩ.

$$V_{GS} = -5{,}000 I_D + (100 \times 10^{-9})(5{,}000 + 10^7)$$
$$V_{GS} \simeq -5{,}000 I_D + 1 \tag{d}$$

The blacked-in area between the two plots of Eqs. (c) and (d) represents the range of $Q$ point variations for $R_S = 5{,}000\ \Omega$.

Note that for $R_S = 500\ \Omega$ the variation in $I_D$ is about 3 mA; when $R_S = 5\ \text{k}\Omega$, however, $I_D$ only changes by about 0.7 mA. It follows that with a larger $R_S$, a more stable operating point can be achieved.

The diagram of Fig. 9.11 is a modified version of the self-bias circuit just described. Both the gate and source are positive with respect to ground; the gate because of $V_{GG}$ and the source because of the voltage drop across $R_S$. For the gate junction to be reverse-biased, $V_{GS}$ must be negative, which means that the source must be at a higher positive potential than the gate. For this circuit, we can therefore write

$$V_{GG} = -R_G I_{GSS} + V_{GS} + I_S R_S \tag{9.7}$$

But

$$I_S = I_D - I_{GSS} \tag{9.5}$$

Therefore

$$V_{GG} = -R_G I_{GSS} + V_{GS} + I_D R_S - I_{GSS} R_S$$
$$V_{GS} = -I_D R_S + I_{GSS}(R_S + R_G) + V_{GG} \tag{9.8}$$

**Example 9.4**

The amplifier of Fig. 9.12a is to be biased for reliable operation over a temperature range of $-25$ to $+75°C$. The following data, which include temperature corrections for the $I_{DSS}$ and $V_P$ specifications, are available:

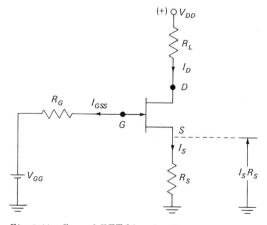

Fig. 9.11  General FET bias circuit.

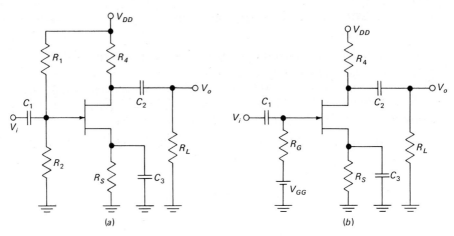

Fig. 9.12  (a) General FET bias circuit; (b) circuit of (a) with gate equivalent circuit.

$\overline{I_{GSS}} = 10$ nA at $+25°C$
Extreme values of $I_{DSS}$: 0.8 and 3 mA
Extreme values of $V_P$: $-3$ and $-10$ V

The amplifier must present an input impedance greater than 10 M$\Omega$ and deliver a peak voltage $V_{om} = 1$ V to $R_L = 30$ k$\Omega$. The values of $R_4$ and $V_{DD}$ are 15 k$\Omega$ and 50 V, respectively. What is the minimum value of $v_{DS}$?

**Solution**

a. The circuit is redrawn in Fig. 9.12b by replacing $R_1$ and $R_2$ with their Thevenin equivalent. The resulting DC circuit is identical to that of Fig. 9.11. The appropriate relationships are

$$V_{GG} = \frac{V_{DD}R_2}{R_1 + R_2} \tag{9.9}$$

$$R_G = R_1 \| R_2 \tag{9.10}$$

b. The AC load $R'_L$ seen by the FET is the parallel combination of $R_L$ and $R_4$:

$$R'_L = (30 \times 10^3) \| (15 \times 10^3) = 10 \text{ k}\Omega$$

c. The drain current must be capable of reaching a peak value $I_{dm}$ equal to

$$I_{dm} = \frac{V_{om}}{R'_L} = \frac{1}{10^4} = 0.1 \text{ mA}$$

d. The quiescent drain current $I_{DQ}$ must be chosen to ensure at least a 0.1-mA swing under worst-case conditions. Since $\underline{I_{DSS} = 0.8}$ mA, a reasonable value for $I_{DQ}$ (quiescent drain current) is 0.4 mA. Thus $I_D$ can swing between extremes of 0.3 and 0.5 mA.

e. To allow for parameter variations due to temperature and device replacement, some range for $I_{DQ}$ must be specified. Generally we want to keep away from the nonlinear regions; therefore we will assume allowable extremes for $I_{DQ}$ of 0.3 and 0.5 mA. Since the AC component of drain current can have a peak value of 0.1 mA, this means that the instantaneous drain current is allowed to vary between 0.2 and 0.6 mA.

f. The extreme transfer characteristics are plotted in Fig. 9.13 using the data supplied and the relationship of Eq. (9.4). The range of allowable values for $I_{DQ}$ is also shown.

g. As indicated in the graph of Fig. 9.10, two parallel load lines are required for any given value of $R_S$. Each of these load lines is drawn here for one of the temperature extremes. In Example 9.3, $R_S$ was known, and we were to determine the range through which the quiescent drain current could be expected to vary. Here we have the reverse situation. Having specified the allowable range for $I_{DQ}$, we must now determine the value of $R_S$ required. For this purpose two parallel load lines are drawn on the characteristic of Fig. 9.13 in such a manner as to enclose an area bounded by our $I_{DQ}$ extremes of 0.3 and 0.5 mA. The voltage-current ratio along either of these load lines corresponds to $R_S$. By computing this ratio between points $X$ and $Y$ we have

$$R_S = \frac{5.3}{0.1 \times 10^{-3}} = 53 \text{ k}\Omega$$

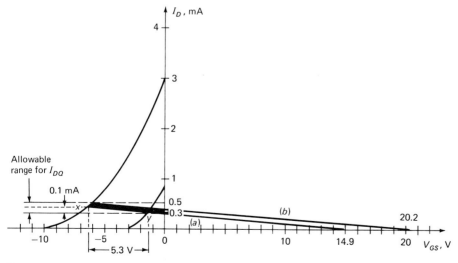

Fig. 9.13   Determination of $R_S$ for Example 9.4.

h. Both load lines $a$ and $b$ represent a plot of

$$V_{GS} = -I_D R_S + I_{GSS}(R_S + R_G) + V_{GG} \tag{9.8}$$

The intercept of the above equation on the $V_{GS}$ axis is obtained by setting $I_D = 0$; this must be done for both extremes, one for a temperature of $-25$ and the other for 75°C. At 25°C, $I_{GSS}$ is negligible; hence the $V_{GS}$ axis intercept is

$$V_{GS1} = V_{GG} \tag{a}$$

At 75°C, $I_{GSS}$ may be assumed to be about 10 times the 25°C value, yielding 100 nA. The $V_{GS}$ axis intercept is therefore

$$V_{GS2} = 10^{-7}(R_S + R_G) + V_{GG} \tag{b}$$

An examination of our graphic construction in Fig. 9.13 indicates that

$$V_{GS1} = 14.9 \text{ V}$$

Therefore

$$V_{GG} = 14.9 \text{ V}$$
$$V_{GS2} = 20.2 \text{ V}$$

Therefore

$$20.2 = 10^{-7}(R_S + R_G) + V_{GG} \tag{b}$$
$$20.2 = 10^{-7}[(53 \times 10^3) + R_G] + 14.9$$
$$R_G \simeq 53 \text{ M}\Omega$$

well above the 10 MΩ required for $R_i$.

i. To determine $R_1$ and $R_2$, we use

$$V_{GG} = \frac{V_{DD} R_2}{R_1 + R_2} \tag{9.9}$$

$$14.9 = \frac{50 R_2}{R_1 + R_2}$$

$$R_1 = 2.36 R_2$$

$$R_G = \frac{R_1 R_2}{R_1 + R_2} = 53 \text{ M}\Omega \tag{9.10}$$

Therefore

$$R_1 = 178 \text{ M}\Omega \quad R_2 = 75 \text{ M}\Omega$$

j. Minimum $v_{DS}$ occurs when the instantaneous drain current is maximum; as indicated in (e), this is 0.6 mA, yielding

$$\underline{v_{DS}} = V_{DD} - \overline{i_D}(R_4 + R_S)$$
$$\underline{v_{DS}} = 50 - (0.6 \times 10^{-3})(68 \times 10^3)$$
$$\underline{v_{DS}} = 9.2 \text{ V}$$

With the selection of $R_G$, $R_1$, $R_2$, and $R_S$ to the nearest commercially available values, the DC circuit design is complete. Close-tolerance resistors should be used, since small changes in component value can affect the $Q$ point.

The JFET biasing circuits so far discussed may also be used for depletion-mode MOSFETs. Since gate leakage current is much smaller in MOSFETs, larger values of $R_G$ may be used. For enhancement-mode operation, though, some modifications are required.

Consider the shunt feedback bias circuit of Fig. 9.14a. The only current through $R_F$ is due to gate leakage, and since for MOSFETs this is quite small, no appreciable voltage is developed across $R_F$. This means that there is no DC voltage between gate and drain; hence $V_{GS} = V_{DS}$. This relationship is plotted on the drain characteristics of Fig. 9.14b. The 1.5-kΩ load line is also plotted between the horizontal intercept $V_{DS} = V_{DD} = -15$ V and the vertical intercept $I_D = V_{DD}/R_L = -10$ mA. The intersection of the load line with the $V_{GS} = V_{DS}$ locus locates the $Q$ point. Note that $R_F$ stabilizes the $Q$ point. For example, should $I_D$ rise, the voltage drop across $R_L$ also rises; since $V_{DG} = 0$, this means that $V_{GS}$ must become less negative, thereby decreasing $I_D$.

Greater biasing flexibility is provided by the circuit of Fig. 9.14c. Here $V_{GS}$ does not have to equal $V_{DS}$; instead the following applies:

$$V_{GS} = \frac{V_{DS} R_G}{R_F + R_G} \tag{9.11}$$

For example, if $R_F = R_G = 50$ MΩ, $V_{GS} = 0.5 V_{DS}$. This yields another locus on the characteristics of Fig. 9.14b, as shown.

In both circuits of Fig. 9.14a and c operating point stabilization is provided by the shunt feedback resistor $R_F$. The input impedance, however, is reduced because of the Miller effect that makes $R_F$ appear divided by $1 - A_v$. For example, if $A_v = -9$, the shunting effect due to $R_F$ is $50 \times 10^6/10 = 5$ MΩ. Since $R_G = 50$ MΩ, $R_i \simeq 4.5$ MΩ. One way to retain the DC feedback while eliminating the AC feedback is to split $R_F$ in two resistances, with a bypass capacitor $C_3$ between the junction and ground, as shown in Fig. 9.14d. This arrangement can also be applied to the circuit of Fig. 9.14a. The total DC resistance between gate and drain is still $R_F$ ($R_F = R_{F1} + R_{F2}$), but the loading effect on the input is now reduced so that $R_i = R_G \| R_{F1}$. On the output side, $R_{F2}$ is in parallel with $R_L$. In our case, since $R_L = 1.5$ kΩ, selecting $R_{F2} = 100$ kΩ

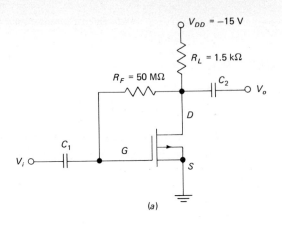

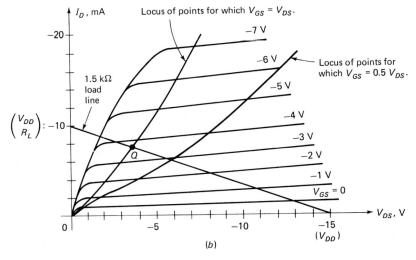

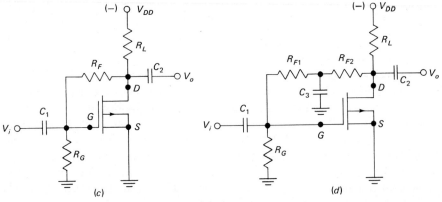

Fig. 9.14  (a) Shunt feedback bias circuit for enhancement mode P channel MOSFET; (b) graphical determination of Q point; (c) modified shunt feedback bias circuit; (d) ac grounding the midpoint of $R_F$ to reduce input loading due to Miller effect.

and $R_{F1} \simeq 50$ MΩ would result in negligible loading at the output and $R_i \simeq 25$ MΩ.

Another MOSFET biasing arrangement is the circuit of Fig. 9.12a, with or without $R_2$. Here stabilization of the operating point results from series (source) feedback as opposed to shunt feedback.

### Problem Set 9.3

9.3.1. The following specifications apply to a typical N-channel JFET: $V_P = -3$ V, $I_{DSS} = 2.5$ mA. Assume $I_{GSS}$ is negligible and construct the transfer characteristic.

9.3.2. If a bias circuit of the type in Fig. 9.9 is used with the device of Prob. 9.3.1, determine the operating point for $R_S = 1$ kΩ and $R_G = 10$ MΩ.

9.3.3. For the circuit of Prob. 9.3.2, determine the quiescent drain current extremes resulting from the following parameter variations: extreme values of $V_P$ are $-3$ and $-6$ V; extreme values of $I_{DSS}$ are 2 and 8 mA.

9.3.4. Assume that the transfer characteristics of Prob. 9.3.3 apply to the JFET in the bias circuit of Fig. 9.12a. Determine the range of $I_{DQ}$ if

    a. $I_{GSS}$ is negligible
    b. $R_1 = 150$ MΩ; $R_2 = 50$ MΩ
    c. $R_S = 5$ kΩ
    d. $V_{DD} = 20$ V

9.3.5. Repeat Prob. 9.3.4 (use the same transfer characteristics) but assume $I_{GSS} = 100$ nA.

9.3.6. Why do you think the self-bias circuit of Fig. 9.8 cannot be used with MOSFETs in the enhancement mode?

9.3.7. For the circuit of Fig. 9.14d and the characteristics of Fig. 9.14b, locate the Q point when $V_{DD} = 10$ V, $R_L = 500$ Ω, $R_{F2} = 100$ kΩ, $R_{F1} = 100$ MΩ, $R_G = 25$ MΩ. Estimate $R_i$.

## 9.6 DYNAMIC CHARACTERISTICS

A model for analyzing small-signal FET amplifiers is shown in Fig. 9.15a. The component values may be obtained from characteristic curves, data-sheet specifications, or experimental measurements. Each component is first discussed; analysis of amplifier circuits follows.

$C_{gs}$ and $C_{gd}$ are depletion-region capacitances in JFETs; in MOSFETs they are due to the insulating oxide layer that separates the gate area from the source and drain. In either case, $C_{gs}$ and $C_{gd}$ are usually in the range of a few picofarads. $C_{ds}$ is the equivalent capacitance across the drain-source channel; it is usually quite low. In general, manufacturers may specify $C_{gs}$ and $C_{gd}$ at a given

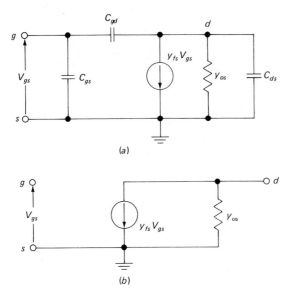

**Fig. 9.15** (a) *Small signal FET model;* (b) *low frequency equivalent of* (a).

$V_{GS}$ and frequency. More often, however, the input capacitance $C_{iss}$ (common source configuration with the output AC short-circuited) is given. For the common source model of Fig. 9.15a, shorting the output places $C_{gs}$ and $C_{gd}$ in parallel; therefore

$$C_{iss} = C_{gs} + C_{gd} \tag{9.12}$$

Another capacitive parameter usually given is $C_{rss}$, the reverse transfer capacitance with the input AC short-circuited. This capacitance is equivalent to $C_{gd}$; hence we may write

$$C_{rss} = C_{gd} \tag{9.13}$$

As an example, the 2N3796 silicon N-channel MOSFET has typical $C_{iss}$ and $C_{rss}$ specifications of 5 and 0.5 pF, respectively. This is equivalent to $C_{gd} = 0.5$ pF and $C_{gs} = 4.5$ pF for the test conditions indicated ($V_{DS} = 10$ V, $V_{GS} = 0$, $f = 1$ MHz) on the data sheets. When using dual gate (tetrode) FETs, the test conditions for each gate connection must be specified.

In addition to the capacitances shown in the model of Fig. 9.15a, there are also equivalent resistances from gate to drain ($r_{gd}$) and from gate to source ($r_{gs}$). These resistances are of the order of 100 MΩ in JFETs and in excess of $10^9$ Ω in MOSFETs; therefore they can be neglected without any appreciable error.

At audio frequencies all capacitances are negligible; the FET model simplifies to that of Fig. 9.15b. Note here that the FET generates an *output* current proportional to the *control* voltage $V_{gs}$. The constant of proportionality $y_{fs}$ is the forward transconductance or forward transadmittance. The symbol $y$ is for admittance, $f$ means forward, and $s$ stands for a short-circuited output. The units of $y_{fs}$ are mhos. In some cases, the symbols $g_m$ or $g_f$ may also be used to denote $y_{fs}$. Note that the $y_{fs}$ generator is similar to the $g_m V_{b'e}$ source in the hybrid-pi model for junction (bipolar) transistors; in either case it is a voltage-controlled current generator.

The forward transadmittance may be determined experimentally by taking the ratio of drain current to $V_{gs}$ with a short-circuited output:

$$y_{fs} = \left. \frac{I_d}{V_{gs}} \right|_{V_{ds}=0} \tag{9.14}$$

If a set of transfer characteristics, such as those of Fig. 9.16, are available, small changes about the operating point may be used to yield

$$y_{fs} = \left. \frac{\Delta I_D}{\Delta V_{GS}} \right|_{V_{DS}=K} \tag{9.14a}$$

Because the transfer characteristics are nonlinear, $y_{fs}$ increases with $I_D$. The data sheets for the 2N3796/7 MOSFET include a graph in which $|y_{fs}|$ (magnitude of forward transfer admittance) is plotted as a function of DC drain current. The magnitude of $y_{fs}$ is specified because in general $y_{fs}$ is a complex number; it has real and imaginary terms due to the presence of resistive and capacitive components. Only at low frequencies can $y_{fs}$ be treated as a pure number (no reactive components). Note that for the 2N3797, $|y_{fs}|$ increases from 1,000 to 3,000 μmho as $I_D$ increases from 0.5 to 5 mA.

An examination of Eq. (9.14a) indicates that $y_{fs}$ is the magnitude of the slope of the $I_D$-$V_{GS}$ transfer characteristic. For JFETs $y_{fs}$ may be obtained analytically by differentiating Eq. (9.4a). The procedure is carried out in Appendix 11.3, and the results are given here:

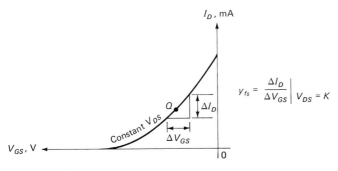

Fig. 9.16  *Graphic determination of* $y_{fs}$.

$$y_{fs} = \frac{2\sqrt{I_D I_{DSS}}}{|V_P|} \qquad (9.15)$$

Note that $y_{fs}$ increases with the square root of DC drain current.

For dual gate devices, three forward transadmittance specifications are possible, one for both gates tied together (usually yields the highest $y_{fs}$) and the other two for each individual gate.

The remaining component in the models of Fig. 9.15 is the output conductance $y_{os}$. This parameter may be determined experimentally by taking the ratio of output current to output voltage when the input is AC short-circuited:

$$y_{os} = \left.\frac{I_d}{V_{ds}}\right|_{V_{gs}=0} \qquad (9.16)$$

It is, of course, possible to treat this parameter as a resistance instead of a conductance; in this case we have

$$r_d = \frac{1}{y_{os}} \qquad (9.17)$$

A set of output characteristics, as in Fig. 9.17, may be used to determine $y_{os}$. Taking small changes about the operating point, we have

$$y_{os} = \left.\frac{\Delta I_D}{\Delta V_{DS}}\right|_{V_{GS}=K} \qquad (9.16a)$$

The output admittance is also, in general, a complex quantity. Data sheets for the 2N3796/7 give a plot of $|y_{os}|$ as a function of DC drain current. Note that as $I_D$ increases, $|y_{os}|$ also increases, which means that the output resistance decreases.

Depending on the application, additional parameters may be specified. In

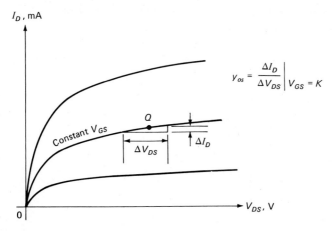

Fig. 9.17  Graphic determination of $y_{os}$.

switching circuits, the drain-source ON resistance determines how well the FET approaches a short circuit when fully turned on. The symbol is $r_{ds(ON)}$, and it is measured with $V_{GS} = 0$ for depletion-mode devices. For enhancement-mode operation, gate-source voltage is required to turn the device on and to measure $r_{ds(ON)}$.

At high frequencies, the input capacitance $C_{iss}$ causes the gain to drop; therefore keeping $C_{iss}$ as low as possible improves the high-frequency response. FETs for use at high frequencies should have a high $y_{fs}/C_{iss}$ ratio—that is, $y_{fs}$ should be as high as possible while $C_{iss}$ is kept as low as possible. The ratio $y_{fs}/C_{iss}$ is, in fact, a high-frequency figure of merit for FETs in general.

### Problem Set 9.4

9.4.1. Using the characteristics for the 2N3796 in Appendix 10, determine $y_{fs}$ and $y_{os}$ at $I_D = 10$ mA, $f = 1$ kHz, $V_{DS} = 10$ V, and $T_A = 25°C$.

9.4.2. Bearing in mind that a FET generates an *output current* proportional to the input *control voltage*, could you classify the FET as approaching an ideal voltage or ideal current amplifier?

9.4.3. Compute the minimum $y_{fs}/C_{iss}$ figure of merit for the 2N3796, using data given in the data sheet for 1 MHz.

9.4.4. A given JFET has $I_{DSS} = 10$ mA, $V_P = -4$ V, and is biased at $I_D = 2.5$ mA. Determine $y_{fs}$.

## 9.7 ANALYSIS OF FET AMPLIFIER CIRCUITS

The basic FET amplifier configurations are discussed in this section. Performance measures, including suitable approximations, are developed. The analysis is limited to low-frequency linear (small-signal) circuits.

### 9.7a The Common Source Configuration

A common source FET amplifier is shown in Fig. 9.18a. The circuit is driven from a generator whose internal resistance ($R_g$) is negligible compared with the input resistance to the FET. The AC equivalent circuit is shown in Fig. 9.18b. Note that $r_d$ instead of $1/y_{os}$ is used ($r_d = 1/y_{os}$).

Basic performance measures for a common source amplifier are now derived. The gate draws no appreciable current; therefore, except for the small current to $R_G$, the generator supplies no current and $A_i$ approaches infinity. The other quantities to consider are as follows.

INPUT RESISTANCE. The signal generator sees $R_G$ in parallel with the resistance looking into the gate which may be considered infinite. Therefore, in general,

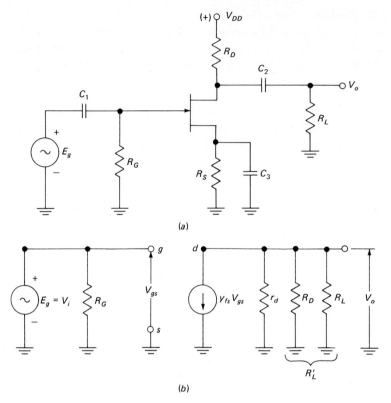

*Fig. 9.18  Analysis of common source amplifier.*

$$R_i = R_G \tag{9.18}$$

*OUTPUT RESISTANCE.* This is the resistance looking into the FET's output with $R_L$ removed and $E_g$ reduced to zero. This means that $V_{gs} = 0$ and the $y_{fs}V_{gs}$ generator produces no current; hence only $r_d$ and $R_D$ remain:

$$R_o = r_d \| R_D \tag{9.19}$$

*VOLTAGE AMPLIFICATION.* This is the ratio of $V_o$ to $V_i$:

$$A_v = \frac{V_o}{V_i} = \frac{V_o}{V_{gs}}$$

But

$$V_o = -y_{fs}V_{gs}(r_d \| R_L')$$

where $R_L' = R_D \| R_L$. Therefore

$$A_v = \frac{-y_{fs}r_d R'_L}{r_d + R'_L} \qquad (9.20)$$

If $r_d \gg R'_L$, $A_v$ simplifies to

$$A_v \simeq -y_{fs}R'_L \qquad (9.20a)$$

Note that when $V_{gs}$ is positive-going, $y_{fs}V_{gs}$ causes $V_o$ to be negative-going, so that $V_o$ and $V_{gs}$ are 180° out of phase and $A_v$ is negative.

### Example 9.5

Estimate $R_i$, $R_o$, and $A_v$ for the circuit in Fig. 9.18a if $R_G = 10\ \text{M}\Omega$, $R_D = 10\ \text{k}\Omega$, $R_L = 15\ \text{k}\Omega$, $y_{fs} = 4{,}000\ \mu\text{mho}$, $y_{os} = 10\ \mu\text{mho}$.

### Solution

a.  $R_i = R_G$     (9.18)
    $R_i = 10\ \text{M}\Omega$

b.  $R_o = r_d \| R_D$     (9.19)
    $$r_d = \frac{1}{y_{os}} = \frac{1}{10 \times 10^{-6}} = 100\ \text{k}\Omega \qquad (9.17)$$
    $R_o = 10^5 \| 10^4 = 9.9\ \text{k}\Omega$

c.  $$R'_L = \frac{R_D R_L}{R_D + R_L} = 6\ \text{k}\Omega$$

Since $6\ \text{k}\Omega \ll 100\ \text{k}\Omega$, $r_d$ may be neglected, and

$$A_v \simeq -y_{fs}R'_L \qquad (9.20a)$$
$$A_v \simeq (-4{,}000 \times 10^{-6})(6 \times 10^3) = -24$$

## 9.7b  The Common Drain Configuration

Also referred to as a source follower, this configuration is analogous to the common collector (emitter follower) bipolar transistor circuit. A typical circuit and its AC equivalent are shown in Fig. 9.19a and b, respectively. Here again the signal generator's internal resistance is neglected, as it is usually small compared with $R_i$. Note that the direction of current flow through the $y_{fs}V_{gs}$ generator is from drain to source, as it was for the common source circuit previously discussed. In this case, however, the drain is grounded and the direction of $y_{fs}V_{gs}$ causes $V_o$ to be positive when $V_i$ is positive. There is, therefore, no phase shift between input and output voltages.

The basic performance measures are as follows.

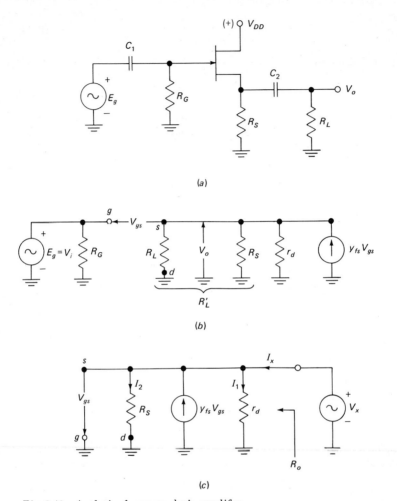

Fig. 9.19 Analysis of common drain amplifier.

CURRENT AMPLIFICATION. This approaches infinity for the same reasons as in common source circuits.

INPUT RESISTANCE. Here again the signal generator sees only $R_G$; therefore

$$R_i = R_G \tag{9.18}$$

VOLTAGE AMPLIFICATION. This is the ratio of $V_o$ to $V_i$:

$$A_v = \frac{V_o}{V_i}$$

But
$$V_i = V_o + V_{gs}$$
and
$$V_o = y_{fs}V_{gs}\frac{r_d R'_L}{r_d + R'_L}$$
where $R'_L = R_S \| R_L$; we now have
$$A_v = \frac{V_o}{V_o + V_{gs}} = \frac{1}{1 + (V_{gs}/V_o)}$$
$$A_v = \frac{1}{1 + [(r_d + R'_L)/(y_{fs}r_d R'_L)]} \tag{9.21}$$
If $r_d \gg R'_L$, $A_v$ simplifies to
$$A_v \simeq \frac{1}{1 + (1/y_{fs}R'_L)} \tag{9.21a}$$
Note that $A_v$ is positive and less than 1.

OUTPUT RESISTANCE. To get $R_o$, $R_L$ is removed, and $E_g$ is replaced with a short circuit, as shown in Fig. 9.19c. $R_o$ is then the ratio of $V_x$ to $I_x$:
$$R_o = \frac{V_x}{I_x} \tag{a}$$
$$I_x = I_1 - y_{fs}V_{gs} + I_2 \tag{b}$$
But
$$V_x = -V_{gs}$$
therefore
$$I_x = \frac{V_x}{r_d} + y_{fs}V_x + \frac{V_x}{R_S} \tag{c}$$
Substituting Eq. (c) into Eq. (a), we have
$$R_o = \frac{1}{(1/r_d) + y_{fs} + (1/R_S)} \tag{9.22}$$
If $r_d \gg R_S$, $R_o$ simplifies to
$$R_o \simeq \frac{1}{y_{fs} + (1/R_S)} = \frac{R_S}{1 + y_{fs}R_S} \tag{9.22a}$$

## Example 9.6

Estimate $R_i$, $A_v$, and $R_o$ for the circuit of Fig. 9.19a if $R_G = 10$ M$\Omega$, $R_S = 10$ k$\Omega$, $R_L = 10$ k$\Omega$, $y_{fs} = 4{,}000$ $\mu$mho, $y_{os} = 10$ $\mu$mho.

## Solution

a. $R_i = R_G$ (9.18)
$R_i = 10$ M$\Omega$

b. $R'_L = R_S \| R_L = 5$ k$\Omega$

$$r_d = \frac{1}{y_{os}} = 10^5 \text{ k}\Omega$$

$$A_v \simeq \frac{1}{1 + (1/y_{fs}R'_L)} \qquad (9.21a)$$

$$A_v \simeq \frac{1}{1 + [1/(4 \times 10^{-3})(5 \times 10^3)]}$$

$$A_v \simeq 0.95$$

c. $R_o \simeq \dfrac{R_S}{1 + y_{fs}R_S}$ (9.22a)

$$R_o \simeq \frac{10^4}{1 + (4 \times 10^{-3})(10^4)}$$

$$R_o \simeq 244 \ \Omega$$

### 9.7c  The Common Gate Configuration

This configuration exhibits the lowest input impedance. It is analogous to the common base bipolar configuration.

A typical common gate amplifier is shown in Fig. 9.20a. The AC equivalent circuit is shown in Fig. 9.20b and further simplified in Fig. 9.20c by converting the current source to an equivalent voltage source (Norton to Thevenin transformation). The generator's internal resistance $R_g$ is included because it may not be negligible compared with $R_i$. No $R_G$ is required, since the gate is grounded.

INPUT RESISTANCE.  $R_i$ may be calculated using the circuit of Fig. 9.20c, as follows:

$$R_i = R_S \| R'_i$$

where

$$R'_i = \frac{V_i}{I_i} = \frac{-V_{gs}}{I_i} \qquad (a)$$

A loop equation around the external circuit yields

Field Effect Transistors and Circuits 355

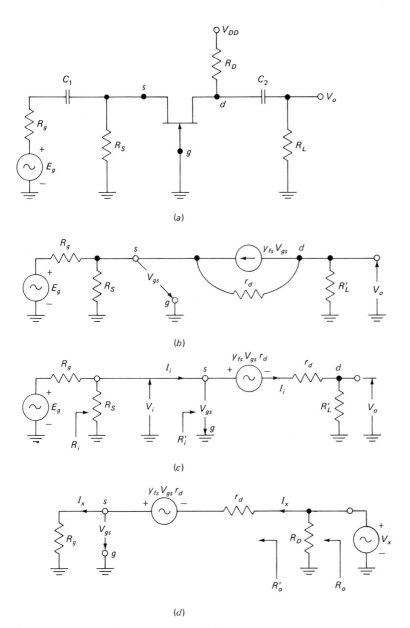

Fig. 9.20 Analysis of common gate amplifier.

$$V_{gs} + y_{fs}V_{gs}r_d + I_i r_d + I_i R'_L = 0$$
$$V_{gs} = \frac{-I_i(r_d + R'_L)}{1 + y_{fs}r_d} \qquad (b)$$

Substitution of Eq. (b) into Eq. (a) yields

$$R'_i = \frac{-V_{gs}}{I_i} = \frac{r_d + R'_L}{1 + y_{fs}r_d}$$

The input resistance is therefore

$$R_i = R_S \left\| \frac{r_d + R'_L}{1 + y_{fs}r_d} \right. \qquad (9.23)$$

If $r_d \gg R'_L$, $R_i$ simplifies to

$$R_i \simeq R_S \left\| \frac{1}{(1/r_d) + y_{fs}} \right. = R_S \left\| \frac{1}{y_{os} + y_{fs}} \right. \qquad (9.23a)$$

VOLTAGE AMPLIFICATION. The output voltage is $I_i R'_L$ and $V_i = -V_{gs}$; therefore

$$A_v = \frac{V_o}{V_i} = \frac{I_i R'_L}{-V_{gs}}$$

Substituting from Eq. (b), we have

$$A_v = \frac{R'_L(1 + y_{fs}r_d)}{r_d + R'_L} \qquad (9.24)$$

If $r_d \gg R'_L$, $A_v$ simplifies to

$$A_v \simeq R'_L \left( \frac{1}{r_d} + y_{fs} \right)$$

Usually $y_{fs} \gg 1/r_d$; therefore

$$A_v \simeq R'_L y_{fs} \qquad (9.24a)$$

OUTPUT RESISTANCE. The signal generator is shorted and $R_L$ removed; using the equivalent circuit of Fig. 9.20d, we write

$$R_o = R_D \| R'_o$$

where

$$R'_o = \frac{V_x}{I_x}$$

A loop equation around the external circuit yields

$$I_x R_g - y_{fs} V_{gs} r_d + I_x r_d = V_x$$

But $V_{gs} = -I_x R_g$; therefore

$$V_x = I_x R_g + y_{fs} I_x R_g r_d + I_x r_d$$

from which

$$R_o' = \frac{V_x}{I_x} = R_g + y_{fs} R_g r_d + r_d$$

and

$$R_o = R_D \| [R_g(1 + y_{fs} r_d) + r_d] \qquad (9.25)$$

If $R_g(1 + y_{fs} r_d) \ll r_d$, $R_o$ simplifies to

$$R_o \simeq R_D \| r_d \qquad (9.25a)$$

**Example 9.7**

Estimate $R_i$, $R_o$, and $A_v$ in the circuit of Fig. 9.20a if $R_g = 0$, $R_S = 10$ k$\Omega$, $R_D = 2$ k$\Omega$, $R_L = 4$ k$\Omega$; $y_{fs} = 4{,}000$ $\mu$mho, $y_{os} = 10$ $\mu$mho.

**Solution**

a. $R_L' = R_D \| R_L = 1{,}333$ $\Omega$ and $r_d = 100$ k$\Omega$; therefore $r_d \gg R_L'$ and

$$R_i \simeq R_S \left\| \frac{1}{y_{os} + y_{fs}} \right. \qquad (9.23a)$$

$$R_i \simeq 10{,}000 \left\| \frac{1}{4{,}010 \times 10^{-6}} \right.$$

$$R_i \simeq 10{,}000 \| 250 = 244 \; \Omega$$

b. $R_g = 0$; therefore

$$R_o \simeq R_D \| r_d \qquad (9.25a)$$
$$R_o = 2{,}000 \| 100{,}000 = 1{,}960 \; \Omega$$

c. $R_L' \ll r_d$; therefore

$$A_v \simeq R_L' y_{fs} = 1{,}333(4 \times 10^{-3}) \qquad (9.24a)$$
$$A_v \simeq 5.3$$

### 9.7d  Maximizing the Input Resistance

One of the FET's advantages over bipolar transistors is its high input resistance. In both common source and common drain configurations, $R_i$ is limited by $R_G$. Because of biasing considerations, however, $R_G$ cannot be too large and in fact prevents us from achieving the large input resistance that FETs are capable of providing. To get around this problem, bootstrapping techniques are often employed.

Consider the common drain circuit of Fig. 9.21. The source resistor has been separated into two components, $R_{S1}$ and $R_{S2}$. The gate resistor $R_G$ is tied to the junction of $R_{S1}$ and $R_{S2}$ instead of being returned to ground. This has negligible effect on the DC bias, but the effective value of $R_G$ is increased due to the bootstrapping action. Since the gate draws no current, the input resistance is

$$R_i = \frac{V_i}{I_G}$$

where

$$I_G = \frac{V_i - V_x}{R_G}$$

$$R_i = \frac{V_i R_G}{V_i - V_x} = \frac{R_G}{1 - (V_x/V_i)} \tag{9.26}$$

Since the current through $R_G$ is usually quite small compared to the source current, $R_{S1}$ and $R_{S2}$ are in series, which allows us to calculate $V_x$ as follows:

$$V_x = V_o \frac{R_{S2}}{R_{S1} + R_{S2}}$$

$$\frac{V_x}{V_i} = \frac{V_o}{V_i} \frac{R_{S2}}{R_{S1} + R_{S2}}$$

$$\frac{V_x}{V_i} = A_v \frac{R_{S2}}{R_{S1} + R_{S2}} \tag{9.27}$$

where $A_v$ is given by Eq. (9.21) or Eq. (9.21a).

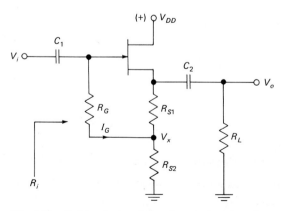

Fig. 9.21  Raising the input impedance by bootstrapping.

### Example 9.8

Estimate $R_i$ for the circuit of Fig. 9.21 if $R_G = 10$ M$\Omega$, $R_{S1} = 500$ $\Omega$, $R_{S2} = 9.5$ k$\Omega$, $R_L = 10$ k$\Omega$, $y_{fs} = 4{,}000$ $\mu$mho, $y_{os} = 10$ $\mu$mho.

### Solution

The load seen by the FET is

$$R'_L = R_L \| (R_{S1} + R_{S2})$$
$$R'_L = 10^4 \| 10^4 = 5 \text{ k}\Omega$$

Note that $R'_L \ll r_d$; hence the voltage amplification is

$$A_v \simeq \frac{1}{1 + (1/y_{fs} R'_L)} \tag{9.21a}$$

$$A_v \simeq \frac{1}{1 + [1/(4 \times 10^{-3})(5 \times 10^3)]} = 0.952$$

To compute $R_i$, the ratio $V_x/V_i$ is required; this is given by Eq. (9.27), as follows:

$$\frac{V_x}{V_i} = A_v \frac{R_{S2}}{R_{S1} + R_2} \tag{9.27}$$

$$\frac{V_x}{V_i} = \frac{0.952(9.5 \times 10^3)}{10^4} = 0.905$$

$$R_i = \frac{R_G}{1 - (V_x/V_i)} \tag{9.26}$$

$$R_i = \frac{10 \times 10^6}{1 - 0.905} = 105 \text{ M}\Omega$$

Note that bootstrapping has substantially increased $R_i$. With $R_G$ returned to ground, the full $V_i$ is across $R_G$, and $I_G = V_i/R_G$. With $R_G$ tied to $V_x$, however, there is less voltage across $R_G$, and hence less current, resulting in a larger effective input resistance. Obviously, the higher $V_x$ is relative to $V_i$, the less current there is through $R_G$ and the larger $R_i$ becomes. To increase the ratio $V_x/V_i$ we might be tempted to make $R_{S1} = 0$, in which case $V_x = V_o$. But now the gate and source are at the same DC potential (the DC voltage drop across $R_G$ is negligible), and hence $V_{GS} = 0$. This provides no gate reverse bias. There is, therefore, a lower limit to $R_{S1}$ based on the minimum DC gate bias that will yield satisfactory operation.

Other bootstrapping connections are shown in Fig. 9.22. The common drain circuit in Fig. 9.22a utilizes an arrangement similar to that used for emitter followers (see Sec. 7.4). In Fig. 9.22b the drain is also bootstrapped to minimize the capacitive loading due to $C_{gd}$.

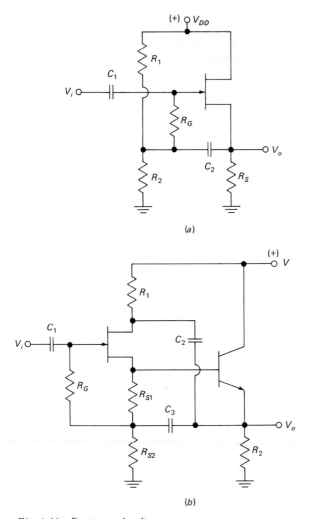

Fig. 9.22 Bootstrap circuits.

## Problem Set 9.5

9.5.1. For each circuit shown in Fig. 9.23, assume $y_{fs} = 1{,}000$ μmho and $r_d = 50$ kΩ. Estimate $A_v$ and $R_i$, making reasonable approximations.

9.5.2. Indicate the FET configuration (CS, CD, or CG) for which the numerical values given below might apply:

a. $R_i = 1{,}000$ Ω
b. $A_v = 15$
c. $R_o = 500$ Ω
d. $A_i = 10^6$

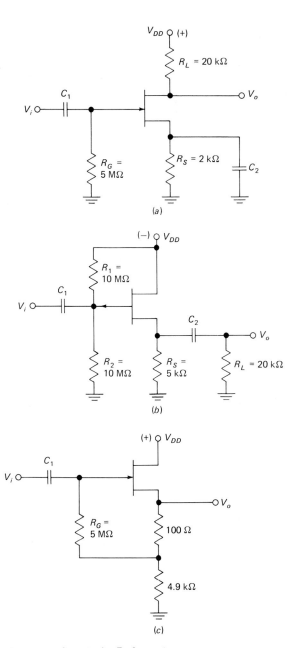

Fig. 9.23 Circuits for Prob. 9.5.1.

e. $A_v = 0.8$
f. $A_v = -10$
g. $A_i = -1$
h. $R_o = 100\ \Omega$

9.5.3. For the circuit of Fig. 9.21 it is desired to have $R_i = 50\ \text{M}\Omega$. The sum of $R_{S1}$ and $R_{S2}$ must equal $10\ \text{k}\Omega$; $R_G = 10\ \text{M}\Omega$, while the other parameters are as given for Example 9.7. Determine $R_{S1}$, $R_{S2}$, and $A_v$.

9.5.4. Why is it desirable to make $R_{S1}$ as small as possible in the circuit of Fig. 9.21? What prevents us from making it too small?

### 9.7e  Frequency Response

The frequency response of FET amplifiers may be obtained using the general techniques developed earlier, using suitable equivalent circuits. An example will be used to illustrate typical methods of solution.

**Example 9.9**

Estimate the midfrequency voltage amplification (in decibels) and $f_L$ in the circuit of Fig. 9.24a. The FET parameters are $y_{fs} = 2{,}000\ \mu\text{mho}$, $r_d = 200\ \text{k}\Omega$, $C_{gs} = 2\ \text{pF}$, $C_{gd} = 1\ \text{pF}$.

**Solution**

a. The midfrequency $A_v$ is determined by treating all capacitors as short circuits; this is done in Fig. 9.24b. Note that $V_i = V_{gs}$. The voltage amplification is therefore

$$A_v = \frac{V_o}{V_i} = \frac{-y_{fs} r_d R'_L}{r_d + R'_L} \tag{9.20}$$

$$A_v = \frac{(-2 \times 10^{-3})(200 \times 10^3)(5 \times 10^3)}{205 \times 10^3} = -9.75$$

and

$$20 \log |A_v| = 19.8\ \text{dB}$$

b. We may assume that $f_L$ is controlled by $C_S$; that is, $C_1$ and $C_2$ behave as short circuits at the frequency where $C_S$ begins to affect the low-frequency response. Whether this assumption is valid or not will have to be verified later. To simplify the algebra, the low-frequency equivalent circuit of Fig. 9.24c has been modified by converting the current generator to a voltage generator (Norton to Thevenin transformation). We can now write

$$V_o = -I_o R'_L$$

$$I_o = \frac{y_{fs} V_{gs} r_d}{R'_L + r_d + Z_s}$$

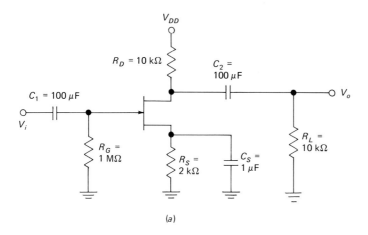

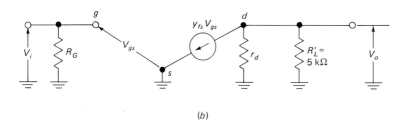

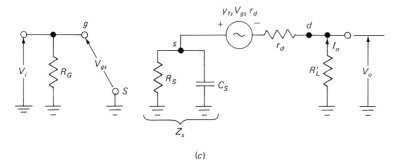

Fig. 9.24  Circuits for Example 9.9.

where

$$Z_s = \frac{R_S[1/(j\omega C_S)]}{R_S + (1/j\omega C_S)}$$

When numerical values are substituted, the result is

$$A_v = \frac{V_o}{V_i} = \frac{V_o}{V_{gs}} = \frac{-9.75(1 + jf/79.5)}{1 + jf/80.5}$$

The plot of $20 \log |A_v|$ is shown in Fig. 9.25. Note that the midfrequency value is 19.8 dB, as obtained earlier, and $f_L = 80.5$ Hz. Although the response flattens out again below 79.5 Hz, eventually $C_1$ and $C_2$ will cause $A_v$ to drop again.

We now go back to determine whether $C_1$ and $C_2$ are significant near $f_L$:

$$X_{C1} = X_{C2} = \frac{1}{2\pi(80.5 \times 10^{-4})} \simeq 20 \, \Omega$$

This is negligible compared to the Thevenin resistance seen by both $C_1$ ($\simeq 10^6 \, \Omega$) and $C_2$ ($\simeq 20 \, \text{k}\Omega$); hence our $f_L$ is valid. It is possible, of course, for the opposite to happen. Here one can calculate the break frequencies due to $C_1$ and $C_2$ assuming $C_S$ to be a perfect bypass, and then verify whether the assumption is valid. It is also possible, if one has the time and the inclination, not to make any assumptions at all and to write an expression for $A_v$ in which all three capacitors are included. This yields a complete (but lengthy) expression from which all break frequencies may be determined.

### Example 9.10

Estimate $f_H$ in Example 9.9 if $R_L$ feeds the input to another, similar circuit.

### Solution

The high-frequency equivalent circuit is shown in Fig. 9.26. $R'_L$ is shunted by the input to the following stage, which, in addition to $R_G$, involves an input capacitance made up of two components, $C_{gs}$ and the Miller equivalent of $C_{gd}$:

$$C_i = C_{gs} + C_{gd}(1 - A_v) \qquad (9.28)$$
$$C_i = 2 \times 10^{-12} + 10^{-12}(1 + 9.75) = 12.75 \text{ pF}$$

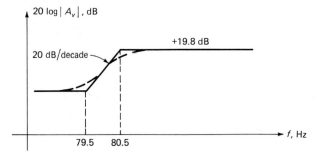

Fig. 9.25  Low frequency response for the circuit of Example 9.9.

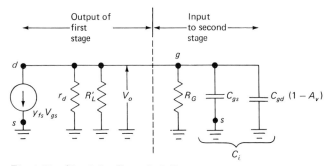

Fig. 9.26   Circuit for Example 9.10.

$f_H$ is the frequency for which the reactance of $C_i$ is equal to the Thevenin resistance it sees

$$f_H = \frac{1}{2\pi C_i (R_G \| R'_L \| r_d)}$$

$$f_H \simeq \frac{1}{2\pi (12.75 \times 10^{-12})(5 \times 10^3)}$$

$$f_H = 2.5 \text{ MHz}$$

## Problem Set 9.6

9.6.1. Determine the break frequencies introduced by $C_1$ and $C_2$ in the circuit of Fig. 9.24a and use them to complete the low-frequency response plot of Fig. 9.25. Assume the circuit is driven from a constant voltage source.

9.6.2. In Example 9.9 it is desired to make $f_L = 200$ Hz. Determine the required value of $C_S$.

9.6.3. Although $C_{gd}$ is usually smaller than $C_{gs}$, it has a greater effect in determining $f_H$. Why?

9.6.4. Under what conditions would $R_G$ in the circuit of Fig. 9.26 significantly affect $f_H$?

## REFERENCES

1. Mulvey, J.: "Semiconductor Device Measurements," Tektronix, Inc., Beaverton, Ore., 1968.
2. Cowles, L. G.: "Transistor Circuits and Applications," Prentice-Hall, Inc., Englewood Cliffs, N.J., 1968.
3. Cowles, L. G.: "Analysis and Design of Transistor Circuits," D. Van Nostrand Company, Inc., Princeton, N.J., 1966.
4. Millman, J., and C. C. Halkias: "Electronic Devices and Circuits," McGraw-Hill Book Company, New York, 1967.

5. Kane, J. F., and D. L. Wollesen: Field Effect Transistor in Theory and Practice, AN 211, Motorola Semiconductor Products Inc., Phoenix, 1966.
6. Ott, H. W.: Biasing the Junction FET, *EEE*, January 1970, pp. 52–57.
7. Reich, S.: MOSFET Biasing Techniques, *EEE*, September 1970, pp. 62–68.
8. Gosling, W.: "Field Effect Transistor Applications," John Wiley & Sons, Inc., New York, 1965.
9. Sevin, L.: "Field-Effect Transistors," McGraw-Hill Book Company, New York, 1965.

# chapter 10

## feedback principles and applications

Any system, whether it is electrical, mechanical, hydraulic, or pneumatic, may be considered to have at least one input and one output. If the system is to perform reliably, we must be able to measure and control the output. Such control is possible through the use of feedback, a process whereby the input is forced to partially depend on the output. This way, should the output deviate from its desired value, the input is modified in such a manner as to yield the desired output.

A simple example may be used to illustrate the nature of feedback. Suppose that we wish to heat a house and that a furnace is available. The heat generated by the furnace is the input, while the temperature of the house is the output. Turning the furnace on represents the application of the input from which the desired output (warm house) should result. Now consider what would happen if the furnace remained on for too long; the temperature would probably rise until it became too hot. Now if there is no one in the house to notice this and do something about it, the input cannot change (furnace is still on) even though the output (temperature of house) is no longer at the desired value. There is therefore no control over the output.

In order to establish control, the input must be modified in response to the output; that is, if it gets too hot, the furnace is turned off, and if it gets too cold, the furnace is turned on. This can be done manually (by having a person continuously monitoring the house temperature and turning the furnace switch on or off) or automatically (by using a thermostat). Although the automatic arrangement is obviously more practical, feedback is being used in both cases to establish control.

Certain applications of feedback have already been discussed, mainly in DC circuits for the stabilization of the operating point. In this chapter we will see how different types of feedback may be used to control and improve the performance of amplifier circuits.

## 10.1  BASIC DEFINITIONS

Basic feedback terminology is introduced here; although it is related to amplifiers, this terminology also applies to more general cases.

Ideally, an amplifier should reproduce at its output an amplified version of the signal fed at its input. All the amplifier is expected to do is to make the input "larger." A real amplifier generally comes short of meeting its expected performance for a variety of reasons. Some typical shortcomings are

1. Changes in amplification from the desired value may be caused by variations in supply voltage, temperature, or other environmental conditions, or aging or replacement of components.
2. Distortion of waveform may be due to nonlinearities in the operating characteristics of the amplifying device or unequal frequency and phase response.
3. The amplifier may introduce noise in addition to that which is already a part of the input signal.

We will see in the sections to follow how, through appropriate applications of feedback, we can improve on the above characteristics. For now, consider the diagram of Fig. 10.1. Here we have an amplifier whose voltage amplification is $A_v$; the input to the amplifier is $V_i$, while the output is $V_o$.[1] For the amplifier proper, the following relationship must apply:

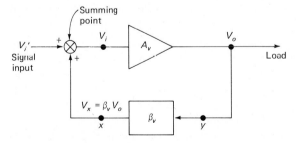

Fig. 10.1  Generalized feedback amplifier.

[1] All voltages are with respect to ground, which, although not shown on the diagram, should be understood to be the common point to which all voltages are referred.

$$A_v = \frac{V_o}{V_i} \tag{10.1}$$

Similarly, the system's overall voltage amplification is the ratio of $V_o$ to the total input signal $V_i'$, as follows:

$$A_v' = \frac{V_o}{V_i'} \tag{10.2}$$

The voltage at point $y$ is $V_o$. The block labeled $\beta_v$ represents a circuit through which a portion of $V_o$ is fed back to the input. The factor $\beta_v$ is the ratio of voltage at point $x$ to voltage at point $y$:

$$\beta_v = \frac{V_x}{V_o} \tag{10.3}$$

Thus $\beta_v$ is dimensionless and generally less than 1. In its simplest form the circuit inside the $\beta_v$ block could be a short circuit between $x$ and $y$, in which case $V_x = V_o$ and $\beta_v = 1$. We could also have a resistive voltage divider, yielding $\beta_v < 1$. In many applications the signal fed back must be shifted in phase; here appropriate $RC$ or $RL$ circuits are used and $\beta_v$ becomes a complex number.

If the connection at $x$ is broken, there is no feedback. The input $V_i'$ to the system is also the input to the amplifier. In this case, the system's overall amplification $A_v'$ is equal to that of the amplifier itself, which is subject to all the variations mentioned earlier. We thus have *no control* because there is *no feedback*.

By closing the feedback loop, a portion of the output voltage $\beta_v V_o$ is combined with $V_i'$ to yield a modified $V_i$. This is accomplished at the *summing point* indicated in Fig. 10.1. The summing point represents a circuit arrangement that produces the algebraic sum of two voltages, in this case $V_i'$ and $\beta_v V_o$. The actual input to the amplifier, also known as the *error* signal, is

$$V_i = V_i' + \beta_v V_o \tag{10.4}$$

The closed-loop amplification is

$$A_v' = \frac{V_o}{V_i'} \tag{10.2}$$

But

$$V_i' = V_i - \beta_v V_o \tag{10.4}$$

therefore

$$A_v' = \frac{V_o}{V_i - \beta_v V_o}$$

$$A'_v = \frac{V_o/V_i}{1 - (\beta_v V_o/V_i)}$$

$$A'_v = \frac{A_v}{1 - \beta_v A_v} \tag{10.5}$$

Equation (10.5) expresses the closed-loop amplification $A'_v$ in terms of the open-loop amplification $A_v$. Note that $A_v$ is divided by $1 - \beta_v A_v$, called the return difference. If there is no feedback, $\beta_v = 0$ and $A'_v = A_v$.

With feedback, there are two main possibilities. If the amplifier produces no phase shift ($V_o$ is in phase with $V_i$), $A_v$ is a positive number. If the feedback circuit produces no phase shift either, then the voltage at point $x$ is in phase with and reinforces $V_i$. This is *positive* or *regenerative* feedback and produces a closed-loop amplification which is greater than the open-loop amplification ($|A'_v| > |A_v|$). If, in fact, $\beta_v A_v = 1$, the return difference is zero and $A'_v$ becomes infinite $[A_v/(1-1) \to \infty]$. Infinite amplification simply means that there is an output without any input! In effect the amplifier supplies its own input $V_i$, derived through the feedback network from $V_o$. The amplifier has become an *oscillator*, that is, a signal generator. Since the output may be present even without an input, there is clearly *no control*. Generally, the effects of positive feedback are opposite to those of negative feedback; however, there are many important applications of positive feedback, such as changing impedance levels and in oscillator circuits, to be discussed later.

When $\beta_v V_o$ (the voltage at point $x$) is of opposite polarity to $V'_i$, the resulting feedback is *negative* or *degenerative*. Here $\beta_v V_o$ subtracts from $V'_i$, causing $V_i$ to be smaller than $V'_i$. The net input to the amplifier is therefore reduced, yielding a smaller $V_o$ for the same $V'_i$. This means that the closed-loop amplification is less than the open-loop amplification ($|A'_v| < |A_v|$). Negative feedback, therefore, reduces the amplification.

In the system of Fig. 10.1, $A_v$ is the voltage amplification of a real amplifier, subject to the parameter variations of all the devices that make up the circuit. Thus $A_v$ can vary, sometimes by quite a bit. Through the appropriate application of negative feedback, however, $A'_v$ can be made quite stable. Suppose, for example, that $A_v$ can vary anywhere between $-1{,}000$ and $-3{,}000$ (the negative sign represents $180°$ phase shift between $V_i$ and $V_o$). This is a 3:1 variation. We now employ 10 percent feedback ($\beta_v = 0.1$) to yield the following closed-loop amplification:

a. When $A_v = -1{,}000$

$$\beta_v A_v = 0.1(-1{,}000) = -100$$

$$A'_v = \frac{A_v}{1 - \beta_v A_v} \tag{10.5}$$

$$A'_v = \frac{-1{,}000}{1-(-100)} = -9.9$$

b. When $A_v = -3{,}000$

$\beta_v A_v = 0.1(-3{,}000) = -300$

$$A'_v = \frac{-3{,}000}{301} = -9.97 \tag{10.5}$$

Note that $A'_v$ varies between $-9.9$ and $-9.97$, a 1.007:1 variation; thus we have established control over our amplification, since a 3:1 variation in $A_v$ is reflected as a 1.007:1 variation in $A'_v$. We pay a price for this control, however, in reduced amplification. If enough feedback is used, in fact, by making $\beta_v A_v \gg 1$, Eq. (10.5) simplifies to

$$A'_v = \frac{A_v}{1 - \beta_v A_v} \simeq -\frac{1}{\beta_v} \tag{10.6}$$

Now $A'_v$ is independent of $A_v$, which means that, as long as $\beta_v A_v \gg 1$, changes in amplifier performance have negligible effect on $A'_v$. $A'_v$ is in fact entirely controlled by the feedback ratio $\beta_v$. In most cases $\beta_v$ can be accurately set (by varying the control on a precision potentiometer) and maintained so that $\beta_v$ may be considered constant. Thus $A'_v$ also remains constant.

## Example 10.1

The following information is available for the block diagram of Fig. 10.1: open-loop voltage amplification: $A_v = -100$; input voltage to the system: $V'_i = 1$ mV. Determine the closed-loop voltage amplification, the output voltage, feedback voltage, input voltage to amplifier, and type of feedback, for

a. $\beta_v = 0$
b. $\beta_v = 0.01$
c. $\beta_v = -0.005$

## Solution

a. $\beta_v = 0$

$A'_v = A_v = -100$
$V_o = A'_v V'_i = -100 \times 10^{-3} = -100$ mV
$V_x = \beta_v V_o = 0$
$V_i = V'_i = 1$ mV
Type of feedback: none

b. $\beta = 0.01$

$$A'_v = \frac{A_v}{1 - \beta_v A_v} \tag{10.5}$$

$$A'_v = \frac{-100}{1 - 0.01(-100)} = -50$$
$$V_o = V'_i A'_v = -50 \times 10^{-3} = -50 \text{ mV}$$
$$V_x = \beta_v V_o = 0.01(-50 \times 10^{-3}) = -0.5 \text{ mV}$$
$$V_i = V'_i + \beta_v V_o \qquad\qquad (10.4)$$
$$V_i = 10^{-3} + (-0.5 \times 10^{-3}) = 0.5 \text{ mV}$$

This is negative feedback because

$$V_i < V'_i; \qquad \text{hence } |A'_v| < |A_v|$$

c. $\beta = -0.005$

$$A'_v = \frac{A_v}{1 - \beta_v A_v} \qquad\qquad (10.5)$$

$$A'_v = \frac{-100}{1 - (-0.005)(-100)} = -200$$
$$V_o = V'_i A'_v = -200 \times 10^{-3} = -200 \text{ mV}$$
$$V_x = \beta_v V_o = -0.005(-0.2) = 1 \text{ mV}$$
$$V_i = V'_i + \beta_v V_o \qquad\qquad (10.4)$$
$$V_i = 10^{-3} + 10^{-3} = 2 \text{ mV}$$

This is positive feedback because

$$V_i > V'_i \quad \text{and} \quad |A'_v| > |A_v|$$

It should be emphasized that what determines whether there is positive or negative feedback is the phase of $V_x$ relative to $V'_i$. Thus negative feedback is possible even if the amplifier's input and output are in phase ($A_v$ a positive number) provided the feedback network introduces 180° phase shift ($\beta_v$ a negative number).

### Example 10.2

In Example 10.1 determine the percent variation in $A'_v$ resulting from a 100 percent increase in $A_v$ when $\beta_v = 0.01$.

### Solution

When $A_v = -100$, $A'_v = -50$. When $A_v$ increases by 100 percent, we have
New $A_v = -200$

$$A'_v = \frac{A_v}{1 - \beta_v A_v} \qquad\qquad (10.5)$$

New $A'_v = \dfrac{-200}{1 - 0.01(-200)} = -66.7$

The change in $A'_v$ is 66.7 − 50 = 16.7, which, when expressed as a percent variation, is

$$\frac{16.7}{50} 100 = 33.3 \text{ percent}$$

Although basic feedback relationships have been developed using voltage amplification, other performance measures, such as current amplification and hybrid transfer functions (for transconductance or transresistance amplifiers), may also be used.

In some cases feedback over more than one amplifier stage may be employed. This is called *overall* feedback, and it has certain advantages over local feedback. To demonstrate this, consider the diagram of Fig. 10.2a. This involves two cascaded amplifiers with identical $A_v = -100$; local feedback with $\beta_v = 0.1$ is used for each amplifier. The closed-loop voltage amplification for each stage is

$$A'_v = \frac{A_v}{1 - \beta_v A_v} = \frac{-100}{1 - 0.1(-100)}$$
$$A'_v = -9.1$$

and for both cascaded stages, the overall $A'_v = -9.1(-9.1) = 82.7$. Should $A_v$ double, for example, the new closed-loop voltage amplification for each stage is

$$A'_v = \frac{-200}{1 - 0.1(-200)} = -9.53$$

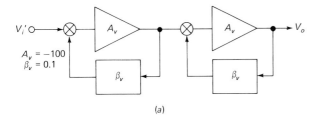

(a)

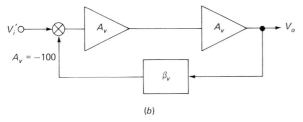

(b)

Fig. 10.2 (a) *Example of local feedback;* (b) *example of overall feedback.*

while the overall $A'_v = -9.53(-9.53) = 90.7$. This would seem to be fairly good stabilization, since doubling of $A_v$ results in the overall $A'_v$ increasing from 82.7 to 90.7, which is only about a 10 percent increase.

Now let us use overall feedback, as in Fig. 10.2b. Since both amplifiers have 180° phase shift, the last output is in phase with the first input. This means that for the feedback to be negative, $\beta_v$ must introduce 180° phase shift. To establish a basis for comparison, let us select a $\beta_v$ that will yield the same closed-loop overall amplification as before, that is, when each stage's $A_v = -100$, the overall $A'_v = 82.7$. Since both stages are cascaded, they can be treated as a single amplifier whose overall $A_v = -100(-100) = 10,000$. We therefore write

$$A'_v = \frac{A_v}{1 - \beta_v A_v} \qquad (10.5)$$

$$82.7 = \frac{10,000}{1 - \beta_v(10,000)}$$

and solve for $\beta_v$:

$$\beta_v = -0.012$$

Now if each stage's $A_v$ doubles, the new overall closed-loop voltage amplification is

$$A'_v = \frac{20,000}{1 - (-0.012)(20,000)}$$
$$A'_v = 83$$

Note that as $A_v$ doubles, $A'_v$ increases from 82.7 to 83; this is less than 0.5 percent compared to the 10 percent obtained using local feedback. The large improvement in gain stabilization is due to the much larger loop amplification (10,000 versus 100) available when overall feedback is used.

### Problem Set 10.1

10.1.1. Give an example of a simple system employing feedback to control some physical parameter that you are familiar with. Try to identify the input, error, output, and feedback signals.

10.1.2. What is the difference between open- and closed-loop amplification? When are these equal?

10.1.3. Given a system of the type in Fig. 10.1 and $A_v = -1,000$, $\beta_v = 0.01$, $V'_i = 10$ mV, determine the amplifier input, output, and feedback voltages. Also determine the closed-loop voltage amplification. What type of feedback is this?

10.1.4. Repeat Prob. 10.1.3 with $\beta_v = -0.01$.

10.1.5. If the amplifier in Prob. 10.1.3 has no phase shift ($A_v = 1{,}000$), what parameter would have to be changed and in what way so that the type of feedback remains the same?

10.1.6. In Example 10.2 how much feedback is required to reduce the closed-loop amplification variation to 1 percent?

10.1.7. Assume $A_v = 500$ for both amplifiers in Fig. 10.2a. Determine $\beta_v$ if the closed-loop amplification for each stage is to be 100. What is the overall $A'_v$?

10.1.8. In Prob. 10.1.7, what is the percent change in the overall $A'_v$ if each stage's $A_v$ drops by 50 percent?

10.1.9. Determine $\beta_v$ in the diagram of Fig. 10.2b if the open-loop $A_v$ for each stage and overall $A'_v$ are the same as in Prob. 10.1.7. What is the percent change in the overall $A'_v$ if each stage's $A_v$ drops by 50 percent?

## 10.2 EFFECTS OF NEGATIVE FEEDBACK ON AMPLIFIER PERFORMANCE

In addition to gain stabilization, negative feedback may be used to improve an amplifier's frequency response, distortion, and noise figure, as well as modify impedance levels.

*FREQUENCY RESPONSE.* In the amplifier of Fig. 10.1, assume low and high cutoff frequencies $f_L$ and $f_H$, respectively. For a 6 dB per octave roll-off, the high-frequency response can be approximated with an equation of the following general form:

$$A_v = \frac{A_{v(\text{mid})}}{1 + jf/f_H} \tag{10.7}$$

where $A_{v(\text{mid})}$ is the midfrequency voltage amplification. The magnitude plot

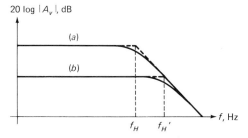

*Fig. 10.3 Frequency-response comparison:* (a) *no feedback;* (b) *negative feedback.*

for $A_v$ is given in curve a of Fig. 10.3. Now if negative feedback is used, the closed-loop amplification is

$$A'_v = \frac{A_v}{1 - \beta_v A_v} \tag{10.5}$$

$$A'_v = \frac{A_{v(\text{mid})}/(1 + jf/f_H)}{1 - \beta_v[A_{v(\text{mid})}/(1 + jf/f_H)]}$$

which becomes

$$A'_v = \frac{A_{v(\text{mid})}/(1 - \beta_v A_{v(\text{mid})})}{1 + jf/[f_H(1 - \beta_v A_{v(\text{mid})})]} \tag{10.8}$$

Equation (10.8) can be rewritten as follows:

$$A'_v = \frac{A'_{v(\text{mid})}}{1 + jf/f'_H} \tag{10.8a}$$

where

$$A'_{v(\text{mid})} = \frac{A_{v(\text{mid})}}{1 - \beta_v A_{v(\text{mid})}} \tag{10.9}$$

and

$$f'_H = f_H(1 - \beta_v A_{v(\text{mid})}) \tag{10.10}$$

Suppose, for example, that $A_{v(\text{mid})} = 1{,}000$, $\beta_v = -0.1$, and $f_H = 1$ MHz. The closed-loop midfrequency amplification is

$$A'_{v(\text{mid})} = \frac{1{,}000}{101} = 9.9 \tag{10.9}$$

while the closed-loop high-frequency cutoff is

$$f'_H = f_H(101) = 101 \text{ MHz} \tag{10.10}$$

Note that as a result of negative feedback, $A_v$ is *reduced* by a factor of 101, while $f_H$ is *increased* by a factor of 101. We are therefore exchanging voltage amplification for bandwidth. This effect is shown in Fig. 10.3.

The extension of bandwidth also takes place on the low-frequency end. In general,

$$f'_L = \frac{f_L}{1 - \beta_v A_{v(\text{mid})}} \tag{10.11}$$

If, for example, $f_L$ is 100 Hz, the closed-loop $f'_L$ is $100/101 = 0.99$ Hz. Therefore

$f_H$ is multiplied while $f_L$ is divided by the same factor as the amplification is reduced.

DISTORTION. Consider the waveforms of Fig. 10.4. Each voltage axis is labeled in accordance with the definitions of Fig. 10.1 except that lowercase symbols are used here to denote instantaneous values. The input and output waveforms are shown in Fig. 10.4a and b, respectively. There is obviously a lot of distortion in the output $v_o$ due to the unequal amplification of upper and lower portions of the signal. Suppose now that we apply negative feedback by deriving an appropriate out-of-phase signal from the output. The feedback signal $v_x$ of Fig. 10.4c is an exact replica of the new output ($v_o'$) of Fig. 10.4e, except for the magnitude reduction and phase inversion. Upon combining $v_x$ and $v_i'$, the modified amplifier input $v_i$ of Fig. 10.4d results. Note that we are predistorting our signal by applying an input that already contains distortion,

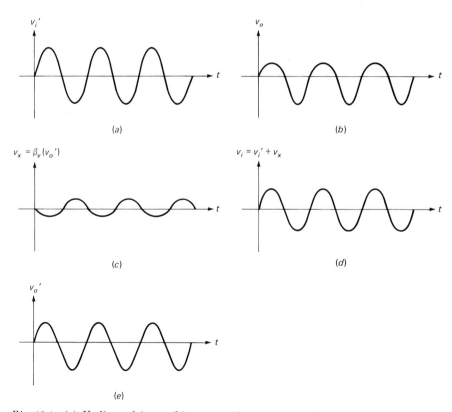

Fig. 10.4 (a) *Undistorted input;* (b) *output without negative feedback;* (c) *negative feedback signal derived from output and phase-inverted;* (d) *modified input: signals of* (a) *and* (c) *combined;* (e) *output with negative feedback—still distorted but not as much as in* (b).

but of the opposite phase to that produced by the amplifier. The new output $v'_o$, shown in Fig. 10.4e, is much less distorted than the previous $v'_o$ in Fig. 10.4b because of the canceling effect between the distortion in the modified input and the distortion due to the amplifier.

To further appreciate the distortion reduction, a mathematical derivation may be developed for the block diagram of Fig. 10.1. Assume that the amplifier, in addition to the desired output, also generates distortion $D$ as follows:

$$V_o = V_i A_v + D \tag{10.12}$$

where

$$V_i = V'_i + \beta_v V_o \tag{10.4}$$

yielding

$$V_o = (V'_i + \beta_v V_o)(A_v) + D$$
$$V_o = \frac{V'_i A_v}{1 - \beta_v A_v} + \frac{D}{1 - \beta_v A_v} \tag{10.12a}$$

The first part of Eq. (10.12a) represents the output signal, reduced because of negative feedback, by the return difference, $1 - \beta_v A_v$. But the distortion $D$ has also been reduced by the same factor so that there appears to be no change in the relative magnitude of signal and distortion! Such is not the case, however, because $V'_i$ can be increased to compensate for the lower amplification, yielding the desired output with reduced distortion.

Assume, for example, that $V'_i = 0.1$ V, $A_v = 100$, and the amplifier produces 10 percent distortion for an output of 10 V. It is important to specify the output level for which the distortion specification applies, since amplifiers generate more distortion as they are driven harder (distortion is a function of output level). With no feedback, $V_o = V'_i A_v = 10$ V plus 10 percent distortion, which is 1 V. Suppose we now apply 20 percent negative feedback ($\beta_v = -0.2$); the amplification is reduced by the return difference $(1 - \beta_v A_v)$, which is 21. It follows that to get the same output as before, the input must be increased by the same factor, that is, $V'_i = 2.1$ V. The total output is now given by Eq. (10.12a), the first part of which represents the desired signal, while the second part is distortion:

$$V_o = \frac{2.1(100)}{21} + \frac{D}{21}$$
$$V_o = 10 + \frac{D}{21}$$

Since the output is 10 V, the distortion specification of 10 percent applies, yielding $D = 1$ V; hence $D/21 = 1/21$ V. For a 10-V output we now have only 1/21-V distortion! The 21-times reduction in distortion has been achieved at a cost of having to increase the input signal also by a factor of 21.

It should be noted that the distortion improvement is possible only when distortion is produced by the amplifier, not when it is already present in the input signal. The same improvement is possible in noise reduction; again this applies to noise generated in the amplifier and not noise that comes in as part of the input signal.

## Problem Set 10.2

10.2.1. An amplifier of the type in Fig. 10.1 has $A_v = 500, f_L = 0, f_H = 10$ kHz. Determine the closed-loop voltage amplification and bandwidth if

    a.    $\beta_v = -0.01$
    b.    $\beta_v = -0.1$

What is the relationship between amplification reduction and bandwidth improvement?

10.2.2. Sketch Bode plots (amplitude) for the open-loop and closed-loop case in Prob. 10.2.1 with $\beta_v = -0.1$.

10.2.3. Determine the gain-bandwidth product $A_v f_H$ in Prob. 10.2.1 for the open-loop case and for each value of $\beta_v$ that is given. What conclusions can you draw about the gain-bandwidth product?

10.2.4. Derive Eq. (10.11).

10.2.5. Can negative feedback be used to reduce distortion or noise that is part of the signal input?

10.2.6. The amplifier of Prob. 10.2.1 produces 5 percent distortion for an output level of 4 V. Determine the RMS input voltage ($V_i'$) and feedback required to yield 1 percent distortion for a 4-V output.

10.2.7. Given an amplifier with $A_v = -100$ which produces 5 percent distortion when the output level is 5 V, show how one can still get 5 V out with 1 percent distortion. What input voltage is required?

10.2.8. Two identical amplifiers are cascaded to yield greater overall amplification. Each amplifier has an open-loop voltage amplification $A_v = -200$. Enough overall negative feedback to reduce the overall $A_v'$ to 50 must be used. The frequency response for each amplifier is flat from DC to 10 kHz, after which it starts to drop at the rate of 20 dB per decade. Sketch the Bode plot for the overall $A_v'$. Determine $\beta_v$ and $f_H'$.

## 10.3 GENERAL FEEDBACK CONNECTIONS

In feedback circuits, it is important to consider how the feedback signal is "derived" from the output and how it is "fed" to the input. Techniques of feeding and deriving as well as their effects on amplification and impedance levels are discussed. The following derivations apply only when the feedback is negative; opposite effects should be expected if the feedback is positive.

Consider the amplifier diagram and signal source of Fig. 10.5a. The output circuitry has been left out for simplicity. Note that $V_i = V_i'$ and there is no feedback. The amplifier generates an output proportional to its input $V_i$. In order to realize the benefits of negative feedback, we must apply a feedback signal that will modify $V_i$. If the circuit is driven from a constant voltage source ($R_g = 0$), the only way a feedback signal can change $V_i$ is if it is applied as a series voltage that adds to or subtracts from $V_i'$. Such an arrangement is shown in Fig. 10.5b, where the feedback signal $V_x$ is said to be series-fed. Note that this circuit is equivalent to the diagram of Fig. 10.1. In either circuit, $V_i = V_i' + V_x = V_i' + \beta_v V_o$ and the voltage amplification is reduced by $1 - \beta_v A_v$.

In order for series-fed feedback to be most effective, the circuit should be driven from a constant voltage source (as shown in Fig. 10.5b) or from a source whose internal resistance $R_g$ is small compared with $R_i$. If, for example, $R_g$ is large compared with $R_i$, $V_i$ will be modified not by the full value of $V_x$

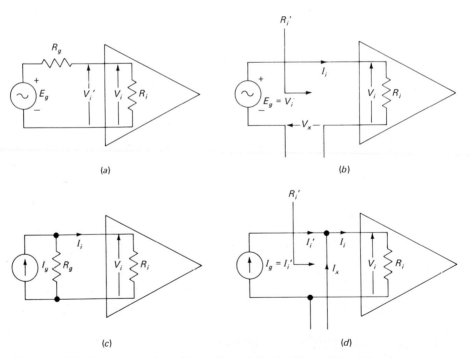

**Fig. 10.5** (a) *Input circuit of amplifier without feedback;* (b) *amplifier driven from a constant voltage source and feedback signal series-fed;* (c) *input circuit of amplifier without feedback;* (d) *amplifier driven from a constant current source and feedback signal shunt-fed.*

but by a smaller amount determined by the relative magnitudes of $R_g$ and $R_i$.

The effects of series feeding on the input resistance are now determined using the diagram of Fig. 10.5b. Without feedback, $V_x = 0$, and the input resistance is

$$R_i = \frac{V_i}{I_i} \tag{10.13}$$

With feedback, the resistance seen by $V_i'$ is

$$R_i' = \frac{V_i'}{I_i} = \frac{V_i - V_x}{I_i}$$

But

$$V_x = \beta_v V_o \tag{10.3}$$

and

$$V_o = V_i A_v \tag{10.1}$$

Therefore

$$R_i' = \frac{V_i - \beta_v V_i A_v}{I_i}$$

Using $V_i = I_i R_i$, we get $\qquad(10.13)$

$$R_i' = \frac{I_i R_i - \beta_v I_i R_i A_v}{I_i}$$

$$R_i' = R_i(1 - \beta_v A_v) \tag{10.14}$$

Thus the input resistance increases by the return difference, the same factor by which the voltage amplification decreases.

Consider now the circuit of Fig. 10.5c. If the amplifier is driven from a constant current source ($R_g = \infty$), $I_i = I_g$. $I_i$ can be modified only if the feedback signal is applied in the form of a shunt current, as shown in Fig. 10.5d. Now $I_i = I_i' + I_x$, and the benefits of negative feedback can be realized.

In order for shunt feedback to be most effective, the amplifier should be driven from a constant current source or a source whose resistance is relatively high ($R_g \gg R_i$). If, for example, we were to drive from a low-resistance source, most of the feedback current $I_x$ would flow through the source instead of $R_i$ and the change in $I_i$ would be minimal.

Without feedback, $I_i = I_i'$ and $A_i' = A_i$, where

$$A_i = \frac{I_o}{I_i} \tag{10.15}$$

With shunt negative feedback, the input current $I_i'$ is reduced, yielding a corresponding reduction in closed-loop current amplification, as follows:

$$A_i' = \frac{I_o}{I_i'} \tag{10.16}$$

But

$$I_i' = I_i - I_x \tag{10.17}$$

and

$$I_o = A_i I_i \tag{10.15}$$

Therefore

$$A_i' = \frac{A_i I_i}{I_i - I_x}$$

We now define a current feedback ratio

$$\beta_i = \frac{I_x}{I_o} \tag{10.18}$$

and

$$A_i' = \frac{A_i I_i}{I_i - \beta_i I_o}$$

$$A_i' = \frac{A_i}{1 - \beta_i A_i} \tag{10.19}$$

If the term $\beta_i A_i \gg 1$, the closed-loop current amplification is

$$A_i' \simeq -\frac{1}{\beta_i} \tag{10.20}$$

which is independent of amplifier parameters. Therefore shunt feeding reduces and stabilizes current amplification in a manner which is analogous to the reduction and stabilization of voltage amplification obtained through series feeding. The general improvements with regard to bandwidth and distortion also apply to $A_i$. Normally shunt feeding does not affect $A_v$.

The effects of shunt feeding on the input resistance may be determined using the diagram of Fig. 10.5d. Without feedback, $I_x = 0$ and $I_i' = I_i$, yielding $R_i' = R_i$. With feedback, we have

$$R_i' = \frac{V_i}{I_i'} \tag{10.21}$$

But

$$I'_i = I_i - I_x \tag{10.17}$$

and

$$I_x = \beta_i I_o \tag{10.18}$$

Therefore

$$R'_i = \frac{V_i}{I_i - \beta_i I_o}$$

and since $I_o = A_i I_i$, $\tag{10.15}$

$$R'_i = \frac{V_i}{I_i - \beta_i A_i I_i}$$

But $V_i = R_i I_i$ $\tag{10.13}$

therefore

$$R'_i = \frac{R_i I_i}{I_i - \beta_i A_i I_i}$$

$$R'_i = \frac{R_i}{1 - \beta_i A_i} \tag{10.22}$$

The reduction in input resistance is due to the shunting effect of the feedback path that carries $I_x$; this is the Miller effect introduced in Chap. 7.

In conclusion, series feeding decreases and stabilizes the voltage amplification by the factor $1 - \beta_v A_v$. The current amplification is not directly affected, but the input resistance is increased by the same factor.

Shunt feeding has no direct effect on $A_v$, but it does reduce and stabilize the current amplification by the current return difference $1 - \beta_i A_i$. The input resistance is reduced by the same factor.

Note that the manner in which the feedback signal is fed at the input affects the input resistance and amplification. The manner in which the feedback signal is derived affects the output resistance.

Block diagrams for each possible combination of feeding and deriving the feedback signal are shown in Fig. 10.6. When the feedback signal is shunt-derived, $V_o$ is simultaneously across $R_L$ and across the input to the feedback network. In the circuit of Fig. 10.6a, this means that $V_x$ is proportional to $V_o$. For example, suppose that $V_o$ attempts to rise. The magnitude of $V_x$ also rises, reducing $V_i$ ($V_x$ is out of phase with $V'_i$). The reduction in $V_i$ returns $V_o$ closer to its original value. The circuit therefore tends to maintain a constant output voltage, which implies a low output resistance. A similar effect takes place in

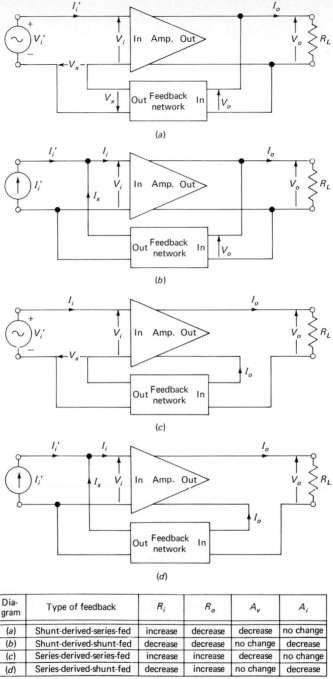

Fig. 10.6 Different types of feedback connections (a, b, c, and d) and effects on amplifier performance (e).

the circuit of Fig. 10.6b. Here the input to the feedback network is also $V_o$, and $I_x$ is therefore proportional to $V_o$. Thus, if $V_o$ tends to rise, $I_x$ also rises, which reduces $I_i$ and $V_i$ ($I_x$ is out of phase with $I_i'$); $V_o$ is therefore decreased toward its original value. Note that the low output resistance is a consequence of the circuit's tendency to maintain a constant output voltage and is not due to any loading on the output from the feedback network, which is assumed negligible. Also, the manner in which the feedback is fed is immaterial; as long as the feedback signal is shunt-derived, the input is modified by a feedback voltage $V_x$ or current $I_x$ that is proportional to $V_o$. The circuit therefore tries to maintain a constant $V_o$, yielding a low output resistance.

When the feedback signal is series-derived, $I_o$ flows through $R_L$ and the input to the feedback network. In both circuits of Fig. 10.6c and d the feedback signal ($V_x$ or $I_x$) is proportional to $I_o$. This means that if $I_o$ tries to rise, the input to the amplifier is reduced accordingly, returning $I_o$ near its original value. The circuit therefore tries to maintain a constant output current, which implies a large output resistance.

In conclusion, an amplifier with shunt-derived negative feedback attempts to maintain a constant output voltage, thereby exhibiting a low output resistance. An amplifier with series-derived negative feedback attempts to maintain a constant output current, thereby exhibiting a high output resistance.

The amplifier of Fig. 10.7a may be used to illustrate the effects of series-derived, series-fed negative feedback. Normally $R_E$ is bypassed and there is no feedback. When $R_E$ is not bypassed, however, the voltage developed across $R_E$ modifies the base-emitter voltage so that the amplifier drive is reduced. The feedback is therefore negative and series-fed.

To further consider this, an equivalent circuit is shown in Fig. 10.7b. This is redrawn to conform with the general representation of Fig. 10.6c; $h_{re}$ is assumed negligible to simplify our calculations. Note that because the output current $I_o$ is also the input current to the feedback network, the feedback is series-derived. $R_g$ must be small compared with $R_i$ for this type of feedback to be effective. Thus we have series-fed, series-derived feedback; its effect on $A_v$ is now determined. From Fig. 10.7c, $V_x = -I_e R_E$ and $V_o = I_o R_L$, but $I_o = -I_c$; therefore $V_o = -I_c R_L$ and

$$\beta_v = \frac{V_x}{V_o} = \frac{-I_e R_E}{-I_c R_L}$$

In general,

$$I_e = I_c \frac{h_{fe} + 1}{h_{fe}}$$

which yields

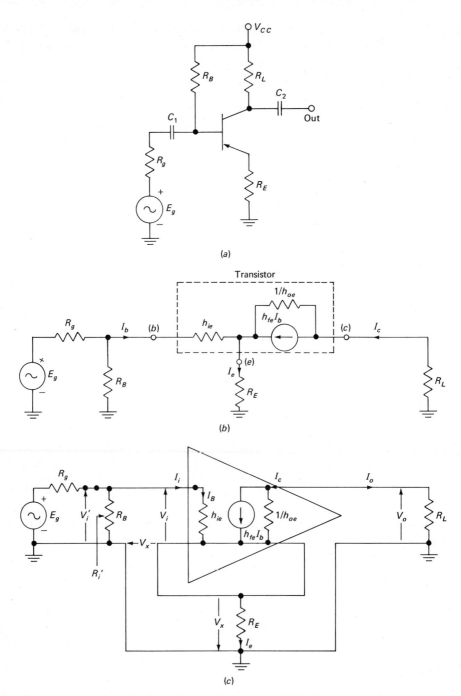

Fig. 10.7 Circuits to illustrate series-derived, series-fed feedback.

$$\beta_v = \frac{h_{fe}+1}{h_{fe}} \frac{R_E}{R_L} \qquad (10.23)$$

Usually $h_{fe} \gg 1$; therefore

$$\beta_v \simeq \frac{R_E}{R_L} \qquad (10.23a)$$

As long as $\beta_v A_v \gg 1$, Eq. (10.6) applies, and

$$A_v' \simeq -\frac{1}{\beta_v} = -\frac{R_L}{R_E} \qquad (10.24)$$

This result was obtained using conventional circuit analysis in Sec. 7.4, Eq. (7.3). Note that $A_v'$ depends only on fixed resistors $R_L$ and $R_E$; transistor parameters such as $h_{fe}$ and $h_{ie}$ do not appear in the final expression.

The input resistance may be derived using the following relationship:

$$R_i' = R_i(1 - \beta_v A_v) \qquad (10.14)$$

$R_i$ is the input resistance when $R_E$ is bypassed, that is, $R_i = h_{ie} \| R_B \simeq h_{ie}$. $A_v$ is the voltage amplification without feedback; it is given by Eq. (6.5b) developed in Sec. 6.2:

$$A_v = \frac{-h_{fe} R_L}{h_{ie}}$$

We therefore have

$$R_i' \simeq h_{ie}\left(1 + \beta_v \frac{h_{fe} R_L}{h_{ie}}\right) \| R_B$$

But

$$\beta_v = \frac{h_{fe}+1}{h_{fe}} \frac{R_E}{R_L} \qquad (10.23)$$

therefore

$$R_i' \simeq h_{ie}\left(1 + \frac{h_{fe}+1}{h_{fe}} \frac{R_E}{R_L} \frac{h_{fe} R_L}{h_{ie}}\right) \| R_B$$

$$R_i' \simeq [h_{ie} + R_E(h_{fe}+1)] \| R_B \qquad (10.25)$$

Note that Eq. (10.25) corresponds to Eq. (7.1) developed in Sec. 7.4.

The current amplification is not directly affected by this type of feedback; hence we can say that $A_i' = A_i$; when $R_L \ll R_o'$, $A_i' \simeq h_{fe}$.

Negative feedback which is series-derived generally increases the output resistance; we will determine $R_o'$ for our circuit using conventional circuit analysis.

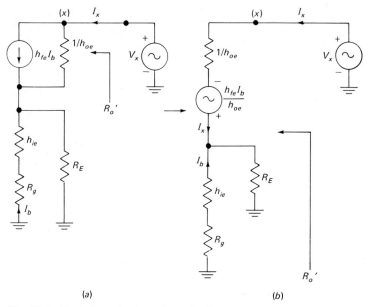

*Fig. 10.8 Equivalent circuits to determine* $R_o$.

The load is removed and the input generator is shorted, yielding the equivalent circuit of Fig. 10.8a. A signal generator $V_x$ is applied at the output; $R_o'$ is simply $V_x/I_x$. The $h_{fe}I_b$ current generator is transformed to its Thevenin equivalent in Fig. 10.8b. Adding all the voltage drops from $x$ to ground yields

$$V_x = I_x \frac{1}{h_{oe}} - \frac{h_{fe}I_b}{h_{oe}} - I_b(h_{ie} + R_g)$$

But

$$I_b = -I_x \frac{R_E}{R_E + h_{ie} + R_g}$$

and $R_o' = V_x/I_x$; therefore

$$R_o' = \frac{1}{h_{oe}} + \frac{h_{fe}R_E}{h_{oe}(R_E + h_{ie} + R_g)} + \frac{R_E(h_{ie} + R_g)}{R_E + h_{ie} + R_g} \tag{10.26}$$

Both the first and last terms in the above expression can be identified with the corresponding resistances in the circuit of Fig. 10.8b. The first term, $1/h_{oe}$, is the output resistance without feedback; the last term is the parallel combination of $R_E$ and $h_{ie} + R_g$. The middle term takes into account the effect of the $h_{fe}I_b/h_{oe}$ generator on the output resistance.

## Example 10.3

For the circuit of Fig. 10.7a, assume $h_{ie} = 1,000 \ \Omega$, $h_{fe} = 100$, $h_{oe} = 5 \times 10^{-5}$ mho, and $h_{re} = 0$. If $R_E = 1 \ \text{k}\Omega$, $R_g = 50 \ \Omega$, $R_L = 5 \ \text{k}\Omega$, and $R_B = 100 \ \text{k}\Omega$, determine $A'_v$, $R'_i$, $A'_i$, and $R'_o$. Make reasonable approximations.

### Solution

The equivalent circuit of Fig. 10.7c applies. The voltage amplification is

$$A'_v \simeq -\frac{R_L}{R_E} \tag{10.24}$$

$$A'_v \simeq -\frac{5,000}{1,000} = -5$$

The input resistance is

$$R'_i = [h_{ie} + R_E(h_{fe} + 1)] \| R_B \tag{10.25}$$
$$R'_i = [1,000 + 1,000(101)] \| 100 \times 10^3$$
$$R'_i = (102 \times 10^3) \| (100 \times 10^3) \simeq 50 \ \text{k}\Omega$$

The current amplification is

$$A'_i \simeq h_{fe} = 100$$
$$A'_i \simeq 100$$

The output resistance is

$$R'_o = \left(\frac{1}{h_{oe}} + R_E\right) \Big\| \left[R_g + h_{ie} + \frac{h_{fe}R_E}{h_{oe}(R_E + h_{ie} + R_g)}\right] \tag{10.26}$$

$$R'_o = [(20 \times 10^3) + 1,000] \Big\| \left[1,050 + \frac{100 \times 1,000}{(5 \times 10^{-5})(2,050)}\right]$$

$$R'_o \simeq 1 \ \text{M}\Omega$$

Note the increase in $R_o$ due to series-derived negative feedback. A circuit to illustrate the effects of shunt-derived, shunt-fed negative feedback is shown in Fig. 10.9a. The circuit is redrawn in Fig. 10.9b to show more clearly how the feedback occurs. $R_g$ must be high compared with $R_i$ for this type of feedback to be effective. Each performance measure is now considered.

Without feedback, the current amplification is

$$A_i = \frac{I_o}{I_i} \tag{10.15}$$

The gate draws no current; hence all $I_i$ flows through $R_G$, yielding

$$V_i = I_i R_G$$

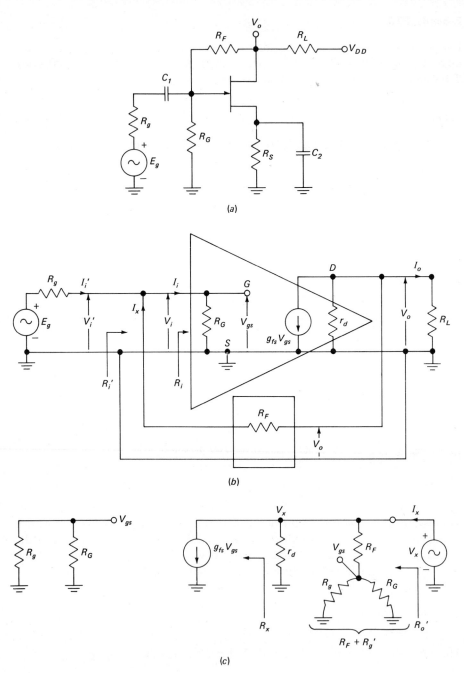

Fig. 10.9  Example of shunt-derived, shunt-fed feedback.

Note that $V_i = V_{gs}$. Provided that $R_L \ll R_o$, we can write
$$I_o \simeq -y_{fs}V_{gs}$$
The current amplification is therefore
$$A_i = \frac{I_o}{I_i} = \frac{-y_{fs}V_{gs}}{V_{gs}/R_G} = -R_G y_{fs}$$
The feedback factor is
$$\beta_i = \frac{I_x}{I_o} \qquad (10.18)$$
$$I_x = \frac{V_o - V_i}{R_F}$$
Using $I_o \simeq -y_{fs}V_{gs} = -y_{fs}V_i$, we get
$$I_x = \frac{V_o + (I_o/y_{fs})}{R_F}$$
and since
$$V_o = I_o R_L$$
$$I_x = \frac{I_o R_L + (I_o/y_{fs})}{R_F}$$
$$\beta_i = \frac{R_L + (1/y_{fs})}{R_F} \qquad (10.18a)$$
The closed-loop current amplification becomes
$$A_i' = \frac{A_i}{1 - \beta_i A_i} = \frac{-R_G y_{fs}}{1 - [(R_L + 1/y_{fs})/R_F](-R_G y_{fs})}$$
$$A_i' = \frac{-1}{[1/(R_G y_{fs})] + [(R_L + 1/y_{fs})/R_F]}$$
If $R_G y_{fs} \gg 1$ and $R_L \gg 1/y_{fs}$, we have
$$A_i' \simeq \frac{-R_F}{R_L} \qquad (10.27)$$
Note that the current amplification has been reduced and can be made independent of device parameters.

The input resistance without feedback is $R_i = R_G$. With shunt-fed feedback, we have
$$R_i' = \frac{R_i}{1 - \beta_i A_i} \qquad (10.22)$$

$$R'_i = \frac{R_G}{1 - [(R_L + 1/y_{fs})/R_F](-R_G y_{fs})}$$

$$R'_i = \frac{1}{[1/R_G] + [(1 + y_{fs}R_L)/R_F]} \qquad (10.28)$$

But

$$A_v \simeq -y_{fs}R_L{}^1 \qquad (9.20a)$$

therefore

$$R'_i = R_G \| R_F/(1 - A_v) \qquad (10.28a)$$

Either Eq. (10.28) or Eq. (10.28a), which will be recognized as the Miller effect equivalent representation, may be used.

The voltage amplification $V_o/V'_i$ is unaffected by the feedback; hence $A'_v \simeq A_v = -y_{fs}R_L$.

As long as $R_g \gg R_G$, the feedback current $I_x$ modifies $I_i$ and hence $V_{gs}$. $I_x$ is proportional to $V_o$; hence if $V_o$ rises, $I_x$ rises, which reduces $I_i$, $V_{gs}$, and $V_o$. The amplifier therefore attempts to maintain a constant output voltage, that is, $R_o$ has decreased. To determine $R'_o$, the circuit of Fig. 10.9c is used. Here $R_L$ is removed and $E_g$ is shorted. We can now write

$$R'_o = (R_F + R'_g) \| r_d \| R_x$$

where $R'_g = R_g \| R_G$ and $R_x$ is the equivalent resistance due to the current generator:

$$R_x = \frac{V_x}{y_{fs} V_{gs}}$$

But

$$V_{gs} = V_x \frac{R_g}{R'_g + R_F}$$

therefore

$$R_x = \frac{V_x(R'_g + R_F)}{y_{fs} V_x R'_g} = \frac{R'_g + R_F}{y_{fs} R'_g}$$

The output resistance is

$$R'_o = (R_F + R'_g) \| r_d \| \frac{R_F + R'_g}{y_{fs} R'_g}$$

$$R'_o = r_d \left\| \frac{R'_g + R_F}{y_{fs} R'_g + 1} \right. \qquad (10.29)$$

---

[1] See Sec. 9.7a for derivation.

If $R'_g \gg R_F$ and $y_{fs}R'_g \gg 1$, we have

$$R'_o \simeq r_d \left\| \frac{1}{y_{fs}} \right. \tag{10.29a}$$

**Example 10.4**

Determine $A'_i$, $R'_i$, and $R'_o$ for the FET amplifier in Fig. 10.9a. The following parameters apply: $y_{fs} = 3{,}000$ $\mu$mho; $r_d = 200$ k$\Omega$; $R_G = 1$ M$\Omega$; $R_L = 10$ k$\Omega$; $R_F = 100$ k$\Omega$; $R_g = 10$ M$\Omega$.

**Solution**

The circuit of Fig. 10.9b will be used. Each performance measure will be considered in turn.

The current amplification is

$$A'_i \simeq -\frac{R_F}{R_L} = -\frac{10^5}{10^4} \tag{10.27}$$

$$A'_i \simeq -10$$

The open-loop voltage amplification is

$$A_v \simeq -y_{fs}R_L \tag{9.20a}$$
$$A_v \simeq (-3{,}000 \times 10^{-6})(10^4) = -30$$

The input resistance is

$$R'_i = R_G \left\| \frac{R_F}{1 - A_v} \right. \tag{10.28a}$$

$$R'_i = 10^6 \left\| \frac{10^5}{31} \right.$$

$$R'_i \simeq 3.2 \text{ k}\Omega$$

The output resistance is

$$R_o \simeq r_d \left\| \frac{1}{y_{fs}} \right. \tag{10.29a}$$

$$R'_o \simeq 200 \times 10^3 \| 333 \simeq 330 \text{ }\Omega$$

The effects of $R_F$ on the output resistance of the common emitter transistor amplifier of Fig. 10.10a are now considered. The output resistance is obtained by shorting $E_g$ and looking into the output terminals. If the circuit is driven from a constant voltage source, $R_g = 0$, so that shorting $E_g$ places the base at ground potential. In this case $R_F$ appears directly across the output and $R'_o =$

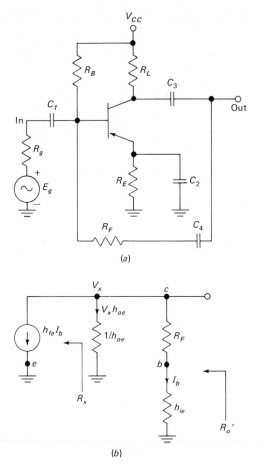

**Fig. 10.10** *Circuits to determine the effect of $R_F$ on output resistance.*

$R_o \| R_F$. The benefits of negative feedback, however, cannot be realized, since shunt feeding is effective only as long as $R_g$ is relatively large.

When the circuit is driven from a constant current source, the equivalent circuit of Fig. 10.10b applies. Here $h_{ie}$ is assumed small compared with $R_B$. This yields

$$R'_o = \frac{1}{h_{oe}} \| R_x \| (R_F + h_{ie})$$

where

$$R_x = \frac{V_x}{h_{fe}I_b}$$

But

$$I_b = \frac{V_x}{R_F + h_{ie}}$$

and

$$R_x = \frac{R_F + h_{ie}}{h_{fe}}$$

yielding

$$R'_o = \frac{1}{h_{oe}} \left\| \frac{R_F + h_{ie}}{h_{fe}} \right\| (R_F + h_{ie}) \tag{10.30}$$

When $R_F \gg h_{ie}$ and $1/h_{oe} \gg R_F$, we have

$$R'_o \simeq \frac{R_F}{h_{fe} + 1} \tag{10.30a}$$

### Example 10.5

Determine the output resistance of the common emitter transistor amplifier in Fig. 10.10a when it is driven from a constant current source. The circuit parameters are $h_{fe} = 50$, $h_{ie} = 1{,}000 \, \Omega$, $h_{oe} = 5 \times 10^{-5}$ mho, $h_{re} \simeq 0$; $R_B = 100 \, k\Omega$, $R_L = 1 \, k\Omega$, $R_F = 10 \, k\Omega$.

### Solution

Since $R_g = \infty$, Eq. (10.30) applies:

$$R'_o = 20{,}000 \left\| \frac{11{,}000}{50} \right\| 11{,}000 \tag{10.30}$$

$$R'_o \simeq 210 \, \Omega$$

## Problem Set 10.3

10.3.1. How is the output resistance of amplifiers affected by negative feedback which is

    a. Series-derived?
    b. Shunt-derived?

10.3.2. How is the input resistance of amplifiers affected by negative feedback which is

    a. Series-fed?
    b. Shunt-fed?

10.3.3. When the feedback signal is series-fed, which performance measures are affected, and in what way?

10.3.4. Repeat Prob. 10.3.3 for shunt feeding.

10.3.5. Which type of feedback stabilizes $A_v$? Which type stabilizes $A_i$?

10.3.6. For the amplifier of Example 10.3, assume $R_E = 2{,}000\ \Omega$ and $R_L = 6{,}000\ \Omega$. Determine $\beta_v$, $R'_i$, and $A'_v$.

10.3.7. Assuming $R_L \ll r_d$, show that, for the FET amplifier of Example 10.4, the closed-loop voltage amplification is

$$A'_v = \frac{(R_L/R_F) - y_{fs}R_L}{1 + (R_L/R_F)}$$

For what ratio $R_L/R_F$ is $A'_v = A_v$?

10.3.8. The transistor in the circuit of Fig. 10.10a has an equivalent $C_{ob} = 4$ pF. Recalling that $C_{ob}$ is the capacitance between base and collector, determine its effect when seen from the output, using the techniques developed to treat $R_F$. The circuit is driven from a 1-MΩ source; $R_B$ also equals 1 MΩ.

10.3.9. Show that, for any amplifier,

$$\frac{A_v}{A_i} = \frac{R_L}{R_i}$$

10.3.10. Using reasonable approximations, estimate the output resistance, voltage amplification, current amplification, and input resistance for the amplifier in Fig. 10.11. The following parameters apply: $h_{fe} = 100$, $h_{ie} = 1{,}000\ \Omega$, $h_{re} = h_{oe} = 0$; $R_F = 20$ kΩ, $R_L = 10$ kΩ.

## 10.4 MULTISTAGE FEEDBACK CIRCUITS

Some specific circuits employing *multistage* (overall) feedback are discussed in this section. As indicated in Sec. 10.1, the advantage of employing feedback across several stages is that, since the open-loop amplification is usually greater than for a single stage, more feedback can be used. This means that greater improvement is possible in performance measures such as stabilization of amplification, distortion reduction, and increased bandwidth.

When amplifier stages are cascaded, their individual phase shifts add; thus for identical stages whose phase shift is 180°, an odd number (3, 5, etc.) of stages must be cascaded to yield overall signal phase inversion. As an example,

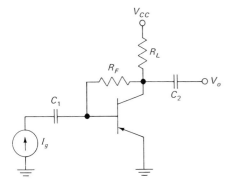

Fig. 10.11  Circuit for Prob. 10.3.10.

the phase shift for $A_v$ in a common emitter stage is usually 180°; therefore two such stages yield an overall phase shift of $180 + 180 = 360°$, which is the same as 0°. Overall feedback taken across these two stages is therefore positive. To achieve negative feedback an additional 180° phase shift must be provided, by either adding one more stage or using a feedback network that provides phase inversion. In practice, feedback over more than three stages is seldom used because of stability problems.

An example of multistage feedback of the series-derived, series-fed type is shown in Fig. 10.12. Here three common emitter stages are cascaded, which results in net phase inversion. The common resistor $R_E$ makes up the feedback

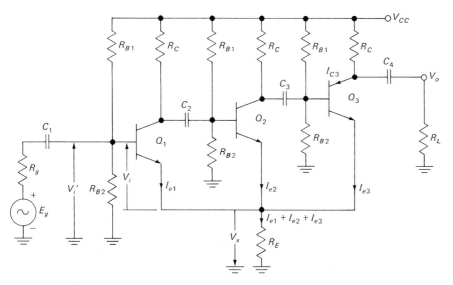

Fig. 10.12  Circuit with multistage feedback.

network, as indicated in the figure. The three cascaded stages may be treated as a single amplifier whose overall open-loop voltage amplification is $A_v$; the voltage feedback ratio is

$$\beta_v = \frac{V_x}{V_o} \tag{10.3}$$

where $V_x = -R_E(I_{e1} + I_{e2} + I_{e3}) \simeq -R_E I_{e3}$ as long as $h_{fe}$ is large enough to result in $I_{e3} \gg I_{e2} \gg I_{e1}$. Similarly, $V_o \simeq -I_{c3} R'_L$ ($R'_L = R_C \| R_L$) and $I_{c3} \simeq I_{e3}$, yielding

$$\beta_v \simeq \frac{-R_E I_{e3}}{-I_{c3} R'_L} \simeq \frac{R_E}{R'_L}$$

$A'_v$ is given by Eq. (10.24), while $R'_i$ is given by Eq. (10.14).

The shunt-derived, shunt-fed feedback circuit of Fig. 10.13 is another example of multistage feedback. Since an odd number of common emitter stages are cascaded, the feedback is negative. Capacitor $C_5$ is used to block DC and may be considered an AC short circuit at midfrequencies.

The circuit may be analyzed using techniques similar to those of Example 10.4. Provided $R_g$ is sufficiently high, $A_i$ can be stabilized, while both $R_i$ and $R_o$ are reduced; $A_v$ is not appreciably affected.

The circuit of Fig. 10.14a employs series-derived, shunt-fed negative feedback. Since two common emitter stages are involved, there is no net phase shift from input to output. For this reason the feedback is derived at the emitter of $Q_2$, which is 180° out of phase with the base of $Q_1$, yielding negative feedback. The circuit is redrawn in Fig. 10.14b to illustrate the feedback connections. For this type of feedback to be effective, $R_g$ should be relatively high. The net result

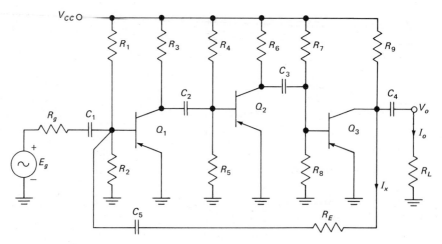

Fig. 10.13  Shunt-derived, shunt-fed multistage feedback circuit.

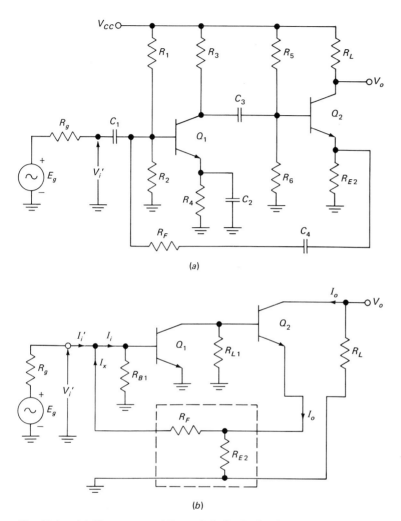

*Fig. 10.14* (a) Two-stage amplifier with feedback; (b) AC equivalent circuit for (a) redrawn to show series-derived, shunt-fed nature of feedback.

should be an increase in $R_o$ and decrease in $R_i$. Also the current amplification is reduced and stabilized. We therefore have a circuit that approaches an ideal current amplifier because of the low input resistance, high output resistance, and stable current amplification.[1]

Another interesting circuit is that of Fig. 10.15. Here the feedback signal is shunt-derived, which lowers the output resistance, and series-fed, which

[1] See Sec. 4.4.

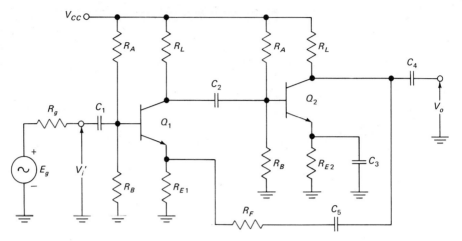

Fig. 10.15  Example of shunt-derived, series-fed feedback across two amplifier stages.

raises the input resistance; the circuit should be driven from a low-impedance source. That the feedback is negative can be determined as follows. Suppose the voltage of the base of $Q_1$ (input) starts to rise; the output at the collector of $Q_2$ also rises. Now, because of the coupling action of $R_F$, the rising output causes the emitter of $Q_1$ to rise. The base drive is the difference between the input signal and the emitter voltage; therefore, as the emitter voltage rises, the actual input to the amplifier (base-emitter voltage) drops. The voltage amplification is reduced and stabilized; coupled with the high input and low output resistance, this means that the circuit approaches an ideal voltage amplifier.[1]

### Problem Set 10.4

10.4.1. What advantages does multistage feedback have over single-stage feedback? Is there any disadvantage?

10.4.2. Why is an odd number of stages usually required when employing negative feedback over several stages, each providing 180° phase shift? Is this always necessary?

10.4.3. The following data apply to the circuit of Fig. 10.12: $h_{fe} = 100$, $h_{ie} = 1$ kΩ, $h_{oe} \simeq h_{re} \simeq 0$ for all transistors; $R_{B1} = 120$ kΩ; $R_{B2} = 60$ kΩ; $R_C = 40$ kΩ; $R_L = 1$ kΩ; $R_g = 0$; $R_E = 5$ Ω. Estimate

    a. The open-loop voltage amplification
    b. $A'_v$
    c. The input resistance seen by $E_g$
    d. $A'_v$ if all $h_{fe}$'s double

[1] See Sec. 4.4.

**10.4.4.** In the circuit diagram of Fig. 10.12, assume $R_L = 600 \ \Omega$. Determine $R_E$ to yield a closed-loop voltage amplification of $-1,000$. What is $\beta_v$?
**10.4.5.** If $h_{ie} = 600 \ \Omega$, $h_{oe} = h_{re} = 0$, $h_{fe} = 100$, $R_{B1} = R_{B2} = 1 \ M\Omega$, in Prob. 10.4.4, estimate $R_i'$.
**10.4.6.** Estimate $A_i'$ and $R_i'$ for the circuit of Fig. 10.13 if $A_i = -20,000$, $R_L = 1,500 \ \Omega$, $R_F = 100 \ k\Omega$, and $R_i = 1 \ k\Omega$.
**10.4.7.** Estimate $A_i'$ in Prob. 10.4.6 when $A_i$ drops by 40 percent.

## 10.5 STABILITY OF FEEDBACK SYSTEMS

It was shown earlier that, with negative feedback, the signal fed back is phase-inverted relative to the input. As long as this condition holds, the closed-loop amplification is less than the open-loop amplification and the system is stable. Stability information can be obtained from the general expression for closed-loop voltage or current amplification. The expression for $A_v'$ may be used, as follows:

$$A_v' = \frac{A_v}{1 - \beta_v A_v} \tag{10.5}$$

There are three main possibilities to consider, each based on a specific value or range of values for the return difference, as follows:

a. $|1 - \beta_v A_v| < 1$: The closed-loop amplification $A_v'$ is greater than the open-loop amplification $A_v$. This is positive feedback.
b. $|1 - \beta_v A_v| > 1$: Here $|A_v'| < |A_v|$; this is negative feedback.
c. $|1 - \beta_v A_v| = 0$: Here the feedback is positive and $A_v' \to \infty$, resulting in oscillations.

A stable system is one in which there are no oscillations, that is, the criterion of (b) is satisfied. Often an amplifier may satisfy this criterion within its midfrequency range, but because of phase shift, the circuit may oscillate at relatively high or low frequencies. Oscillators are treated in Chap. 14.

At low frequencies, coupling capacitors and transformers reduce the gain and produce phase shift. At high frequencies, shunt capacitances of the active device and other components also reduce the gain and produce phase shift. These phase shifts and gain reductions mean that $A_v$ is, in general, a complex number—that is, it has both a real and imaginary part, the imaginary part being due to reactive components in the system. The criterion for oscillation can therefore be expressed as follows:

$$1 - \beta_v A_v = 0$$
$$\beta_v A_v = 1 \tag{10.31}$$
$$\beta_v A_v = 1 + j0 \tag{10.31a}$$
$$\beta_v A_v = 1/\underline{0°} \tag{10.31b}$$

Thus if the magnitude of $\beta_v A_v$ is 1 (0 dB) and its phase 0°, the circuit is unstable. To illustrate this point, consider the magnitude and phase plots of Fig. 10.16. The graphs represent the transfer function:

$$\beta_v A_v = \frac{-1{,}000}{(1 + jf/10^6)^3}$$

whose magnitude is 1,000 (60 dB) at midfrequencies and drops off at 60 dB per decade beyond $f = 1$ MHz. The total phase shift at midfrequencies is 180° (because of the negative sign). The denominator is insignificant until we approach the corner frequency of $10^6$ Hz.

At midfrequencies, $\beta_v A_v = -1{,}000$, a negative real number; the quantity $1 - \beta_v A_v$ is therefore equal to 1,001 which is real, positive, and greater than 1. This means that $|A_v'|$ is less than $|A_v|$ and the feedback is negative.

As the frequency increases toward 1 MHz, the phase is no longer 180° but changes because of the $(1 + jf/10^6)^3$ term in the denominator. Eventually the phase becomes zero, yielding positive feedback. If $|\beta_v A_v| < 1$ when the phase is zero, the closed-loop amplification is greater than the open-loop amplification but the system does not oscillate. If, for example, $\beta_v A_v = 0.8$ when the phase is zero, then the quantity $1 - \beta_v A_v = 0.2$ and $A_v' = 5 A_v$.

For the function of Fig. 10.16, the frequency of zero phase is about 2 MHz. At this frequency, $|\beta_v A_v|$ is greater than 1, and hence the system is unstable. To

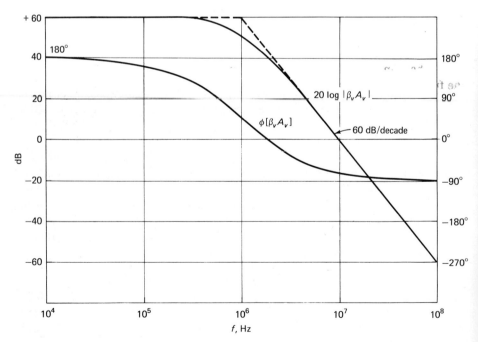

Fig. 10.16 Magnitude and phase plot for $\beta_v A_v = -1{,}000/(1 + jf/10^6)^3$.

make the system stable, we could either reduce $|\beta_v A_v|$ so that it is less than 0 dB when the phase is zero, or we could add phase compensation (usually an $RC$ network) so that the total phase shift does not become zero until $|\beta_v A_v|$ drops below 0 dB.

When using gain reduction to stabilize a system, a safety margin is usually necessary so that if $\beta_v A_v$ should change (due to component aging or replacement), the system will remain stable. In our case, a 30-dB gain reduction at 2 MHz would make the system stable and provide a gain margin of about 10 dB.

If phase compensation is chosen, a phase-shifting network must be used that will cause the phase response to remain above the 0° axis until after the magnitude curve drops below 0 dB (around 10 MHz). We would therefore need about 90° (and preferably more) of phase lead near $f = 10$ MHz for the system to be stable. An amplifier is considered unstable if there is one or more frequency for which the criterion $\beta_v A_v = 1 + j0$ is satisfied. This applies even if the amplifier is not intended for use at the frequency of instability. For example, a high-gain audio amplifier which is unstable at 100 kHz will oscillate at 100 kHz even though only audio signals (below 15 kHz) are applied at the input. This is further discussed in Chap. 14; for now it is sufficient to say that a well-designed amplifier should be stable at all frequencies.

In general, as more amplifier stages are cascaded, the total phase shift increases and instability problems are more likely to occur. In fact, it is rather difficult to design stable amplifiers involving feedback around three or more stages.

Two measures of an amplifier's stability are the gain and phase margins. Gain margin represents by how much $20 \log |\beta_v A_v|$ is below 0 dB when the phase is 0°. Phase margin represents by how much the phase of $\beta_v A_v$ is above 0° at the frequency for which $20 \log |\beta_v A_v| = 0$ dB.

## Example 10.6

An amplifier has the following open-loop voltage amplification:

$$A_v = \frac{-1{,}000}{(1 + jf/10^5)^3}$$

Determine whether the amplifier is stable, including gain and phase margins, for:

a. $\beta_v = 0.1$
b. $\beta_v = 0.005$

## Solution

Magnitude and phase plots for $\beta_v A_v$ are sketched in Figs. 10.17 and 10.18 for (a) and (b), respectively.

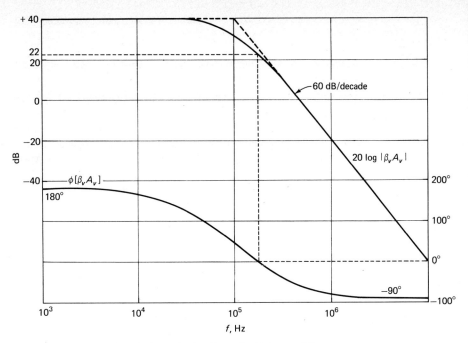

Fig. 10.17  Magnitude and phase plot for Example 10.6, part (a).

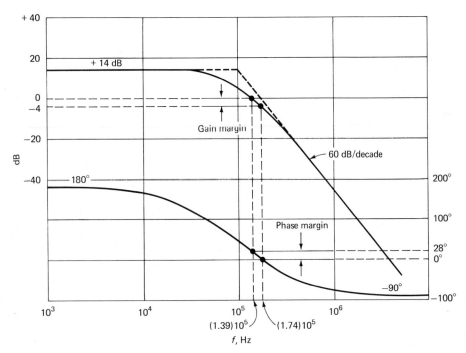

Fig. 10.18  Magnitude and phase plot for Example 10.6, part (b).

a. $\beta = 0.1$: At the frequency of zero phase shift, $20 \log |\beta_v A_v| \simeq 22$ dB; the system is therefore unstable.
b. $\beta = 0.005$: At the frequency of zero phase shift ($f \simeq 1.74 \times 10^5$ Hz), the gain, from Fig. 10.20, is $-4$ dB. The phase at the frequency for which $20 \log |\beta_v A_v| = 0$ dB is around 28°. The system is therefore stable. The gain margin is 4 dB and the phase margin is 28°. In practice, however, it may be desirable to have a gain margin greater than 4 dB.

**Example 10.7**

For the Bode plots of Fig. 10.19, determine

a. The transfer function $\beta_v A_v$
b. How much feedback is being used if $20 \log |A'_v| = 60$ dB at midfrequencies
c. The reduction in midfrequency voltage amplification due to negative feedback
d. The low-frequency cutoff without feedback
e. The low-frequency cutoff with feedback
f. Whether the system is stable
g. Whether the system remains stable after the midfrequency $A_v$ is reduced by 40 dB

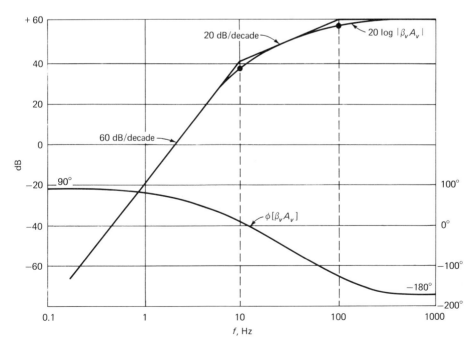

*Fig. 10.19  Magnitude and phase plots for Example 10.7.*

## Solution

a. The transfer function $\beta_v A_v$ can be constructed from the information on the graph. At midfrequencies the phase is $-180°$; this introduces a negative sign. Below 10 Hz the response rises with frequency at 60 dB per decade, which implies a $(jf)^3$ factor in the numerator. Above 10 Hz the response continues to rise, but only at 20 dB per decade, which indicates a $(1 + jf/10)^2$ factor in the denominator. Finally at 100 Hz the response flattens out, implying a denominator factor of $1 + jf/100$. Our expression is therefore of the following general form:

$$\beta_v A_v = \frac{-K(jf)^3}{(1 + jf/10)^2(1 + jf/100)}$$

$K$ is obtained by evaluating the expression at any frequency; at $f - 1$ Hz, $20 \log |\beta_v A_v| = -20$ dB; therefore $\beta_v A_v = 0.1$. This yields $K = 0.1$. Note that although the phase response is given here, it could be determined from the transfer function just constructed, which required only the plot of $|\beta_v A_v|$ and the midfrequency phase.

b. If $20 \log |A'_v| = 60$ dB, $A'_v = -10^3$; at midfrequencies $20 \log |\beta_v A_v| = 60$ dB, and therefore $\beta_v A_v = -1{,}000$, or $A_v = -1{,}000/\beta_v$. We can now write

$$A'_v = \frac{A_v}{1 - \beta_v A_v} \qquad (10.5)$$

$$A'_v = \frac{1{,}000/\beta_v}{1 - \beta_v A_v}$$

$$-1{,}000 = \frac{-1{,}000/\beta_v}{1 + 1{,}000}$$

$$\beta_v = 0.001$$

c. The midfrequency voltage amplification has been reduced from $10^6$ to 1,000. This is a 60-dB drop.

d. The transfer function for $\beta_v A_v$ is down 3 dB at 100 Hz; since $\beta_v$ is a constant, $|A_v|$ is also down 3 dB at 100 Hz.

e. With feedback the frequency response is extended by the factor $1 - \beta_v A_v \simeq 1{,}000$; therefore $f'_L = 100/1{,}000 = 0.1$ Hz.

f. The phase is $0°$ at $f = 10$ Hz; the gain here is about $+34$ dB. The amplifier is therefore unstable and should oscillate at 10 Hz.

g. At the frequency of zero phase shift (10 Hz), the new value of $20 \log |\beta_v A_v|$ is $-6$ dB, and the system is stable. The gain margin is 6 dB, and the phase margin is $30°$ (near $f = 15$ Hz).

Another approach for analyzing the stability of a system was developed by Nyquist. The criterion is based on the condition that feedback is negative if

$$|1 - \beta_v A_v| > 1 \tag{a}$$

as developed earlier. Now, in general, $\beta_v A_v$ is a complex number and may be expressed either as a magnitude and phase (polar form) or as the sum of real and imaginary parts (rectangular form). Using the rectangular form, we can write

$$\beta_v A_v = a + jb \tag{b}$$

where both $a$ and $b$ may be functions of frequency. The condition for negative feedback can therefore be expressed as follows:

$$\begin{aligned}|1 - (a + jb)| &> 1 \\ |(1 - a) - j(b)| &> 1 \\ \sqrt{(1 - a)^2 + b^2} &> 1 \\ (1 - a)^2 + b^2 &> 1\end{aligned} \tag{c}$$

The above equation describes the area outside a circle whose radius is 1 and which is centered at $a = 1$ and $b = 0$, that is, the point $(1, j0)$ in Fig. 10.20. Points outside the *unit circle* represent negative feedback and a stable system. Points inside the unit circle represent positive feedback.

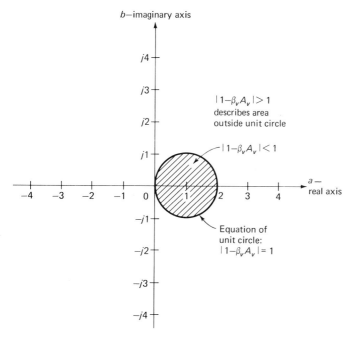

**Fig. 10.20** Use of the Nyquist stability criterion. Note: $\beta_v A_v = a + jb$.

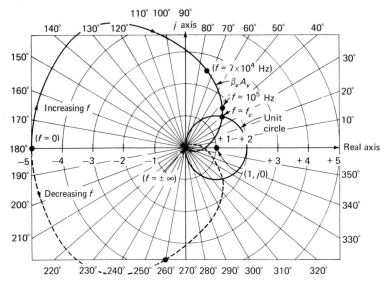

Fig. 10.21 Polar plot of $\beta_v A_v = -5/(1 + jf/10^5)^3$.

To illustrate Nyquist's criterion, consider Fig. 10.21. Here, in addition to the unit circle, a polar plot of $\beta_v A_v$ has been constructed by calculating the magnitude and phase of $\beta_v A_v$ at several frequencies. For example, given

$$\beta_v A_v = \frac{-5}{(1 + jf/10^5)^3}$$

at $f = 10^5$ Hz, $\beta_v A_v = -5/(1 + j1)^3 = 1.76\underline{/45°}$. We plot this by locating the magnitude 1.76 along the 45° line. This is repeated for as many frequencies between zero and infinity as are necessary to generate the curve. Note that at zero frequency, the magnitude is 0 and the phase is $-90°$. The dashed curve represents the plot of $\beta_v A_v$ for negative frequencies ($f = 0$ to $f = -\infty$); this plot is the mirror image of that for positive frequencies.

Nyquist's stability criterion states that if the $\beta_v A_v$ plot does not encircle the point $(1, j0)$, the system is stable. For our example, the feedback is negative for frequencies below $f_c$. At $f_c$, the $\beta_v A_v$ locus crosses the unit circle and the feedback becomes positive. For frequencies between $f_c$ and infinity the feedback remains positive but the magnitude of $\beta_v A_v$ is always less than 1 [the point $(1, j0)$ is not encircled]. Hence, even though there is positive feedback, the system is stable.

### Example 10.8

Discuss, using Nyquist's criterion, the stability for each plot of $\beta_v A_v$ in Fig. 10.22.

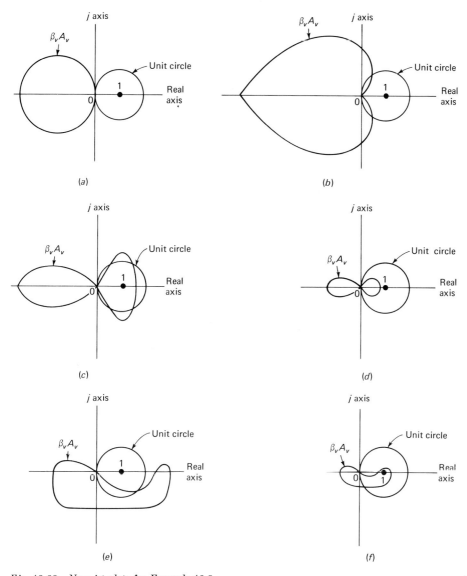

**Fig. 10.22** Nyquist plots for Example 10.8.

## Solution

a. The $\beta_v A_v$ locus remains outside the unit circle; therefore the system is stable.
b. The $\beta_v A_v$ locus does enter the unit circle, but the point $(1, j0)$ is not encircled; hence even though there is positive feedback at certain frequencies, the system remains stable.

c,d. The system in (c) is unstable because the point $(1, j0)$ is encircled. To make the system stable, the gain could be reduced until the $\beta_v A_v$ locus no longer encloses $(1, j0)$. This results in the stable plot of (d).

e,f. In (e) the point $(1, j0)$ is not encircled, and hence the system appears stable. If the gain were reduced, however, the plot would shrink as in (f) and the system would become unstable. Therefore the system whose $\beta_v A_v$ locus is shown in (e) is conditionally stable, the condition being that the gain must remain above a certain value.

## Problem Set 10.5

10.5.1. Why must amplifiers be stable even outside their midfrequency range?

10.5.2. Given the following values for $\beta_v A_v$, determine whether the feedback is negative or positive. If positive, is the feedback sufficient for oscillations to take place?

   a. $3/\underline{0°}$
   b. $3/\underline{180°}$
   c. $0.5 + j0$

10.5.3. Given $\beta_v A_v = -100/(1 + jf/5{,}000)^3$, sketch magnitude and phase plots. Is the system stable? If so, determine gain and phase margins; if not, determine frequency of oscillation.

10.5.4. Given the $\beta_v A_v$ magnitude plot of Fig. 10.23 and assuming a 180° midfrequency phase shift, sketch the phase response. What conclusions can you make regarding stability?

10.5.5. For the plot of Prob. 10.5.4, determine $\beta_v$ if $20 \log |A_v| = 60$ dB in the midfrequency range.

10.5.6. What are the open-loop $f_L$ and $f_H$ in Prob. 10.5.4? What are the closed-loop values if $\beta_v$ from Prob. 10.5.5 applies?

10.5.7. It is desired to provide phase compensation in a feedback loop using the network of Fig. 10.24. Sketch magnitude and phase plots for the

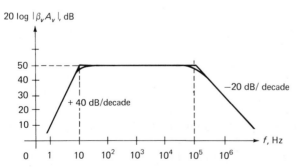

Fig. 10.23 Magnitude plot for Prob. 10.5.4.

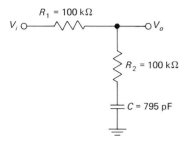

Fig. 10.24  Circuit for Prob. 10.5.7.

transfer function $V_o/V_i$. What is the DC attenuation? What is the high-frequency attenuation? What is the phase shift at $f = 1{,}000$ Hz?

10.5.8. Sketch a Nyquist plot for

$$\beta_v A_v = \frac{-100}{(1 + jf/10^4)(1 + jf/10^6)}$$

Is the system stable?

## REFERENCES

1. Joyce, M. V., and K. K. Clarke: "Transistor Circuit Analysis," Addison-Wesley Publishing Company, Inc., Reading, Mass., 1961.
2. Cowles, L. C.: "Analysis and Design of Transistor Circuits," D. Van Nostrand Company, Inc., Princeton, N.J., 1966.
3. Millman, J., and C. C. Halkias: "Electronic Devices and Circuits," McGraw-Hill Book Company, New York, 1967.
4. Cutler, P.: "Semiconductor Circuit Analysis," McGraw-Hill Book Company, New York, 1964.
5. Gibbons, J. F.: "Semiconductor Electronics," McGraw-Hill Book Company, New York, 1966.
6. Fitchen, F. C.: "Transistor Circuit Analysis and Design," 2d ed., D. Van Nostrand Co., Inc., Princeton, 1966.

# chapter 11

# DC circuits

There are many systems utilizing signals that change very slowly with time; typical applications include analog computation, bioelectric measurements, power supply regulators, and so on. To process these signals, one must employ circuits whose frequency response is flat down to DC. This precludes the use of interstage-coupling elements such as capacitors and transformers, because these components attenuate very low frequencies and completely block DC.

There are two basic techniques for amplifying signals that change very slowly. One is to use direct-coupled (dc)[1] amplifiers; the other requires "chopping" the DC signal so as to change it to an AC signal, which is then amplified using conventional AC amplifiers and reconstructed at the output. Both of these techniques are discussed in the sections to follow.

## 11.1 DIRECT-COUPLED AMPLIFIERS

As the notation implies, dc amplifiers are directly connected; that is, no coupling capacitors or transformers are used. When this is done, care must be taken to ensure that the DC levels of each stage are compatible with those of other circuits to which the stage is connected.

As an example, consider the two identical amplifier stages of Fig. 11.1a.

---

[1] We define dc to mean direct-coupled, while DC specifies any frequency so low that it may be considered zero.

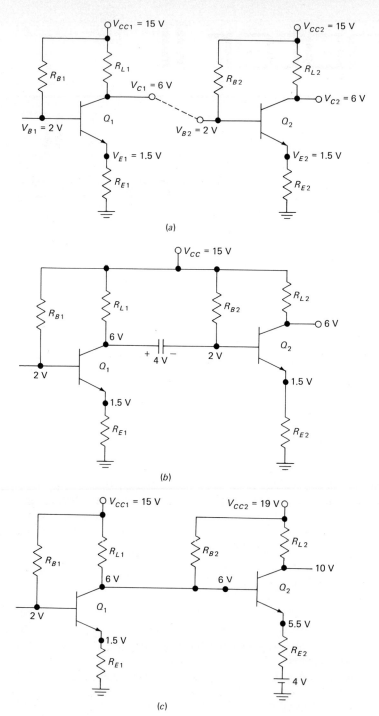

Fig. 11.1 Interstage coupling techniques.

Assume that $V_{CC1} = V_{CC2} = 15$ V; $V_{B1} = V_{B2} = 2$ V; $V_{E1} = V_{E2} = 1.5$ V; and $V_{C1} = V_{C2} = 6$ V. Each stage is therefore biased at $V_{BE} = 0.5$ V and $V_{CB} = 4$ V. Also there is a potential difference of 4 V between the collector of $Q_1$ and base of $Q_2$.

For AC operation a capacitor is connected between the collector of $Q_1$ and the base of $Q_2$ as shown in Fig. 11.1b. This way the AC signal is coupled from $Q_1$ to $Q_2$, but the capacitor charges to 4 V DC so that $V_{C1}$ is still 6 V and $V_{B2}$ is still 2 V. The DC levels for either stage remain independent of the other.

For DC operation, the capacitor is omitted, but we cannot tie the collector of $Q_1$ directly to the base of $Q_2$ unless the voltages at these two points are the same. One possible solution is to leave $Q_1$ as it is, but "lift" $Q_2$ by 4 V relative to ground so that $V_{B2}$, $V_{E2}$, and $V_{C2}$ are all 4 V higher than they were earlier. This may be accomplished by connecting a 4-V battery in series with $R_{E2}$ and by using a 19-V supply for $V_{CC2}$, as shown in Fig. 11.1c. We now have $V_{B2} = 6$ V and $V_{C2} = 10$ V. This causes $V_{C1}$ to equal $V_{B2}$, enabling us to tie the two points directly to each other. Note that raising base, emitter, and collector voltages by the same amount does not alter any of the potential differences; that is, $V_{BE2}$ is still 0.5 V, $V_{CB2}$ is still 4 V, and therefore the operating point is unchanged.

The arrangement of Fig. 11.1c has a number of obvious disadvantages. Because $V_{CC1}$ and $V_{CC2}$ are not the same and because of the 4-V battery in series with $R_{E2}$, a total of three different supplies are needed. The 4-V battery could be replaced with a 4-V Zener diode, or it could simply be eliminated and $R_{E2}$ made larger. This would increase the input resistance to $Q_2$ and lower the gain because of the increased negative feedback. Another disadvantage, however, is that $V_{CC2}$ must be larger than $V_{CC1}$, and if a third stage is added, $V_{CC3}$ must be larger than $V_{CC2}$. For a multistage amplifier, one might require impractically large DC voltage supplies.

A circuit arrangement that solves the above problems is shown in Fig. 11.2. Here alternate polarity transistors are cascaded, and the DC voltages are so adjusted that each of the three transistors is operating at exactly the same point. For example, there is 0.5-V forward bias across each base-emitter junction and 6-V reverse bias across each collector-base junction. Also note that only one external supply is needed.

We must now consider an additional problem peculiar to dc amplifiers. This has to do with small changes in the operating point of a particular stage; as discussed earlier, changes in $Q$ point may be due to temperature or power supply fluctuations as well as aging of components. Since there is nothing to block DC, these changes are amplified by each successive stage and appear across the output.

The DC input voltage required to bring the output voltage back to its original (no-signal) level is a measure of performance for dc amplifiers and is called voltage offset, $E_{os}$. In the same fashion an offset current $I_{os}$ is used to denote

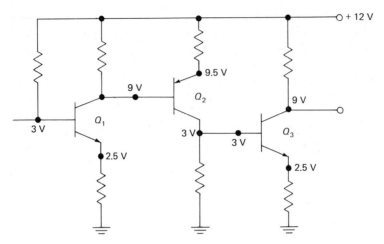

Fig. 11.2  *Alternating polarity transistor direct-coupled stages.*

the DC input current required to bring the output current back to its original level. Both the offset voltage and offset current tend to change (drift) with time and temperature; dc amplifiers generally include a potentiometer adjustment for canceling such offsets. The manufacturer also usually specifies, in addition to $E_{os}$ and $I_{os}$, the expected drift with time and temperature.

In dc amplifiers the weakest signal that can be amplified is determined not so much by noise (as in AC amplifiers) but by the drift in voltage and current offsets.

The various techniques discussed in earlier chapters for stabilizing the Q point (Chap. 8 for bipolar transistors; Chap. 9 for FETs) are also used with dc amplifiers. These include the use of negative feedback and nonlinear components such as diodes and thermistors.

Another useful and interesting dc circuit is the *compound* or *Darlington* connection shown in Fig. 11.3. Here the emitter current of $Q_1$ is the base current of $Q_2$; assuming equal $h_{fe}$'s for both transistors, we can write

$I_c = I_{c1} + I_{c2}$
$I_c = I_{b1}h_{fe} + I_{b2}h_{fe}$

But $I_{b2} = I_{e1}$; therefore

$I_c = I_{b1}h_{fe} + I_{e1}h_{fe}$

and

$I_{e1} = I_{b1}(h_{fe} + 1)$

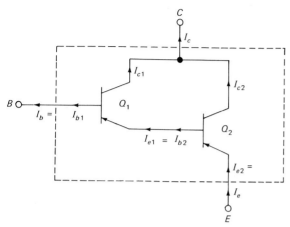

Fig. 11.3  Darlington connection.

yielding

$$I_c = I_{b1}h_{fe} + I_{b1}(h_{fe} + 1)(h_{fe})$$
$$I_c = I_{b1}(2h_{fe} + h_{fe}^2)$$

Usually

$$h_{fe}^2 \gg 2h_{fe}$$

and

$$I_c \simeq I_{b1}h_{fe}^2 \tag{11.1}$$

The two-transistor combination therefore behaves as a single transistor with a much larger $h_{fe}$; if, for example, each transistor's $h_{fe} = 100$, the Darlington connection behaves as a single transistor whose $h_{fe} = 100^2 = 10{,}000$. In addition to the high $h_{fe}$, there is also the advantage of emitter feedback, since the resistance looking into the base of $Q_2$ is in series with the emitter of $Q_1$. This helps stabilize the gain as well as increasing the input resistance to the base of $Q_1$. Darlington pairs are often manufactured on a single chip, which helps in the matching of their characteristics. With the single package, only three leads are externally accessible; these may be treated as the collector, base, and emitter of a single, very high $h_{fe}$ transistor.

## Problem Set 11.1

11.1.1. Why can capacitors or transformers not be utilized to couple very low frequency signals?

11.1.2. What are some of the disadvantages of direct coupling?

### 418   Solid State Electronic Circuits

11.1.3. Why must direct-coupled stages have a very high degree of Q point stabilization?

11.1.4. What limits the smallest DC signal that a dc amplifier can satisfactorily amplify?

11.1.5. Do you think a 4-V Zener diode could be used in place of the capacitor in the circuit of Fig. 11.1b to allow amplification of DC signals? If so, how would you connect it?

11.1.6. The following information applies to the Darlington pair of Fig. 11.3: $I_{e1} = 0.1$ mA, $I_{c2} = 15$ mA. Determine the overall $h_{fe}$ if both transistors have identical parameters.

11.1.7. The input voltage offset of a given direct-coupled amplifier can be expected to drift up to 20 $\mu$V per day. If the offset was adjusted to zero 3 days ago and $A_v = 100$, what is the maximum output voltage that can be expected without any input signal?

## 11.2  THE DIFFERENTIAL AMPLIFIER

The dc amplifiers discussed in Sec. 11.1 had a single-ended input; that is, one of the input terminals was grounded. The differential amplifier is a special type of dc amplifier (although it can also be AC-coupled) in which both input terminals are floating (can take on any potential) with respect to ground.

The output of an ideal differential amplifier is proportional only to the *difference* between the signals at each input terminal. For the diagram of Fig. 11.4a, we can write

$$V_o = A_v(V_2 - V_1) \tag{11.2}$$

where $V_1$ and $V_2$ are the potentials, relative to ground, at input terminals 1 and 2, respectively.

A typical bipolar transistor differential amplifier circuit is shown in Fig. 11.4b. The output, taken between the two collectors, may be expressed as follows:

$$V_o = I_{c2}R_{L2} - I_{c1}R_{L1} \tag{11.3}$$

where $I_{c1}$ and $I_{c2}$ are the collector currents of $Q_1$ and $Q_2$, respectively.

The drift problem is significantly reduced by the use of differential amplifiers. To show how this happens, assume that $Q_1$ and $Q_2$ are very closely matched, so that each transistor's characteristics ($I_{CBO}$, $h_{FE}$, $V_{BE}$) and temperature dependence are virtually the same. Suppose now that $I_{C1}$ should rise, possibly because of an increase in temperature. This means that the voltage between the collector of $Q_1$ and ground drops. Now as long as $Q_2$ is at the same temperature as $Q_1$ and its characteristics have the same temperature dependence as those of $Q_1$, the same thing should happen here—that is, the collector voltage

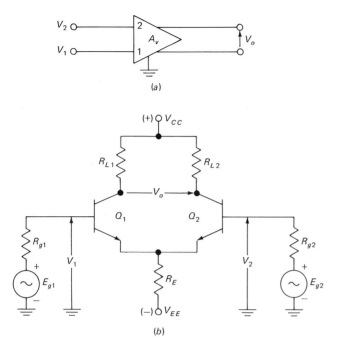

Fig. 11.4 *Differential amplifier symbol* (a) *and circuit* (b).

of $Q_2$ should drop by the same amount as at $Q_1$. Thus there is no change in $V_o$; the individual drifts of $Q_1$ and $Q_2$ cancel and do not appear across the output. It is, of course, impossible to perfectly match $Q_1$ and $Q_2$ in every detail; therefore drift is not completely eliminated. However, it is certainly reduced by a large factor compared to the single-ended type of amplifier. The drift reduction can be maximized using planar transistors, because of their highly uniform characteristics. It is also possible to obtain matched transistors mounted on a single header; this helps keep both at the same temperature for better tracking.

The $Q$ point stability of the differential amplifier can be quite good. In single-stage amplifiers, the emitter current of only one transistor flows through $R_E$. In a differential amplifier, both $I_{E1}$ and $I_{E2}$ flow through $R_E$; therefore any changes in emitter currents produce approximately twice as large a change in emitter voltage as in a single-stage amplifier. This negative DC feedback cuts the transistor drive, thereby offsetting the increase in current.

The inputs to a differential amplifier may be classified into two categories: differential mode and common mode. Both may be simultaneously present. The great virtue of differential amplifiers, in fact, lies in being able to amplify the differential-mode signal while *at the same time* rejecting the common-mode signal.

To see how this happens, consider the simple strain-gage bridge shown in

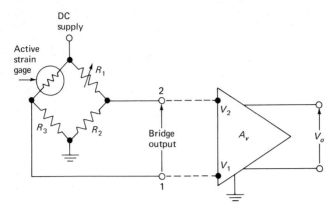

*Fig. 11.5  Example of a strain-gage bridge feeding a differential amplifier.*

Fig. 11.5. The signal is taken between points 2 and 1 and fed to the differential amplifier. The differential-mode input can be expressed as follows:

Differential-mode input:  $V_2 - V_1$ (11.4)

Now in many applications it is possible for interference signals to be picked up so that they appear simultaneously at point 2 and point 1, relative to ground. Examples of such undesirable interference are 60 Hz power line hum or stray fields from nearby electric machinery. As long as the same interference is present at both input terminals of the differential amplifier, it is common mode. The common-mode input is defined as the average value of $V_1$ and $V_2$:

Common-mode input:  $\dfrac{V_1 + V_2}{2}$ (11.5)

An ideal differential amplifier should provide a large gain to differential-mode inputs and reject completely (zero gain) all common-mode inputs. A measure of performance for differential amplifiers is, in fact, the ratio of differential gain to common-mode gain; the result is called the common-mode rejection ratio ($CMRR$):

$$CMRR = \left|\dfrac{K_{DM}}{K_{CM}}\right| \qquad (11.6)$$

where $K_{DM}$ and $K_{CM}$ are the differential- and common-mode voltage amplifications, respectively. The ideal $CMRR$ is infinite, but in practice this can only be approached. Usually the $CMRR$ is expressed in decibels, as follows:

$$CMRR \text{ (in dB)} = 20 \log \left|\dfrac{K_{DM}}{K_{CM}}\right| \qquad (11.6a)$$

Typical values for high-quality differential amplifiers exceed 100 dB.

We can therefore conclude that two measures of performance for differential amplifiers are drift and $CMRR$. The drift should be minimized, while the $CMRR$ is maximized.

The output of an ideal differential amplifier is given by Eq. (11.2). For a practical differential amplifier, $V_2$ and $V_1$ do not experience equal amplification; hence we must write

$$V_o = K_2 V_2 - K_1 V_1 \tag{11.7}$$

It is also possible to express the output in terms of differential-mode and common-mode components, as follows:

$V_o$ = (common-mode input $\times$ common-mode gain) + (differential input $\times$ differential gain)

$$V_o = \frac{V_1 + V_2}{2} K_{CM} + (V_2 - V_1)(K_{DM}) \tag{11.8}$$

$$V_o = \frac{V_1 K_{CM}}{2} + \frac{V_2 K_{CM}}{2} + V_2 K_{DM} - V_1 K_{DM}$$

$$V_o = \left(\frac{K_{CM}}{2} + K_{DM}\right)(V_2) - \left(K_{DM} - \frac{K_{CM}}{2}\right)(V_1) \tag{11.8a}$$

Direct comparison of Eqs. (11.7) and (11.8a) yields

$$K_2 = K_{DM} + \frac{K_{CM}}{2} \tag{a}$$

$$K_1 = K_{DM} - \frac{K_{CM}}{2} \tag{b}$$

Simultaneous solution of Eqs. (a) and (b) yields

Differential gain: $\quad K_{DM} = \dfrac{K_1 + K_2}{2}$ \hfill (11.9)

Common-mode gain: $\quad K_{CM} = K_2 - K_1$ \hfill (11.10)

### Example 11.1

A differential amplifier has an output given by the following expression:

$$V_o = 49 V_2 - 50 V_1$$

The two inputs are $V_1 = 5$ mV and $V_2 = 10$ mV. Determine

a. The total output using the expression given above
b. The common-mode input
c. The differential input

d. The common-mode gain
e. The differential gain
f. The common-mode output
g. The differential output
h. The total output expressed as the sum of (f) and (g)
i. The $CMRR$

**Solution**

a. $V_o = 49V_2 - 50V_1$
$V_o = 49(10 \times 10^{-3}) - 50(5 \times 10^{-3})$
$V_o = 240$ mV

b. Common-mode input $= \dfrac{V_1 + V_2}{2}$ (11.5)

$= \dfrac{15 \times 10^{-3}}{2} = 7.5$ mV

c. Differential input $= V_2 - V_1$ (11.4)
$= 5$ mV

d. Common-mode gain $K_{CM} = K_2 - K_1$ (11.10)
By inspection of the expression for $V_o$, $K_2 = 49$ and $K_1 = 50$;

$K_{CM} = 49 - 50 = -1$

e. Differential gain $K_{DM} = \dfrac{K_2 + K_1}{2}$ (11.9)

$= \dfrac{49 + 50}{2} = 49.5$

f. Common-mode output = common-mode input × common-mode gain
$= (7.5 \times 10^{-3})(-1) = -7.5$ mV

g. Differential output = differential input × differential gain
$= (5 \times 10^{-3})(49.5) = 247.5$ mV

h. $V_o = (-7.5 \times 10^{-3}) + (247.5 \times 10^{-3}) = 240$ mV

i. $CMRR = \left| \dfrac{KDM}{KCM} \right|$ (11.6)

$= \left| \dfrac{49.5}{-1} \right| = 49.5$

$CMRR$ (in dB) $= 20 \log 49.5 = 34$ dB

Having established that the more closely matched the transistors, the higher the $CMRR$, we now consider the effect of $R_E$ on the performance of the differential amplifier of Fig. 11.4b. We will show that negative feedback through $R_E$ improves the $CMRR$ and that, ideally, $R_E$ should be as large as possible.

You will recall that an unbypassed emitter resistance reduces the voltage amplification because of negative feedback. Thus there is greater gain when $R_E$ is bypassed. $R_E$ is effectively bypassed if the current through it does not change.

Let us assume an input which has only a differential component. This means that $V_1$ and $V_2$ are equal but opposite in phase; thus there is no common-mode input. If $V_1$ is of such a polarity as to increase conduction through $Q_1$, the emitter current of $Q_1$ increases. Since $V_2$ is of opposite phase to $V_1$, the emitter current of $Q_2$ must decrease. Now the current through $R_E$ is the sum of the two emitter currents. If the increase in $I_{e1}$ is equal to the decrease in $I_{e2}$, it follows that the current through $R_E$ remains the same. But a constant current through $R_E$ means no feedback (it is as if $R_E$ were bypassed); hence the gain is high! In other words, the *differential* gain is high.

Now let us apply an input that has only a common-mode component. Such is the case if $V_1 = V_2$; the differential-mode component is zero. If the amplifier were ideal, the output would also equal zero. In practice, however, we know that because of small differences in the characteristics of $Q_1$ and $Q_2$, a net difference in potential is developed between the two collectors. The circuit must minimize this difference. Since $V_1 = V_2$, the currents of both transistors are equally affected; if $I_{e1}$ rises, so does $I_{e2}$. This means that the current through $R_E$ cannot remain constant, but must change by the sum of $\Delta I_{e1}$ and $\Delta I_{e2}$. When the current through $R_E$ changes, there is negative feedback and a consequent reduction in gain.

The effect of $R_E$ is therefore to provide negative feedback (lower gain) when the inputs are common mode and no feedback (higher gain) when the inputs are differential mode. Obviously, the larger $R_E$, the better this function can be realized.

### Example 11.2

Determine the $CMRR$ in the circuit of Fig. 11.4b for $R_E = 0$. Assume $h_{fe1} = 49$; $h_{fe2} = 50$; $R_{L1} = R_{L2} = 1$ k$\Omega$; $h_{ie1} = h_{ie2} = 1$ k$\Omega$; $R_{g1} = R_{g2} = 0$. Neglect other transistor parameters.

### Solution

When $R_E = 0$, the equivalent circuit of Fig. 11.6a applies, and we can write

$$V_o = I_{c2}R_{L2} - I_{c1}R_{L1} \tag{11.3}$$
$$V_o = h_{fe2}I_{b2}R_{L2} - h_{fe1}I_{b1}R_{L1} \tag{11.3a}$$

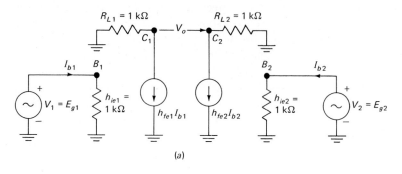

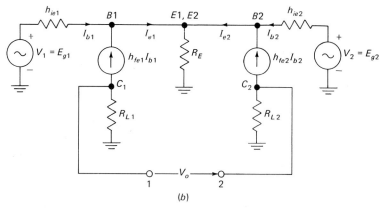

**Fig. 11.6** *Differential amplifier equivalent circuits.*

To determine the $CMRR$, the output is calculated first for a common-mode and then for a differential-mode input.

a. A common-mode input requires $V_1 = V_2$; if we assume $V_1 = 1$ mV, $V_2 = 1$ mV, the common-mode input [Eq. (11.5)] is 1 mV.
b. Since $h_{ie1} = h_{ie2} = 1$ kΩ, both base currents are equal:

$$I_{b1} = I_{b2} = \frac{10^{-3}}{10^3} = 1 \ \mu A$$

Substitution into Eq. (11.3a) yields

$$V_o = (50 \times 10^{-6})(10^3) - (49 \times 10^{-6})(10^3) = 1 \text{ mV}$$

c. The common-mode gain is

$$K_{CM} = \frac{\text{common-mode output}}{\text{common-mode input}} = \frac{10^{-3}}{10^{-3}} = 1$$

d.  To determine the differential-mode gain, we assume $V_2 = 1$ mV, $V_1 = -1$ mV, which represents a differential input [Eq. (11.4)] of 2 mV. Using Eq. (11.3a), we have

$$V_o = 50\left(\frac{10^{-3}}{10^3}\right)(10^3) - 49\left(\frac{-10^{-3}}{10^3}\right)(10^3)$$
$$V_o = 99 \text{ mV}$$

e.  The differential gain is

$$K_{DM} = \frac{\text{differential output}}{\text{differential input}} = \frac{99 \times 10^{-3}}{2 \times 10^{-3}} = 49.5$$

f.  The $CMRR$ is

$$CMRR = \left|\frac{K_{DM}}{K_{CM}}\right| = \frac{49.5}{1} = 49.5$$
$$20 \log CMRR \simeq 34 \text{ dB}$$

**Example 11.3**

Repeat Example 11.2 for the case when $R_E = 1$ k$\Omega$.

**Solution**

When $R_E = 1$ k$\Omega$, the equivalent circuit is as shown in Fig. 11.6b. We now write two loop equations, as follows:

$$V_1 = I_{b1}h_{ie1} + R_E[I_{b1}(h_{fe1} + 1) + I_{b2}(h_{fe2} + 1)] \quad (a)$$
$$V_2 = I_{b2}h_{ie2} + R_E[I_{b2}(h_{fe2} + 1) + I_{b1}(h_{fe1} + 1)] \quad (b)$$

a.  Using the same common-mode input as before and substituting numerical values, we have

$$V_1 = 10^{-3} = I_{b1}(1{,}000) + 1{,}000 I_{b1}(50) + 1{,}000 I_{b2}(51) \quad (c)$$
$$V_2 = 10^{-3} = I_{b2}(1{,}000) + 1{,}000 I_{b2}(51) + 1{,}000 I_{b1}(50) \quad (d)$$

Simultaneous solution of Eqs. (c) and (d) yields

$$I_{b1} = I_{b2} = 9.8 \text{ nA}$$

b.  The common-mode output is

$$V_o = 50(9.8 \times 10^{-9})(10^3) - 49(9.8 \times 10^{-9})(10^3) \quad (11.3a)$$
$$V_o = -9.8 \text{ }\mu\text{V}$$

c.  The common-mode gain is

$$K_{CM} = \frac{-9.8 \times 10^{-6}}{1 \times 10^{-3}} = -9.8 \times 10^{-3}$$

d. With the same differential-mode input as before and substituting numerical values, we have

$$V_1 = -10^{-3} = 51{,}000 I_{b1} + 51{,}000 I_{b2} \qquad (e)$$
$$V_2 = 10^{-3} = 50{,}000 I_{b1} + 52{,}000 I_{b2} \qquad (f)$$

Simultaneous solution of Eqs. (e) and (f) yields

$$I_{b1} = -1.01 \ \mu A \quad \text{and} \quad I_{b2} = 0.99 \ \mu A$$

e. The differential output is

$$V_o = 50(0.99 \times 10^{-6})(10^3) - 49(-1.01 \times 10^{-6})(10^3) \qquad (11.3a)$$
$$V_o = 99 \ \text{mV}$$

f. The differential gain is

$$K_{DM} = \frac{99 \times 10^{-3}}{2 \times 10^{-3}} = 49.5$$

g. The $CMRR$ is

$$CMRR = \left| \frac{K_{DM}}{K_{CM}} \right| = \left| \frac{49.5}{-9.8 \times 10^{-3}} \right|$$
$$CMRR = 5{,}050$$
$$20 \log CMRR \simeq 74 \ \text{dB}$$

Note the improvement in $CMRR$; when $R_E = 1$ k$\Omega$, the $CMRR$ is 74 dB, about 40 dB higher than for $R_E = 0$. Further improvement is possible if $R_E$ is increased; however, $R_E$ is limited by the available DC supply.

The examples just carried out assumed all parameters to be matched except $h_{fe}$. If any of the other quantities involved, such as $h_{ie}$, generator resistance, biasing resistors, and so on, are not matched, the $CMRR$ will be affected.

A circuit arrangement designed to provide the benefits of a large effective $R_E$ without the large DC voltage drop is shown in Fig. 11.7a. The schematic given is for the Fairchild μA 730 integrated circuit. Differential amplifiers lend themselves to integrated-circuit manufacturing techniques because of the very high degree of uniformity in parameters that can be achieved. Here $R_E$ from the circuit of Fig. 11.4b has been replaced by $Q_5$ and its biasing resistors. To see the advantages of using $Q_5$ instead of an emitter resistor, assume that 6 V DC are available for dropping between point $A$ and ground. If the two emitter currents add up to 1 mA, the allowable $R_E$ is $6/10^{-3} = 6$ k$\Omega$. The same 6 V and 1 mA, however, can be used to bias $Q_5$; this means that the dynamic resistance from point $A$ to ground is now the output resistance of a common emitter amplifier,

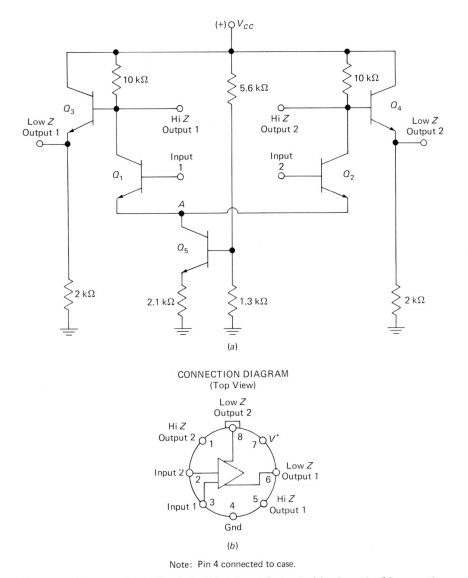

Fig. 11.7 *Differential amplifier ($\mu A$ 730) integrated circuit:* (a) schematic; (b) connection diagram. (*Courtesy of Fairchild Semiconductor.*)

which can reach 20 k$\Omega$ or more. We therefore achieve the benefits of a larger $R_E$ than would be possible if an actual resistor were used.

There are two sets of outputs available in this amplifier. If a large output impedance is required, the output is taken between the collectors of $Q_1$ and $Q_2$.

This is the same as in the circuit of Fig. 11.4b. For a low output impedance, the output is taken between the emitters of $Q_3$ and $Q_4$. Both $Q_3$ and $Q_4$ are emitter followers with essentially unity voltage amplification. Therefore the voltage appearing at the collector of $Q_1$ also appears at the emitter of $Q_3$; $Q_3$ and $Q_4$ act as impedance-matching devices.

## Problem Set 11.2

11.2.1. What are some of the advantages of differential amplifiers?

11.2.2. Why are matched transistors required in differential amplifiers? What happens to the drift if the transistors are not matched?

11.2.3. Describe the difference between common-mode and differential-mode signals.

11.2.4. For a differential amplifier of the type in Fig. 11.4b, determine the differential- and common-mode inputs when $V_1 = 3$ mV and

    a. $V_2 = 3$ mV
    b. $V_2 = -3$ mV
    c. $V_2 = 5$ mV
    d. $V_2 = -5$ mV

11.2.5. What relationship must exist between $V_1$ and $V_2$ if there is to be no common-mode input?

11.2.6. What relationship must exist between $V_1$ and $V_2$ if there is to be no differential input?

11.2.7. For an ideal differential amplifier, what should the common-mode gain be? What prevents us from achieving this in practice?

11.2.8. What does $CMRR$ refer to?

11.2.9. What type of interference signals are normally rejected by differential amplifiers?

11.2.10. Given a differential amplifier whose output can be expressed as

$$V_o = 1{,}000 V_2 - 999 V_1$$

determine

    a. Common-mode gain
    b. Differential gain
    c. $CMRR$ (in decibels)

11.2.11. If $V_1 = 10$ $\mu$V and $V_2 = -8$ $\mu$V in Prob. 11.2.10, determine

    a. Common-mode output
    b. Differential output

11.2.12. In the circuit of Fig. 11.4b, what is the function of $R_E$? Why should $R_E$ be as large as possible, and what prevents us from making it too large?

11.2.13. Determine the $CMRR$ (in decibels) for the circuit of Fig. 11.4b if $R_{g1} = R_{g2} = 0\ \Omega$; $R_E = 2$ k$\Omega$; $h_{fe1} = 99$, $h_{fe2} = 100$; $R_{L1} = R_{L2} = 1.5$ k$\Omega$; $h_{ie1} = h_{ie2} = 2{,}000\ \Omega$. Neglect all other transistor parameters.

11.2.14. In Fig. 11.7a a circuit diagram is given in which $Q_5$ and its biasing resistors replace $R_E$. What are the advantages of doing this?

11.2.15. In the circuit of Fig. 11.7a, do you think that $Q_3$ and $Q_4$ need to be matched? Explain your answer.

## 11.3  CHOPPER-STABILIZED AMPLIFIERS

It has been shown that gradual changes in input offset voltage and current (drift) represent a basic limitation to the amplification of low-level, low-frequency signals. The drift reduction possible with differential amplifiers has also been discussed. In this section another technique is presented that can provide significant improvement in the drift performance of DC amplifiers. This technique is called *chopping*, and the resultant amplifier is said to be chopper-stabilized.

You will recall that a direct-coupled amplifier cannot distinguish internal DC-level shifts from externally applied DC signals. One way to remove this difficulty is to change the slowly varying (DC) signal into an AC signal. The AC signal can be amplified using standard $RC$ or transformer-coupled amplifiers. The amplifier's output is then reconstructed to yield a DC signal that is an amplified version of the input.

To see how this is done, consider the simplified block diagram of Fig. 11.8a. The input signal is applied between point $X$ and ground. The switch is in position $A$ for a brief period of time, yielding a positive output at point $C$; it is then switched to position $B$, yielding a negative output at $C$. The switching action must take place at a relatively fast rate, at least 10 times faster than the input-signal frequency. Typical switching devices that may be used are mechanical vibrators, diodes, bipolar transistors, field effect transistors, or photoconductors.

Note that the information contained in the input signal is retained in the "envelope" of the chopped signal. The chopped signal itself has no DC level; hence it can be amplified by the standard $RC$ or transformer-coupled amplifier following the chopper circuit. The amplifier's output is still "chopped," but, of course, its amplitude is greater. The recovery of the input signal may be accomplished by one of two methods. An output-switching circuit (Fig. 11.8b) can be used to perform the opposite operation of the input chopper. The two circuits have to be switched in synchronism so that when the input chopper is at $A$, the output chopper is at $A'$, and at the instant that the input is switched to $B$, the output is switched to $B'$. This is referred to as synchronous detection. Synchronous switching of input and output chopper is achieved in practice by driving both switching devices from the same source.

Another possible arrangement for recovering the input signal is shown in

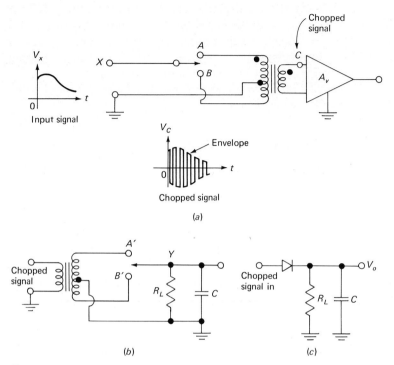

*Fig. 11.8* Chopper-stabilized amplifier: (a) input chopper and amplifier; (b) output recovered by synchronous detection; (c) output recovered by rectification and filtering.

Fig. 11.8c. Here the amplifier output is rectified and filtered; this operation is identical to that employed in power supplies (see Sec. 3.2c). The time constant of the $R_L C$ network must be long enough to prevent appreciable decay of the output signal during the nonconducting time of the diode.

A block diagram for a chopper-stabilized amplifier is shown in Fig. 11.9. Here the DC (low-frequency) components of $V_i$ are fed directly to the chopper amplifier, while $C_1$ prevents the higher-frequency AC components from doing the same. $C_2$ couples the AC components to the AC amplifier and blocks the DC components. Thus there are two amplifiers—the chopper-stabilized amplifier for the low frequencies and the AC amplifier for the higher-frequency components. The amplified signals are combined in the AC amplifier to yield the composite $V_o$.

### Problem Set 11.3

11.3.1. What does chopping refer to?
11.3.2. Why are chopper-stabilized amplifiers used?

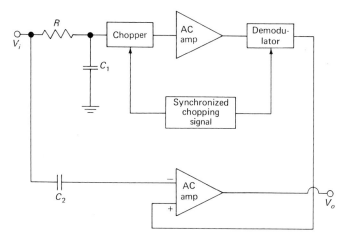

*Fig. 11.9  Block diagram of chopper-stabilized amplifier with separate AC and DC amplification.*

11.3.3. What relationship should exist between the chopping frequency and the signal frequency?

11.3.4. What is synchronous detection?

## 11.4  OPERATIONAL AMPLIFIERS

Operational amplifiers (abbreviated OA) are high-gain amplifiers whose low-frequency response is flat down to DC and whose high-frequency response rolls off at a controlled uniform rate, usually 6 dB per octave. Also typical of the OA is the high input impedance, low output impedance, and 180° phase shift. The OA may be used to perform mathematical operations (hence the name *operational*) or simulate physical processes. These functions can be carried out with a high degree of accuracy through the application of negative feedback. When sufficient negative feedback is used, the closed-loop performance becomes virtually independent of the OA parameters and is entirely controlled by the passive components in the feedback loop. Thus OAs find wide application in analog computers, instrumentation, and control systems.

Because a low-frequency response down to DC is required, OAs utilize direct-coupled, differential, or chopper-stabilized stages. Depending on the application, discrete bipolar or field effect transistors as well as integrated circuits are employed.

A standard OA symbol is shown in Fig. 11.10. The − and + signs do not mean that $V_2$ and $V_1$ are, respectively, negative and positive voltages. Instead these signs denote the inverting input ($V_2$) and noninverting input ($V_1$). This means that a signal applied between the inverting (−) input and ground appears

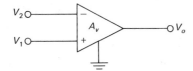

Fig. 11.10 Operational amplifier (OA) symbol.

phase-inverted at the output. The output is proportional to the difference between the two inputs, as follows:

$$V_o = A_v(V_2 - V_1) \tag{11.2}$$

where $A_v$, the open-loop voltage amplification, is a negative real number at DC. The above will be recognized as the standard expression for differential inputs. The difference $V_2 - V_1$ is also called the error voltage. For single-ended inputs, $V_1$ is simply tied to ground, and only $V_2$ is used. Because input and feedback currents are algebraically added at the inverting input, this point is also referred to as the summing point.

An ideal OA is one in which $A_v = -\infty$, $R_i = \infty$ Ω, and $R_o = 0$ Ω. Although this cannot be achieved in practice, the ideal OA is a convenient reference against which real OAs may be compared. The infinite input impedance means that no input current is required; thus the ideal OA is voltage-controlled. Similarly, zero output impedance means that $V_o$ is not dependent on the load resistance connected across the output. In addition, the ideal OA has a flat frequency response from DC to infinity. There are also no current or voltage offsets and hence no drift problems. For differential inputs, the $CMRR$ is infinite.

Let us now consider typical performance measures for real OAs. These are specified by the supplier, using fairly standard terminology. The open-loop voltage amplification is usually given for DC; it is a function of load resistance and temperature. Typical DC magnitudes of $A_v$ range from around 10,000 to in excess of $10^8$.

The frequency at which $|A_v|$ drops to 1 is also given; this is the unity gain bandwidth. It ranges from a low of 100 kHz to over 100 MHz for specially designed wideband units. When large output signals are involved, however, circuit nonlinearities affect the frequency response in a manner that cannot be predicted from small-signal unity gain bandwidth specifications. For this purpose, the slewing rate, a measure of the amplifier's ability to produce large and fast output voltage changes, is usually specified. Slewing rates for ultra-wideband OAs may reach 200 V/μs.

The input impedance is made up of an equivalent resistance shunted by an equivalent capacitance. There is more than one way to specify $Z_i$, but generally the essential information is given in one form or another. As an example, a single-ended-input chopper-stabilized OA might have an input-impedance specification of 1 MΩ‖400 pF. Differential stages, on the other hand, require two $Z_i$ specifications, one for differential- and the other for common-mode inputs. Differential input resistances range from less than 1 MΩ for some bipolar transistor stages to more than 10,000 MΩ for FET stages.

In working with OAs, care must be taken not to exceed the input and output voltage ranges specified by the manufacturer. For a typical differential stage, a maximum voltage might be 5 V between the two inputs and ±10 V at either input. The output voltage range might be around ±10 V. Exceeding these levels results in nonlinear operation (saturation) with the consequent distortion and general degradation of performance.

Specifications for input voltage offset and input current offset, including the drift, are usually provided. At times there is a built-in or external adjustment for voltage and current offsets. Drifts of 10 μV/°C and less than 10 nA/°C are typical for general purpose differential OAs. With chopper stabilization, these can be reduced to the range of 1 μV/°C and 10 pA/°C or less. Other information that may be available includes $CMRR$, power requirements, noise, output impedance, and, of course, price.

To prevent oscillations in OAs, it is important that stray capacitances are minimized by keeping leads as short as possible. All power supply connections should be capacitively bypassed to prevent signal feedback through the power supply's internal impedance. Similarly, the amplifier's input and output should be kept as far apart as possible for maximum isolation. Proper shielding from stray pickup may be required, especially with extremely low level signals. Connections between the different grounds (power supply, signal source, etc.) should be carefully laid out to minimize the generation of emfs. Taking these precautions, at times an art in itself, can prevent many problems associated with the use of OAs.

### 11.4a  Basic Relationships

Consider the functional diagram of Fig. 11.11. For simplicity, a number of details, such as external power connections and voltage offset zero adjustment,

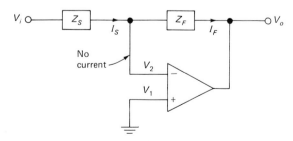

*Fig. 11.11* Basic OA with series and feedback impedances: $A'_v = V_o/V_i = -Z_F/Z_S$.

have been omitted. Also note that the noninverting input $V_1$ has been tied to ground, yielding single-ended input operation.

$V_i$ is the input voltage, while $Z_S$ and $Z_F$ are series and feedback impedances, respectively. These impedances may involve simple resistances or other passive components; they are externally connected and sometimes variable. The great usefulness of OAs is that the closed-loop performance can be made to depend almost entirely on $Z_S$ and $Z_F$. For the amplifier proper, Eq. (11.2) applies, and since $V_1 = 0$, we can write

$$V_o = A_v V_2$$

or

$$V_2 = \frac{V_o}{A_v} \tag{11.2a}$$

Now as long as the input resistance is infinite (or very high), no current flows into the amplifier; hence the current through $Z_S$ is equal to the current through $Z_F$:

$$I_S = I_F$$
$$\frac{V_i - V_2}{Z_S} = \frac{V_2 - V_o}{Z_F}$$
$$V_o = \frac{Z_F}{Z_S}(V_2 - V_i) + V_2 \tag{11.11}$$

Substituting Eq. (11.2a) above yields

$$V_o = \frac{Z_F}{Z_S}\left(\frac{V_o}{A_v} - V_i\right) + \frac{V_o}{A_v}$$
$$V_o\left(1 - \frac{Z_F}{Z_S A_V} - \frac{1}{A_v}\right) = -\frac{Z_F}{Z_S} V_i$$

and the closed-loop voltage amplification becomes

$$A'_v = \frac{V_o}{V_i} = -\frac{Z_F}{Z_S[1 - Z_F/(Z_S A_v) - 1/A_v]} \tag{11.12}$$

Generally $A_v$ is very large, and the denominator term approaches $(Z_S)(1)$; therefore

$$A'_v = \frac{V_o}{V_i} = -\frac{Z_F}{Z_S} \tag{11.12a}$$

Thus as long as the open-loop $A_v$ is very large, the closed loop $A'_v$ is independent of $A_v$ and is determined entirely by the ratio of feedback and series impedances. Note that since $A_v$ is large, the quantity $V_2 = V_o/A_v$ becomes very small and may be considered zero. The inverting input terminal ($V_2$) is in fact referred to as a *virtual* ground, not because it is tied to ground but because its voltage is virtually the same as at ground.

For a physical picture of what actually happens, assume that initially $V_i$ and hence $V_2$, $V_o$, $I_S$, and $I_F$ are all zero. Now if $V_i$ rises, both $I_S$ and the voltage at the summing point tend to rise. If there were no feedback ($I_F = 0$), all $I_S$ would flow into the amplifier's input impedance (which is very high). Thus $V_2$ would rise, and $V_o$ would drop (because of 180° phase shift). Even with feedback, the output is prevented from immediately responding to changes in $V_i$ because of capacitive effects; hence initially $V_2$ does tend to rise. But as soon as $V_o$ can respond, the potential difference across $R_F$ increases, producing an increase in $I_F$. The rise in $I_S$ is now matched by an equal rise in $I_F$ so that there is little or no current flowing into the amplifier's input resistance, and $V_2$ remains close to zero. Hence, any attempt of $V_2$ to rise initiates action that reduces $V_2$ back to its near-ground level.

### 11.4b  The Inverter Connection

This will be discussed by considering the circuit of Fig. 11.12. Here $V_o$ is 180° out of phase with $V_i$; the amplification magnitude is controlled by an appropriate selection of $R_F$ and $R_S$. In this case, we have

$$V_o = V_i \frac{-R_F}{R_S} \tag{11.12b}$$

$$V_o = -1\left(-\frac{50 \times 10^3}{10 \times 10^3}\right) = 5 \text{ V}$$

Thus both amplification and phase inversion are achieved.

### 11.4c  The Adder Circuit

Consider the circuit of Fig. 11.13. We will assume $R_F = 50$ kΩ and all $R_S = 10$ kΩ. Also $V_{i1} = 0.5$ V, $V_{i2} = -1$ V, and $V_{i3} = -0.8$ V. The derivations

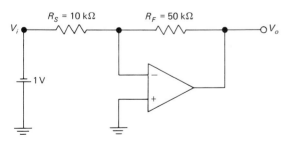

Fig. 11.12  The OA inverter: $A'_v = -R_F/R_S = -5$.

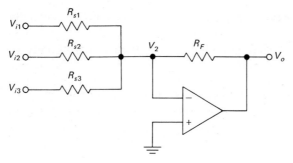

**Fig. 11.13** OA adder circuit: $V_o = (-R_F/R_S)(V_{i1} + V_{i2} + V_{i3})$, when $R_{S1} = R_{S2} = R_{S3} = R_S$.

leading to Eq. (11.12b) may be extended using the principle of superposition. The total output is the algebraic sum of each output due to only one input at a time. For example, if $V_{i2}$ and $V_{i3}$ are momentarily assumed to equal zero,[1] the contribution to $V_o$ from $V_{i1}$ is

$$V_{o1} = V_{i1} \frac{-R_F}{R_{S1}}$$

$$V_{o1} = \frac{0.5(-50 \times 10^3)}{10 \times 10^3} = -2.5 \text{ V}$$

Similarly, the individual contributions to $V_o$ from $V_{i2}$ and $V_{i3}$ are

$$V_{o2} = V_{i2} \frac{-R_F}{R_{S2}}$$
$$V_{o2} = -1(-5) = 5 \text{ V}$$
$$V_{o3} = V_{i3} \frac{-R_F}{R_{S3}}$$
$$V_{o3} = -0.8(-5) = 4 \text{ V}$$

The total output is

$$V_o = V_{o1} + V_{o2} + V_{o3}$$
$$V_o = -2.5 + 5 + 4 = 6.5 \text{ V}$$

Note that $V_o$ is proportional to the algebraic sum of the three inputs. Since $V_{i1} + V_{i2} + V_{i3} = -1.3$ V and $V_o = 6.5$ V, the *scale factor* is $6.5/-1.3 = -5$; this is simply $A'_v$. By selecting $R_{s1} \neq R_{s2} \neq R_{s3}$, it is possible to obtain *weighted* addition; that is, each input is multiplied by a different scale factor. In general, we can write

$$V_o = V_{i1} \frac{-R_F}{R_{S1}} + V_{i2} \frac{-R_F}{R_{S2}} + V_{i3} \frac{-R_F}{R_{S3}} + \cdots \quad (11.12c)$$

[1] $R_{s2}$ and $R_{s3}$ are now effectively in parallel and shunt the path from $V_2$ to ground. This effect, however, is negligible because $V_2$ is a virtual ground.

### 11.4d  The Subtractor Circuit

The operation of subtraction can be accomplished using the circuit of Fig. 11.14. Both inverting and noninverting inputs are required; hence a differential OA, as opposed to the single-ended input type, must be used. In general, we can write

$$V_o = A_v(V_2 - V_1) \tag{11.2}$$

Assuming a high impedance looking into the noninverting input, we find

$$V_i = V_{i2}\frac{R_2}{R_1 + R_2} \tag{a}$$

Equation (11.11) may be rewritten by replacing $Z_F$ and $Z_S$ with $R_2$ and $R_1$, respectively, to yield

$$V_o = \frac{R_2}{R_1}(V_2 - V_{i1}) + V_2$$

$$V_2 = \frac{V_{i1}R_2 + V_o R_1}{R_1 + R_2} \tag{11.11a}$$

Substituting Eqs. (11.11a) and (a) into Eq. (11.2) yields

$$V_o = \frac{R_2(V_{i2} - V_{i1})}{R_1 - [(R_1 + R_2)/A_v]} \tag{11.13}$$

Usually $A_v$ is large, and the above equation simplifies to

$$V_o = \frac{R_2}{R_1}(V_{i2} - V_{i1}) \tag{11.13a}$$

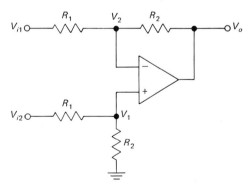

Fig. 11.14  OA subtractor circuit: $V_o = (R_2/R_1)(V_{i2} - V_{i1})$.

If $R_2 = R_1$, the output is exactly equal to the difference between the two inputs. If $V_{i1} = 1$ V and $V_{i2} = 3$ V in Fig. 11.14, we then have

$$V_o = V_{i2} - V_{i1} = 3 - 1 = 2 \text{ V}$$

If $R_2 \neq R_1$, a scale factor is introduced; if, for example, $R_2 = 5R_1$, then the scale factor is 5, and $V_o = 5(V_{i2} - V_{i1}) = 5(2) = 10$ V.

### 11.4e  Constant Voltage and Constant Current Sources

The output impedance of OAs can be very low; hence a constant voltage source may be approached, as shown in Fig. 11.15a. Here, as $R_L$ is varied, $V_o$ remains virtually constant. Obviously, if $R_L$ is as low as $R_o$, this is no longer true.

A constant current source is approached if $R_L$ is connected in place of the feedback resistor, as shown in Fig. 11.15b. The current through $R_L$ is the same as the current through $R_S$, which is $I \simeq V_i/R_S$. Therefore, as long as $V_i$ and $R_S$ are constant, $I$ does not change. This is the requirement for a constant current source (no change in $I_o$ as $R_L$ is varied). Note that in this circuit $R_L$ must be floating with respect to ground.

**Example 11.4**

Determine $V_o$ in the circuit of Fig. 11.15a if $V_i = -3$ mV, $R_S = 1$ k$\Omega$, $R_F = 20$ k$\Omega$, and $R_L = 10$ k$\Omega$.

**Solution**

Since the circuit approaches a constant voltage source, $R_L$ has no effect on $V_o$. The closed-loop voltage amplification is $-R_F/R_S = -20$ and $V_o = -20(-3 \times 10^{-3}) = 60$ mV.

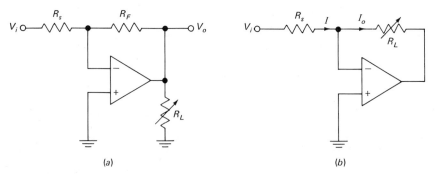

*Fig. 11.15* Use of an OA as a constant voltage (a) and constant current (b) source. In (a), $V_o$ is independent of $R_L$; in (b), $I_o$ is independent of $R_L$.

## Example 11.5

Determine $I_o$ in the circuit of Fig. 11.15b if $V_i = 1$ V, $R_S = 1$ k$\Omega$, and

a. $R_L = 10$ k$\Omega$
b. $R_L = 20$ k$\Omega$

### Solution

In either case, $I_o$ is given by $V_i/R_s$:

$$I_o = \frac{1}{10^3} = 1 \text{ mA}$$

Therefore a constant current is maintained through $R_L$.

### 11.4f The Noninverter Connection

Consider now the circuit of Fig. 11.16. Here $V_i$ is applied to the noninverting input, resulting in $V_o$ being in phase with $V_i$. The input impedance seen by $V_i$ is also increased. To determine the closed-loop $A'_v$, we use the fact that $V_2$ is very close to $V_1$; hence the potential at the junction of $R_1$ and $R_2$ is approximately $V_i$. Since $I_1 = I_2$, we can write

$$\frac{V_i}{R_1} = \frac{V_o - V_i}{R_2}$$

$$V_o = V_i \left(1 + \frac{R_2}{R_1}\right) \tag{11.14}$$

$$A_v = \frac{V_o}{V_i} = 1 + \frac{R_2}{R_1} \tag{11.14a}$$

### 11.4g The Voltage Follower

If $R_2 = 0$ and $R_1 = \infty$ in the circuit of Fig. 11.16, the circuit of Fig. 11.17 results. This is a voltage follower whose closed-loop $A'_v$ is positive and approaches

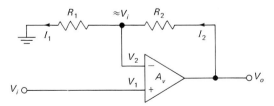

**Fig. 11.16** The OA noninverter circuit: $A'_v = V_o/V_i = 1 + R_2/R_1$.

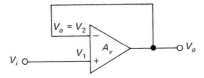

Fig. 11.17 The OA follower circuit:
$V_o = V_i[A_v/(A_v - 1)] \simeq V_i$.

unity, as long as $A_v$ is very large. If $A_v$ is not large enough, an accurate expression for $V_o$ may be derived as follows:

$$V_o = (V_2 - V_1)(A_v) \tag{11.2}$$

But $V_1 = V_i$ and $V_2 = V_o$; therefore

$$V_o = (V_o - V_i)(A_v)$$
$$V_o(1 - A_v) = -V_i A_v$$
$$V_o = V_i \frac{A_v}{A_v - 1} \tag{11.15}$$

The voltage follower is characterized by very high input and very low output impedance.

### 11.4h  Offset Voltage

Offset voltage may be measured using the circuit of Fig. 11.18. This circuit is similar to the noninverter of Fig. 11.16 except that $V_i$ is replaced with a voltage generator $E_{os}$ connected between the noninverting input and ground. The output is then monitored using a high input impedance voltmeter. The rela-

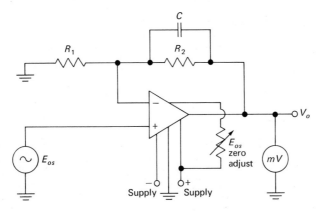

Fig. 11.18  Circuit to measure offset voltage.

tionship between voltage offset and $V_o$ is determined using Eq. (11.14), rewritten as follows:

$$V_o = E_{os}\left(1 + \frac{R_2}{R_1}\right) \tag{11.14b}$$

Normally a potentiometer adjustment is provided for nulling the offset voltage. Since $V_o$ is read directly on the meter, adjusting the potentiometer until $V_o$ is minimum minimizes $E_{os}$.

If the OA is subjected to temperature gradients so that one of the differential transistors is at a slightly different temperature than the other, there is a large drift in offset voltage due to the unbalance in the transistors. When a voltage offset adjustment is made, a capacitor is also often connected across $R_2$ to reduce the amplifier's high-frequency gain. This effectively attenuates high-frequency noise which might otherwise mask the DC offset. The DC gain is, of course, not affected by the capacitor.

### 11.4i Offset Current

In addition to voltage offset, there is also an equivalent offset current through each differential input. The circuit of Fig. 11.19a may be used to measure $I_{os}$. Here, power supply and other connections have been omitted for simplicity. Capacitors may be connected across $R_1$ and $R_2$, again to attenuate high-frequency noise.

If $R_1$ is shorted, $V_1$ is at ground potential, and so is $V_2$. We can therefore write

$$I_{os2} = \frac{V_o}{R_2} \tag{11.16}$$

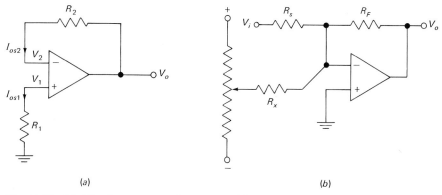

Fig. 11.19 (a) Circuit to measure offset current; (b) circuit to cancel offset current.

If $R_2$ is shorted, we have a circuit similar to the voltage follower of Fig. 11.17, and we can write

$$I_{os1} = \frac{V_1}{R_1} \quad\quad\quad (11.16a)$$

But $V_1 \simeq V_o$ (unity gain); hence

$$I_{os1} = \frac{V_o}{R_1} \quad\quad\quad (11.16b)$$

To determine each offset current, a measurement of $V_o$ is therefore required. The total offset current, also referred to as the input bias current, is

$$I_{os} = |I_{os2} - I_{os1}| \quad\quad\quad (11.17)$$

The technique shown in Fig. 11.19b may be used to cancel out current offset. $R_x$ should be several hundred megohms, and the adjustment should be made after the voltage offset has been canceled.

### 11.4j  Common-mode Rejection Ratio

The $CMRR$ is a measure of the ability to amplify differential-mode signals and reject common-mode signals. Ideally, a signal which is common to both inputs is not amplified; hence, the $CMRR$ is infinite. In practice, errors are to be expected because of the finite value of the $CMRR$. For the circuit of Fig. 11.20, we can write

$$V_o = A_v(\text{differential input}) + A_v\left(\frac{\text{common-mode input}}{CMRR}\right) \quad\quad\quad (11.18)$$

With an infinite $CMRR$ the total output is due only to a differential input. In this case, unwanted signals, such as noise, pickup from ground loops, etc., which are common to both inputs are rejected. For example, if a 2-mV differential input produces an output of 10 V and a 2-mV common-mode input produces an output

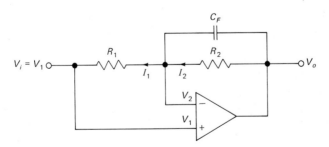

Fig. 11.20  Circuit to measure CMRR.

of 10 mV, the differential gain is $10/(2 \times 10^{-3}) = 5{,}000$, and the common-mode gain is $(10 \times 10^{-3})/(2 \times 10^{-3}) = 5$. The $CMRR$ is simply $5{,}000/5 = 1{,}000$, or 60 dB. A common-mode input, in this case, would have to be 1,000 times greater than a differential input to yield the same output.

In Fig. 11.20, the differential input is $V_2 - V_1$, while the common-mode input is $V_1$. An expression for the output can therefore be written as follows:

$$V_o = A_v(V_2 - V_1) + \frac{A_v V_1}{CMRR} \tag{11.18a}$$

But $V_2 = I_1 R_1$, and since $I_1 = I_2$,

$$V_2 = I_2 R_1$$

where

$$I_2 = \frac{V_o - V_2}{R_2}$$

Therefore

$$V_2 = \frac{(V_o - V_2)(R_1)}{R_2} \tag{a}$$

Substituting Eq. (a) into Eq. (11.18a) and solving for $CMRR$, we get

$$CMRR = \frac{V_1(R_1 + R_2)}{V_o(R_1 + R_2)/A_v + R_1(V_1 - V_o)} \tag{11.19}$$

Usually $A_v$ is large enough that the first term in the denominator is relatively insignificant, yielding

$$CMRR = \frac{V_1(R_1 + R_2)}{R_1(V_1 - V_o)} \tag{11.19a}$$

$V_1$ and $V_2$ can be measured in the circuit of Fig. 11.20, and the $CMRR$ is then calculated using Eq. (11.19a). Capacitor $C_F$ is again used to limit the OA high-frequency response.

Another technique for measuring $CMRR$ is shown in Fig. 11.21. The

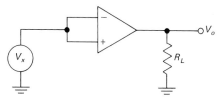

*Fig. 11.21* Circuit to measure CMRR.

signal $V_x$ is simultaneously applied to both inputs; hence there is no differential input. Since the gain for common-mode signals is fairly low, no feedback is used. In Eq. (11.18a), $V_2 = V_1 = V_x$, and we have

$$V_o = \frac{A_v V_x}{CMRR}$$

$$CMRR = \frac{A_v V_x}{V_o} \tag{11.20}$$

### 11.4k  Frequency Response Considerations

The high-frequency response of operational amplifiers (OAs) is usually intentionally limited to prevent stability problems from arising. The main consideration is for the gain to fall below 0 dB long before the phase shift crosses the 0° axis.

A typical open-loop frequency response for an OA is given in Fig. 11.22. If $A_o$ is the low frequency $A_v$, we can write

$$A_v = \frac{A_o}{1 + jf/f_H} \tag{11.21}$$

which is the standard expression for a response with a 6 dB per octave roll-off beyond $f_H$. If negative feedback is used as in Fig. 11.11, the closed-loop gain falls, and the frequency response is extended. Reactive elements in the series or feedback paths may also be used to obtain specific frequency response characteristics. Some of these applications are differentiation and integration; these will be taken up in Chap. 15.

### Problem Set 11.4

11.4.1. Name four important characteristics of operational amplifiers.
11.4.2. What is a virtual ground?

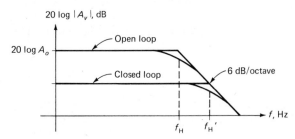

*Fig. 11.22  Frequency-response comparison for open-loop and closed-loop OA circuit.*

11.4.3. What is unity gain bandwidth? What is slewing rate? Compare these two performance measures.

11.4.4. For the circuit of Fig. 11.11, assume $V_i = 1$ V, $Z_F = 40$ kΩ, and $Z_S = 10$ kΩ; determine $V_o$.

11.4.5. Determine $V_o$ in the circuit of Fig. 11.13 if $V_{i1} = 1$ mV, $V_{i2} = -1$ mV, $R_{s1} = R_{s2} = 10$ kΩ, and $R_F = 100$ kΩ. Assume $V_{i3}$ is not used. What operation is the circuit performing?

11.4.6. Repeat Prob. 11.4.5 with $R_{s1} = 10$ kΩ and $R_{s2} = 1$ kΩ.

11.4.7. Explain how the circuit of Fig. 11.15a simulates a constant voltage source.

11.4.8. Explain how the circuit of Fig. 11.15b simulates a constant current source.

11.4.9. For the noninverter circuit of Fig. 11.16, it is desired to have a closed-loop voltage amplification of 100. Determine the relationship between $R_1$ and $R_2$.

11.4.10. How much feedback (value of $\beta_v$) is being employed in the circuit of Fig. 11.17? What is $A'_v$? Is there phase inversion?

11.4.11. If $E_{os} = 1$ mV in the circuit of Fig. 11.18 and $R_2 = 20R_1$, determine $V_o$.

11.4.12. An operational amplifier has an open-loop voltage amplification of $-10^5$. With a common-mode input of 1 mV, $V_o = -10$ mV. Determine the $CMRR$ in decibels.

11.4.13. Given an OA whose DC open-loop voltage amplification is 100 dB and whose $f_H$ (high-frequency 3-dB cutoff) is 100 Hz, determine the unity gain bandwidth. Assume 6 dB per octave roll-off beyond $f_H$.

## 11.5 VOLTAGE REGULATION

A simple Zener diode voltage regulator was discussed in Chap. 3. The performance of such circuits can be improved by using negative feedback in conjunction with direct-coupled amplifiers with single-ended or differential inputs. Operational amplifiers (OAs) employing discrete components or integrated circuits may also be used.

  A number of typical circuits are analyzed in this section, using reasonable approximations. As long as $h_{FE}$ is sufficiently high, transistor emitter and collector currents are assumed to be equal. Since transistors in these circuits operate in the active region, typical $V_{BE}$ values are 0.2 and 0.6 V for Ge and Si, respectively. $I_{CBO}$ is neglected when dealing with Si transistors.

  The objective of a voltage regulator, it will be recalled, is to maintain as constant an output voltage as possible, even if the load current or input voltage happens to change, within a given range. In addition to ripple reduction factor ($RRF$) and voltage regulation, which were introduced in Chap. 3, other parameters may be specified. The output resistance, for example, is a measure of how constant $V_o$ remains as $I_o$ changes. The ideal voltage regulator would have $R_o = 0$; however, in practice this can only be approached. Other specifications may include the temperature dependence of $V_o$ as well as the current limits of the unit. To protect the regulator against accidental short circuits, the current may be automatically limited to a safe value.

### 11.5a  Shunt Regulator

The Zener diode regulator has limitations of range. The load current range for which regulation is maintained is the difference between maximum allowable and minimum required diode current. As an example, if a Zener diode requires a minimum current of 10 mA and is limited to a maximum of 1 A (to prevent excessive dissipation), the range is $1 - 0.01 = 0.99$ A. If the load current variations exceed 0.99 A, regulation may be lost.

The shunt regulator circuit of Fig. 11.23 uses a transistor to amplify the Zener diode current, therefore extending the Zener's current range by a factor equal to the transistor's $h_{FE}$. Except for this modification, the circuit operates as discussed in Chap. 3. $R_1$ supplies the Zener current.

### Example 11.6

For the circuit of Fig. 11.23, determine the following:

a. Nominal output voltage
b. Value of $R_1$
c. Load current range
d. Maximum transistor power dissipation
e. Value of $R_S$ and its power dissipation

The relevant information is as follows:

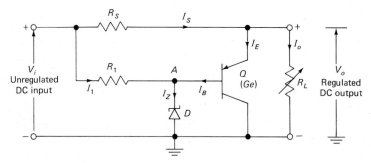

Fig. 11.23  *Shunt regulator circuit.*

$V_i$ = constant 8 V
D: 6.3 V, 200 mW; requires 5-mA minimum current
Q: $V_{EB} = 0.2$ V, $h_{FE} = 49$, $I_{CBO} \simeq 0$

## Solution

a. The nominal output voltage is the sum of the transistor's $V_{EB}$ and the Zener voltage:

$$V_o = 0.2 + 6.3 = 6.5 \text{ V}$$

b. $R_1$ must supply 5 mA to the Zener:

$$R_1 = \frac{8 - 6.3}{5 \times 10^{-3}} = 340 \, \Omega$$

c. The maximum allowable Zener current is

$$P/V = 0.2/6.3 = 31.8 \text{ mA}$$

The load current range is the difference between minimum and maximum current through the shunt path provided by the transistor. At junction $A$, we can write

$$I_B = I_Z - I_1;$$

$I_1$ is a constant 5 mA; therefore

$$\underline{I_B} = \underline{I_Z} - I_1 = (5 \times 10^{-3}) - (5 \times 10^{-3}) = 0$$
$$\overline{I_B} = \overline{I_Z} - I_1 = (31.8 \times 10^{-3}) - (5 \times 10^{-3}) = 26.8 \text{ mA}$$

The transistor's emitter current is $(h_{FE} + 1)(I_B)$. $I_B$ ranges from a minimum of around zero to a maximum of 26.8 mA; therefore the load current range is $(h_{FE} + 1)(26.8 \times 10^{-3}) = 50(26.8 \times 10^{-3}) = 1.34$ A. The Zener alone could provide a maximum range of 26.8 mA.

d. The maximum transistor power dissipation occurs when the current is maximum. Using $I_E \simeq I_C$, we have

$$P_D = V_o I_E = 6.5(1.34) = 8.7 \text{ W}$$

e. $R_S$ must pass 1.34 A to supply current to the transistor and $R_L$:

$$R_S = \frac{V_i - V_o}{1.34} = \frac{8 - 6.5}{1.34} = 1.12 \, \Omega$$

The power dissipated by $R_S$ is

$$I_S^2 R_S = (1.34^2)(1.12) \simeq 2 \text{ W}$$

### 11.5b Series Regulator

Further improvements in performance are possible using the series regulator of Fig. 11.24a. Here the load voltage $V_o$ is also the voltage across the sampling resistance $R_S$. The name *sampling* resistance is used because $V_s$ is a "sample" of $V_o$. If $V_o$ rises or drops, $V_s$ also rises or drops. The amplifier is of the differential input type; one input is $V_s$, while the other is $V_r$, a constant reference voltage. The reference may be a Zener diode or a dry cell. The amplifier output is proportional to the difference between its two inputs; this difference is called the error signal and may be expressed as follows:

$$\epsilon = V_s - V_r \tag{11.22}$$

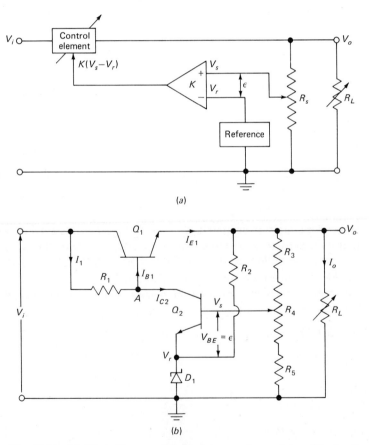

*Fig. 11.24* (a) Block diagram of feedback-type series regulator; (b) circuit diagram of feedback-type series regulator.

The amplifier multiplies the error by a constant $K$, yielding an appropriate signal to the control element. The function of the signal is to increase or decrease the effective resistance of the control element, depending on whether $V_o$ is too high or too low, respectively. For example, if $V_o$ should increase, the error voltage $\epsilon$ also increases. The amplifier's output therefore increases; this causes the resistance of the control element to increase, therefore reducing the load current and load voltage. Note that the original *increase* in $V_o$ caused the circuit to act in such a manner as to *decrease* $V_o$. It is because of this *negative feedback* that we are able to achieve *control* over the load voltage.

A circuit that operates on the above principle is shown in Fig. 11.24b. $R_3$, $R_4$, and $R_5$ make up the sampling network. The current through the sampling network must be sufficiently large that, whatever current the base of $Q_3$ draws, the loading on the voltage divider is negligible and $V_s$ remains an accurate sample of $V_o$.

It is possible to simply eliminate $R_3$ and $R_5$ and leave only $R_4$, but there is the danger that in setting the potentiometer one will go to either extreme, in which case the transistor may be damaged. For example, if the wiper is brought all the way to the lower extreme and there is no $R_5$, the base is at ground potential. The emitter, on the other hand, is above ground by the Zener voltage; this reverse-biases the emitter-base junction, and since reverse $V_{BE}$ ratings are rather low, the transistor could be damaged. For this reason, fixed resistor $R_5$ is used to prevent the wiper from being brought all the way to ground. Without $R_3$, the wiper on $R_4$ could be brought all the way to the top, which would forward-bias the collector-base junction, therefore losing transistor action. For this reason, fixed resistor $R_3$ is used to prevent the base from ever rising above the collector. None of these resistors are critical; $R_4$ is variable to allow setting of the output voltage.

The sample voltage $V_s$ is applied to the base of $Q_2$, while the reference voltage $V_r$, which is provided by Zener diode $D_1$, is applied to the emitter. $R_2$ sets the Zener current; in addition to ensuring that the Zener operates in the breakdown region, the Zener's temperature coefficient, which is both voltage- and current-dependent, may be controlled by an appropriate selection of current through $R_2$. This way the Zener's temperature coefficient (usually positive) may be used to cancel the transistor's $V_{BE}$ temperature dependence (usually negative).

The error voltage is $V_s - V_r$, or simply $V_{BE}$. $Q_2$ yields an output current ($I_{C2}$) which is proportional to this "error." At junction $A$ the following expression may be written

$$I_{C2} + I_{B1} = I_1 \tag{a}$$

If $I_1$ is constant, changes in $I_{C2}$ yield equal but opposite changes in $I_{B1}$; that is, if $I_{C2}$ rises, $I_{B1}$ drops. Changes in $I_{B1}$ are amplified by $Q_1$ to yield corresponding changes in $I_{E1}$ and $I_o$. As the current through it changes, $Q_1$ appears as a variable resistor whose resistance depends on the control current $I_{B1}$. If $I_{B1}$ drops, the resistance of $Q_1$ increases, allowing less current to flow through the load.

It is important to note that this type of corrective action is effective only as long as $I_1$ remains reasonably constant. This is to ensure that $I_{B1}$ changes mainly in response to variations in $I_{C2}$ (which are due to the error signal). If $I_1$ is not constant, it is possible for $I_{B1}$ (and hence $I_o$ and $V_o$) to change in response to $I_1$ variations which do not reflect conditions in the load circuit.

### Example 11.7

The following data apply to a circuit of the type in Fig. 11.24b:

$$\begin{aligned}
&\text{Unregulated input:} && V_i = 15 \text{ to } 20 \text{ V} \\
&\text{Desired output:} && V_o = 10 \text{ V} \\
&\text{Load current requirements:} && I_o = 0 \text{ to } 0.1 \text{ A} \\
&&& D_1 = 6.8 \text{ V} \\
&&& \underline{h_{FE1}} = 15 \\
&&& \overline{h_{FE1}} = 50
\end{aligned}$$

Both transistors are the silicon planar type with negligible $I_{CBO}$. Determine

a. Extreme values of $R_L$
b. Maximum value of $I_{B1}$
c. Suitable value for $R_1$
d. Extreme values of $I_1$
e. Value of $R_2$ if $D_1$ requires a minimum of 5 mA
f. Worst-case power dissipation for $Q_1$ under normal operating conditions
g. Worst-case power dissipation for $Q_2$ under normal operating conditions

### Solution

a. Since $R_L = V_o/I_o$, we can write

$$\underline{R_L} = \frac{V_o}{\overline{I_o}} = \frac{10}{0.1} = 100 \text{ }\Omega$$

The minimum load current is zero; therefore the load resistance is anywhere from 100 Ω to an open circuit.

b. The base current of $Q_1$ is $I_{E1}(h_{FE1} + 1)$, where $I_{E1}$ is the sum of load current $I_o$, current through the sampling network, and current through $R_2$. Usually $I_o$ is the largest component involved; therefore

$$\overline{I_{E1}} \simeq \overline{I_o} = 0.1 \text{ A}$$

and

$$\overline{I_{B1}} \simeq \frac{\overline{I_{E1}}}{h_{FE1}} = \frac{0.1}{15} = 6.66 \text{ mA}$$

c. $R_1$ should supply the maximum current required by the base of $Q_1$. This is 6.66 mA; however, it is a good practice to increase it by about 50 percent to provide a safety margin. We therefore select $R_1$ on the basis that its current must never drop below 10 mA:

$$R_1 = \frac{V_i - V_A}{I_1}$$

$$R_1 = \frac{V_i - (V_o + V_{EB1})}{I_1}$$

$$R_1 = \frac{15 - 10.6}{10 \times 10^{-3}} = 440 \text{ }\Omega$$

d. The extreme values of $I_1$ are determined using the extreme values of $V_i$:

$$\underline{I_1} = \frac{\underline{V_i} - V_A}{R_1} = \frac{15 - 10.6}{440} = 10 \text{ mA}$$

$$\overline{I_1} = \frac{\overline{V_i} - V_A}{R_1} = \frac{20 - 10.6}{440} = 21.4 \text{ mA}$$

e. $R_2 = \dfrac{V_o - V_r}{I_{D1}} = \dfrac{10 - 6.8}{5 \times 10^{-3}}$

$R_2 = 640 \text{ }\Omega$

f. Worst-case power dissipation for $Q_1$ occurs when both the voltage and current are maximum. Normally the collector-base junction dissipates the most power, since its voltage is much greater than $V_{BE}$; however, the total transistor dissipation involves both junctions. This can be approximated as follows:

Voltage across $Q_1$:      $V_{CE1} = V_i - V_o$
Current through $Q_1$:     $I_{C1} \simeq I_{E1} \simeq I_o$

$$\overline{P_{D1}} \simeq (\overline{V_i} - V_o)(\overline{I_o})$$
$$\overline{P_{D1}} \simeq (20 - 10)(0.1) = 1 \text{ W}$$

g. $Q_2$ does not handle large currents as $Q_1$ does; therefore its power dissipation is relatively low:

$P_{D2} = (V_{CE2})(I_{C2})$

Both $V_A$ and $V_r$ are relatively constant; therefore $V_{CE2}$ is also constant:

$V_{CE2} = V_A - V_r = V_o + V_{EB1} - V_r$
$V_{CE2} = 10 + 0.6 - 6.8 = 3.8 \text{ V}$

The maximum value of $I_{C2}$ is approximately equal to $\overline{I_1} = 21.4$ mA; we therefore have

$$\overline{P_{D2}} \simeq V_{CE2}\overline{I_{C2}}$$
$$\overline{P_{D2}} \simeq 4(21.4 \times 10^{-3}) = 81.5 \text{ mW}$$

Note that the current through $R_1$ in Example 11.7 does not remain constant; in fact, it can vary between 10 and 21.4 mA, a total range of 11.4 mA. The regulator would perform much better if $I_1$ were regulated. This will be discussed shortly.

### Example 11.8

Determine $R_1$ and extreme values of $I_1$ in the circuit of Example 11.7 if $Q_1$ is replaced with the compound connection of Fig. 11.25. Assume both transistors have a minimum $h_{FE}$ of 15 and negligible cutoff currents.

### Solution

The current through $R_1$ should be selected about 50 percent higher than the maximum current required by the base of $Q_{12}$. For worst-case design, the minimum $h_{FE}$ of 15 is used, yielding a composite $h_{FE} \simeq 15(15) = 225$ for both $Q_{11}$ and $Q_{12}$. This yields

$$\overline{I_{B12}} \simeq \frac{\overline{I_o}}{225} = \frac{0.1}{225} = 0.445 \text{ mA}$$

We now select $I_1 \simeq 0.65$ mA; $R_1$ is calculated as follows:

$$R_1 = \frac{V_i - V_A}{I_1}$$

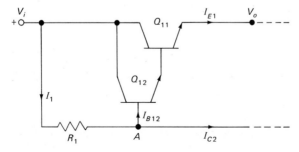

Fig. 11.25  Replacing $Q_1$ in the circuit of Fig. 11.24b with a Darlington pair.

where

$V_A = V_o + V_{EB11} + V_{EB12}$
$V_A = 10 + 0.6 + 0.6 = 11.2$ V
$R_1 = \dfrac{15 - 11.2}{0.65 \times 10^{-3}} \simeq 5.85$ k$\Omega$

The extreme values of $I_1$ occur when $V_i$ is at its extremes. The minimum $I_1$ is 0.65 mA (as indicated above), while the maximum is

$\overline{I_1} = \dfrac{\overline{V_i} - V_A}{R_1} = \dfrac{20 - 11.2}{5.85 \times 10^{-3}}$
$\overline{I_1} = 1.5$ mA

The variation in $I_1$ is $1.5 - 0.65 = 0.85$ mA.

### 11.5c  Series Regulator with Current Preregulator

The circuit of Fig. 11.26 is an improved version of that in Fig. 11.24b. Besides $Q_1$ being replaced with a Darlington pair, $R_1$ has also been replaced with a current regulator circuit. The function of $D_2$, $R_6$, $R_7$, and $Q_3$ is to establish and maintain a constant $I_1$. The circuit works this way: $I_1$ is the collector current of $Q_3$, and hence it is also approximately equal to $I_{E3}$. The voltage at the base of $Q_3$

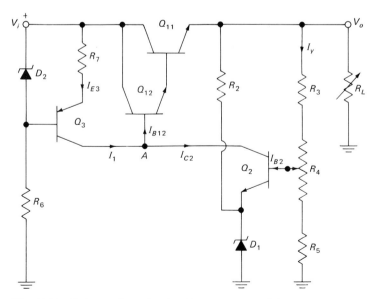

*Fig. 11.26*  Series regulator circuit with current preregulator.

relative to $V_i$ is held at a constant level by $D_2$; current through $R_6$ is selected to keep $D_2$ in breakdown and to yield the proper temperature coefficient. Should $I_1$ rise, $I_{E3}$ will also rise, increasing the voltage across $R_7$. This reduces $V_{EB3}$, which in turn reduces $I_{B3}$ and $I_1$. Thus $I_1$ is regulated and remains fairly constant even if there are changes in the unregulated input.

One disadvantage of this circuit is that a larger input voltage is required to supply the various voltage drops between $V_i$ and $V_o$. In this case $V_i$ must supply $V_o$ plus the two $V_{EB}$ drops of $Q_{11}$ and $Q_{12}$ (which takes us to point $A$), plus the collector-base bias for $Q_3$ (which takes us to the base of $Q_3$), plus the Zener voltage for $D_2$.

## Problem Set 11.5

11.5.1. What is the function of $Q$ in the circuit of Fig. 11.23?

11.5.2. In the circuit of Fig. 11.23, assume $V_i$ can vary between 6 and 8 V, $R_S = 1\ \Omega$, $V_D = 4.7$ V, and the transistor is a germanium type with $h_{FE} = 49$. $R_L$ can vary anywhere between an open circuit and 10 $\Omega$; $R_1 = 500\ \Omega$. Determine

    a. The nominal $V_o$
    b. The extreme values of load current
    c. The extreme value of $I_S$
    d. The extreme values of transistor emitter and base currents
    e. The extreme values of $I_1$
    f. The extreme values of Zener current
    g. The maximum Zener power dissipation
    h. The maximum transistor dissipation

11.5.3. Assume that the maximum load current in the circuit of Fig. 11.23 is 1 A. If the remaining parameters of Example 11.6 apply, determine the lowest value of $V_i$ for which regulation is possible.

11.5.4. Consider the block diagram of Fig. 11.24a. If $V_o$ should drop below its desired value, how should the resistance of the control element change to provide corrective action? What happens to the magnitude of error voltage $\epsilon$ when $V_o$ drops?

11.5.5. Given the circuit of Fig. 11.24b, how is the output voltage sampled? What considerations determine the values of sampling-network resistors? Why do we need $R_3$? Why do we need $R_5$?

11.5.6. What are the functions of $R_2$ and $D_1$ in the circuit of Fig. 11.24b? What temperature coefficient should $D_1$ have?

11.5.7. If $V_o$ in the circuit of Fig. 11.24b starts to rise, what happens to each of the following?

a. $V_s$
b. $V_r$
c. $I_{B2}$
d. $I_{C2}$
e. $I_1$
f. $I_{B1}$
g. $I_{E1}$
h. Effective resistance of $Q_1$
i. $I_o$
j. $V_o$

11.5.8. Repeat Prob. 11.5.7 if $V_o$ starts to drop.

11.5.9. Why is it desirable that $I_1$ in the circuit of Fig. 11.24b remain constant? What causes $I_1$ to vary?

11.5.10. Repeat Example 11.7 if all parameters remain the same except

$\overline{h_{FE1}} = 100 \quad \underline{h_{FE1}} = 60$
$D_1 = 3.3$ V
$I_o = 0$ to 1 A

11.5.11. Which transistor must dissipate maximum power in the Darlington connection of Fig. 11.25?

11.5.12. For the circuit of Fig. 11.26, suppose $I_1$ begins to rise. What happens to the following?

a. $V_{D2}$
b. $V_{R7}$
c. $V_{BE3}$
d. $I_{B3}$
e. $I_1$

## REFERENCES

1. Cowles, L. G.: "Analysis and Design of Transistor Circuits," D. Van Nostrand Company, Inc., Princeton, N.J., 1966.
2. Millman, J., and C. C. Halkias: "Electronic Devices and Circuits," McGraw-Hill Book Company, New York, 1967.
3. Texas Instruments Incorporated: "Transistor Circuit Design," McGraw-Hill Book Company, New York, 1963.
4. "General Electric Transistor Manual," 7th ed., Syracuse, 1964.
5. Operational Amplifiers, pts. I, II, III, and IV, *Analog Devices*, Cambridge, Mass., 1965.
6. Hilbiber, D. F.: A New DC Transistor Differential Amplifier, *IRE Trans. Circuit Theory*, December 1961, pp. 434–437.
7. Middlebrook, R. D.: "Differential Amplifiers," John Wiley & Sons, Inc., New York, 1964.
8. "Handbook of Operational Amplifier Active RC Networks," Burr-Brown Research Corporation, Tucson, 1966.
9. "Handbook of Operational Amplifier Applications," Burr-Brown Research Corporation, Tucson, 1963.

10. "Applications Manual for Computing Amplifiers for Modelling, Measuring, Manipulating, and Much Else," Philbrick Researches, Inc., Boston, 1966.
11. Weber, H. F.: A Regulated Power Supply Using a Reference Amplifier, *Application Note* 181, Motorola Semiconductor Products Inc., Phoenix, 1966.
12. Breece, H. T.: Current Limiting for Transistor Series Voltage Regulators, *Application Note* SMA-18, RCA Semiconductor and Materials Division, Somerville, N.J., 1963.

# chapter 12

# large-signal amplification

## 12.1 GENERAL CONSIDERATIONS

Large-signal amplification involves circuits in which voltage and current variations extend over a large portion of the amplifying device's operating region, as opposed to small-signal amplification, in which variations about the $Q$ point are typically small. This chapter deals with applications for which large signals result in large power, such as in transistor audio power amplifiers. We will also limit our discussions to those frequencies for which reactive effects are not very significant, typically in the audio range.

Although semiconductors have many advantages over vacuum tubes (one is the absence of any filament power requirement), much high-power RF transmitting equipment still utilizes vacuum-tube power stages simply because high-power, high-frequency transistors have not been available. The state of the art, however, is constantly improving, and high-frequency semiconductor power capabilities are continuing to increase. Field effect transistors have not yet matched the power capabilities of bipolar transistors; for this reason, only bipolar transistor circuits are treated here. Power ratings of FETs are also continuing to increase, however, leading to greater utilization of these devices in power circuits.

Usually, an amplifying system must produce sufficient power to drive a load; the load may be some type of electromechanical or electromagnetic transducer, such as a loudspeaker, servomotor, or transmitting antenna. Such a system often employs several successive stages of amplification. The initial stages are designed to minimize noise and optimize gain, while the last stage must produce significant power with low distortion. In many cases, a *driver* stage may be used to provide the input levels required by the power stage to achieve optimum performance.

There are many differences between small- and large-signal amplifiers, as we shall soon see. These differences exist because the performance requirements are not the same. Mainly, power amplifiers are characterized by how much power they can handle, the distortion they produce, and their efficiency. Frequency response may be a consideration if, for example, a high-fidelity sound system is involved; when the amplifier is used to drive a servomotor, however, it may not be as critical. Size and weight may also be important, if the system is to be airborne, as in an aircraft, or much more critically, a space vehicle. Maximization of power gain, although desirable, is not usually a consideration, because the conditions that would favor it are detrimental to a higher-priority requirement, the maximization of output power. This will soon be discussed in more detail.

### 12.1a Output Power

To obtain the largest possible power output from any device, the output voltage and current variations must be maximized. Although a given power level may be achieved through an infinite variety of voltage and current combinations, it is the limitations inherent in the amplifying device that dictate the maximum voltage and current levels and hence the maximum power. Since transistors are basically low-voltage devices (there are exceptions to this, since it is now possible to obtain transistors that can operate with voltages in excess of 1,000 V), we usually require large current swings, and hence low values of load resistance, to achieve large power levels. An idealized graph for the output voltage and current relationships of a typical amplifying device is shown in Fig. 12.1. If a $Q$ point is chosen midway along the AC load line, symmetrical maximum voltage and current swings are possible. Assuming a sinusoidal signal, the maximum available output power is

$$P_{o(\text{max})} = V_o I_o \tag{12.1}$$

where $V_o$ and $I_o$ are the maximum RMS output voltage and current, respectively:

$$P_{o(\text{max})} = \left(0.707 \frac{\Delta V_o}{2}\right)\left(0.707 \frac{\Delta I_o}{2}\right)$$

$$P_{o(\text{max})} = \frac{\Delta V_o \Delta I_o}{8} \tag{12.1a}$$

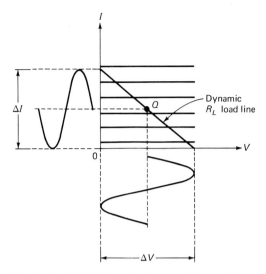

*Fig. 12.1  Idealized graph of output voltage and current swings for an amplifying device.*

where $\Delta V_o$ and $\Delta I_o$ represent the maximum output voltage and current change, respectively.

The dynamic resistance of the load may be expressed as the ratio of voltage change to current change:

$$R_L = \frac{\Delta V_o}{\Delta I_o} \tag{12.2}$$

### Example 12.1

A given ideal power transistor may be operated anywhere within output voltage and current extremes of 10 V and 2 A, respectively. Select a load line to yield maximum output power, and locate the $Q$ point for maximum symmetrical swing. Assuming a sinusoidal signal, determine the maximum output power and the load resistance required.

### Solution

A load line joining $V = 10$ V and $I = 2$ A is chosen; the $Q$ point is halfway between these voltage and current extremes:

$Q$ point:   $V = 5$ V     $I = 1$ A

The voltage and current swings are

$\Delta V_o = 10$ V     $\Delta I_o = 2$ A

The maximum output power is

$$P_{o(\text{max})} = \frac{\Delta V_o \, \Delta I_o}{8} = \frac{10(2)}{8} = 2.5 \text{ W} \tag{12.1a}$$

The load required is

$$R_L = \frac{\Delta V_o}{\Delta I_o} = \frac{10}{2} = 5 \, \Omega \tag{12.2}$$

The above value of $R_L$ is typically low. If we were to consider maximizing the power gain in a real amplifier, it is unlikely that it could be realized with such a low value of load resistance. For example, a typical output resistance for a common emitter amplifier (which exhibits high power gain) might be 20 k$\Omega$. To achieve maximum power gain, a load would be required whose resistance is of the order of 20 k$\Omega$, but if the output power-handling capabilities are to be maximized, the load should be 5 $\Omega$. Hence, when dealing with large-signal amplifiers, we usually sacrifice power gain in order to achieve a *large power* capability.

### 12.1b  Transistor Parameters

Since we are now dealing with large voltage and current swings, small-signal AC parameters such as $h_{ie}$ and $h_{fe}$ may not be valid, because they were obtained assuming very small changes about the operating point. Much more valid for large-signal analysis are static parameters, normally given on data sheets or easily estimated from characteristic curves. Because both $R_L$ and $A_v$ are usually small, it is reasonable to neglect the $h$ parameters dealing with output admittance and reverse voltage transfer ($h_o$ and $h_R$); the effects of these parameters become less significant as $R_L$ decreases. Hence, the following static parameters are of the greatest significance at the moment:

*Common Emitter*

Forward current transfer ratio: $\quad h_{FE} = \dfrac{I_C}{I_B} \qquad (12.3)$

Input resistance: $\quad h_{IE} = \dfrac{V_{BE}}{I_B} \qquad (12.4)$

*Common Base*

Forward current transfer ratio: $\quad h_{FB} = \dfrac{-I_C}{I_E} \qquad (12.5)$

Input resistance: $\quad h_{IB} = \dfrac{V_{EB}}{I_E} \qquad (12.6)$

All the above parameters depend on the point at which they are measured. Since $R_L$ is relatively small, a good approximation to the common emitter current amplification is $h_{FE}$; similarly, $h_{FB}$ is the approximate value of common base current amplification. The input resistance is $h_{IE}$ and $h_{IB}$ for common emitter and common base, respectively.

The parameter $h_{FE}$ can be estimated using the transfer characteristics. As an example, consider the 2N2142 germanium power transistor whose data sheet is reproduced in Appendix 10. At $V_{CE} = 2$ V and $I_C = 2$ A, the corresponding value of base current is 50 mA (see "Collector current versus Base current" graph). This yields

$$h_{FE} = \frac{2}{50 \times 10^{-3}} = 40$$

The common emitter input resistance may be estimated using the "Base-emitter voltage versus Base current" graph. At the point in question, $V_{BE} = 0.52$ V and $h_{IE} = 0.52/(50 \times 10^{-3}) = 10.4\ \Omega$.

If a different point is chosen, say $V_{CE} = 2$ V and $I_C = 3$ A, $h_{FE} \simeq 3/(90 \times 10^{-3}) = 33.3$ and $h_{IE} \simeq 0.6/(90 \times 10^{-3}) = 6.66\ \Omega$. Thus both $h_{FE}$ and $h_{IE}$ drop at higher current levels.

### 12.1c  Distortion

The minimization of distortion is another matter of concern. When small signals are involved, device characteristics are fairly linear, and small-signal models are valid. In such cases, distortion is not much of a problem. As signal swings get larger, however, operation in the nonlinear regions cannot be avoided, and distortion becomes an important factor, especially in high-fidelity sound systems. For this purpose, equivalent circuit models are of limited use, and graphic analysis techniques are usually employed.

Distortion may arise in the input circuit of a transistor amplifier because of the nonlinear $VI$ characteristics of the forward-biased base-emitter diode. This is demonstrated graphically in Fig. 12.2b for the common emitter circuit of Fig. 12.2a, which is driven from a voltage source. This type of distortion can be minimized by driving from a high internal resistance (current source) so as to linearize the total resistance in the input circuit. Such an arrangement results in an undistorted base current waveform.

Distortion in the output circuit is due to *alpha crowding*, which is a reduction of $h_{FE}$ at high current levels. At high current densities, the lifetime of carriers injected into the base decreases, and hence fewer reach the collector. The common emitter output characteristics of Fig. 12.3a illustrate this point. At high collector currents, the curves tend to "crowd" each other; hence the ratio of

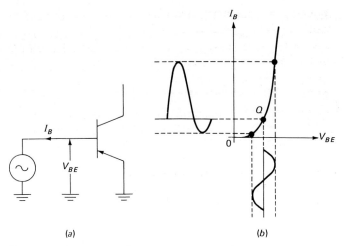

*Fig. 12.2* Distortion of base current waveform caused by nonlinear base-emitter diode characteristics.

collector to base current is smaller. Figure 12.3b demonstrates how the upper half of the current waveform receives less amplification than the lower half.

Since alpha crowding distorts the signal in a manner opposite to the input circuit's nonlinearity, cancellation of the two types of distortion is possible. For this reason, common emitter amplifiers are usually driven from a low-resistance (voltage) source in order to produce the type of input distortion required to cancel alpha-crowding effects in the output circuit.

When we are dealing with common base circuits, the distortion problem

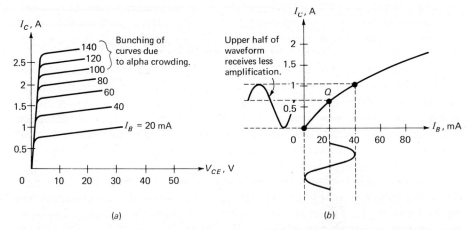

*Fig. 12.3* (a) *Common emitter output characteristics of power transistor;* (b) *waveform distortion due to alpha crowding.*

is handled differently. As an example, suppose that a common emitter amplifier has an $h_{FE}$ of 50 which, at high current levels, drops to 35. Such a drop would be quite noticeable, since the upper parts of our waveform receive only $35/50 = 70$ percent of the maximum amplification. If the same transistor is used in a common base configuration, the current amplification is $h_{FB} = -h_{FE}/(h_{FE} + 1) = -0.98$ when $h_{FE} = 50$. When $h_{FE}$ drops to 35, the corresponding value of $h_{FB}$ is $-35/36 = -0.972$. Although $h_{FE}$ has dropped by 30 percent, $h_{FB}$ is down by less than 1 percent. It is for this reason that distortion due to alpha crowding is significant mainly in common emitter stages, but negligible in common base circuits.

Since there is no significant output distortion in common base circuits, it follows that any distortion due to the nonlinear base-emitter circuit cannot be compensated as with common emitter amplifiers. We must therefore minimize the input distortion by driving with a current source whose high internal resistance "swamps out" the smaller, nonlinear resistance of the emitter-base diode.

To summarize, distortion considerations dictate that common emitter stages should be driven from low-resistance (voltage) sources, while common base stages should be driven from high-resistance (current) sources. More will be said on other types of distortion when specific circuits are discussed.

### 12.1d Efficiency

Efficiency may be defined as the ratio of output (AC) power to input (DC) power:

$$\eta = \frac{P_o}{P_{\text{DC}}} \tag{12.7}$$

The DC power supplied to the amplifier is partially converted to AC signal power; the remainder, $P_D$, is dissipated as heat by the active and passive components in the circuit. We can therefore write

$$P_D = P_{\text{DC}} - P_o \tag{12.8}$$

In an efficient system, $P_o$ approaches $P_{\text{DC}}$, and there is little $P_D$ left to dissipate. For an inefficient system, however, there are many drawbacks; a few are listed here:

1. With mobile equipment, more power must be supplied by the battery; hence the battery's life is reduced.
2. If an AC-DC power supply is used, a unit with greater power output may be required; this usually means greater cost, heavier and bulkier equipment, and quite likely more heat problems.

3. The amplifying device must dissipate more heat. This is a definite limiting factor to the maximum power the device can handle; overheating results in degradation of performance and eventual device destruction.

The design of circuits with high efficiency is therefore a primary objective; the more successful we are in improving efficiency, the less of a problem heat removal becomes.

**Example 12.2**

A transistor audio amplifier delivers an output power of 10 W with an efficiency of 45 percent. Determine

a. The DC power input
b. The power that must be dissipated as heat

**Solution**

a. $\eta = \dfrac{P_o}{P_{DC}}$ (12.7)

$0.45 = \dfrac{10}{P_{DC}}$

$P_{DC} = \dfrac{10}{0.45} = 22.2 \text{ W}$

b. $P_D = P_{DC} - P_o$ (12.8)
$P_D = 22.2 - 10 = 12.2 \text{ W}$

The amplifier therefore requires 22.2 W; of this, 10 W is converted to signal power, while the remaining 12.2 W is wasted as heat.

## Problem Set 12.1

12.1.1. List three basic performance measures for power amplifiers.
12.1.2. Indicate, for each criterion listed above, why it is significant in power amplifiers, but relatively unimportant when small signals are involved.
12.1.3. Why is graphic analysis useful when dealing with large signals?
12.1.4. Assume that a power transistor can be safely operated between the following extremes:

Output voltage: 0 to 15 V
Output current: 0 to 3 A

Determine

a. The maximum output power
b. The $Q$ point and value of $R_L$ required to achieve (a)

12.1.5. Calculate the power that must be dissipated by the transistor in Prob. 12.1.4 when no signal is present.

12.1.6. A power transistor whose output swings between 0 and 24 V produces 12 W of signal power. Determine the value of $R_L$, current swing, and $Q$ point.

12.1.7. Estimate $h_{IB}$ and $h_{FB}$ for the 2N2142 at $V_{CE} = 2$ V and
a. $I_C = 1$ A
b. $I_C = 2$ A

12.1.8. A 50-Ω generator is used to drive a power amplifier. From purely distortion considerations, what transistor configuration would you use?

12.1.9. Why are the effects of alpha-crowding distortion not significant in common base circuits?

12.1.10. If $h_{FB} = -0.95$ and then drops by 1.5 percent, what is the percent drop in $h_{FE}$?

12.1.11. A given transistor power amplifier delivers an RMS current of ½ A to its 8-Ω load while dissipating 4 W of power. Determine the efficiency.

12.1.12. How much power does a 70 percent efficient amplifier dissipate while delivering 20 W to a 32-Ω load? How much DC power must be supplied?

## 12.2 OPERATING REGION

The ratings and characteristics of a particular device set the limits of its operation; circuit design must, of necessity, start with a definition of these limits. A *rating* refers to the maximum value of a quantity which, if exceeded, may damage the device; *characteristics*, on the other hand, are parameters of the device for which reliable operation is to be expected. We will illustrate the development of an operating region for the common emitter output characteristics of Fig. 12.4:

1. The voltage ($V_{CE}$) cannot exceed the breakdown rating of the transistor. A worst-case design would ensure that $V_{CE}$ is always less than $BV_{CEO}$. Thus an upper limit is set to the voltage, indicated as $V_{CE(MAX)}$ in the figure.
2. The current ($I_C$) cannot be so large that significant alpha-crowding distortion takes place; the manufacturer may also specify an absolute maximum value for $I_C$. The upper limit on $I_C$ is indicated in the figure as $I_{C(MAX)}$.
3. The transistor must not operate in cutoff, or clipping of the waveform results. Cutoff occurs somewhat below $I_B = 0$. This is because when $I_B = 0$, collector current ($I_{CEO}$) still flows and the transistor is not really off until $I_C$ drops closer to zero. This can be done by reverse-biasing the base by a small amount, which yields $I_{CEX}$, the lowest possible value of collector current

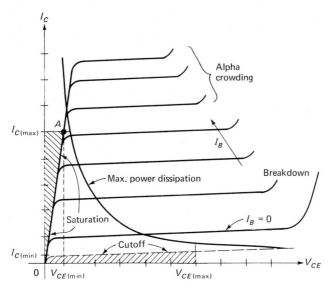

*Fig. 12.4 Allowable region of operation for common emitter output characteristics.*

in this configuration. Thus transistor cutoff places a lower limit on $I_C$; this is indicated on the graph as $I_{C(\text{MIN})}$.

4. The transistor must not operate in saturation, or clipping occurs. In the saturation region, the collector-base junction is actually forward-biased, and a small saturation voltage $V_{CE(\text{SAT})}$ is developed which is close to but slightly greater than zero. $V_{CE(\text{SAT})}$ can be minimized by using a transistor with low ON resistance. This requires a heavily doped collector region. In the graph, saturation resistance $r_{(\text{ON})}$ is given by the voltage-current ratio along line $OA$. The saturation resistance is more significant in silicon than germanium transistors. Because of $r_{(\text{ON})}$ and its associated voltage drop, there is a lower limit to $V_{CE}$; this is indicated as $V_{CE(\text{MIN})}$.

5. The transistor must be kept within a specified temperature range to avoid damage. If the transistor cannot dissipate the heat that it generates, it will overheat and destroy itself. Basically, heat is generated because of the current through and voltage across the device; the product of these represents power. Because the collector-base junction handles the large voltage and current swings that represent output power, it follows that it is in the region near this junction that most of the heat is generated. The base-emitter junction generates some heat as well, but it is usually negligible in comparison.

The heat produced by the collector-base junction must escape, or the junction temperature continues to rise until self-destruction occurs. Hence the basic limitation to power-handling capability is the junction temperature: $T_{j(\text{max})}$. Silicon units can withstand maximum junction temperatures of 175°C or higher; the limit is around 100°C for germanium transistors.

The maximum power that can be dissipated is therefore a direct consequence of $T_{j(\text{max})}$. Based on the device's $T_{j(\text{max})}$, the allowable $P_{D(\text{MAX})}$ is established; a power hyperbola can then be constructed on the output characteristics, as shown in the figure.

An additional phenomenon of concern when designing transistor power circuits is secondary breakdown. This is usually a destructive condition, and hence it must be avoided. It will be recalled that avalanche breakdown involves a significant rise in $I_C$, which is not necessarily destructive, provided it is limited to a safe value. If $I_C$ is allowed to increase further, however, localized melting in small areas near the junction may result in a short circuit from collector to emitter and a sudden drop in $V_{CE}$. This effect, known as secondary breakdown, is not entirely defined by voltage and current values. Because melting of the semiconductor requires energy, both power and time are involved. Manufacturers of power transistors usually provide information on the safety limits of their devices to ensure that secondary breakdown is avoided.

In conclusion, an acceptable region for device operation is the area which lies within the boundaries set by minimum and maximum voltage, minimum and maximum current, and maximum power.

### Problem Set 12.2

12.2.1. What considerations designed to avoid transistor damage are involved when specifying an operating region?

12.2.2. What prevents us from achieving zero voltage in a transistor? Zero current?

12.2.3. A typical power transistor is rated at 2 A, 20 V, and can safely dissipate 10 W. It has a saturation resistance of 0.5 Ω and a cutoff current of 100 mA at 20 V. Construct, on a graph of $V$ versus $I$, the allowable operating region. Determine the minimum voltage.

12.2.4. Determine the maximum undistorted power output available for Prob. 12.2.3, assuming the transistor is operated with a sinusoidal signal within the device ratings, but neglecting saturation and cutoff. What value of $R_L$ is required, and what is the $Q$ point?

## 12.3 THERMAL CONSIDERATIONS

When a transistor must handle substantial amounts of power, the dissipation of heat is a matter that requires attention. Unless care is taken to remove the heat, a condition known as *thermal runaway* may result. As the temperature rises, $I_C$ rises, which may result in more power dissipation. More power dissipation causes further heat generation, which increases $I_C$ still further. The process may thus continue feeding on itself until self-destruction occurs. It is possible to design circuits in which thermal runaway cannot occur; for the moment, however, we will concern ourselves with the problems of heat removal.

As has already been pointed out, the transistor rating that sets a limit to the power it can dissipate is the maximum junction temperature: $T_{j(\text{max})}$. When two points are at a different temperature, heat flows from the higher-temperature point to the lower. The rate of heat flow depends on the temperature difference and the nature of the medium between the two points. Since this process is very similar to current flow through a conductor, certain analogies have been developed to facilitate the analysis.

Figure 12.5 illustrates the similarities in behavior of electric and thermal circuits. In an electric circuit, potential difference produces current flow; in a thermal circuit, heat flow is due to the difference in temperature between two points. Ohm's law for thermal circuits can be stated as

$$\theta_{xy} = \frac{T_x - T_y}{P_D} \tag{12.9}$$

where $T_x - T_y$ is the temperature difference in degrees Celsius between points $x$ and $y$; $P_D$ is the heat flow (power), in watts, from $x$ to $y$, and $\theta_{xy}$ is the thermal resistance, in degrees Celsius per watt, between points $x$ and $y$. The actual thermal resistance depends on the ability of the medium to transmit heat between the points in question. Since heat may be transmitted by conduction, convection, or radiation, there are many factors that determine thermal resistance.

In semiconductor devices, heat generated at a PN junction flows out toward the case and from the case to the surrounding atmosphere. There is an effective thermal resistance from junction to case ($\theta_{jc}$) and from case to air ($\theta_{ca}$). Because junction temperature cannot be measured directly, manufacturers usually provide a power-temperature derating curve similar to that shown in Fig. 12.6a. This curve tells us that we can safely dissipate 75 W as long as $T_c$ (case temperature) is 25°C or less. If the case is hotter than 25°C, the transistor's power-dissipation capabilities decrease.

Consider now the thermal circuit of Fig. 12.6b. As long as heat flows from the junction to the case, there is a thermal drop across $\theta_{jc}$ equal to the product

Fig. 12.5  Electric-thermal circuit analogies: (a) electric circuit; (b) thermal circuit.

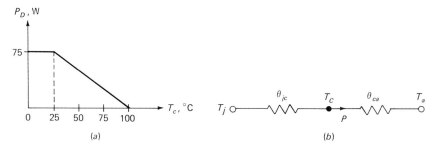

*Fig. 12.6* (a) *Typical transistor power-temperature derating curve;* (b) *thermal circuit from junction to air.*

of power ($P_D$) and thermal resistance ($\theta_{jc}$). The units of thermal drop are degrees Celsius; hence the junction temperature is

$$T_j = T_c + \theta_{jc} P_D \tag{12.10}$$

Note that the junction is at the highest temperature and heat flows away from it. If $T_c = T_j$, the power is zero; thus if the case cannot be maintained at a lower temperature than the junction, no heat flows from junction to case, and no power can be dissipated. For the curve of Fig. 12.6a, this occurs when the case is at 100°C; hence the maximum allowable junction temperature is also 100°C. We can therefore conclude that this is a germanium transistor.

The junction-case thermal resistance may be obtained from the power derating curve by noting that $T_j = 100°C$ anywhere along that curve. If we take a specific value of case temperature and read the allowable value of power dissipation, all the information needed to calculate $\theta_{jc}$ is available. When $T_c = 25°C$, $P_D = 75$ W; since $T_j = 100°C$,

$$\theta_{jc} = \frac{T_j - T_c}{P_D} = \frac{100 - 25}{75} \tag{12.10}$$
$$\theta_{jc} = 1 \text{ °C/W}$$

Obviously, the lower the junction-case thermal resistance, the more power can be dissipated. Power transistors are manufactured to yield low values of $\theta_{jc}$, thus facilitating the removal of heat.

The thermal resistance from case to ambient depends on the path available for heat to escape from the transistor case. If the case is simply exposed to the surrounding air, heat removal is poor, and a large case-air thermal resistance results. Typically, $\theta_{ca}$ might be around 800 °C/W, which precludes power-dissipation levels greater than a few milliwatts.

To reduce $\theta_{ca}$, a *heat sink* is used, which is essentially a large heat-conducting area in close thermal contact with the case. Since heat radiation increases with area, the objective is to maximize the area so as to minimize thermal resistance. A plot of thermal resistance versus area for a flat-sheet, aluminum heat sink is shown in Fig. 12.7. The thermal resistance depends not only on the area but on the mounting (vertical or horizontal), color (heat sinks are usually painted black to maximize heat radiation), thickness, and air velocity (the thermal resistance can be substantially lowered if air is constantly blown past the heat sink). Heat sinking may be achieved by mounting a power transistor directly on a metal chassis; commercially available heat sinks, however, are more efficient, since a large surface area is available in a relatively small volume.

The mounting of power transistors on heat sinks requires some care. Usually, the collector is electrically connected to the case to provide good thermal conductivity from junction to case. To mount the transistor case on a heat sink, however, an insulating washer (mica or Teflon) should be used, unless we want the collector to be at the same electric potential as the heat sink (ground). The insulating washer should be coated on each side with silicone grease, a thermal compound used to provide good thermal conductivity. Figure 12.8 shows a technique for mounting power transistors and a table of typical thermal resistances for different types of insulating washers. It should be noted that using silicone lubricant on both sides of the washer cuts the thermal resistance by approximately one-half. A heat sink, such as the Motorola MS-10, whose thermal characteristics are shown in Fig. 12.9$b$, exhibits a thermal resistance (from heat sink to air) of approximately 3 °C/W. Note the reduction in thermal resistance under forced air flow (Fig. 12.9$c$).

**Example 12.3**

Determine the maximum allowable power dissipation for the transistor whose power derating curve is given in Fig. 12.6$a$. Assume forced air flow of 0.3 lb/min

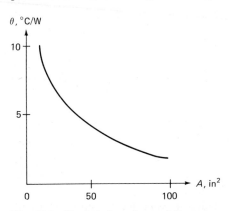

*Fig. 12.7* Typical dependence of thermal resistance $\theta$ of a flat-sheet aluminum heat sink on area of one side, A.

*Large-signal Amplification* 471

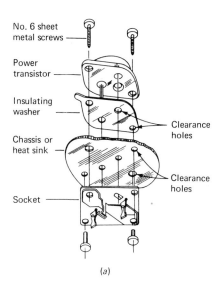

(a)

| Kit no. | Insulating washer | Typical mounting thermal resistance ($\theta_{cs}$) °C/W (includes contact resistance) | |
|---|---|---|---|
| | | Dry | With DC4* |
| —— | No insulator | 0.20 | 0.10 |
| MK-10 | Teflon | 1.45 | 0.80 |
| MK-15 | Mica | 0.80 | 0.40 |
| MK-20 | Anodized Aluminum | 0.40 | 0.35 |

*DC4 is Dow Corning No. 4 Silicone Lubricant. The use of the DC4 or equivalent is highly recommended especially for high power applications. The grease should be applied in a thin layer on both sides of the washer. When transistors are replaced in the sockets a new layer of the grease should be added.

(b)

*Fig. 12.8* (a) *Method for mounting power transistors;* (b) *typical values of thermal resistance for different types of insulating washers.* (*Courtesy of Motorola Inc.*)

and that the transistor is mounted on an MS-10 heat sink with a mica washer coated with silicone grease on both sides. The air temperature surrounding the transistor–heat-sink combination can be maintained at 30°C.

## Solution

Using the equivalent circuit of Fig. 12.10, the total resistance from junction to air is

$$\theta_{ja} = \theta_{jc} + \theta_{c-hs} + \theta_{hs-a} \tag{12.11}$$

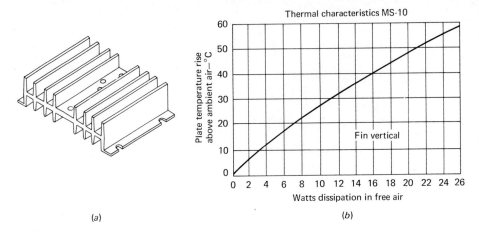

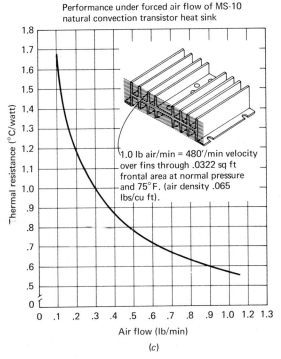

Fig. 12.9  (a) *Motorola MS-10 heat sink;* (b) *thermal characteristics for MS-10;* (c) *thermal resistance under forced air flow conditions.* (Courtesy of Motorola Inc.)

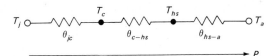

Fig. 12.10  *Thermal circuit for Example 12.3.*

$\theta_{jc}$ is the thermal resistance from junction to case; it was determined earlier to be 1 °C/W. $\theta_{c-hs}$ is the thermal resistance from case to heat sink, and it is due to the insulating washer; the value can be read from Fig. 12.8b as 0.4 °C/W. $\theta_{hs-a}$ is the heat-sink thermal resistance; for an air flow of 0.3 lb/min, its value can be read from the curve of Fig. 12.9c as 1 °C/W. We therefore have a total resistance from junction to air of

$$\theta_{ja} = 1 + 0.4 + 1 \qquad (12.11)$$
$$\theta_{ja} = 2.4 \text{ °C/W}$$

The maximum allowable junction temperature is 100°C. Since the ambient temperature is 30°C, a maximum thermal drop between junction and air of 70°C is allowed:

$$T_j = T_a = 70°C$$
$$P_D = \frac{T_j - T_a}{\theta_{ja}} \qquad (12.9a)$$
$$P_D = \frac{70}{2.4} = 29.2 \text{ W}$$

The thermal equivalent circuits so far discussed are somewhat incomplete. A *capacitive* effect is also present that results in *thermal time constants*. This simply means that when the level of power dissipation suddenly changes, the various temperatures (including the junction) do not immediately respond but change exponentially with time. A thermal time constant can thus be defined as the time required for temperature to rise to 63 percent of its final value after the initial application of a power pulse. A time delay equal to five time constants is normally required for the temperature to reach its ultimate value. It may thus be possible to temporarily exceed the continuous maximum power-dissipation limit if the time is too short for the junction temperature to rise above $T_{j(\text{max})}$.

## Problem Set 12.3

12.3.1. A diode dissipates 2 W. It has a junction-case thermal resistance of 0.5 °C/W. The case is at 60°C, while the ambient temperature is 35°C. Find

    a. The case-air thermal resistance
    b. The junction temperature

12.3.2. If we can maintain the ambient temperature in Prob. 12.3.1 at 25°C, what is the new junction temperature?

12.3.3. What is the maximum safe power dissipation for a transistor mounted in air at 25°C if $\theta_{ja} = 0.8$ °C/mW and $T_{j(\text{max})} = 175$°C?

12.3.4. What heat-sink thermal resistance is required for 10-W dissipation by a transistor with $\theta_{jc} = 0.8$ °C/W at an ambient temperature of 30°C if the junction temperature must be maintained below 100°C? Assume that a Teflon washer with silicone grease on both sides is used between case and sink.

12.3.5. A transistor manufacturer provides the following data: allowable power dissipation at 25°C case temperature: 150 W; derate linearly to a maximum junction temperature of 175°C.

    a. What type of material is the transistor made of?
    b. What is the junction-case thermal resistance?
    c. The transistor is mounted on an MS-10 heat sink using a mica washer coated with silicone grease on both sides; the maximum ambient temperature is 80°C. What is the maximum power that the transistor can safely dissipate?
    d. Using the above heat-sinking configuration, what is the junction temperature when the ambient temperature is 70°C and the power dissipation 5 W?

12.3.6. A germanium transistor in a common emitter configuration is operated without signal at $V_{CE} = 10$ V and $I_C = 1$ A. If $\theta_{jc} = 2$ °C/W, determine the maximum allowable case-air thermal resistance when the ambient temperature is 25°C.

## 12.4 AMPLIFIER CIRCUITS

Amplifier circuits may be classified in terms of that portion of the cycle for which the active device conducts. A class A amplifier is one in which the active device is biased to conduct for the full 360° of a cycle. This is typical of amplifier circuits so far discussed and results in an output which is a reasonable reproduction of the input. We will discuss class A power amplifiers first; other types will be taken up later.

### 12.4a Class A Circuits

Consider the idealized output characteristics for a direct-coupled class A amplifier, as shown in Fig. 12.11a. The output circuit for such an amplifier is shown in Fig. 12.11b. We will neglect the limitations of cutoff and saturation and assume, for now, that output voltage and current swings down to zero are possible; similarly, we will assume that the voltage, current, and power-dissipation ratings have been satisfied.

A 25-Ω load line is drawn on the output characteristics; in this simple circuit, the 25-Ω load line applies for both DC and AC analysis. To obtain maximum usable voltage and current swings, a $Q$ point midway along the load line

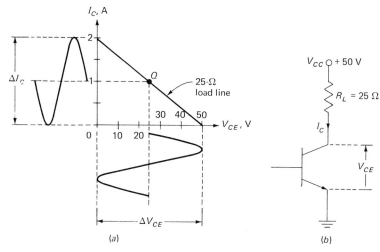

*Fig. 12.11  Idealized output voltage and current swings and circuit for a class A direct-coupled amplifier.*

is chosen, at $V_{CE} = 25$ V and $I_C = 1$ A. The maximum available output power (for a sinusoidal input) is

$$P_{o(\text{max})} = \frac{\Delta V_o \, \Delta I_o}{8} \qquad (12.1a)$$

$$P_{o(\text{max})} = \frac{50(2)}{8} = 12.5 \text{ W}$$

Whether or not the full 12.5 W of output power is obtained depends on the input drive. If no signal is applied at the input, the output just sits at the $Q$ point, and there is no signal output.

To determine the efficiency, we assume maximum output power and compute the DC power that the 50-V source must supply. Since power is the product of voltage and current, the graphic approach of Fig. 12.12 may be helpful. In practice, the collector current accounts for almost all the supply current; hence the instantaneous power is the product of 50 V and $i_C$, yielding the waveform of Fig. 12.12c. The average power supplied by the 50-V source is therefore 50 W. The efficiency is

$$\eta = \frac{P_o}{P_{\text{DC}}} \qquad (12.7)$$

$$\eta = \frac{12.5}{50} = 25 \text{ percent}$$

Note that 25 percent is the *maximum* efficiency, based on maximum AC power output. In cases where less than the full output voltage and current

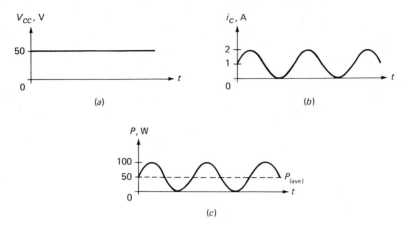

Fig. 12.12 *Determination of power by the graphic multiplication of voltage and current.*

swings are utilized, the AC output power is less, while the DC input power is the same; hence the efficiency is significantly lower.

We can conclude that, for the class A circuit just discussed, a DC power input of 50 W is always required; this is true whether maximum AC power is being produced or none at all. When there is no AC output, the full 50 W is dissipated; the transistor's voltage and current are 25 V and 1 A, and hence it dissipates 25 W. The load resistor dissipates the remaining 25 W.

There is usually no guarantee that a signal will always be present; we must therefore be sure that the transistor can dissipate whatever power it has to under worst-case conditions, that is, with no signal. It can generally be stated that a direct-coupled class A transistor amplifier must be capable of dissipating an amount of power equal to at least twice its AC output power capability.

Because of its relatively high power dissipation and low efficiency, direct-coupled class A amplifiers are limited to milliwatt power levels.

### Example 12.4

Given the circuit and characteristics of Fig. 12.13, determine $R_E$ and $R_B$ to yield a $Q$ point of 25 V and 1.25 A. A germanium transistor with negligible $V_{BE}$ drop is used. Draw the DC load line on the characteristics, and determine the value of emitter bypass capacitor for operation down to 200 Hz.

### Solution

The $Q$ point is located on the output characteristics, from which $I_B$ can be estimated at 35 mA. The voltage drop across $R_L$ is 20(1.25) = 25 V; $V_{CE}$ is also

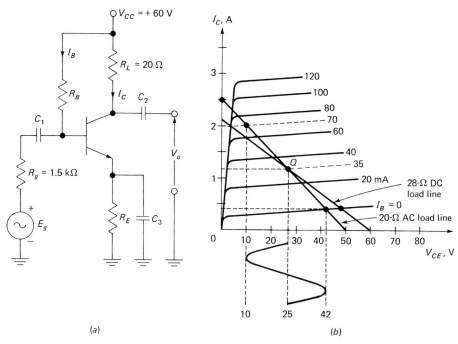

*Fig. 12.13  Circuit and characteristics for Examples 12.4 and 12.5.*

25 V, and hence there is 10 V across $R_E$. $R_E$ is, of course, required for temperature stabilization of the $Q$ point.

$$R_E \simeq \frac{10}{1.25} = 8 \ \Omega$$

Since $R_L + R_E = 28 \ \Omega$, the 28-$\Omega$ DC load line is drawn through the $Q$ point and the 60-V horizontal intercept. $R_B$ must supply 35 mA to the base; hence its value can be determined as follows:

$$R_B = \frac{60 - V_B}{35 \times 10^{-3}}$$

$$V_B \simeq V_E = I_E R_E = 10 \ \text{V}$$

$$R_B = \frac{60 - 10}{35 \times 10^{-3}} = 1.43 \ \text{k}\Omega$$

To adequately bypass $R_E$, $C_3$ must have a reactance no greater than 0.1 $R_E$ at 200 Hz:

$$\frac{1}{2\pi f C_3} = 0.8 \ \Omega$$

$$C_3 = \frac{1}{(6.28)(200)(0.8)} = 1{,}000\ \mu F$$

**Example 12.5**

For the circuit of Example 12.4, determine

a. Maximum available output power
b. Worst-case transistor power dissipation
c. Maximum efficiency
d. Average DC power dissipated by $R_L$

**Solution**

a. The AC load line is constructed to pass through the $Q$ point on the characteristics of Fig. 12.13b with a voltage-current ratio equal to the AC load resistance. Since $R_E$ is bypassed, the AC load is $R_L = 20\ \Omega$. How much of a signal swing we can handle is now limited by cutoff or saturation. On the cutoff side we can go as far as $I_B = 0$ or even below; however, we will limit our swing to $I_B = 0$, yielding a net 35-mA variation from the quiescent value of $I_B$. If we swing 35 mA on the other side of the $Q$ point, the peak base current is 70 mA; this is still safely away from the saturation region. We therefore assume a maximum base current swing of 35 mA on each side of the $Q$ point. An undistorted base current waveform is to be expected, due to the relatively large internal resistance of the signal generator.

The corresponding output voltage and current swings are read from the characteristics as

$\Delta V_{CE} = 42 - 10 = 32$ V
$\Delta I_C = 2 - 0.4 = 1.6$ A

The maximum available AC output power is

$$P_{o(\max)} \simeq \frac{32(1.6)}{8} = 6.4\ W \qquad (12.1a)$$

b. Worst-case power dissipation occurs halfway along the DC load line, where $V_{CE} = 30$ V and $I_C = 1.07$ A, yielding $P_D = 30(1.07) = 32.2$ W. Note that this is significantly greater than twice the AC power out; this is because we have limited the signal swing to avoid excessive waveform distortion.

c. The 60-V DC source supplies an average $I_C = 1.25$ A and an average $I_B = 35$ mA. The total average current is 1.285 A, yielding

$P_{DC} = 60(1.285) = 77$ W

The efficiency is

$$\eta = \frac{P_o}{P_{DC}} \tag{12.7}$$

$$\eta = \frac{6.4}{77} = 8.3 \text{ percent}$$

d.  The DC load current is 1.25 A; hence the DC load power is $1.25^2(20) = 31.25$ W.

The poor efficiency of the direct-coupled class A amplifier is primarily due to the large amount of DC power that the load must dissipate. The transformer-coupled class A circuit of Fig. 12.14a eliminates this problem because the transformer keeps DC out of the load. Ideally, the DC current through the transformer primary does not develop any power at all; in practice, there is a small DC resistance which does absorb some power, although not nearly to the same extent of a direct-coupled load.

Let us assume that the transistor in the circuit of Fig. 12.14a is ideal in every respect, but that a 2.5-A current limit is necessary. The DC conditions must be established first. If the transformer is ideal, it has no DC resistance and hence no DC voltage drop. Thus all 50 V from the supply is dropped across the transistor. The $Q$ point is at $V_{CE} = 50$ V and some value of $I_C$ determined by the input drive. If $I_{C(MAX)} = 2.5$ A, we may want to bias at $I_C = 1.25$ A to allow maximum swing on each side of the $Q$ point. The DC load is 0 Ω; hence the DC load line is vertical, along $V_{CE} = 50$ V, as shown on the graph of Fig. 12.14b.

The AC load line is established by joining the $Q$ point with the 2.5-A collector current intercept, as shown on the graph. This yields a load line whose

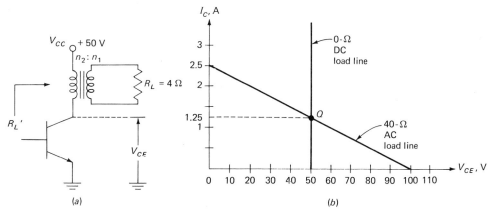

Fig. 12.14  Circuit and graphic construction for transformer-coupled class A amplifier.

ohmic value is $50/1.25 = 40\,\Omega$ and which cuts the $V_{CE}$ axis at 100 V. The 40 $\Omega$ represents the dynamic resistance $R'_L$ seen from the transformer primary. Since the actual load $R_L$ is equal to 4 $\Omega$, the transformer's turns ratio must be selected to yield the proper impedance transformation. Generally we can write

$$\frac{R'_L}{R_L} = \left(\frac{n_2}{n_1}\right)^2 \tag{12.12}$$

where $n_2/n_1$ is the transformer's turns ratio;

$$\frac{40}{4} = \left(\frac{n_2}{n_1}\right)^2$$

$$\frac{n_2}{n_1} = \frac{\sqrt{10}}{1} = \frac{3.16}{1}$$

This transformer can be specified using the notation 40/4.

It may be proper to question how $V_{CE}$ can reach 100 V when the DC supply is only 50 V. This point calls for an understanding of basic transformer action. First, the DC voltage across the ideal transformer primary always equals zero. This is true no matter how much current flows. Voltage appears only when the current varies. Suppose that a signal is applied, causing $I_C$ to rise; a voltage is developed across the transformer primary equal to the product of the changing component of $I_C$ and the transformer's dynamic resistance. Thus, when $I_C$ has risen from its quiescent level to the maximum 2.5 A, the changing component of $I_C$ is 1.25 A, and the voltage is $1.25(40) = 50$ V. This voltage is positive on the $V_{CC}$ side of the transformer primary. At this instant

$$V_{CC} = V_{R'_L} + V_{CE}$$
$$50 = 50 + V_{CE}$$
$$V_{CE} = 0$$

During the other half of the cycle, however, $I_C$ drops; this means that the changing component of $I_C$ reverses polarity, and hence the voltage across $R'_L$ also changes polarity. When $I_C$ reaches its minimum value, we have

$$V_{CC} = V_{R'_L} + V_{CE}$$
$$50 = -50 + V_{CE}$$
$$V_{CE} = 100 \text{ V}$$

The circuit can therefore provide a total usable voltage swing that is twice the supply voltage; care must be taken, of course, to ensure that the transistor's voltage breakdown rating is not exceeded. The maximum load power is

$$P_{o(\text{max})}: \quad \frac{100(2.5)}{8} = 31.25 \text{ W} \tag{12.1a}$$

The average power from the DC supply (for maximum load power) can be determined using the technique of Fig. 12.12. The peak power is $50(2.5) = 125$ W, and the minimum power is 0. Since the curve is symmetric, the average power is halfway between minimum and peak, 62.5 W. Under these conditions, maximum efficiency is achieved:

$$\eta = \frac{P_o}{P_{DC}} = \frac{31.25}{62.5} = 50 \text{ percent} \tag{12.7}$$

The higher efficiency, compared to direct coupling, is a direct consequence of using a transformer to eliminate waste of DC power in the load.

### Example 12.6

The class A amplifier of Fig. 12.15a is to feed an 8-$\Omega$ load at a fixed frequency of 400 Hz. A data sheet for the 2N2139 transistor is available in Appendix 10. The transformer is 75 percent efficient, has a DC primary resistance of 1.2 $\Omega$, and can handle up to 10 W; its frequency response extends down to 150 Hz, and the ratio of primary to secondary impedance is 16:8. The transistor is mounted on a heat sink which provides an effective case to ambient thermal resistance of 0.8 °C/W. The maximum ambient temperature is 50°C.

a. Determine the $Q$ point, allowing 6-V DC drop across $R_E$ and the transformer's primary DC resistance.
b. Draw the AC load line and determine the maximum signal power that can be delivered to the load.
c. Compute $R_E$ and $R_B$.
d. Draw the DC load line and maximum power-dissipation hyperbola, and compare your operating region with the manufacturer's "Safe operating areas."

### Solution

a. If 6 V is to be dropped across $R_E$ and the transformer's primary DC resistance, the quiescent value of $V_{CE}$ is 16 V. (Since this is a PNP transistor, $V_{CE} = -16$ V; however, all voltages are treated as positive here to keep the calculations as simple as possible.) The total AC voltage swing can be twice this, that is, 32 V. This is safely within the transistor's $BV_{CEO}$ rating of 45 V. Since the dynamic resistance looking into the transformer primary is 16 $\Omega$, there is a maximum $I_C$ swing of $32/16 = 2$ A. The quiescent collector current is therefore established at 1 A.

b. The AC load line is drawn on the graph of Fig. 12.15b through the $Q$ point with a voltage-current ratio of 16 $\Omega$. The horizontal and vertical intercepts are 32 V and 2 A, respectively.

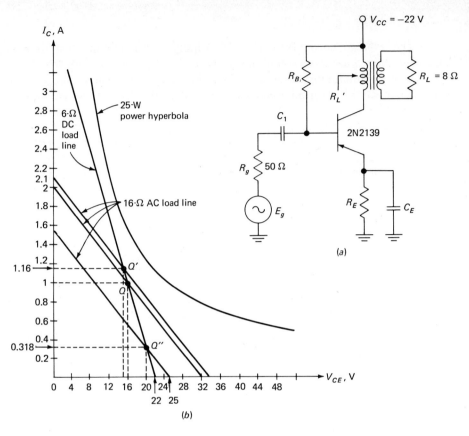
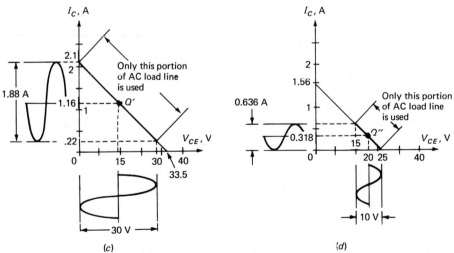

Fig. 12.15  Circuit and graphic constructions for Examples 12.6 and 12.7.

Maximum signal power occurs when all the available swing is utilized. If we could neglect saturation and cutoff, the power to $R'_L$ would be

$$P'_{o(\text{max})} = \frac{\Delta V'_o \Delta I'_o}{8} = \frac{32(2)}{8} = 8 \text{ W} \tag{12.1b}$$

where $\Delta V'_o$ and $\Delta I'_o$ are the voltage and current swings for $R'_L$. Since the transformer is 75 percent efficient, $R_L$ does not receive the total power delivered to $R'_L$; instead, the following applies:

$$P_o = \eta_T P'_o \tag{12.13}$$

where $\eta_T$ is the transformer efficiency. This yields

$$P_o = 0.75 P'_o = 0.75(8) = 6 \text{ W}$$

Because of saturation and cutoff, however, we should expect $P_o$ to be somewhat less than 6 W.

c. The DC drop across the transformer primary is the product of the 1-A current and 1.2-Ω DC resistance—1.2 V. This leaves 4.8 V for $R_E$:

$$R_E = \frac{V_E}{I_E} \simeq \frac{4.8}{1} = 4.8 \text{ Ω}$$

To determine $R_B$, $I_B$ is required. If we locate the $Q$ point on the "Collector characteristics" in the data sheet, the corresponding value of $I_B$ is read at about 28 mA.[1] The voltage across $R_B$ is

$$V_{R_B} = V_{CC} - V_{BE} - V_{R_E}$$

$V_{BE}$ can be estimated from the "Base-emitter voltage versus Base current" curves. For $I_B \simeq 28$ mA, $V_{BE} \simeq 0.5$ V. Although these curves are given for $V_{CE} = 2$ V, they can be applied, with minimal error, to other values of $V_{CE}$, yielding

$$V_{R_B} = 22 - 0.5 - 4.8 = 16.7 \text{ V}$$
$$R_B = \frac{V_{R_B}}{I_B} \simeq \frac{16.7}{28 \times 10^{-3}} = 600 \text{ Ω}$$

d. The DC resistance seen by the transistor is due to $R_E$ and the transformer; hence $R_{DC} = 1.2 + 4.8 = 6$ Ω. The 6-Ω DC load line is shown on the graph of Fig. 12.15b.

---

[1] Our $V_{CE} = -16$ V; the horizontal axis on the characteristics, however, is labeled in terms of the percent of rated $BV_{CES}$. $BV_{CES}$ is given elsewhere in the data sheet as $-60$ V; hence the horizontal ordinate of the $Q$ point is $(16/60)(100) = 26.7$ percent. The intersection of this horizontal ordinate and the vertical ordinate ($I_C = 1$ A) is the $Q$ point; the base current is then read as 28 mA.

The maximum allowable transistor power dissipation can be determined using Eq. (12.9a):

$$P_D = \frac{T_j - T_a}{\theta_{ja}} \tag{12.9a}$$

$T_{j(\text{max})}$ is given in the data sheet as 100°C; the highest ambient temperature is 50°C. The junction-ambient thermal resistance is the sum of $\theta_{jc}$ (given in the data sheet as 1.2 °C/W) and $\theta_{ca}$ (0.8 °C/W). We can therefore write

$$P_D = \frac{100 - 50}{1.2 + .8} = \frac{50}{2}$$
$$P_D = 25 \text{ W}$$

The maximum power hyperbola has been constructed on the graph. Our $Q$ point is below the curve; hence safe operation should result, provided the $Q$ point does not shift to a higher power-dissipation region. It is important to note that $Q$ point shifts due to temperature or transistor replacement occur along the DC load line. In our case, an increase in $I_C$ would definitely increase the power to be dissipated by the transistor.

Examination of the "Safe operating areas" provided in the data sheet indicates that our operating region is well within the maximum reliability area and that destructive voltage breakdown will not occur.

## Example 12.7

Estimate $Q$ point shifts due to temperature and transistor replacement in Example 12.6. What is the available power in each case?

### Solution

The $Q$ point shifts are determined by estimating the maximum and minimum values of collector current. The following expression for $I_C$ will be used:

$$I_C = \alpha \left[ \frac{V_{CC}R_B/R_1 + I_{CBO}R_B - V_{BE}}{R_E + R_B(1 - \alpha)} \right] + I_{CBO} \tag{8.4}$$

Note that $\alpha = -h_{FB}$; also $R_B = R_1$. Assuming that all resistors and $V_{CC}$ remain constant,

$$\overline{I_C} = -\overline{h_{FB}} \left[ \frac{V_{CC} + \overline{I_{CBO}}R_B - V_{BE}}{R_E + R_B(1 + \overline{h_{FB}})} \right] + \overline{I_{CBO}} \tag{a}$$

and

$$I_C = -h_{FB} \left[ \frac{V_{CC} + \overline{I_{CBO}} R_B - \overline{V_{BE}}}{R_E + R_B(1 + h_{FB})} \right] + \overline{I_{CBO}} \quad (b)$$

Before proceeding, we must determine the extreme values of $h_{FB}$, $I_{CBO}$, and $V_{BE}$. Since temperature is a factor, the highest case temperature is computed first:

$$\overline{T_c} = \overline{T_a} + \overline{P_D \theta_{ca}}$$
$$\overline{T_c} = 50 + 16(0.8)$$
$$\overline{T_c} \simeq 63°C$$

a. $h_{FB}$: $h_{FB}$ is linearly related to $h_{FE}$; hence the maxima and minima of each occur at the same time. The maximum $h_{FE}$ at 25°C is 60; this applies for $V_{CE} = 2$ V and $I_C = 0.5$ A. Although we are operating at a higher current and voltage level, $h_{FE}$ tends to drop with current and rise with voltage. Hence a maximum $h_{FE}$ of 60 at 25°C is a safe assumption. The "$h_{FE}$ versus Temperature" graph indicates that $h_{FE}$ rises with temperature for $I_C = 0.5$ A, but drops with temperature for $I_C = 2$ A. To be on the safe side, we assume $h_{FE}$ to rise in accordance with the $I_C = 0.5$ A curve; this yields

$$\overline{h_{FE}} \text{ (when } T_c = 63°C) = 115 \text{ percent} \times \overline{h_{FE}} \text{ at } 25°C$$
$$\overline{h_{FE}} = 1.15(60) = 69$$
$$\overline{h_{FB}} = \frac{-\overline{h_{FE}}}{\overline{h_{FE}} + 1} = \frac{-69}{70} = -0.985$$

The minimum $h_{FE}$ is 30 when $I_C = 0.5$ A and 15 when $I_C = 2$ A. These values apply for $T_c = 25°C$. Although our specific value of $I_C$ is 1 A, using the 2-A value provides a greater safety margin:

$$\underline{h_{FE}} = 15$$
$$\underline{h_{FB}} = \frac{-\underline{h_{FE}}}{\underline{h_{FE}} + 1} = \frac{-15}{16} = -0.94$$

b. $V_{BE}$: A maximum $V_{BE}$ of 1.2 V is given in the data sheet. Again, to be on the safe side, we will use this value even though it is unlikely that $V_{BE}$ would ever be this high in our case:

$$\overline{V_{BE}} = 1.2 \text{ V}$$

$V_{BE}$ generally drops with temperature at the rate of $\simeq 2.5$ mV/°C. The 25°C value of $V_{BE} \simeq 0.5$ V; hence at 63°C (our highest case temperature), $V_{BE}$ must have dropped by $2.5(63 - 25) = 95$ mV, to yield $\underline{V_{BE}} = 0.5 - 0.095 \simeq 0.4$ V. Actually, the $V_{BE}$ variation hardly matters here, since both the $V_{R_E}$ and $V_{R_B}$ drops are much larger than $V_{BE}$.

c. $I_{CBO}$: Since $I_{CBO}$ is very small, the simplest assumption is $I_{CBO} = 0$. The maximum value of $I_{CBO}$ is given as 5 mA; this applies when the voltage is maximum and $T_c = 71°C$. Here we are operating at less than the maximum voltage and less than the specified temperature. It is difficult to make a voltage adjustment, because the voltage dependence of $I_{CBO}$ is not specified. The temperature correction, however, can be easily made. The "$I_{CBO}$ versus Temperature" curve indicates that $I_{CBO}$ doubles for every 10°C rise; hence $I_{CBO}$ should be about 2.5 mA at 61°C (10°C below $T_c = 71°C$). Our maximum case temperature is 63°C. Therefore a safe estimate for $\overline{I_{CBO}}$ is 2.7 mA.

Substituting the appropriate values in (a) and (b), we get

$$\overline{I_C} = 0.985 \left[ \frac{15 + (2.7 \times 10^{-3})(600) - 0.4}{4.8 + 600(0.015)} \right] = 2.7 \times 10^{-3}$$

$$\overline{I_C} = 1.16 \text{ A}$$

$$\underline{I_C} = 0.94 \left[ \frac{15 + 0 - 1.2}{4.8 + 600(0.06)} \right]$$

$$\underline{I_C} = 0.318 \text{ A}$$

The $Q$ points corresponding to $\overline{I_C}$ and $\underline{I_C}$ are located on the DC load line in Fig. 12.15b; these are $Q'$ and $Q''$ respectively. The slope of the AC load line is always the same, but it must cross the $Q$ point, and hence different AC load lines for $Q'$ and $Q''$ are constructed.

The available power is now computed for each $Q$ point. Saturation and cutoff are neglected for simplicity:

$Q'$: $V_{CE} \simeq 15$ V, $I_C = 1.16$ A  (see Fig. 12.15c)
Maximum $\Delta V$:  $2(15) = 30$ V
Maximum $\Delta I$: $2(2.1 - 1.16) = 1.88$ A

$$P'_o = \frac{30(1.88)}{8} = 7.05 \text{ W}$$

$P_o = \eta_T P'_o = 0.75(7.05) = 5.3$ W  (12.13)

$Q''$: $V_{CE} \simeq 20$ V, $I_C = 0.318$ A  (see Fig. 12.15d)
Maximum $\Delta V$: $2(25 - 20) = 10$ V
Maximum $\Delta I$:  $2(0.318) = 0.636$ A

$$P'_o = \frac{10(0.636)}{8} = 0.795 \text{ W}$$

$P_o = \eta_T P'_o = 0.75(0.795) = 0.597$ W  (12.13)

## Problem Set 12.4

12.4.1. Given the circuit of Fig. 12.16 in which $V_{CC} = 15$ V, $R_E = 0 \, \Omega$, $R_L = 1$ k$\Omega$, $h_{FE} = 50$, determine the maximum AC power available and the

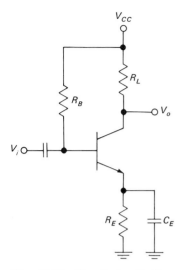

Fig. 12.16  Circuit for Problem Set 12.4.

value of $R_B$. Construct a graph of output voltage versus output current, locate the $Q$ point, and draw the load line. Neglect $V_{BE}$ and $I_{CBO}$.

12.4.2. Repeat Prob. 12.4.1 when $R_E = 1$ k$\Omega$. Draw the AC and DC load lines. Select the $Q$ point halfway along the DC load line.

12.4.3. The following data apply to the circuit of Fig. 12.16: $V_{CC} = 40$ V, $R_E = 5$ $\Omega$, $h_{FE} = 20$, $R_B = 500$ $\Omega$. Both $V_{BE}$ and $I_{CBO}$ are negligible. Determine the $Q$ point and value of $R_L$ that will produce maximum AC power output.

12.4.4. What is the worst-case power dissipation for the transistor in Prob. 12.4.3? The average power supplied by $V_{CC}$ when the output power is maximum? The maximum efficiency?

12.4.5. Using the data and the value of $R_L$ calculated for Prob. 12.4.3, determine the $Q$ point when $h_{FE}$ drops to 15 and when it rises to 30. What is the maximum AC power output in each case?

12.4.6. A 15-V supply is available for the circuit of Fig. 12.16; $h_{FE} = 40$, and a 5-V drop across $R_E$ is required. If $V_{BE} = 0.7$ V and $I_{CBO}$ is negligible, determine the values of $R_L$, $R_E$, and $R_B$ required to produce 1 W of AC power output to $R_L$. Construct a graph showing the load line and $Q$ point.

12.4.7. For the circuit of Fig. 12.17, assume $V_{CC} = 50$ V, $R_E = 0$, the maximum collector current rating is 2 A, $BV_{CEO} = 150$ V, and $I_{C(MIN)} = V_{CE(MIN)} = 0$.

   a. Determine the maximum AC power deliverable to $R_L$ if the transformer is lossless.

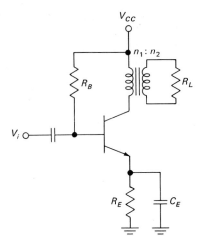

Fig. 12.17  Circuit for Problem Set 12.4.

    b. What must $n_1/n_2$ be if $R_L = 10\ \Omega$ and maximum power out is desired?

    c. If the transformer has an efficiency of 80 percent and a turns ratio $n_1/n_2 = 2:1$, what value of $R_L$ is needed to receive 4 W under conditions of maximum output voltage swing?

12.4.8.  For the circuit of Fig. 12.17, $R_E = 0$, $n_1/n_2 = 5$, and $R_L = 4\ \Omega$. If $V_{CE(\text{MAX})} = 150$ V, what is the maximum available power (neglect saturation and cutoff) that the transistor can deliver to the transformer primary? What is the Q point?

### 12.4b  Class B Circuits

    The class A amplifier, because of its poor efficiency, is used mainly when large amounts of power are not required, typically for driver stages that feed a last "power" stage.

    A more practical circuit for power stages is the class B push-pull amplifier of Fig. 12.18a. We will discuss this circuit first by assuming ideal transistors in which $V_{BE}$ and all cutoff currents are zero. The problems of real components will be introduced later. The circuit operates on the push-pull principle in which $Q_1$ and $Q_2$ conduct during opposite halves of the AC cycle, each for 180°.

    Consider the input voltage waveform of Fig. 12.18b. This is the voltage applied to the primary of $T_1$. Before $t_o$, the input voltage is zero; hence the base of $Q_1$ and the base of $Q_2$ are both at ground potential. Neither transistor conducts, and therefore $I_{B1} = I_{B2} = I_{C1} = I_{C2} = I_{E1} = I_{E2} = 0$. There is no

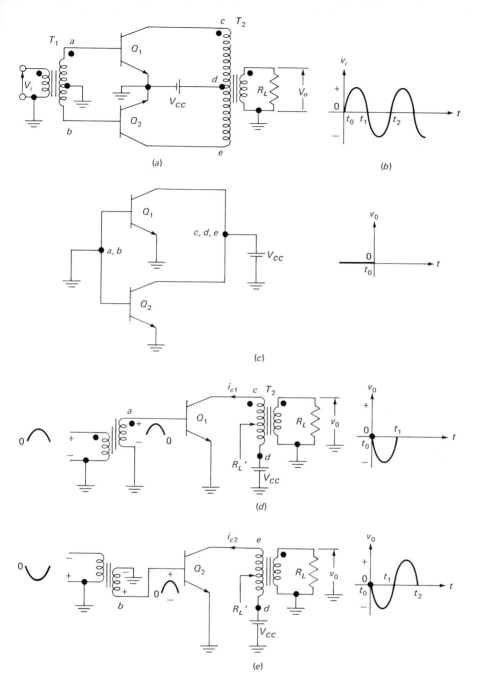

Fig. 12.18  (a) Class B push-pull amplifier circuit; (b) input voltage waveform; (c) equivalent circuit when $v_i = 0$ ($t < t_o$); (d) equivalent circuit when $v_i$ is positive-going; $Q_2$ does not conduct; (e) equivalent circuit when $v_i$ is negative-going; $Q_1$ does not conduct.

voltage across either primary winding of $T_2$ and no output. An equivalent circuit and graph of $v_o$ for $t < t_o$ are given in Fig. 12.18c.

At $t_o$, the input voltage $v_i$ begins to rise; this results in a positive-going voltage at the base of $Q_1$ and a negative-going voltage at the base of $Q_2$, relative to ground. The base-emitter junction of $Q_1$ is forward-biased, causing $Q_1$ to conduct; $Q_2$, however, is cut off, since its base-emitter junction is reverse-biased. A simplified equivalent circuit for this condition is given in Fig. 12.18d. The collector current of $Q_1$, which follows the input-signal variations, develops a voltage across the primary winding of $T_2$; this voltage is proportional to $i_{c1}$ and the reflected resistance $R'_L$. The actual output voltage is taken across $R_L$ (the secondary of $T_2$); hence its magnitude depends on the turns ratio of $T_2$. While $Q_1$ conducts, $Q_2$ is off and no current flows through the lower half of input and output transformers.

At $t_1$, the instantaneous input voltage $v_i$ reverses polarity, and the equivalent circuit of Fig. 12.18e applies. $Q_2$ now conducts, while $Q_1$ turns off. An output voltage is developed across $R_L$, as indicated on the accompanying graph.

Note that each transistor conducts for only one-half the time. The input center-tapped transformer $(T_1)$ is a convenient device for producing opposite polarity signals at the bases of $Q_1$ and $Q_2$. The output transformer combines the signals from $Q_1$ and $Q_2$, thus producing a composite output across $R_L$ which is an amplified version of the input signal. There is no DC current through either transformer, therefore no saturation of the magnetic core. Since the AC signal does not "ride" on a DC level, power dissipation is substantially lower than for the class A amplifier.

To consider these aspects more fully, the load line method of Fig. 12.19 will be used. Here $R'_L = 32$ Ω and $V_{CC} = 16$ V. $R'_L$ is the dynamic resistance reflected across each half of the output transformer's primary winding. Since the resistance ratio is proportional to the turns ratio squared, it follows that the total dynamic resistance reflected across both halves of the transformer's primary winding is $4R'_L$ (twice the turns, 4 times the resistance). Transformer $T_2$ is therefore specified as $128CT/R_L$, to indicate that when the secondary is loaded with a specific value of $R_L$, the total resistance reflected on the primary side is 128 Ω. $CT$ simply stands for center tap.

With no signal, $I_{C1} = 0$ and $V_{CE1} = 16$ V; this is the Q point for $Q_1$ and is so indicated in the figure. Note that the Q point represents zero power; hence the transistor is not required to dissipate any power when there is no signal, a distinct advantage over the class A amplifier. The maximum current is $V_{CC}/R'_L = 16/32 = 0.5$ A. When $Q_1$ conducts, $I_{C1}$ rises from zero toward 0.5 A while $V_{CE}$ drops from 16 V toward zero; the waveforms shown on the graph apply when the full swing is utilized. During the next half-cycle, $I_{C2}$ rises from zero toward 0.5 A while $V_{CE2}$ drops from 16 toward 0 V. The maximum signal power available is

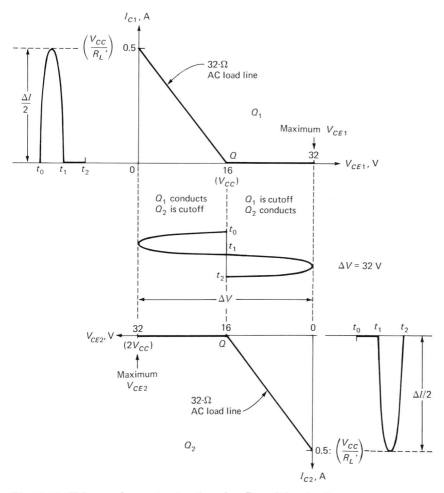

*Fig. 12.19* Voltage and current swings in a class B amplifier circuit.

$$P_{o(\text{max})} = \frac{\Delta V \, \Delta I}{8} \tag{12.1a}$$

$$P_{o(\text{max})} = \frac{2V_{cc}(2V_{cc}/R'_L)}{8}$$

$$P_{o(\text{max})} = \frac{V_{cc}^2}{2R'_L} \tag{12.14}$$

which in our case works out to $16^2/64 = 4$ W.

The maximum voltage across either transistor occurs during cutoff. For example, at the instant of maximum $Q_1$ conduction, $V_{CE1} = 0$ and the voltage between points $d$ and $c$ (see Fig. 12.18a) is $V_{CC}$. At the same time, $Q_2$ is not conducting, but a voltage is developed between points $e$ and $d$ (due to the coupling action of the transformer) such that $v_{ed} = v_{dc} = V_{CC}$. The voltage from collector to emitter of $Q_2$ is equal to $v_{ed} + V_{CC} = 2V_{CC}$. The same situation applies to $Q_1$; that is, $V_{CE1} = 2V_{CC} = 32$ V at the instant of maximum $Q_2$ conduction. This is indicated on the graphs of Fig. 12.19.

Maximum power occurs halfway along the load line; hence maximum instantaneous power dissipation for each transistor is

$$P_{d(\max)} = \frac{V_{CC}}{2} \frac{V_{CC}}{2R'_L}$$

$$P_{d(\max)} = \frac{V_{CC}^2}{4R'_L} \qquad (12.15)$$

which in our case is 2 W.

We can now make some preliminary conclusions regarding class B push-pull amplifiers. These conclusions are based on the assumption of ideal transistors and maximum voltage and current swings:

a. *Output power:* The maximum available total output power is given by Eq. (12.14).
b. *Reverse voltage across each transistor:* The maximum instantaneous reverse voltage across each transistor occurs during its nonconducting half-cycle and is equal to twice the supply voltage.
c. *Power dissipation for each transistor:* The peak instantaneous power dissipation for each transistor occurs halfway along the load line and is given by Eq. (12.15). Note that $p_{d(\max)}$ is one-half of $P_{o(\max)}$.

## Problem Set 12.5

12.5.1. An ideal amplifier is capable of delivering 4 W to the load it feeds. What is the highest value of instantaneous power dissipation for each transistor in

    a. Class A operation
    b. Class B operation

12.5.2. Given a supply voltage of 15 V and $R_E = 0$, what voltage rating must a transistor have for

    a. Class A direct-coupled amplifier
    b. Class A transformer-coupled amplifier
    c. Class B transformer-coupled, push-pull amplifier

12.5.3. A class B amplifier of the type in Fig. 12.18a and $V_{CC} = 40$ V and $T_2$: 40CT/4. Assuming $R_L = 4\ \Omega$ and ideal transistors and transformers,

a. Determine $R'_L$
b. Locate the $Q$ point and draw the AC load line
c. Find the transistor voltage and current with
   1. No signal
   2. Maximum signal in conduction
   3. Maximum signal in cutoff
d. Determine the maximum AC power deliverable to $R_L$
e. Find the peak instantaneous device dissipation
f. Determine the minimum breakdown voltage rating for each transistor

We shall now concentrate on transistor power-dissipation considerations. Although peak instantaneous power dissipation occurs midway along the load line, it is not always necessary that each transistor have a power rating equal to $p_{d(\max)}$. If, in fact, the signal period $T$ is much shorter than the transistor's thermal time constant $\tau_j$, the junction temperature cannot follow the instantaneous power variations; instead it responds only to the average power.

For example, if the lowest frequency to be amplified is 20 Hz and $\tau_j$ is 15 ms, it follows that $T = \frac{1}{20} = 50$ ms. Since $T$ is greater than $\tau_j$, the transistor's junction temperature responds to the peak power dissipation, $p_{d(\max)}$. If, on the other hand, the lowest frequency is 2,000 Hz, $T = 0.5$ ms, which is short compared to 15 ms. Now the junction temperature cannot follow $p_d$, and the only protection required is against the average power dissipation $P_D$.

In practice, a typical sound signal may include frequencies low enough to result in $T \gg \tau_j$, but it is unlikely that these frequencies involve a significant amount of power. Usually, most of the power is concentrated in the midfrequencies with the low- and high-frequency content making up a small portion of the whole spectrum. Audio amplifiers used for sound reproduction are usually tested for maximum power output at a standard frequency of 1,000 Hz; they are not normally expected to deliver maximum output power at either the low or the high end of their frequency range. We can therefore conclude that in the majority of cases the junction temperature responds to $P_D$ (average) instead of $p_{d(\max)}$ (peak). This can make quite a difference in the transistor and heat sinking required, as we shall soon see.

It should be emphasized that a transistor circuit designed to dissipate $P_D$ can be damaged if low-frequency signals whose period is comparable to $\tau_j$ are used, since the junction temperature can then follow the instantaneous $p_d$. Similarly, damage can result if the circuit is tested with high-frequency signals at maximum output power. Although a discussion of this aspect is beyond the scope of this text, we can briefly state that at higher frequencies, the load is no longer resistive but also involves reactance with its inherent phase shift. The reactive load may shift transistor operation to a higher power region, where overheating results.

We will now consider the power dissipation for the voltage and current swings of Fig. 12.19. Graphs of $v_{ce1}$, $i_{c1}$, and $p_{d1}$ versus time are shown in Fig. 12.20. The $p_{d1}$ graph represents the instantaneous power dissipation for $Q_1$; each point on this graph is the product of $v_{ce1}$ and $i_{c1}$ at a specific instant. The highest value of $p_{d1}$, which is equal to 2 W, occurs at the halfway point on the AC load line. Note that during the interval from $t_1$ to $t_2$, $Q_1$ does not conduct, and hence $p_{d1} = 0$. A similar graph, of course, applies to $Q_2$, with the exception that the nonconduction interval is from $t_o$ to $t_1$. The average dissipation $P_D$ is obviously much less than $p_{d(\max)}$. Since it cannot be computed exactly without the use of

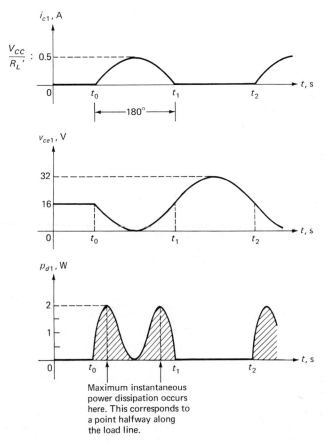

Fig. 12.20 *Instantaneous power dissipation obtained by graphic multiplication of current and voltage.*

TABLE 12.1.

|  | $K = 1$ | $K = 0.8$ |
|---|---|---|
| $\Delta V$ | 32 V | 0.8(32) = 25.6 V |
| $\Delta I$ | 1 A | 0.8(1) = 0.8 A |
| $P_o = \dfrac{\Delta V \, \Delta I}{8}$ | 4 W | 2.56 W |
| $p_{d(\max)}$ | 2 W | 2 W |

calculus, the derivation is carried out in Appendix 11.4, and the results are given here:

$$P_D = \frac{0.068 V_{cc}^2}{R'_L} \tag{12.16}$$

when full swing is utilized. For $V_{CC} = 16$ V and $R'_L = 32\,\Omega$, $P_D$ for each transistor is $(0.068)(16^2)(32) = 0.544$ W. We can therefore get a total output power of 4 W with transistors that can dissipate only 0.544 W each! A class A amplifier would require an 8-W transistor to deliver the same power. We should not get overly optimistic, however, since the above applies only as long as full swing is utilized, a condition which is not always guaranteed to occur.

What happens when less than the full swing is utilized? Obviously, less output power is delivered; power dissipation for the transistor is also affected. To consider this further we will define a new parameter $K$, which represents the ratio of actual swing to full swing. If full swing is utilized, $K = 1$; if no signal at all is applied, $K = 0$, and so on. Figure 12.21 represents a situation in which $K = 0.8$. Table 12.1 compares various parameters for $K = 1$ and $K = 0.8$. $P_D$ for any value of $K$ is derived in Appendix 11.4 and given here:

$$P_D = \frac{K V_{cc}^2}{R'_L}\left(\frac{1}{\pi} - \frac{K}{4}\right) \tag{12.17}$$

When $K = 1$, Eq. (12.17) simplifies to Eq. (12.16). For $K = 0.8$, $P_D = 0.755$ W, which is higher than the 0.544 W obtained for $K = 1$. To show more clearly the dependence of $P_D$ on $K$, Eq. (12.17) is plotted in Fig. 12.22. To use this curve, the vertical-axis reading must be multiplied by the actual value of $V_{CC}^2/R'_L$. Examination of this curve indicates that worst-case $P_D$[1] occurs when $K$ is between 0.6 and 0.7; an actual differentiation carried out in Appendix 11.4 yields a more accurate value of $K = 0.636$. Since there is usually no way of knowing what $K$ will be in any particular case (it can vary from 0 to 1 depending on the applied

[1] Worst-case $P_D$ is the highest value of average power dissipation.

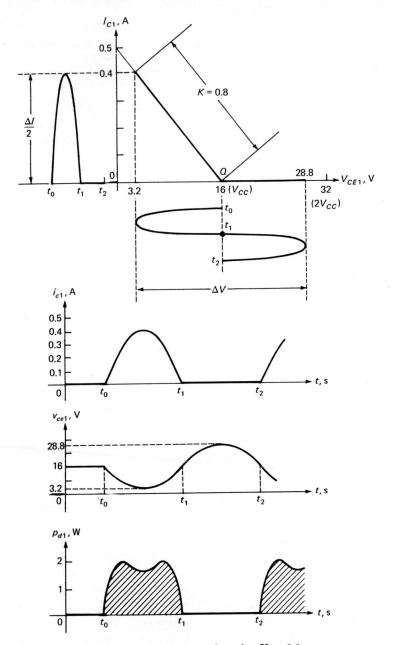

Fig. 12.21  Voltage, current, and power plots when K = 0.8.

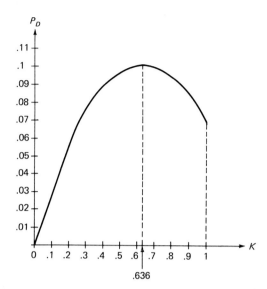

Fig. 12.22 Average power dissipation for each transistor in the circuit of Fig. 12.18a as a function of K. $P_D = K(V_{cc}^2/R'_L)(1/\pi - K/4)$. Multiply vertical axis readings by actual value of $V_{cc}^2/R'_L$.

signal), we must assume worst case for design purposes. For $K = 0.636$, $P_D = 0.1015 V_{cc}^2/R'_L$. For our earlier example in which $V_{cc} = 16$ V and $R'_L = 32$ Ω, worst-case $P_D$ works out to 0.812 W, as opposed to 0.544 W for $K = 1$.

## Problem Set 12.6

12.6.1. An ideal class B push-pull amplifier has $V_{cc} = 15$ V, $R'_L = 15$ Ω. Assume $K = 1$. Determine

    a. The Q point and draw the AC load line
    b. The voltage and current with

        1. No signal
        2. Maximum signal in conduction
        3. Maximum signal in cutoff

    c. Maximum AC power output
    d. The peak instantaneous transistor power dissipation
    e. The average transistor dissipation
    f. The minimum transistor voltage breakdown rating
    g. The output transformer specification if $R_L = 4$ Ω
    h. The transistor power-dissipation rating ($\tau_j \gg T$)

**498** Solid State Electronic Circuits

12.6.2. Repeat Prob. 12.6.1, assuming $K = 0.5$.

12.6.3. What is the relationship between the total AC output power and $p_{d(\max)}$ when $K = 1$? Compare with class A.

12.6.4. What is the relationship between the total output power and worst-case $P_D$? For what value of $K$ does worst-case $P_D$ occur?

Efficiency is one of the main advantages of class B amplifiers. Since a large amount of quiescent power is not required, it follows that there should be an improvement over class A circuits. To pursue this further, the power supplied by the external source will be determined; the graphic approach of Fig. 12.23 applies when $K < 1$, which is the most general case. Each of the waveforms in Fig. 12.23 is now described.

*DC SUPPLY VOLTAGE.* This is always equal to a constant, $V_{CC}$, regardless of how much current is drawn.

*DC SUPPLY CURRENT.* It is assumed here that no signal is applied prior to $t_o$; hence neither transistor draws current until $t_o$. Beyond $t_o$ each transistor

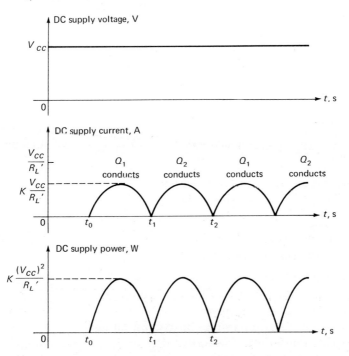

Fig. 12.23  Power supply waveforms for class B amplifier circuit.

conducts on alternate half-cycles. The power supply must provide current to both transistors and hence the *full-wave rectified* waveform. In general, the maximum instantaneous current is $KV_{CC}/R'_L$.

DC SUPPLY POWER. Since power is the product of voltage and current, this graph is simply the product of the voltage and current graphs just described. The peak current is $KV_{CC}/R'_L$; hence the peak power is $KV_{CC}^2/R'_L$.

The average value of any full-wave rectified sinusoid is 0.636 of the peak value. Hence the average power supplied by the DC source is

$$P_{DC} = 0.636 \frac{KV_{CC}^2}{R'_L} \tag{12.18}$$

The AC power to the load is

$$P_o = \frac{K^2 V_{CC}^2}{2R'_L} \tag{12.14a}$$

and the efficiency is

$$\eta = \frac{P_o}{P_{DC}} \tag{12.7}$$

$$\eta = \frac{K^2 V_{CC}^2/(2R'_L)}{0.636 K V_{CC}^2/R'_L} = 0.785K \tag{12.19}$$

It is evident that the highest efficiency occurs when $K = 1$, at which point $\eta = 78.5$ percent. This is the highest *theoretical* efficiency of a class B amplifier based on a sinusoidal signal, ideal transistors, and full utilization of voltage and current swings. Needless to say, this cannot be achieved in practice.

Distortion in class B push-pull circuits will be considered next. Distortion may be due to a number of different factors. As long as large signals are involved, we should expect distortion because of the transistors' nonlinear transfer characteristics. Such nonlinearities are present in the input circuit (the relationship between $V_{BE}$ and $I_B$ is nonlinear) and the output circuit ($h_{FE}$ is a function of the operating point). In class A circuits, these nonlinearities cause one-half of the output waveform to receive more or less amplification than the other half; in class B circuits, if the transistors have identical characteristics, both halves of the waveform are equally distorted. This has the effect of eliminating even harmonics; the net distortion is then due mainly to third-order harmonics.

Distortion produced by core saturation of the output transformer is also minimized in class B push-pull circuits. This is because current flow through each half of the primary winding is in opposite directions; hence the core receives no net DC magnetization, and operation in the saturation region is avoided.

Differences in the characteristics of the two transistors of a class B circuit may produce second-order distortion because of the unequal amplification received by each half of the signal waveform. The effects of unequal $h_{FE}$ may also escalate near the cutoff frequency. If $f_{h_{FE}}$ (the frequency at which $|h_{FE}|$ is down 3 dB from the midfrequency value) is not the same for both transistors, distortion in the frequency cutoff region increases. This is further complicated by unequal phase shift for each transistor in this frequency range. It is therefore quite important that both transistors be matched as closely as possible. Negative feedback is invariably used in good high-fidelity amplifiers to reduce the overall distortion.

An additional source of distortion in class B circuits is due to the "crossover" period when one transistor stops conducting before the other one is able to start conducting. The result is the "dead zone" shown in Fig. 12.24. The dead zone occurs because neither transistor can conduct until its base-emitter threshold voltage has been exceeded. The effect is less pronounced as the signal gets larger but must be corrected if faithful signal reproduction is to be achieved. Usually this is done by providing a small amount of forward bias to both transistors, even under no-signal conditions. This yields class AB operation, where each device conducts for slightly more than 180°; now no signal is required to overcome the threshold voltage, and hence no dead zone results. Figure 12.25 illustrates class AB biasing using composite characteristics for the input circuit.

The diagram of Fig. 12.26 shows a modified class B push-pull amplifier circuit in which the voltage-divider action of $R_1$ and $R_2$ produces the necessary

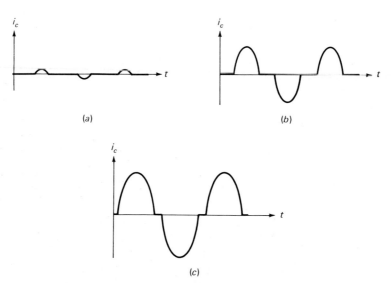

*Fig. 12.24* Crossover distortion when the signal is relatively (a) *small;* (b) *medium;* (c) *large.*

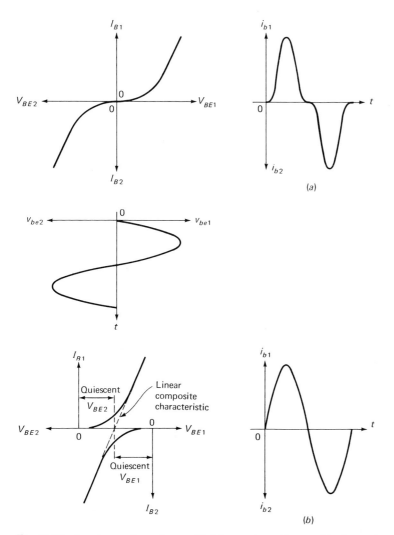

*Fig. 12.25 Input current waveforms with* (a) *no quiescent base-emitter forward bias;* (b) *quiescent base-emitter forward bias.*

bias to overcome each transistor's base-emitter threshold voltage. Often $R_2$ is temperature-dependent (thermistor) to stabilize the bias against temperature variations. An NTC thermistor is used to cause a reduction in forward base-emitter drive with temperature; this offsets the collector current's tendency to rise as the temperature increases. The addition of $R_2$ results in some loss of signal, since the input voltage developed across the secondary of $T_1$ must divide itself between the transistor's base-emitter junction and the parallel combination

**502** Solid State Electronic Circuits

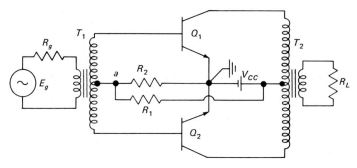

*Fig. 12.26* Transformer-coupled push-pull class AB amplifier circuit with forward bias to base-emitter junctions provided by $R_1$-$R_2$ voltage divider.

of $R_1$ and $R_2$. It would appear at first glance that AC bypassing of $R_2$ would solve this problem. This is not the case, however. If a bypass capacitor is connected across $R_2$, the polarity of voltage developed during any given half-cycle would tend to reverse-bias the opposite transistor when its turn came to conduct. As an example, suppose $Q_1$ is conducting; the capacitor charges up with a fast time constant because of the base-emitter junction's low resistance. Thus point $a$ acquires a negative DC voltage with respect to ground. When $Q_1$ stops conducting, the base of $Q_2$ is tied to the negative potential of point $a$, which reverse-biases $Q_2$, preventing it from conducting. The capacitor can only discharge with a long time constant, because neither base-emitter junction is conducting; hence the only available path is through $R_2$ and $R_1$. A bypass capacitor therefore cannot be used.

It may be advantageous to have a small unbypassed resistance in each emitter lead; this provides both DC and AC negative feedback. The DC feedback helps stabilize the quiescent collector current; the AC feedback improves the frequency response and reduces distortion.

Class B push-pull stages need not be driven from a center-tapped transformer. All that is really required are two inputs, each 180° out of phase with the other. The circuit of Fig. 12.27 illustrates one technique for accomplishing this. There are many other circuits that can be used, each with its own special characteristics. The RC coupling used here is usually less costly and yields better low-frequency response than transformer coupling.

$Q_1$ is a split-load phase inverter; it operates class A and produces two outputs, one each across $R_4$ and $R_3$. These opposite-phase outputs drive the class B push-pull circuit that follows. $R_3$ and $R_4$ are chosen to produce equal magnitude voltages. A small resistance may be placed in series with $C_3$ to raise the effective internal resistance at the emitter of $Q_4$ to make it comparable to that seen at the collector.

$C_2$ and $C_3$ couple the AC signals to $Q_2$ and $Q_3$, respectively. $D_1$ and $D_2$

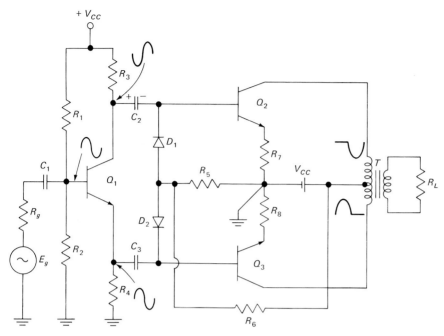

*Fig. 12.27* Class B push-pull amplifier driven from split-load phase inverter.

allow $C_2$ and $C_3$ to discharge during the half-cycle that $Q_2$ and $Q_3$ do not conduct. To illustrate this point, consider what happens when the voltage at the collector of $Q_1$ is positive-going: $Q_2$ conducts, $D_1$ is reverse-biased, and $C_2$ is charged, as shown in the figure. During the next half-cycle (when the collector voltage of $Q_1$ drops), $Q_2$ turns off, and $C_2$ discharges through the now-conducting $D_1$ and the low resistance of $R_5$. Without $D_1$ this could not happen, and $C_2$ would retain most of its charge, preventing $Q_2$ from conducting when its turn came once again. $D_2$ performs the same function as $D_1$; it is reverse-biased when $Q_3$ conducts, but comes on during the nonconducting half-cycle, allowing $C_3$ to quickly discharge.

$R_7$ and $R_8$ provide negative feedback for DC stability, frequency-response, and distortion improvement. $R_5$ and $R_6$ provide the forward bias needed to minimize crossover distortion.

## Example 12.8

The circuit of Fig. 12.26 is to be used as a servo amplifier at a fixed frequency of 400 Hz. A minimum of 6 W is required by the 600-$\Omega$ load. The supply is 12 V, and the maximum ambient temperature is 65°C. Both transistors are 2N2139. The internal resistance of the driving generator is 50 $\Omega$. Assume 75 percent efficient transformers with no DC resistances. Determine

a. The transformer specifications required to achieve the desired power. The input transformer is to be chosen on the basis of maximum power transfer, rather than minimization of distortion.
b. $R_2$, $R_1$, and the heat sinking required.
c. Power gain.
d. Efficiency.

**Solution**

a. The power available to the primary of $T_2$ must be greater than the 6 W required by the load. Since the transformer efficiency is 75 percent, the primary should receive $6/0.75 = 8$ W. $V_{CC} = 12$ V; hence the maximum theoretical useful voltage swing is 24 V. Based on this, the required current swing is computed as follows:

$$\text{Power to } T_2 = \frac{\Delta V \, \Delta I}{8}$$

$$8 = \frac{24 \, \Delta I}{8}$$

$$\Delta I = \frac{64}{24} = 2.66 \text{ A}$$

Each transistor must supply $\frac{1}{2} \Delta I$, which means that the maximum collector current should be at least 1.33 A. Since in practice neither the full voltage swing (because of saturation) nor the full current swing (because of cutoff) can be obtained, we should provide for a slightly larger current, say 1.5 A,

$$I_{C(\text{MAX})} = 1.5 \text{ A}$$

$$R'_L = \frac{V_{CC}}{I_{C(\text{MAX})}}$$

$$R'_L = \frac{12}{1.5} = 8 \, \Omega$$

The 2N2139 has a $BV_{CEO}$ rating of 45 V; hence no problem should result from our 24-V swing. Our maximum $I_C$ of 1.5 A is also well within the transistor's 3-A rating. The output transformer should be able to handle at least 8 W; the resistance across the full primary should be $4R'_L = 32 \, \Omega$, yielding an impedance ratio of $32CT/600$.

To select an input transformer on the basis of maximum power transfer, $h_{IE}$ is required. Examination of the "Collector characteristics–Common emitter" configuration for the 2N2139 (see Appendix 10) yields $I_B \simeq 55$ mA when $I_C = 1.5$ A. This value is read near the saturation region, since it is here that maximum current occurs. The "Base-emitter voltage versus Base current" graph yields $V_{BE} \simeq 0.65$ V when $I_B = 55$ mA. From this we can estimate $h_{IE}$:

$$h_{IE} \simeq \frac{0.65}{55 \times 10^{-3}} \simeq 12 \, \Omega$$

Since $h_{IE}$ depends on the operating point, the above is only a rough estimate of the effective input resistance. To achieve maximum power transfer from signal generator to the circuit, the secondary winding for each half of $T_1$ must reflect a resistance of 12 $\Omega$ plus $R_1 \| R_2$ when driven from a 50-$\Omega$ source. Since $R_1 \| R_2$ is normally small, the impedance specification for $T_1$ is $50/(48CT)$.

b. $R_1$ and $R_2$ are chosen to minimize crossover distortion. Since these are germanium transistors, a base-emitter quiescent voltage of 0.1 V is adequate. The parallel combination of $R_1$ and $R_2$ should be small enough not to waste much AC signal, but not so small that excessive current is drawn from the supply (which lowers the efficiency). Since $h_{IE} \simeq 12 \, \Omega$, we choose $R_2 = 3 \, \Omega$. Under no-signal conditions neither base of $Q_1$ or $Q_2$ draws appreciable current; hence the DC voltage at point $a$ in Fig. 12.26 can be computed as follows:

$$V_a = 0.1 = \frac{V_{cc} R_2}{R_1 + R_2} = \frac{12(3)}{3 + R_1}$$

$0.3 + 0.1 R_1 = 36$

$R_1 = 357 \, \Omega$

The quiescent current drawn from the power supply by the biasing resistors is $12/(357 + 3) = 33.3$ mA.

The heat sinking required must be based on the worst-case power dissipation. We must first decide whether protection against the peak instantaneous or average dissipation is required. Since the amplifier will operate only at 400 Hz, the signal period is $T = \frac{1}{400} = 2.5$ ms. The transistor has a thermal time constant of about 50 ms (see the last page of data sheet for the 2N2139), which means that the junction will not be able to follow the instantaneous variations of $p_d$, but will respond to the average power instead ($T \ll \tau_j$). The worst-case $P_D$ occurs for $K = 0.636$ and is equal to $0.1015 V_{cc}^2 / R_L' = 0.1015(144)/8 = 1.83$ W for each transistor. To provide some margin of safety, a 2-W worst-case dissipation will be assumed. The 2N2139's junction-case thermal resistance is 1.2 °C/W maximum; $T_{j(max)} = 100$°C. Hence

$$\theta_{ja} \leq \frac{T_{j(max)} - T_{a(max)}}{P_D}$$

$$\theta_{ja} \leq \frac{100 - 65}{2} = 17.5 \text{ °C/W}$$

$$\theta_{ca} = \theta_{ja} - \theta_{jc}$$
$$\theta_{ca} \leq 17.5 - 1.2 = 16.3 \text{ °C/W}$$

Heat sinking must therefore be provided for each transistor so that the total thermal resistance from case to ambient is less than 16.3 °C/W. If both transistors are mounted on the same heat sink, twice the power has to be dissipated, and a heat sink with one-half the above thermal resistance is required.

c. To determine the power gain $P'_i$, the AC power to the secondary of $T_1$ must be computed. This is simply twice the AC power to each transistor and the biasing resistors:

$$P'_i = 2I_b^2 R_i$$
$$P'_i = 2(0.707 I_{bm})^2 \left( h_{IE} + \frac{R_1 R_2}{R_1 R_2} \right)$$
$$\simeq 2[0.707(55 \times 10^{-3})]^2 (12 + 3)$$
$$\simeq 45 \text{ mW}$$

The AC power $P_i$ to the primary of $T_1$ is

$$P_i = \frac{P'_i}{\eta} \qquad (12.20)$$

$$P_i = \frac{45 \times 10^{-3}}{0.75} = 60 \text{ mW}$$

The power gain is

$$G = \frac{P_o}{P_i} \qquad (4.11)$$

$$= \frac{6}{60 \times 10^{-3}} = 100$$

$$G \text{ (in dB)} = 20 \text{ dB}$$

d. The efficiency is greatest when the maximum output power of 6 W is produced. The DC input power is

$$P_{DC} = I_{DC} V_{CC}$$

where $I_{DC}$ includes DC current supplied to the biasing resistors and the average value of the full-wave rectified collector current waveform. As long as the DC current drawn by each base is small compared to that through $R_2$, we can write

$$P_{DC} \simeq \left( \frac{V_{CC}}{R_1 + R_2} + 0.636 I_{C(MAX)} \right) (V_{CC})$$

$$P_{\text{DC}} \simeq \left[\frac{12}{360} + 0.636(1.5)\right] \quad (12)$$

$P_{\text{DC}} \simeq 11.85 \text{ W}$

The efficiency is

$$\eta = \frac{6}{11.85} = 50.5 \text{ percent} \tag{12.7}$$

Note that the efficiency is substantially lower than the 78.5 percent which is theoretically possible. This is due to the power losses in $T_1$, $T_2$, $R_1$, and $R_2$.

## Problem Set 12.7

12.7.1. Why is magnetic-core saturation of the output transformer more likely to occur in class A circuits than in class B?

12.7.2. What is crossover distortion, and how can it be minimized?

12.7.3. Why can the $R_1$-$R_2$ combination in Fig. 12.26 *not* be bypassed for AC?

12.7.4. When is crossover distortion most pronounced?

12.7.5. The output characteristics of Fig. 12.28 are to be used in conjunction with the circuit of Fig. 12.26. Neglect the $R_1$-$R_2$ combination, and assume $V_{CC} = 20$ V, $R'_L = 10 \ \Omega$, $R_L = 5 \ \Omega$, and the maximum power

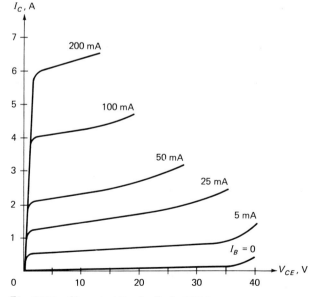

Fig. 12.28 Characteristics for Prob. 12.7.5.

dissipation per transistor (suitably heat-sinked) is 10 W. Assume ideal transistors ($V_{BE} = 0$, $I_{CBO} = 0$, $V_{CE(SAT)} = 0$).

a. Draw the AC load line and power-dissipation hyperbola, and show the Q point on the output characteristics.
b. What is the maximum instantaneous power dissipation per transistor?
c. What is the worst-case average power dissipation per transistor?
d. Assuming an 80 percent efficient output transformer, what is the maximum signal power deliverable to $R_L$?
e. What is the maximum DC power input to the circuit?
f. Specify the output transformer.

12.7.6. Why must diodes $D_1$ and $D_2$ be used in the circuit of Fig. 12.27?

## 12.5 COMPLEMENTARY SYMMETRY CIRCUITS

The standard class B push-pull amplifier requires two out-of-phase signals; these may be supplied through a center-tapped transformer or phase inverter. Complementary symmetry circuits need only one phase. Their chief requirement is a pair of closely matched, oppositely doped transistors. Until recently it was difficult to manufacture such transistors, but now they are available at relatively low cost.

The basic complementary circuit of Fig. 12.29a operates class B. The input is capacitively coupled, and $R_L$ is direct-coupled. With no signal, neither transistor conducts, and the current through $R_L$ equals 0. When the signal is positive-going, $Q_1$ is turned off while $Q_2$ conducts; the resulting current through $R_L$ develops a negative-going voltage at point a relative to ground. When the signal is negative-going, $Q_2$ goes off while $Q_1$ turns on; current flows through $R_L$ in such a direction as to make point a positive with respect to ground. Note that there is no DC current through $R_L$; hence an electromagnetic load, such as a loudspeaker, can be connected directly without introducing saturation problems.

The circuit of Fig. 12.29a produces crossover distortion unless a DC bias to overcome the threshold voltage for each base-emitter junction is provided. This can be achieved as in the circuit of Fig. 12.29b, in which the voltage developed across $R_2$ forward-biases both base-emitter junctions. $R_2$ is normally small so as not to produce any significant loss in drive to $Q_2$ relative to $Q_1$. This circuit requires only one DC supply and is commonly referred to as the "totem pole" configuration.

Without signal, neither transistor conducts, and point a is at approximately $\frac{1}{2}V_{CC}$ relative to ground. If $V_{CC} = 12$ V, $V_a = 6$ V and C charges to 6 V with the polarity shown. A positive-going input causes $Q_1$ to conduct and

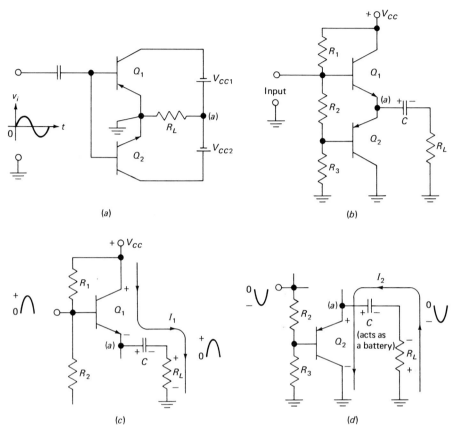

Fig. 12.29  (a) Basic complementary circuit; (b) totem pole configuration, including forward biasing resistors; (c) equivalent circuit for (b) when the input signal is positive-going; (d) equivalent circuit for (b) when the input signal is negative-going.

$Q_2$ to turn off. Current now flows through the load, which develops a positive voltage across $R_L$, as indicated in Fig. 12.29c. When the input goes negative, $Q_1$ turns off while $Q_2$ conducts. Since $Q_1$ is off, no current can flow from $V_{CC}$ through $Q_1$, but the capacitor acts as a battery, discharging as indicated in Fig. 12.29d. The voltage across $R_L$ contains no DC component and is in phase with the input; since $Q_1$ and $Q_2$ operate as emitter followers (with their inherent negative feedback), the load is driven from a low-impedance, low-distortion source.

Figure 12.30 shows the schematic of a direct-coupled power amplifier capable of producing 7.5 to 10 W into a 16-Ω load. $TR1$ operates class A. The two output transistors $TR4$ and $TR5$ are of the same polarity and operate class B. The complementary pair of driver transistors $TR2$ and $TR3$ also operate

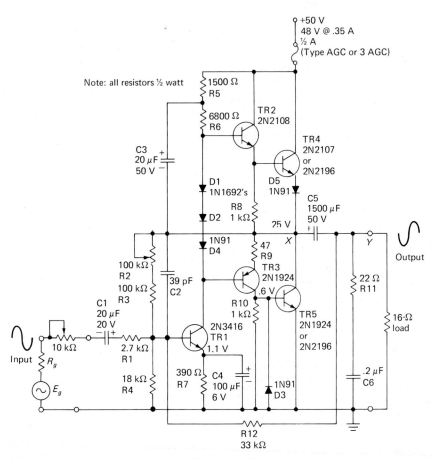

*Fig. 12.30* 7.5- to 10-W amplifier. (*Courtesy of General Electric Company.*)

class B. This type of circuit is referred to as quasi-complementary symmetry. It has the advantage of requiring only the driver transistors to be complementary matched, instead of the higher-power output transistors. A negative-going input is inverted by $TR1$, causing $TR3$ to go off and $TR2$ to conduct. The $TR2$-$TR4$ Darlington-connected transistors operate as emitter followers and cause current to flow through the load in such a direction as to make point $Y$ positive with respect to ground. $C5$ is charged with the polarity shown. A positive-going input is inverted by $TR_1$, causing $TR_2$ to go off and $TR_3$ to conduct. $TR_3$ inverts the signal, yielding a positive input to $TR_5$; this causes $TR_5$ to conduct ($C5$ supplies the current while $TR_2$ and $TR_4$ are off) yielding a negative output voltage at $Y$. Note that there is a net 180° phase shift from input to output.

$R12$ provides overall shunt-derived, shunt-fed negative feedback to improve the frequency response and reduce distortion. To ensure the effectiveness of shunt feeding, a 2.7-kΩ resistor ($R1$) is placed in series with the input path. This way, regardless of the value of $R_g$ or the setting of the 10-kΩ volume control, there is always an effective resistance of at least 2.7 kΩ in series with the signal generator. The input impedance is reduced because of the Miller effect. $C1$ is simply a coupling capacitor to prevent DC from flowing into the signal generator circuit.

The variable resistance $R2$ provides a method for setting the DC voltage at point $X$ equal to one-half of the 50-V supply. Diodes $D1$, $D2$, and $D4$ provide forward bias to $TR2$ and $TR3$ so as to reduce crossover distortion. Diodes are used instead of resistors because their temperature dependence can be closely matched to that of the transistors ($V_{BE}$ drops at the rate of approximately 2.5 mV/°C). The $TR2$-$TR3$ combination is biased at around 1 mA, while the $TR4$-$TR5$ combination is biased at around 10 mA to minimize crossover distortion.

$R9$ (47 Ω) provides both DC and AC local negative feedback for DC stabilization and reduced distortion. $R8$ (1 kΩ) is required to develop an output voltage for emitter follower $TR2$. $D3$ shunts any increase in $TR3$'s collector current due to temperature away from $TR5$; since $D3$ is a germanium diode, its reverse current can follow closely that of $TR3$, which is also made of germanium, thus obtaining good temperature compensation.

The positive-going portion of the output signal is produced by the $TR2$-$TR4$ emitter-follower combination, which has less gain than the $TR3$-$TR5$ combination. For this reason, positive feedback is provided by $C3$ to increase the amplification received by the positive half-cycles. The positive feedback does not create oscillation problems, because there is substantial overall negative feedback ($R12$, $R2$, and $R3$, $C2$) and local negative feedback ($TR2$ and $TR4$) to compensate.

$C2$ is chosen to provide the exact amount of overall negative feedback for optimum square-wave response (maximum rise time and minimum overshoot). The $R11$-$C6$ network helps cancel the inductive reactance of the load (loudspeaker) at high frequencies. $D5$ reduces the off current of $TR4$.

The circuit performance is described by the distortion and frequency response graphs of Fig. 12.31. Although these graphs were obtained for a 16-Ω resistive load, no significant loss of power results for slightly lower or larger loads. The circuit efficiency is between 47 and 60 percent, and the noise figure is better than 98 dB.

## 12.6 OTHER POWER CIRCUITS

Class C amplification is often used when larger amounts of power are required, as in the output stage of a radio transmitter. While class B circuits use transistors that conduct for ½ cycle, class C involves conduction angles of less than 180°.

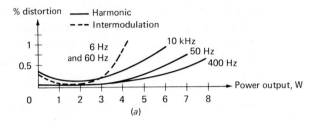

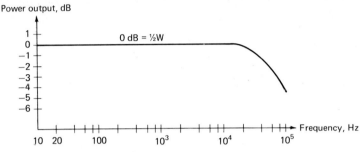

Fig. 12.31 (a) *Distortion;* (b) *frequency response for circuit of Fig. 12.30.* (*Courtesy of General Electric Company.*)

This produces significant distortion in the output, but higher practical efficiencies than class B (around 70 percent). Although such techniques are unsuitable for audio amplification, it is possible, through the use of tuned circuits, to recover the fundamental component of the input signal and filter out the remaining harmonics. This is indeed done and results in practical, high-efficiency circuits. Because their application is generally limited to radio frequencies, class C circuits are discussed in Chap. 13.

### Problem Set 12.8

12.8.1. What is one advantage of complementary symmetry circuits?
12.8.2. Why is negative feedback used in audio amplifiers?
12.8.3. What advantages are provided by emitter-follower stages?
12.8.4. Why are diodes instead of resistors used in the circuit of Fig. 12.30 to minimize crossover distortion?
12.8.5. What transistor stage in the circuit of Fig. 12.30 provides the greatest voltage amplification?

## REFERENCES

1. Phillips, A. B.: "Transistor Engineering and Introduction to Integrated Semiconductor Circuits," McGraw-Hill Book Company, New York, 1962.

2. Kiver, M. S.: "Transistors," McGraw-Hill Book Company, New York, 1962.
3. Malvino, A. P.: "Transistor Circuit Approximations," McGraw-Hill Book Company, New York, 1968.
4. Amos, S. W.: "Principles of Transistor Circuits," Hayden Book Company, Inc., New York, 1965.
5. Kahn, M., and J. M. Doyle: "The Synthesis of Transistor Amplifiers," Holt, Rinehart and Winston, Inc., New York, 1970.
6. Cowles, L. G.: "Transistor Circuits and Applications," Prentice-Hall, Inc., Englewood Cliffs, N.J., 1968.
7. Cutler, P.: "Semiconductor Circuit Analysis," McGraw-Hill Book Company, New York, 1964.
8. Texas Instruments Incorporated: "Transistor Circuit Design," McGraw-Hill Book Company, New York, 1963.
9. "General Electric Transistor Manual," 7th ed., Syracuse, 1964.
10. Daugherty, D. G., and R. A. Greiner: Low Distortion and High Thermal Stability in Transistor Audio Power Amplifiers, *IEEE Trans. Audio*, March-April 1964, pp. 26–29.
11. Blaser, L., and H. Franco: Push-Pull Class AB Transformerless Power Amplifiers, *IEEE Trans. Audio*, January-February 1963, pp. 6–14.
12. Clark, M. A.: Power Transistors, *Proc. IRE*, June 1958, pp. 1185–1204.
13. Reich, B.: Measurement of Transistor Thermal Resistance, *Proc. IRE*, June 1958, pp. 1204–1207.
14. Nelson, J. T., and J. E. Iwersen: Measurement of Internal Temperature Rise of Transistors, *Proc. IRE*, June 1958, pp. 1207–1208.
15. Von Recklinghansen, D. R.: Class B Amplifier Dissipation, Instantaneous and Steady State, *IEEE Trans. Audio*, July–August 1965, pp. 83–87.
16. Baker, L.: Power Dissipation in Class B Amplifiers, *IRE Trans. Audio*, September–October 1962, pp. 139–145.
17. Buchsbaum, W. H.: Push-Pull Class B Transistor Power-Output Circuits, *Electronics World*, November 1960.
18. Inbar, G. F.: Thermal and Power Considerations in Class B Transistorized Amplifiers, *IEEE Trans. Audio*, July–August 1965, pp. 88–95.
19. Lin, H. C.: Quasi-complementary Transistor Amplifier, *Electronics*, September 1956, pp. 173–175.
20. Marcus, P., and L. Zide: Solid State Designs for Hi-Fi Amplifiers, *Electronics World*, June 1965, pp. 49–54.
21. Greenburg, R.: Determining Maximum Reliable Load Lines for Power Transistors, *Motorola Semiconductor* AN 137.
22. ———: Selecting Commercial Power Transistor Heat Sinks, *Motorola Semiconductor* AN 135.

# chapter 13

# RF circuits

Radio frequencies (RF) represent a large range of frequencies useful in various types of radio transmission. The range covered extends from less than 100 kHz to over 1 GHz.[1] This chapter deals with techniques applicable below 500 MHz. Beyond this frequency some rather specialized techniques are required which are beyond the scope of this book. The main discussion deals with small-signal amplifiers; however, some large-signal functions such as mixing and modulation are also introduced.

## 13.1 SMALL-SIGNAL RF AMPLIFIERS

RF amplifiers are normally tuned; that is, they employ resonant $LC$ circuits to achieve frequency-selective amplification. For this reason, properties of resonant circuits form an important part of any study of RF amplifiers.

Communication systems make extensive use of tuned amplifiers. For example, the receiving antenna in a communications receiver may pick up many different signals simultaneously. A good receiver should select only the desired frequency band and reject all others. This *selectivity* is provided by tuned circuits.

---

[1] Frequency ranges may also be expressed as follows: LF (low frequencies): 30 to 300 kHz, MF (medium frequencies): 300 kHz to 3 MHz, HF (high frequencies): 3 to 30 MHz, VHF (very high frequencies): 30 to 300 MHz, UHF (ultrahigh frequencies): 300 MHz to 3 GHz.

In communication systems, the RF amplifier is that stage (or stages) of amplification that processes the signal at the transmission frequencies, that is, the signal that is received or transmitted by the antenna.

Standard AM (amplitude modulation) broadcasts cover the frequency range from 535 kHz to 1.605 MHz. Each station is allotted a 10-kHz *band* within this range for the transmission of its program. In the receiver, it is the function of the RF amplifier to select and amplify the frequency band corresponding to the station whose broadcast we wish to receive. Thus the RF amplifier must be tunable (variable in frequency) so that we can tune in different stations. In *superheterodyne* receivers, the amplified RF signal is shifted in frequency to a fixed lower frequency, called the intermediate frequency (IF). This is 455 kHz in AM broadcast receivers. Here the signal receives additional amplification before it is further processed to yield the final sound from a loudspeaker. The reason for the shift in signal frequency is that more efficient amplification is possible at the lower IF. The IF amplifier is fixed tuned, as opposed to the RF amplifier, which must be capable of receiving a narrow band of frequencies anywhere within the broadcast band. In most inexpensive AM receivers, no RF stage is provided; selectivity here is achieved mainly in the IF amplifier.

The frequency range for frequency modulation (FM) broadcasts is 88 to 108 MHz. Each station is allotted a 200-kHz band. This is wider than standard AM bands, but it is the price we must pay for the benefits of FM.[1] Nearby stations are kept at least 400 kHz apart to minimize interference. The IF amplifier in broadcast FM receivers is fixed tuned at 10.7 MHz. The RF amplifier must obviously be able to select and amplify any 200-kHz band within the 88- to 108-MHz range.

Television broadcasting involves both the VHF range (channels 2 to 6: 54 to 88 MHz; channels 7 to 13: 174 to 216 MHz) and UHF range (channels 14 to 83: 470 to 890 MHz). Each channel is 6 MHz wide. Note that television channels are wider than AM or FM radio bands. This is because more information is transmitted over a TV channel (sound plus picture) and thus a larger bandwidth is required.

### 13.1a  Performance Measures

Tuned amplifiers are mainly characterized by their sensitivity, bandwidth, gain, and noise figure. Normally there is a trade-off between gain and bandwidth so that if bandwidth is increased, gain decreases, and vice versa. A typical set of specifications might be as follows:

Center frequency:   $f_o = 200$ MHz
Bandwidth (3 dB):   $B = 10$ MHz

---

[1] One of the advantages of FM over AM is greater noise immunity.

Power gain at $f_o$:    $G_o = 25$ dB
Noise figure at $f_o$:   $NF = 4$ dB

Other measures of performance may include the amplifier's stability and dynamic range. Stability has to do with the tendency of high-frequency circuits to oscillate; a stable circuit is one which does not oscillate under normal operating conditions. Dynamic range is a measure of the spread between the weakest signal that can be amplified, which is limited by noise, and the strongest signal, which is limited by the distortion-producing nonlinearities in the active device. Dynamic range is very important in applications where wide variations in signal level are encountered.

### 13.1b   Basic Circuits

A qualitative introduction to basic tuned amplifiers is given here. More detailed analytical treatment is provided in the sections to follow.

Consider the circuit of Fig. 13.1a. Here we have a common emitter amplifier suitably biased by $R_1$, $R_2$, $R_3$, and $V_{CC}$. $R_3$ is bypassed for AC by $C_2$. As far as DC is concerned, $L$ has negligible effect and the collector is directly tied to $V_{CC}$. $C_c$ are DC-blocking capacitors with negligible reactance at the frequency of operation.

The gain of the amplifier is determined by the transistor's parameters and the dynamic (AC) load seen by the collector. Here the collector sees $R_L$ shunted by $L$ and $C_1$, as shown in Fig. 13.1b. For simplicity, all losses, including the inductor's, are lumped in $R_L$. The parallel $LC_1R_L$ circuit has impedance characteristics as shown in Fig. 13.1c. At the resonant frequency $f_o$, the impedance is highest (it is also purely resistive—no reactive component); also note that $|Z|$ drops on each side of $f_o$. Since gain is proportional to load impedance (within certain limits), the amplifier gain is highest at $f_o$. At frequencies away from $f_o$, the impedance seen by the collector drops; therefore the gain also drops. The amplifier's gain, in fact, when plotted as a function of frequency, is essentially identical with the impedance curve of Fig. 13.1c.

There are two main items of interest in the frequency response of Fig. 13.1c. One is $f_o$, the resonant or *center* frequency. This is determined by $L$ and $C_1$. For example, making $C_1$ variable would enable us to vary $f_o$. Our response curve could therefore be shifted to a higher or lower $f_o$, depending on whether $C_1$ was reduced or increased, respectively.

The bandwidth is a function of the circuit $Q$.[1] A high $Q$ means a narrow bandwidth, while a lower $Q$ results in a wide bandwidth. The $Q$ can be increased by reducing the resistive losses or increasing the energy stored in the

---

[1] $Q$ is a figure of merit that expresses a circuit's ability to store energy, as in the magnetic field of a coil, relative to the power dissipated in the circuit's resistive elements.

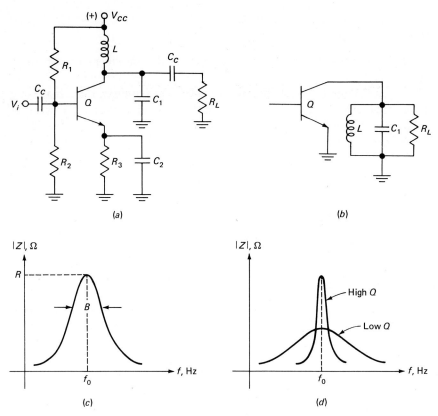

*Fig. 13.1* (a) *Amplifier with tuned output;* (b) *equivalent load seen by transistor in* (a); (c) *impedance as a function of frequency for the tuned circuit in* (b); (d) *impedance curves for high and low Q.*

$LC$ circuit. Response curves, both centered at $f_o$, are compared for high and low $Q$ in Fig. 13.1d.

The circuit just discussed is a single-tuned, fixed-frequency amplifier. Based on the application, other tuned circuits may be used. To achieve greater selectivity, there may also be a tuned circuit at the amplifier's input.

The circuit of Fig. 13.2 involves two tuned circuits; each has a variable capacitor to change $f_o$. The two capacitors are *ganged;* that is, they can be simultaneously varied by rotating a common shaft. This means that whenever one wishes to receive a different frequency signal, only one control needs to be changed; the control varies $f_o$ for both tuned circuits.

The input signal is picked up by the antenna and coupled through transformer $T_1$ to the transistor. The transformer is *tapped* to provide a proper impedance match to the base. This way the tank ($LC$ circuit) impedance can be made

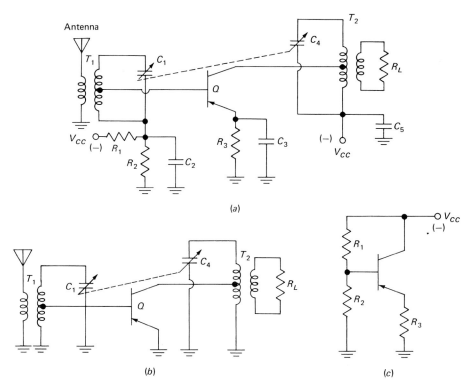

**Fig. 13.2** (a) *RF amplifier with tuned input and output;* (b) *AC circuit for* (a); (c) *DC circuit for* (a).

very high to achieve a high $Q$, but the transistor can be fed from a lower equivalent-source impedance to achieve maximum power transfer. At the frequency of operation, $C_2$, $C_3$, and $C_5$ behave as short circuits, yielding the AC equivalent circuit of Fig. 13.2b. Note that both transformers are wound around an *air* core, that is, a nonmagnetic core. The air core is used at high frequencies because the inductance required is usually small and can be achieved without a magnetic core; the core losses (energy lost through eddy currents and hysteresis) are also quite small, yielding a high $Q$.

The DC circuit, shown in Fig. 13.2c, involves a fairly straightforward biasing arrangement and will not be discussed here. At high frequencies the DC supply may present an effective impedance through which signals could be coupled from output to input, causing various problems, including oscillations. To prevent such undesirable coupling, the DC supply is AC-bypassed (decoupled) using $C_5$.

In general, the center frequency $f_o$ of a tuned circuit can be changed using variable capacitors or inductors. Variable capacitors work on the principle that capacitance is a function of the common area between capacitor plates. By rotating a shaft, we alter the capacitor plates' relative position, thus varying the net capacitance. It is also possible in certain cases to use reverse-biased semiconductor diodes as variable capacitors (varactors). Here the diode's capacitance can be varied by changing the applied voltage; as the reverse bias increases, the depletion region widens, and the capacitance decreases. Variable inductors or transformers are also used. The inductance change may be achieved by varying the position of a powdered iron core around which the coil is wound. When the core is pulled out, the magnetic field completes most of its path through air, which has low permeability, and hence the inductance is small. As the core is pushed in, the inductance increases.

The bandwidth is set during the initial alignment procedure and is not usually changed during normal operation. In the circuits so far discussed, the bandwidth is determined by the circuit $Q$. It is also possible to control the bandwidth and general shape of the response curve by varying the coupling between tuned circuits. This aspect is more fully explored in a later section.

## Problem Set 13.1

13.1.1. What is selectivity?
13.1.2. Why are tuned circuits extensively used in RF amplifiers?
13.1.3. What is sensitivity?
13.1.4. What does stability refer to?
13.1.5. What is meant by dynamic range?
13.1.6. What frequency range must an RF amplifier cover in an

    a. AM broadcast receiver
    b. FM broadcast receiver

13.1.7. What is an IF amplifier, and why is it used?
13.1.8. Explain how frequency-selective amplification is achieved in the amplifier of Fig. 13.1a.
13.1.9. How can the center frequency of a tuned amplifier's response be varied?
13.1.10. How does a variable capacitor work?
13.1.11. What is the general relationship between bandwidth and $Q$?
13.1.12. What is the function of $C_5$ in the circuit of Fig. 13.2?

### 13.1c Tuned Circuits

Properties of tuned circuits must be well understood before tuned amplifiers can be thoroughly discussed. A review of these properties is given here.

Shown in Fig. 13.3 is the series and parallel representation for any re-

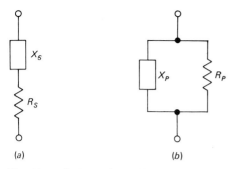

Fig. 13.3 Series and parallel representation of a real reactance.

actance, capacitive or inductive. The series representation in Fig. 13.3a is discussed first. The $Q$, at any frequency, is defined as follows:

$$Q = \frac{X_S}{R_S} \qquad (13.1)$$

where $X_S$ is the reactance magnitude in ohms, and $R_S$ is the equivalent series resistance, also in ohms. $R_S$ accounts for the energy losses. With coils, these are due to wire resistance and core losses; there may also be some energy lost through radiation at high frequencies. $X_S$ is directly related to the stored energy (magnetic field for a coil, electric field for a capacitor). A high $Q$ involves a high ratio of stored to dissipated energy.

The impedance of the series circuit can be written as follows:

$$Z_S = R_S \pm jX_S \qquad (13.2)$$

where the positive sign is for inductance and the negative sign for capacitance. For the parallel circuit in Fig. 13.3b, the impedance is

$$Z_P = \frac{(R_P)(\pm jX_P)}{R_P \pm jX_P}$$

which can be expanded to

$$Z_P = \frac{R_P X_P^2}{R_P^2 + X_P^2} \pm j \frac{X_P R_P^2}{R_P^2 + X_P^2} \qquad (13.3)$$

A practical circuit can be simulated at any given frequency using a series or parallel equivalent circuit, so that either Eq. (13.2) or Eq. (13.3) may be used. This means that

$$Z_S = Z_P \qquad (13.4)$$

from which

$$R_S = \frac{R_P X_P^2}{R_P^2 + X_P^2} \tag{13.5}$$

$$X_S = \frac{X_P R_P^2}{R_P^2 + X_P^2} \tag{13.6}$$

The $Q$, which is defined by Eq. (13.1), can now be rewritten for a parallel circuit using Eqs. (13.5) and (13.6):

$$Q = \frac{X_S}{R_S} = \frac{R_P}{X_P} \tag{13.7}$$

Equations (13.5) and (13.6) may be rearranged to yield the following:

$$R_P = \frac{X_S^2 + R_S^2}{R_S} \tag{13.5a}$$

$$X_P = \frac{R_S^2 + X_S^2}{X_S} \tag{13.6a}$$

The above transformations are valid only at the frequency for which they are calculated. Typical coil $Q$'s range to a maximum of about 400 for specially designed units with very low core losses. Usually, though, coil $Q$'s are somewhat lower. Capacitors, on the other hand, have fairly high $Q$'s, so that normally their losses can be neglected. This is especially true in $LC$ circuits, where the coil's losses predominate. Because losses in general increase with frequency, neither $R_S$ nor $R_P$ is constant; instead they are a function of frequency, and so is the $Q$. For this reason $Q$ specifications are incomplete unless the frequency is also given.

### Example 13.1

A 10-mH inductor has an effective series resistance of 50 Ω at 50 kHz. Determine the coil $Q$ at 50 kHz.

### Solution

Using the series representation of Fig. 13.3a, we have

$X_S = 2\pi f L_S = (6.28)(50 \times 10^3)(10 \times 10^{-3})$
$X_S = 3{,}140 \; \Omega$

$$Q = \frac{X_S}{R_S} = \frac{3{,}140}{50} = 62.8 \tag{13.1}$$

### Example 13.2

For large values of $Q$, show that $X_S \simeq X_P$ in the two circuits of Fig. 13.3.

## Solution

The relationship between $X_S$ and $X_P$ is

$$X_S = \frac{X_P R_P^2}{R_P^2 + X_P^2} \qquad (13.6)$$

This may be rewritten as

$$X_S = \frac{X_P}{1 + (X_P/R_P)^2}$$

But

$$Q = \frac{R_P}{X_P} \qquad (13.7)$$

Therefore

$$X_S = \frac{X_P}{1 + (1/Q^2)}$$

As long as $1/Q^2 \ll 1$, the denominator approaches 1 and

$$X_S \simeq X_P \qquad (13.8)$$

Equation (13.8) may be used with less than 5 percent error if $Q \geq 5$.

## Example 13.3

Determine the equivalent parallel circuit for the coil in Example 13.1.

## Solution

Since $Q > 5$, $X_S \simeq X_P$ and $L_S \simeq L_P$. Therefore $L_P = 10$ mH. The equivalent parallel resistance is

$$R_P = QX_P \qquad (13.7)$$
$$R_P = 62.8(3{,}140) = 197 \text{ k}\Omega$$

## Example 13.4

Show that, as long as $Q \geq 5$, $R_P \simeq Q^2 R_S$.

## Solution

$$R_P = QX_P \qquad (13.7)$$

For $Q \geq 5$, $X_P \simeq X_S$ and

$$R_P \simeq QX_S$$

But
$$X_S = QR_S \tag{13.1}$$

Therefore
$$R_P \simeq Q(QR_S)$$
$$R_P \simeq Q^2 R_S \tag{13.9}$$

**Example 13.5**

Determine $R_P$ in Example 13.3 using Eq. (13.9).

**Solution**

$$R_P \simeq Q^2 R_S \tag{13.9}$$
$$R_P \simeq (62.8^2)(50) = 197 \text{ k}\Omega$$

**Example 13.6**

A 100-pF capacitor has a $Q$ of 1,000 at 1 MHz. Determine $R_S$ and $R_P$, and draw the equivalent series and parallel circuits.

**Solution**

The equivalent series resistance is

$$R_S = \frac{X_S}{Q} \tag{13.1}$$

$$R_S = \frac{1}{(6.28 \times 10^6)(100)(10^{-12})(1,000)} = 1.59 \ \Omega$$

Since the $Q$ is very high, the equivalent parallel resistance is found using

$$R_P \simeq Q^2 R_S \tag{13.9}$$
$$R_P \simeq (1,000^2)(1.59) = 1.59 \text{ M}\Omega$$

Both equivalent circuits are shown in Fig. 13.4.

**Example 13.7**

A 250-$\mu$H inductor has a $Q$ of 300 at 1 MHz. Determine $R_S$ and $R_P$ at this frequency, and draw both equivalent circuits.

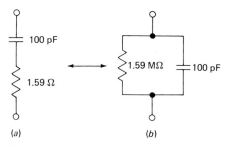

Fig. 13.4 The circuit of (a) is equivalent to the circuit of (b) at a frequency of 1 MHz (Example 13.6).

## Solution

The equivalent circuits are shown in Fig. 13.5. Since the $Q$ is high, $L_S$ and $L_P$ are essentially equal; the resistances are

$$R_S = \frac{X_S}{Q} \tag{13.1}$$

$$R_S = \frac{(6.28 \times 10^6)(250 \times 10^{-6})}{300} = 5.25 \; \Omega$$

$$R_P \simeq Q^2 R_S \tag{13.9}$$

$$R_P \simeq (300^2)(5.25) = 475 \; \text{k}\Omega$$

## Example 13.8

The coil of Example 13.7 is series-loaded with a 12-$\Omega$ resistor. Determine the new $Q$ and $R_P$.

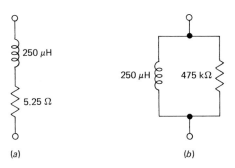

Fig. 13.5 The circuit of (a) is equivalent to the circuit of (b) at a frequency of 1 MHz (Example 13.7).

## Solution

Here we must make a distinction between the unloaded $Q$, which is 300, and the loaded $Q$, which we must calculate. The coil's $R_S$ is 5.25 Ω; if we add 12 Ω in series, the new $R_S$ is $R'_S = 12 + 5.25 = 17.25$ Ω, and the loaded $Q$ is

$$Q_L = \frac{X_S}{R'_S} \tag{13.1}$$

$$Q_L = \frac{(6.28 \times 10^6)(250 \times 10^{-6})}{17.25} = 91$$

The new $R_P$ is

$$R'_P \simeq R'_S Q^2 \tag{13.9}$$
$$R'_P \simeq 17.25(91^2) = 143 \text{ k}\Omega$$

Note that adding resistance in series reduces both the $Q$ and $R_P$.

### Example 13.9

The capacitor of Example 13.6 is shunt-loaded with a 3.3-MΩ resistor. Determine the new $R_P$ and loaded $Q$.

## Solution

The capacitor's $R_P$ is 1.59 MΩ. Shunt loading yields a new $R_P$ whose value is the parallel combination of 3.3 and 1.59 MΩ:

$$R'_P = 1.07 \text{ M}\Omega$$

The loaded $Q$ is

$$Q_L = \frac{R'_P}{X_P} \tag{13.7}$$

$$Q_L = (1.07 \times 10^6)(6.28 \times 10^6)(100)(10^{-12}) = 670$$

Note that shunt loading has reduced the capacitor's $Q$ from 1,000 to 670.

### Example 13.10

Simplify the circuit in Fig. 13.6a to the type in Fig. 13.3b; what is the overall $Q$?

## Solution

The series combination of $X_S$ and $R_S$ has a $Q$ of $X_S/R_S = 100$. This yields an $R'_P \simeq Q^2 R_S = (100^2)(10) = 100$ kΩ. $R'_P$ is in parallel with $R_L$, yielding an overall $R_P$ of 100 kΩ∥100 kΩ = 50 kΩ as shown in Fig. 13.3b. The overall $Q$ is

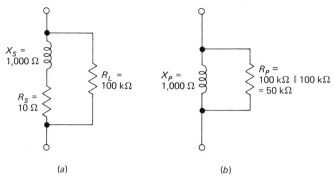

*Fig. 13.6* The circuit of (a) is equivalent to the circuit of (b) at the frequency for which the coil's reactance is 1,000 Ω (*Example 13.10*).

$$Q = \frac{R_P}{X_P} = \frac{50 \times 10^3}{1,000} = 50$$

We now consider a series $RLC$ circuit driven from a voltage source, as shown in Fig. 13.7a. The impedance of this circuit is a function of frequency. At low frequencies the impedance is high and mainly capacitive. At high frequencies it is also high and mainly inductive. At the resonant frequency $f_o$, the inductive and capacitive reactances are equal in magnitude; therefore they cancel (because they have opposite signs). The impedance is purely resistive and at its minimum value, that is, $Z_o = R$. Since $|I| = |E|/|Z|$, it follows that $I$ is maximum when $Z$ is minimum, and vice versa. The current, as a function of frequency, is shown in Fig. 13.7b. Note that at $f_o$ the current is at its maximum value.

$$I_o = \frac{E}{R} \tag{13.10}$$

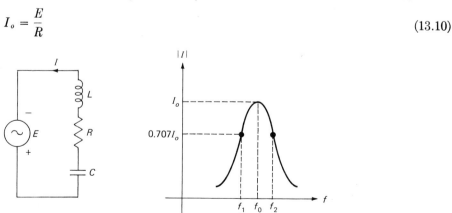

*Fig. 13.7* (a) *Series RLC circuit;* (b) *current magnitude as a function of frequency for the circuit of* (a).

Since inductive and capacitive reactances are equal at the resonant frequency, we have

$$2\pi f_o L = \frac{1}{2\pi f_o C}$$

$$f_o = \frac{1}{2\pi \sqrt{LC}} \qquad (13.11)$$

The 3-dB (half-power) bandwidth is based on the two frequencies for which the current magnitude is 0.707 of the maximum current:[1]

$$B = f_H - f_L \qquad (13.12)$$

A relationship between $B$ and $Q$ is developed in Appendix 11.5; the results are

$$B = \frac{f_o}{Q} \qquad (13.13)$$

**Example 13.11**

Given the series $RLC$ circuit of Fig. 13.8a, determine

a. Resonant frequency
b. Total resistance at $f_o$
c. Current at $f_o$
d. Voltage across $R_L$ at $f_o$
e. Circuit $Q$ at $f_o$
f. Bandwidth
g. Also sketch a phasor diagram showing all voltages in the circuit at $f_o$

**Solution**

a. $f_o = \dfrac{1}{2\pi \sqrt{LC}} \qquad (13.11)$

$$f_o = \frac{1}{6.28 \sqrt{(795 \times 10^{-6})(31.8 \times 10^{-12})}} = 1 \text{ MHz}$$

b. At $f_o$, the inductive and capacitive reactances cancel, and the total circuit resistance is

$$R_t = R_g + R_S + R_L \qquad (a)$$

The coil's $R_S$ is computed as follows:

---

[1] A current reduction to 0.707 implies a power reduction to $0.707^2 = \frac{1}{2}$; one-half power corresponds to 3 dB.

$$R_S = \frac{X_S}{Q} \tag{13.1}$$

$$R_S = \frac{(6.28 \times 10^6)(795 \times 10^{-6})}{50} = 100 \ \Omega$$

$$R_t = 50 + 100 + 100 = 250 \ \Omega \tag{a}$$

c. The resonant current is

$$I_o = \frac{E}{R} = \frac{1}{250} = 4 \text{ mA} \tag{13.10}$$

d. The output voltage across $R_L$ at $f_o$ is

$$V_o = I_o R_L = (4 \times 10^{-3})(100) = 400 \text{ mV}$$

e. The *circuit* $Q$ at $f_o$ is

$$\text{Circuit } Q = \frac{X_S}{R_t} \tag{13.1}$$

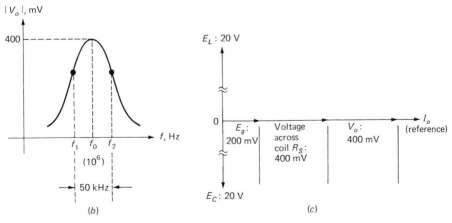

Fig. 13.8 (a) *Circuit for Example 13.11;* (b) *output voltage magnitude as a function of frequency for the circuit of* (a); (c) *phasor diagram at* $f_o$.

where $R_t$ is the *total* resistance in the circuit at $f_o$. Either the inductive or capacitive reactance may be used, because at $f_o$ they are equal:

$$X_S = QR_S \qquad (13.1)$$

where $Q$ is the coil $Q$:

$$X_S = 50(100) = 5{,}000 \; \Omega$$

We can now write

$$\text{Circuit } Q = \frac{X_S}{R_t} = \frac{5{,}000}{250} = 20$$

f. The 3-dB bandwidth is

$$B = \frac{f_o}{Q} = \frac{10^6}{20} = 50 \text{ kHz} \qquad (13.13)$$

The response curve for $V_o$ is shown in Fig. 13.8b. Note that $B$ is not determined by the coil $Q$ but by the circuit $Q$ which includes the losses in $R_g$ and $R_L$ as well as $R_S$.

g. The various voltages at $f_o$ are computed here and shown, with their proper phase relationship, on the phasor diagram of Fig. 13.8c; the current $I_o$ has been chosen as the reference.

Voltage across $R_g$: $\quad I_o R_g = (4 \times 10^{-3})(50) = 200 \text{ mV}$
Voltage across coil $R_S$: $\quad I_o R_S = (4 \times 10^{-3})(100) = 400 \text{ mV}$
Voltage across $R_L$: $\quad I_o R_L = (4 \times 10^{-3})(100) = 400 \text{ mV}$

The above voltages are in phase with each other and with the current $I_o$.

Voltage across $L$: $\quad I_o X_L = (4 \times 10^{-3})(5{,}000) = 20 \text{ V}$
Voltage across $C$: $\quad I_o X_C = (4 \times 10^{-3})(5{,}000) = 20 \text{ V}$

The coil voltage leads $I_o$ by 90°, while the capacitor voltage lags $I_o$ by 90°, as shown in the figure. Since the coil and capacitor voltages are equal and 180° out of phase with each other, their sum is zero. The generator voltage $E_g$ therefore supplies only the resistive voltage drops, as follows:

$$E_g = (200 \times 10^{-3}) + (400 \times 10^{-3}) + (400 \times 10^{-3}) + j20 - j20 = 1 \text{ V}$$

Consider now the parallel resonant circuit of Fig. 13.9a. The impedance of this circuit is small and inductive at low frequencies, small and capacitive at high frequencies, and large and purely resistive at the resonant frequency $f_o$. At resonance, in fact, the effects of $L$ and $C$ cancel, and the circuit impedance is simply $R$. The resonant frequency is given by Eq. (13.11) provided all losses in the coil are lumped in $R$.

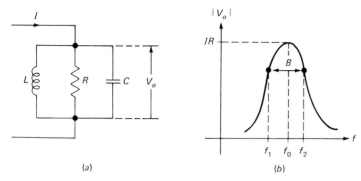

Fig. 13.9 (a) Parallel resonant circuit; (b) frequency response for the circuit of (a).

If the circuit is driven from a constant current source $I$, the output voltage, which is the product of $I$ and the total circuit impedance, is then as shown in Fig. 13.9b. Here, at $f_o$, $V_o = IR$; the bandwidth is also a function of $Q$ and is given by Eq. (13.13).

## Example 13.12

Given the parallel $RLC$ circuit of Fig. 13.10a, determine

a. The resonant frequency
b. The total resistance at $f_o$
c. Current through $R_L$ at $f_o$
d. $V_o$ at $f_o$
e. Circuit $Q$ at $f_o$
f. Bandwidth

## Solution

a. Although $L$ has series resistance, the $Q$ is high enough that $L_S \simeq L_P$, and Eq. (13.11) applies:

$$f_o = \frac{1}{2\pi \sqrt{(31.8 \times 10^{-6})(795 \times 10^{-12})}} \tag{13.11}$$

$$f_o = 1 \text{ MHz}$$

b. The total resistance at $f_o$ is the parallel combination of $R_g$, $R_L$, and the coil's $R_P$. The coil's reactance is

$$X_S = 2\pi f L_S$$
$$X_S = (6.28 \times 10^6)(31.8 \times 10^{-6}) = 200 \text{ }\Omega$$

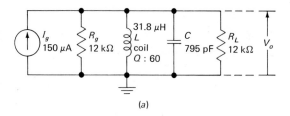

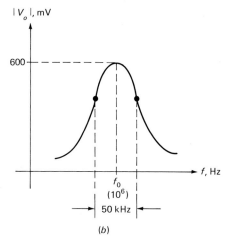

**Fig. 13.10** (a) *Circuit for Example 13.12;* (b) *frequency response for the circuit of* (a).

The $Q$ is high enough that $X_S \simeq X_P$; $R_P$ is

$$R_P = QX_P \tag{13.7}$$
$$R_P = 60(200) = 12 \text{ k}\Omega$$

The total resistance is

$$R = R_g \| R_L \| R_P$$
$$R = 12 \times 10^3/3 = 4 \text{ k}\Omega$$

c. At $f_o$, the coil and capacitor currents are equal and 180° out of phase[1] with each other; therefore their sum is zero. What happens physically is that after the initial transient, current is continuously exchanged between $L$ and $C$, and the generator has to supply only the resistive components of current needed by $R_g$, $R_L$, and $R_P$. $I_g$ therefore divides equally among the three 12-k$\Omega$ paths available. Thus each resistance receives one-third of the total, and $I_o = \frac{1}{3}(150 \times 10^{-6}) = 50 \text{ }\mu\text{A}$.

---

[1] This is true only as long as all the losses in the coil are lumped with $R_p$.

d.  $V_o = I_o R_L$
    $V_o = (50 \times 10^{-6})(12 \times 10^3) = 600$ mV

e.  The circuit $Q$ is the ratio of the total resistance $R$ to the reactance:

$$\text{Circuit } Q = \frac{R}{X_P} \tag{13.7}$$

where $R = 4$ k$\Omega$ and $X_P \simeq X_S = 200$ $\Omega$:

$$\text{Circuit } Q = \frac{4,000}{200} = 20$$

f.  $$B = \frac{f_o}{Q} = \frac{10^6}{20} = 50 \text{ kHz} \tag{13.13}$$

The circuit frequency response is shown in the graph of Fig. 13.10b.

Quite often resonant circuits must work with fixed source and load resistances that do not result in the desired overall $Q$ and bandwidth. To achieve control over the $Q$, various impedance transformation techniques are used, some of which are now discussed.

Consider the circuit of Fig. 13.11a. $X_1$ and $X_2$ can represent inductive or capacitive reactances. $R_L$ is the load. As long as the $Q$ is sufficiently high, the transformation of Fig. 13.11b is valid. To establish the relationship between $R_S$ and $R_L$, we let $Q_1$ equal the $Q$ of the $X_2$-$R_L$ combination and write

$$Q_1 = \frac{R_L}{X_2}$$

But

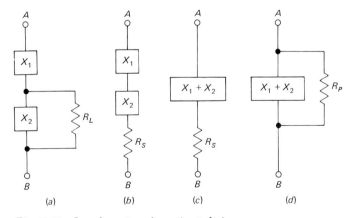

Fig. 13.11  Impedance transformation technique.

$$R_L \simeq R_S Q_1^2 \quad (13.9)$$

Therefore

$$R_S = \frac{R_L}{(R_L/X_2)^2} = \frac{X_2^2}{R_L} \quad (a)$$

Now $X_1$ and $X_2$ are in series and may be combined as in Fig. 13.11c. Let $Q_2$ equal the Q of the $X_1 + X_2$ and $R_S$ combination; we can then write

$$Q_2 = \frac{X_1 + X_2}{R_S}$$

The circuit of Fig. 13.11c can now be transformed to that of Fig. 13.11d by writing

$$R_P \simeq Q_2^2 R_S \quad (13.9)$$
$$R_P = \frac{(X_1 + X_2)^2}{R_S}$$

Substituting $R_S$ from Eq. (a), we have

$$R_P = \left(\frac{X_1 + X_2}{X_2}\right)^2 R_L$$
$$R_P = R_L \left(1 + \frac{X_1}{X_2}\right)^2 \quad (13.14)$$

If $X_1$ and $X_2$ are uncoupled inductive reactances, $X_1 = 2\pi f L_1$ and $X_2 = 2\pi f L_2$; this allows us to write

$$R_P = R_L \left(1 + \frac{L_1}{L_2}\right)^2 \quad (13.15)$$

If $X_1$ and $X_2$ are capacitive reactances, we have $X_1 = 1/(2\pi f C_1)$ and $X_2 = 1/(2\pi f C_2)$, which yields

$$R_P = R_L \left(1 + \frac{C_2}{C_1}\right)^2 \quad (13.15a)$$

Note that in both the above cases, the transformations are independent of frequency.

**Example 13.13**

Determine the equivalent parallel capacitance and resistance from $A$ to $B$ if both capacitors are lossless in the circuit of Fig. 13.12a.

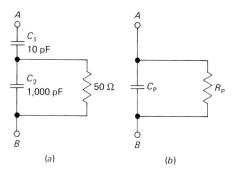

Fig. 13.12 Impedance transformation for Example 13.13.

## Solution

As long as the $Q$ is sufficiently high, the two capacitors may be treated as if they were in series, yielding

$$C_P \simeq \frac{C_1 C_2}{C_1 + C_2} = 9.9 \text{ pF}$$

The equivalent $R_P$ is

$$R_P = 50\left(1 + \frac{1{,}000}{10}\right)^2 = 500 \text{ k}\Omega \qquad (13.15a)$$

The 50-$\Omega$ resistance has therefore been "transformed" to 500 k$\Omega$.

## Example 13.14

An amplifier with an output resistance of 18.5 k$\Omega$ must be coupled, through a tuned circuit, to the next stage, whose input resistance is 500 $\Omega$. The frequency response must be centered at 1 MHz and produce a 100-kHz bandwidth. Assume a 100-$\mu$H inductor with a $Q$ of 300 is available. Design a suitable circuit.

## Solution

The block diagram is shown in Fig. 13.13a. The tuned circuit is the interstage coupling network that provides selective amplification. In Fig. 13.13b the output of amplifier 1 and input to amplifier 2 are shown as the source and load, respectively, for the tuned circuit.

Initially, we might consider using the circuit of Fig. 13.13c for our interstage coupling network. Since the coil $Q$ is high, the effect of its series resistance on $f_o$ is negligible and

$$f_o = \frac{1}{2\pi \sqrt{LC}} \tag{13.11}$$

$$10^6 = \frac{1}{2\pi \sqrt{(100 \times 10^{-6})(C)}}$$

from which $C = 250$ pF. To achieve the desired bandwidth, a circuit $Q$ is required as follows:

$$Q = \frac{f_o}{B} \tag{13.13}$$

$$Q = \frac{10^6}{100 \times 10^3} = 10$$

The coil has a $Q$ of 300; therefore an equivalent $R_P$ can be calculated:

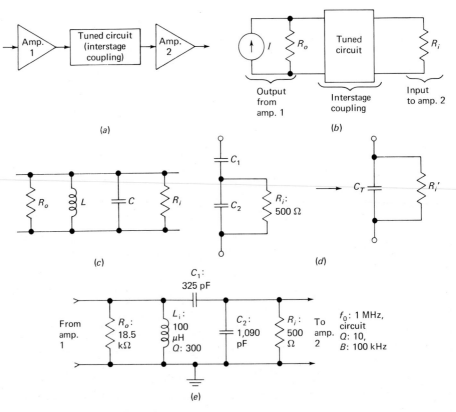

Fig. 13.13  Circuits for Example 13.14.

$$R_P = QX_P \qquad (13.7)$$
$$X_P = 2\pi f L_P = (6.28 \times 10^6)(100 \times 10^{-6}) = 628 \; \Omega$$
$$R_P = 300(628) = 1.88 \; \text{M}\Omega$$

The total circuit resistance is

$$R_T + R_o \| R_P \| R_i = (18.5 \times 10^3) \| (1.88 \times 10^6) \| 500$$
$$R_T = 485 \; \Omega$$

The circuit $Q$ is therefore

$$\text{Circuit } Q = \frac{R_T}{X_P}$$
$$\text{Circuit } Q = \frac{485}{628} = 0.775$$

which is unsatisfactory, since the required $Q$ is 10.

The circuit of Fig. 13.13c yields a very low $Q$, because the 500-$\Omega$ input resistance to amplifier 2 is connected directly across the tank circuit. It is possible, however, to tap $L$ or $C$ so that $R_i$ may be transformed to a much higher value. This is done in Fig. 13.13d, where $C$ is separated into $C_1$ and $C_2$ and the 500-$\Omega$ load is connected only across $C_2$. To determine $R'_i$ (the transformed value of $R_i$), note that an overall $Q$ of 10 can be achieved if the total circuit resistance is

$$R'_T = QX_P$$
$$R'_T = 10(628) = 6{,}280 \; \Omega$$

But $R'_T = R_o \| R_P \| R_i$, and since $R_o \ll R_P$, $R'_T \simeq R_o \| R'_i$,

$$R'_T = \frac{R_o R'_i}{R_o + R'_i}$$
$$6{,}280 = \frac{18{,}500 R'_i}{18{,}500 + R'_i}$$
$$R'_i = 9.5 \; \text{k}\Omega$$

The transformed value of $R_i$ is

$$R'_i = R_i \left(1 + \frac{C_2}{C_1}\right)^2 \qquad (13.15a)$$
$$9{,}500 = 500 \left(1 + \frac{C_2}{C_1}\right)^2$$
$$C_2 = 3.36 C_1$$

But the series combination of $C_1$ and $C_2$ must still equal 250 pF; therefore

$$250 \times 10^{-12} = \frac{C_1 C_2}{C_1 + C_2} = \frac{C_1(3.36 C_1)}{C_1 + 3.36 C_1}$$

from which $C_1 = 325$ pF and $C_2 = 1,090$ pF. With the calculation of $C_1$ and $C_2$ the circuit design is complete, as shown in Fig. 13.13e. It should be noted, however, that the transformation of $R_i$ to 9.5 k$\Omega$ is valid only as long as the $Q$ of the $C_2$-$R_i$ combination is fairly high. In this case, the reactance of $C_2$ at $f_o$ is

$$X_2 = \frac{1}{(6.28 \times 10^6)(1.09 \times 10^{-9})} = 146 \ \Omega$$

yielding a $Q$ of $R_i/X_2 = 500/146 = 3.43$. This is rather low; therefore some experimental adjustment of the relative values of $C_1$ and $C_2$ will be required to obtain the desired bandwidth. This is normal procedure in most cases anyway.

With tightly coupled transformers, impedance transformation can be achieved by selecting an appropriate turns ratio. A single inductor can also be tapped to yield an impedance transformation proportional to the number of turns squared. In the circuit of Fig. 13.14, the value of $Z$, when seen across the entire $n_2$ turns, is multiplied by $(n_2/n_1)^2$. For example, if $Z$ is a pure 50-$\Omega$ resistance and $n_2/n_1 = 10$, then the 50 $\Omega$ is stepped up by a factor of $10^2 = 100$ and appears as 5,000 $\Omega$ when seen from the $y$ side.

If $Z$ were due to a capacitance, though, it would appear divided by $(n_2/n_1)^2$, because the reactance of a capacitor is inversely proportional to the capacitance. If $C = 100$ pF and $n_2/n_1 = 10$, then the capacitance seen at $y$ is 1 pF. Similarly, any impedance connected to the $y$ side appears divided by $(n_2/n_1)^2$ when seen from the $x$ side. In many cases more than one tap may be needed on the coil to properly transform the various impedances that may be connected across it.

Certain precautions when using capacitors, inductors, or transformers at high frequencies are in order. For example, all inductances exhibit distributed capacitance between coil windings. This results in an equivalent capacitance across the inductance which resonates at $f_o$. If a 10-$\mu$H inductor has a distributed capacitance of 10 pF and the $Q$ is sufficiently high that the resistive losses have no influence on $f_o$, then $L$ is self-resonant at

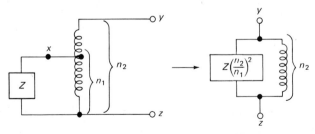

Fig. 13.14  Impedance transformation in a tightly coupled transformer.

$$f_o = \frac{1}{2\pi \sqrt{(10 \times 10^{-6})(10 \times 10^{-12})}}$$
$$f_o \simeq \text{MHz}$$

Below 16 MHz, the coil behaves as an inductance, but above 16 MHz it becomes capacitive and useless as an inductor. At exactly 16 MHz the impedance is purely resistive and fairly high.

Capacitors can also be self-resonant. The leads on a capacitor's body have an equivalent inductance, usually in the nanohenry range; the longer the lead length, the greater the inductance. This inductance is in series with the capacitance and forms a series resonant circuit at $f_o$. For example, a 0.01-$\mu$F capacitor with 10-nH series inductance is self-resonant at

$$f_o = \frac{1}{2\pi \sqrt{10^{-8}(0.01 \times 10^{-6})}}$$
$$f_o \simeq 16 \text{ MHz}$$

Below 16 MHz, the component still behaves as a capacitor, but above 16 MHz, it becomes inductive. At 250 MHz, this component would exhibit a total *inductive* reactance of about 16 Ω. To keep the self-resonant frequency of capacitors high, their leads are usually kept as short as possible. In fact, the soldering of capacitors with very short leads without burning the capacitor, nearby components, or one's fingers is quite an art in itself!

*COUPLED TUNED CIRCUITS.* A wider bandwidth, flatter passband, and greater selectivity are possible when two tuned circuits are *coupled*, as shown in Fig. 13.15a. Here energy is coupled from the transformer's primary to its secondary circuit through the common magnetic field, as indicated by the mutual inductance $M$. Although both sides of the transformer are shown to be tuned, there are applications where sufficient selectivity is possible with only one side tuned.

Coupled tuned circuits involve relatively "loose" coupling; therefore relationships involving turns ratios cannot be used. An equivalent circuit for a loosely coupled transformer, neglecting resistive and capacitive components, is shown in Fig. 13.15b. The mutual inductance is related to the primary and secondary inductances ($L_P$ and $L_S$) as follows:

$$M = k \sqrt{L_P L_S} \tag{13.16}$$

where $k$ is the coupling coefficient. For tightly coupled circuits, $k$ approaches 1, but in tuned coupled circuits, values of $k$ around 0.01 are typical. The dotted ends on the transformer diagram indicate that points $x$ and $y$ are in phase.

Before discussing specific coupled circuits, it might be worthwhile to consider the effect of coupling on the circuit frequency response. Shown in Fig.

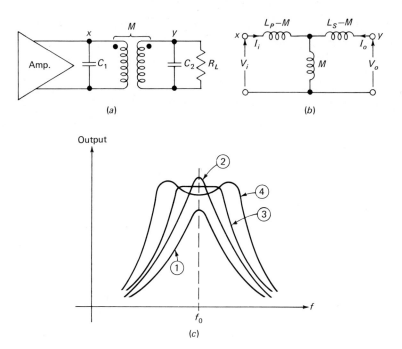

*Fig. 13.15  Coupling between tuned circuits may be achieved with a transformer as in (a); (b) equivalent circuit for the transformer in (a); (c) frequency response for different degrees of coupling: 1—undercoupled, 2—critically coupled, 3—transitionally coupled, 4—overcoupled.*

13.15c is the frequency response for different degrees of coupling. A tuned circuit of the type in Fig. 13.15a is assumed in which both primary and secondary circuits are tuned to the same frequency.

Curve 1 is for an undercoupled circuit; here the response is similar to that of a single tuned circuit. As the coupling is increased, the output increases until curve 2 is reached; this is critical coupling which yields the highest output. A higher $k$ yields curve 3, referred to as a transitionally coupled response. This is the flattest response that can be obtained. Increasing the coupling beyond this point yields the overcoupled response of curve 4; here a wider bandwidth is possible than with transitional coupling. The overcoupled response, though, is double-humped, not flat; the departure from a flat response is called ripple, usually specified in decibels.

Two tuned circuits may be coupled without using a transformer. The transformer equivalent circuit of Fig. 13.15b suggests one possibility, as in Fig. 13.16a. By suitable network transformations, the $T$ network made up of $L_A$, $L_B$, and $L_C$ may be replaced with an equivalent $\pi$ network using $L_1$, $L_2$, and $L_3$ as in Fig. 13.16b. Here $L_3$ provides the coupling between the $L_1C_1$ and $L_2C_2$ circuits.

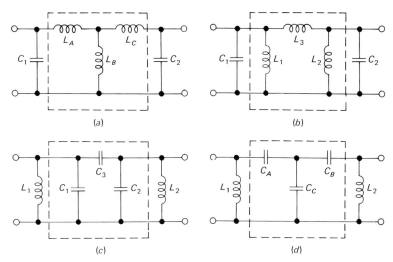

**Fig. 13.16** *Techniques for coupling tuned circuits.*

But the coupling can also be achieved with capacitor $C_3$, yielding the circuit of Fig. 13.16c. Finally, the $\pi$ network involving $C_1$, $C_2$, and $C_3$ in Fig. 13.16c may be transformed to an equivalent $T$ network made up of $C_A$, $C_B$, and $C_C$, as shown in Fig. 13.16d.

Any of the coupling circuits of Fig. 13.16 may be used; the choice is usually based on which configuration yields the most practical component values for the required response. Both $C_1$ and $C_2$ as well as $L_1$ and $L_2$ may be tapped to provide proper source and load impedance matching as required.

When very wide bandwidths are required, a technique known as stagger tuning may be used. Here a number of cascaded stages are tuned to slightly different frequencies; the total frequency response, in decibels, is the sum of the responses for each stage. This is illustrated in Fig. 13.17 for two cascaded stages, one tuned to $f_1$ and the other to $f_2$. Note that the total response is much wider and flatter than that for either amplifier stage.

### Problem Set 13.2

13.2.1. A 250-$\mu$H inductor has a $Q$ of 150 at 2 MHz. Determine $R_S$ and $R_P$, and draw both equivalent circuits.

13.2.2. What happens to the $Q$ of a coil when $R_S$ is increased? When $R_P$ is increased?

13.2.3. If more energy is removed from a circuit, what happens to the $Q$?

13.2.4. Determine the $Q$ of a capacitor whose reactance is 1,000 $\Omega$ and $R_P$ = 1 M$\Omega$. What is the equivalent $R_S$?

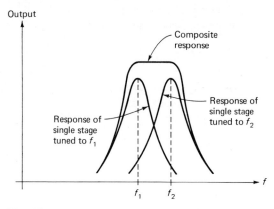

Fig. 13.17 Response for two stagger-tuned stages; one stage is tuned at $f_1$, the other at $f_2$, yielding the composite response shown.

13.2.5. What is the $Q$ of a circuit that stores energy but does not lose any energy? What are $R_S$ and $R_P$ for such a circuit?

13.2.6. Determine the loaded $Q$ in Prob. 13.2.1 if a 50-Ω load is connected in series with the coil. What is the new $R_P$?

13.2.7. A 2-pF capacitor with negligible losses is shunt-loaded with a 100-kΩ resistor at 100 MHz. What is the loaded $Q$?

13.2.8. An inductor with a $Q$ of 100 has an effective $R_S = 10$ Ω. What is the coil's $R_P$?

13.2.9. Under what conditions is $X_S \simeq X_P$?

13.2.10. In a series $RLC$ circuit, indicate whether each of the items below is maximum or minimum at $f_o$:

   a. Impedance
   b. Current through $R$
   c. Voltage across $R$

13.2.11. Repeat Prob. 13.2.10 for a parallel $RLC$ circuit.

13.2.12. In what way is the bandwidth of a tuned circuit affected by $Q$?

13.2.13. A 600-Ω generator produces an open-circuit output voltage of 1,000-mV RMS. The generator feeds a series $RLC$ circuit in which $R = 600$ Ω and $L = 1$ mH. It is desired to obtain maximum output across $R$ at 1 MHz; the coil's $Q$ at this frequency is 50. Determine

   a. The value of $C$
   b. The total circuit impedance at $f_o$
   c. The current at $f_o$
   d. The output voltage (across $R$) at $f_o$
   e. The bandwidth

**13.2.14.** In the circuit of Prob. 13.2.13, display all voltages at $f_o$ on a phasor diagram. Does the sum equal the applied voltage of 1,000 mV?

**13.2.15.** An amplifier with an output resistance of 100 k$\Omega$ is loaded with a parallel $RLC$ tuned circuit where $R = 100$ k$\Omega$, $L = 100$ $\mu$H with a $Q$ of 30 at 10 MHz, and $C$ must be selected to yield maximum response at 10 MHz.

    a. Determine $C$
    b. What is the circuit $Q$ at $f_o$?
    c. What is the bandwidth?

**13.2.16.** In Prob. 13.2.15, assume that a bandwidth of 1 MHz is required. What is the circuit $Q$ needed to achieve this bandwidth? Replace $C$ with two series capacitors, $C_1$ and $C_2$, and connect $R$ across $C_2$. Select $C_1$ and $C_2$ to satisfy the bandwidth requirement.

**13.2.17.** Repeat Prob. 13.2.16, but this time replace $L$ with two inductors $L_1$ and $L_2$ instead of $C$.

**13.2.18.** Draw a phasor diagram for all currents in Prob. 13.2.15 at $f_o$. Assume the amplifier's output can be simulated with a 1-$\mu$A current source shunted by the 100-k$\Omega$ output resistance. What is the voltage across the parallel tuned circuit at $f_o$?

**13.2.19.** Assume that in Example 13.14 the driving source impedance is 50 $\Omega$ instead of 18.5 k$\Omega$. To retain the same performance, the arrangement of Fig. 13.18 is proposed. $C_1$ and $C_2$ are as calculated in Example 13.14. Assume $L_1 + L_2 = 100$ $\mu$H with a $Q$ of 300. Determine $L_1$ and $L_2$.

**13.2.20.** A 0.02-$\mu$F capacitor is found to be self-resonant at 10 MHz. What is its equivalent series inductance? What causes this inductance? How can it be minimized?

**13.2.21.** What is the equivalent reactance, and is it inductive or capacitive, for the component in Prob. 13.2.20 at 100 MHz?

**13.2.22.** A 100-$\mu$H inductor has a total distributed capacitance of 10 pF. What is its self-resonant frequency? How does the inductor behave for frequencies above its self-resonant frequency?

**13.2.23.** Determine the turns ratio $n_2/n_1$ in the circuit of Fig. 13.14 if $Z$ is 50 $\Omega$ and if it is desired to transform it to 10,000 $\Omega$ on the $y$ side.

**13.2.24.** The coil of Fig. 13.14 has an equivalent $R_P$ of 10 k$\Omega$ when seen from the $y$ side. If $n_1/n_2 = 1/4$, what is the resistance seen at $x$ due to $R_P$?

**13.2.25.** Consider the circuit of Fig. 13.19. The inductor is 10 $\mu$H with an unloaded $Q$ of 50 and 4-pF distributed capacitance. Determine $C$ to resonate $L$ at 10.7 MHz. What is the coil's $R_P$?

**13.2.26.** In Prob. 13.2.25, assume $R_g = 50$ $\Omega$, $R_i = 2,000$ $\Omega$, $n_2/n_1 = 5$, $n_3/n_2 = 2$. What is the total resistance across $L$? What is the loaded $Q$? The bandwidth?

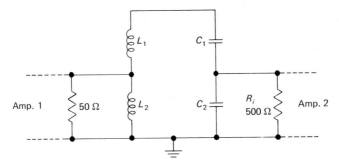

*Fig. 13.18  Circuit for Prob. 13.2.19.*

13.2.27. In Prob. 13.2.26, what is the equivalent resistance seen by $R_g$?

13.2.28. In Prob. 13.2.26, what is the equivalent resistance seen by $R_i$?

13.2.29. The following specifications apply to the circuit of Fig. 13.19: $L = 10$ μH with a $Q$ of 150 and 4-pF distributed capacitance at 10.7 MHz, $f_o = 10.7$ MHz, $B = 200$ kHz, $R_i = 2{,}000$ Ω, $R_g = 5$ kΩ. The equivalent resistance seen by $R_i$ (looking into $n_2$ turns of $L$) must be 2,000 Ω. Determine $C$, $n_3/n_2$, and $n_2/n_1$. What is the equivalent resistance seen by $R_g$ (looking into $n_1$ turns of $L$)?

13.2.30. Sketch the response curves for tuned circuits that are

    a. Undercoupled
    b. Critically coupled
    c. Transitionally coupled
    d. Overcoupled

13.2.31. What is ripple? Which coupled response exhibits minimum ripple? Maximum ripple?

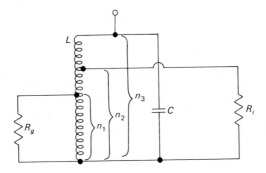

*Fig. 13.19  Circuit for Prob. 13.2.25.*

13.2.32. Which coupled response yields the widest bandwidth?
13.2.33. What is the 3-dB bandwidth increase over a single tuned circuit for a critically coupled, double-tuned circuit?
13.2.34. What is the coupling coefficient, in terms of $Q$, for critical coupling?
13.2.35. What is stagger tuning, and when is it used?

### 13.1d Circuits Employing Bipolar Transistors

The bipolar transistor has been described using the hybrid-pi model in Sec. 6.3 and 7.6. It was shown there that at low frequencies the various capacitances could be disregarded. We must now consider these capacitances and develop ways in which their effects can be predicted.

A hybrid-pi model is shown in Fig. 13.20a. This model simulates the small-signal behavior of a transistor biased at a suitable operating point. The parameters are determined either experimentally or from manufacturer's data sheets; in any case they remain constant only as long as the operating point does not change.

Normally, a source and load are connected at the transistor input and output, respectively. Both source and load may involve reactive as well as resistive elements; transformer coupling may also be used to achieve the proper impedance levels. Once all these components are added, the circuit becomes fairly complex and unless certain simplifications are made, very difficult to solve. One approach is to transform the circuit of Fig. 13.20a to that in Fig. 13.20b. The transformation procedure has been carried out elsewhere,[1] and only the results are given here. This way, if the parameters in Fig. 13.20a are known, their values can be substituted in the equations below to yield the parameters for the circuit in Fig. 13.20b. Although time-consuming, the procedure is straightforward:

$$R_1 = r_{bb'} + r_{b'e} \left[ \frac{r_{b'e} + r_{bb'}}{r_{b'e} + r_{bb'} + \omega^2 (C_{b'e} + C_{b'c})^2 (r_{bb'} r_{b'e}{}^2)} \right] \tag{13.17}$$

$$C_1 = \frac{C_{b'e} + C_{b'c}}{\left(1 + \dfrac{r_{bb'}}{r_{b'e}}\right)^2 + \omega^2 r_{bb'}{}^2 (C_{b'e} + C_{b'c})^2} \tag{13.18}$$

$$R_2 = \frac{1}{\dfrac{1}{r_{ce}} + \dfrac{1}{r_{b'c}} + g_m \dfrac{\dfrac{1}{r_{b'c}}\left(\dfrac{1}{r_{bb'}} + \dfrac{1}{r_{b'e}}\right) + \omega^2 C_{b'e} C_{b'c}}{\left(\dfrac{1}{r_{bb'}} + \dfrac{1}{r_{b'e}}\right)^2 + \omega^2 C_{b'e}{}^2}} \tag{13.19}$$

---

[1] See chap. 5 of Reference 14 at the end of this chapter.

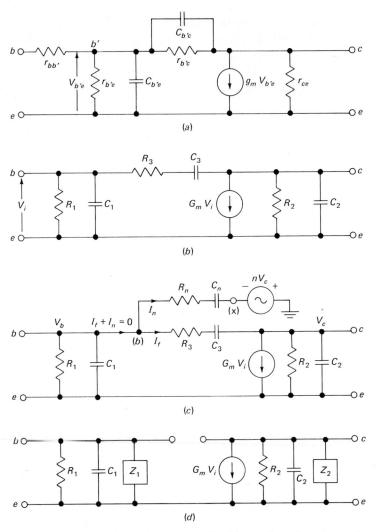

Fig. 13.20 (a) Transistor hybrid-pi model; (b) transformed version of the circuit in (a); (c) unilateralized circuit; (d) unilateralized circuit including loading on input and output due to $R_3$, $C_3$, $R_n$, and $C_n$.

$$C_2 = C_{b'c}\left[1 + \frac{g_m\left(\dfrac{1}{r_{bb'}} + \dfrac{1}{r_{b'e}}\right)}{\left(\dfrac{1}{r_{bb'}} + \dfrac{1}{r_{b'e}}\right)^2 + \omega^2 C_{b'e}^2}\right] \qquad (13.20)$$

$$R_3 = r_{bb'}\left(1 + \frac{C_{b'e}}{C_{b'c}}\right) + \frac{1 + \dfrac{r_{bb'}}{r_{b'e}}}{r_{b'c}\omega^2 C_{b'c}^2} \tag{13.21}$$

$$C_3 = \frac{C_{b'c}}{1 + \dfrac{r_{bb'}}{r_{b'e}} - \dfrac{r_{bb'}}{r_{b'c}}\dfrac{C_{b'e}}{C_{b'c}}} \tag{13.22}$$

$$G_m = \frac{g_m}{\sqrt{\left(1 + \dfrac{r_{bb'}}{r_{b'e}}\right)^2 + [r_{bb'}\omega(C_{b'e} + C_{b'c})]^2}} \tag{13.23}$$

Consider the path involving $R_3$ and $C_3$ between points $c$ and $b$ in Fig. 13.20$b$. These components account for the internal feedback present in all transistors. Because of this feedback, the input and output circuits are not isolated; instead the output circuit is influenced by the source at the input, while the input circuit is influenced by the load. This not only makes the analysis more complicated, but also may cause the circuit to oscillate. If both input and output circuits are tuned, the alignment procedure is more difficult, because every time the input circuit is tuned, the output circuit is affected; this necessitates retuning the output, which in turn affects the input, and so on.

Ideally the amplifier should be unilateral—that is, it should be possible for a signal to be transmitted from the input to the output but not from the output to the input. This would be the case if, instead of the $R_3$-$C_3$ network, we had an open circuit in its place. Then there would be perfect isolation between input and output.

One method for unilateralizing[1] a transistor is shown in Fig. 13.20$c$. $I_f$ is the current due to internal feedback. A passive network, made up of $R_n$ and $C_n$, is externally connected to supply an additional current $I_n$ equal in magnitude but opposite in phase to $I_f$. This is achieved by connecting point $x$ to a voltage source which is 180° out of phase with respect to $V_c$. This source holds point $x$ at $-nV_c$ V relative to ground; the $n$ is simply a constant. How the 180°-out-of-phase voltage is achieved is unimportant for now (a transformer is one way). What matters is that $I_f$ must equal $-I_n$ so that there is no net current leaving point $b$, as follows:

$$I_f = -I_n$$
$$\frac{V_b - V_c}{R_3 + (1/j\omega C_3)} = -\frac{V_b - (-nV_c)}{R_n + (1/j\omega C_n)}$$

---

[1] A *neutralized* amplifier is one in which only the internal reactive feedback is canceled; a *unilateralized* amplifier is one in which both resistive and reactive effects are canceled.

Usually $V_b \ll V_c$ ($V_b$ is the input, while $V_c$ is the transistor output voltage); therefore

$$\frac{-V_c}{R_3 + (1/j\omega C_3)} \simeq -\frac{nV_c}{R_n + (1/j\omega C_n)}$$

$$R_n + \frac{1}{j\omega C_n} = n\left(R_3 + \frac{1}{j\omega C_3}\right)$$

$$R_n + \frac{1}{j\omega C_n} = nR_3 + \frac{1}{j\omega C_3/n}$$

Therefore

$$R_n = nR_3 \tag{13.24}$$

and

$$C_n = \frac{C_3}{n} \tag{13.25}$$

Once the circuit is unilateral, the input and output may be treated independently, as shown in Fig. 13.20d. $Z_1$ and $Z_2$ account for the loading due to $R_3$, $C_3$, $R_n$, and $C_n$. These impedances can usually be neglected, as they are generally large compared with the rest.

A typical tuned amplifier is shown in Fig. 13.21a. $C_A$ is a coupling capacitor, while $C_B$ and $C_C$ are bypass capacitors. They all have negligible reactance at the operating frequency. $R_A$, $R_B$, and $R_E$ set and stabilize the operating point. The choice of operating point is made on the basis of desired performance (maximum gain, low noise, and so on). $C_D$ plus any other reflected or stray capacitance form a parallel resonant circuit with the transformer's inductance to provide maximum gain at the desired frequency. The transformer primary is tapped for proper impedance transformation. The transformer is tightly coupled ($k \simeq 1$); therefore impedances are transformed in proportion to the turns ratio squared.

The AC equivalent circuit is shown in Fig. 13.21b. The $R_n$-$C_n$ network is connected between the transistor base and the undotted end (point $x$) of the transformer secondary. The voltage at $x$ is 180° out of phase with the voltage at the collector, and the ratio of these two voltages corresponds to the parameter $n$ used in Eqs. (13.24) and (13.25):

$$n = \frac{V_x}{V_c} \tag{13.26}$$

**Example 13.15**

The circuit of Fig. 13.21a is to be designed with a center frequency of 1 MHz and a bandwidth of 20 kHz. The transformer primary has an inductance of 50 μH

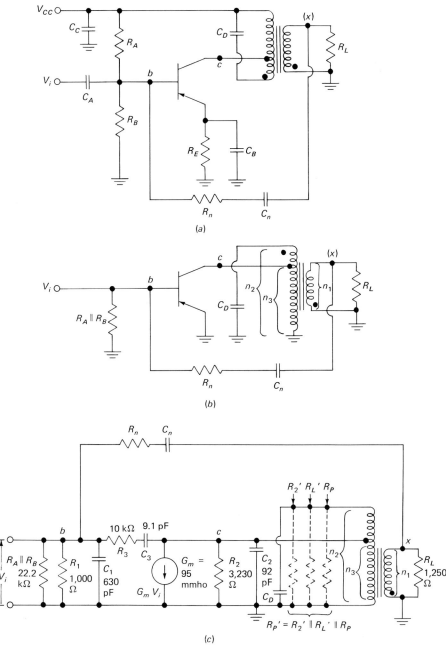

Fig. 13.21 Circuits for Example 13.15. $R'_2$: value of $R_2$ reflected across entire transformer primary: $R'_2 = R_2(n_2/n_3)^2$; $R'_L$: value of $R_L$ reflected across the entire transformer primary: $R'_L = R_L(n_2/n_1)^2$; $R_P$: transformer losses expressed in terms of a parallel resistance: $R_P = Q_U X$.

shunted by 100 pF and an unloaded $Q$ of 150. The load is 1,250 $\Omega$, purely resistive. A power match between load and transistor output is to be realized. The transistor has the following parameters:

$$r_{bb'} = 100\ \Omega \qquad g_m = 120\ \text{mmho} \qquad r_{b'e} = 1,000\ \Omega,$$
$$r_{b'c} = 5\ \text{M}\Omega \qquad C_{b'e} = 1,000\ \text{pF} \qquad C_{b'c} = 10\ \text{pF},$$
$$r_{ce} = 100\ \text{k}\Omega$$

The biasing resistors have been selected to yield the proper operating point; they are $R_A = 68$ k$\Omega$, $R_B = 33$ k$\Omega$, $R_E = 1.5$ k$\Omega$. Determine the following:

a. All parameters in the equivalent circuit of Fig. 13.20b
b. The loaded $Q$ for the output tuned circuit and the total parallel equivalent resistance required to achieve it
c. The value of $R_L$ reflected across the transformer primary and the turns ratio $n_2/n_1$ required to achieve it
d. The value of $R_2$ reflected across the entire transformer primary and the turns ratio $n_2/n_3$ required
e. $C_n$ and $R_n$
f. $C_D$

**Solution**

a. The parameters in the equivalent circuit of Fig. 13.20b are calculated by substituting the information given above into Eqs. (13.17) through (13.23). The calculations are not carried out here, but the results are given:

$$R_1 = 1,000\ \Omega \qquad C_1 = 630\ \text{pF} \qquad R_2 = 3,230\ \Omega$$
$$C_2 = 92\ \text{pF} \qquad R_3 = 10\ \text{k}\Omega \qquad C_3 = 9.1\ \text{pF}$$
$$G_m = 95\ \text{mmho}$$

A complete equivalent circuit is shown in Fig. 13.21c.

b. The loaded $Q$ is determined on the basis of the required bandwidth. This is

$$Q_L = \frac{f_o}{B} = \frac{10^6}{20 \times 10^3} \tag{13.13}$$

$$Q_L = 50$$

The transformer primary reactance at $f_o$ is

$$X_L = (6.28 \times 10^6)(50 \times 10^{-6}) = 3,140\ \Omega$$

To achieve a loaded $Q$ of 50, the total parallel resistance must be

$$R'_P = QX = 50(3,140) = 15.7\ \text{k}\Omega$$

The 15.7 k$\Omega$ is made up of the transformer's own $R_P$, $R'_L$, and $R'_2$ (values of $R_L$

and $R_2$ reflected across the entire transformer primary) as shown in Fig. 13.21c.

c. For a power match, $R'_L$ should equal the parallel combination of $R'_2$ and $R_P$, as follows:

$$R'_L = R'_2 \| R_P$$

but

$$R'_L \|(R'_2 \| R_P) = 15.7 \text{ k}\Omega$$
$$R'_L \| R'_L = 15.7 \text{ k}\Omega$$
$$R'_L = 31.4 \text{ k}\Omega$$

Since $R_L = 1{,}250 \ \Omega$ and $R'_L = 31.4 \text{ k}\Omega$, the turns ratio $n_2/n_1$ is

$$\left(\frac{n_2}{n_1}\right)^2 = \frac{R'_L}{R_L} = \frac{31.4 \times 10^3}{1{,}250}$$

$$\frac{n_2}{n_1} \simeq \sqrt{25} = 5$$

d. The coil's $R_P$ is

$$R_P = Q_U X = 150(3{,}140) = 47.1 \text{ k}\Omega$$

The parallel combination of $R_P$ and $R'_2$ is equal to $R'_L = 31.4 \text{ k}\Omega$; therefore

$$\frac{R_P R'_2}{R_P + R'_2} = 31.4 \times 10^3$$

from which $R'_2 = 94.5 \text{ k}\Omega$. But

$$R_2 \left(\frac{n_2}{n_3}\right)^2 = R'_2$$

therefore

$$\frac{n_2}{n_3} = \sqrt{\frac{94.5 \times 10^3}{3.23 \times 10^3}} = 5.4$$

e. To determine $C_n$ and $R_n$, the parameter $n$ is required. Note that $n$ is the ratio of $V_x$ to $V_c$:

$$n = \frac{V_x}{V_c} \qquad (13.26)$$

$$V_x = \frac{n_1}{n_3} V_c$$

$$n = \frac{n_1}{n_3} = \frac{n_1}{n_2} \frac{n_2}{n_3} = \frac{1}{5}(5.4)$$

$$n = 1.08$$

$$C_n = \frac{C_3}{n} = \frac{9.1 \times 10^{-12}}{1.08} \tag{13.25}$$

$$C_n = 8.4 \text{ pF}$$
$$R_n = nR_3 = 1.08(10 \times 10^3) \tag{13.24}$$
$$R_n = 10.8 \text{ k}\Omega$$

f. The total tuning capacitance required is

$$C = \frac{1}{4\pi^2 f_o^2 L} \tag{13.11}$$

$$C = \frac{1}{(40 \times 10^{12})(50 \times 10^{-6})} = 500 \text{ pF}$$

The transformer itself has 100 pF of distributed capacitance; $C_2$, when transformed to its equivalent value across the entire transformer, is

$$C_2' = C_2 \left(\frac{n_3}{n_2}\right)^2 = (92 \times 10^{-12}) \left(\frac{1}{5.4}\right)^2$$
$$C_2' = 3.15 \text{ pF}$$

The additional capacitance required is

$$C_D = (500 - 100 - 3.15)(10^{-12})$$
$$C_D \simeq 400 \text{ pF}$$

Usually some stray capacitance is unavoidable; therefore a practical choice might be $C_D = 360$ pF shunted by a variable 8- to 50-pF capacitor.

### Example 13.16

Estimate the power gain in the circuit of Example 13.15.

### Solution

The power gain is the ratio of output to input power. At the input we have

$$P_i = \frac{V_i^2}{R_i}$$

At the output,

$$P_o = (I_o')^2 R_L'$$

where $I_o'$ is the current through $R_L'$. Since $R_L'$ is equal to the remaining parallel resistances, $I_o'$ at resonance is simply one-half $G_m V_i$:

$$P_o = \left(\frac{G_m V_i}{2}\right)^2 R'_L$$

The input resistance is $R_1 \| R_A \| R_B = 1{,}000 \| 22{,}200 = 955\ \Omega$; the power gain is therefore

$$G = \frac{P_o}{P_i} = \frac{G_m^2 V_i^2 R'_L}{4 V_i^2 / R_i}$$
$$G = \tfrac{1}{4} G_m^2 R'_L R_i = \tfrac{1}{4}(95^2)(10^{-6})(31.4 \times 10^3)(955)$$
$$G = 67{,}500 = 48.3\ \text{dB}$$

Other techniques for canceling internal feedback are often used. In Fig. 13.22, an inductor $L_n$ is connected between collector and base; capacitor $C_c$ is used to block DC (without $C_c$ the collector and base would be practically at the same DC potential) but is essentially a short circuit at the operating frequency. $L_n$ is chosen to have approximately the same reactance at $f_o$ as the internal capacitance $C_{ob}$:[1]

$$L_n = \frac{1}{\omega_o^2 C_{ob}} \tag{13.27}$$

The current through $L_n$ is of opposite phase to the internal feedback current; therefore cancellation takes place. Note that only the capacitive feedback is neutralized and that the resistive feedback remains. Therefore the circuit is not completely unilateralized.

In Fig. 13.23, another neutralization circuit is shown. Here $C_A$ is chosen so that its reactance at $f_o$ is much lower than the inductor's reactance, in which case

$$I \simeq \frac{V_1}{j\omega_o L}$$

and

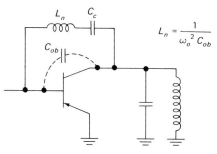

Fig. 13.22  *Neutralization technique.*

[1] $C_{ob} \simeq C_{b'c}$.

$$V_2 = I\,\frac{1}{j\omega_o C_A} = \frac{V_1}{(j\omega_o L)(j\omega_o C_A)}$$

But $\omega_o L = 1/(\omega_o C_T)$ (at the resonant frequency); therefore

$$V_2 = \frac{V_1}{j[1/(\omega_o C_T)](j\omega_o C_A)} = -\frac{V_1 C_T}{C_A} \qquad (a)$$

Now we have a source of voltage $V_2$ which is $180°$ out of phase relative to $V_1$. If we assume $V_1 \gg V_b$, the internal feedback current is

$$I_f \simeq \frac{V_1}{1/(j\omega_o C_{ob})} \qquad (b)$$

and the current through $C_n$ is

$$I_n \simeq \frac{V_2}{1/(j\omega_o C_n)} \qquad (c)$$

For neutralization to take place, $I_f = -I_n$:

$$V_1 j\omega_o C_{ob} = -V_2 j\omega_o C_n$$
$$V_1 C_{ob} = -V_2 C_n$$

But

$$V_2 = \frac{-V_1 C_T}{C_A} \qquad (a)$$

therefore

$$V_1 C_{ob} = V_1 C_n \frac{C_T}{C_A}$$

$$C_n = \frac{C_A}{C_T} C_{ob} \qquad (13.28)$$

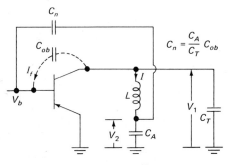

Fig. 13.23  *Neutralization technique.*

## Example 13.17

Determine the neutralizing capacitance required in the circuit of Fig. 13.23 if $C_{ob} = 1$ pF, $C_T = 100$ pF, $f_o = 5$ MHz.

**Solution**

At $f_o$, the tuning capacitor's reactance is

$$X_{C_T} = \frac{1}{(6.28)(5 \times 10^6)(100)(10^{-12})}$$

$$X_{C_T} \simeq 320 \ \Omega$$

$L$ will also have a reactance of around 320 $\Omega$; the reactance of $C_A$ should be much less than this, no more than one-tenth, say 30 $\Omega$. The value of $C_A$ is therefore

$$C_A = \frac{1}{2\pi f_o X} = \frac{1}{(6.28)(5 \times 10^6)(30)}$$

$$C_A \simeq 0.001 \ \mu\text{F}$$

The neutralizing capacitor can now be calculated:

$$C_n = \frac{C_A}{C_T} C_{ob} \tag{13.28}$$

$$C_n = \left(\frac{10^{-9}}{10^{-10}}\right)(10^{-12})$$

$$C_n = 10 \text{ pF}$$

The neutralization techniques so far discussed have the disadvantage of being frequency-dependent; that is, perfect neutralization can be achieved only at one frequency. For narrow-band amplifiers this can still be adequate, although variations in transistor parameters (due to replacement or $Q$ point shifts) may require changing the values of neutralizing components.

Mismatching is often used instead of neutralization to prevent instability. Here source and load are deliberately not matched to the device's input and output impedances, respectively. Although mismatching precludes obtaining maximum power transfer, it is often preferred because it is essentially independent of frequency and transistor parameters, which is an advantage in wideband amplifiers. Mismatching also provides some isolation between input and output circuits so that the input impedance can be made almost independent of the load.

### Problem Set 13.3

13.3.1. The following parameters are given for a transistor at 5 MHz: $r_{bb'} = 150 \ \Omega$, $r_{b'e} = 1.2$ k$\Omega$, $C_{b'e} = 1{,}000$ pF, $C_{b'c} = 5$ pF, $r_{b'c} = 10^6 \ \Omega$, $r_{ce} =$

75 kΩ, $g_m$ = 80 mmho. Determine the equivalent parameters for the circuit in Fig. 13.20b.

13.3.2. Why is internal feedback from collector to base in a transistor usually undesirable?

13.3.3. What is meant by a unilateralized amplifier?

13.3.4. Using the results of Prob. 13.3.1 and $n = 1$, what is the value of $R_n$ and $C_n$ in the circuit of Fig. 13.20c?

13.3.5. What do you think would happen if the secondary winding in the transformer of Fig. 13.21a were reversed (dot near $x$)?

13.3.6. Using the data from Prob. 13.3.1 in the circuit of Fig. 13.21a,

a. Determine the loaded $Q$ for the output tuned circuit if $f_o = 5$ MHz and a 3-dB bandwidth of 250 kHz is required.

b. Determine the total parallel equivalent resistance required to achieve the specifications in (a) assuming a total inductance of 40 μH.

c. If the 40-μH inductance has an unloaded $Q$ of 100, what portion of the resistance in (b) must be due to loading?

13.3.7. If $R_L = 50$ Ω in Prob. 13.3.6, determine the turns ratio $n_2/n_1$ (as defined in Fig. 13.21b) for maximum power transfer.

13.3.8. Using the data of Probs. 13.3.6 and 13.3.7, determine the turns ratio $n_2/n_3$. What value of $C_D$ is required if the transformer's distributed capacitance (across $n_2$ turns) is 5 pF?

13.3.9. Determine the neutralizing components $R_n$ and $C_n$ required in the circuit of Prob. 13.3.6 (the data of Probs. 13.3.7 and 13.3.8 also apply).

13.3.10. Estimate $R_i$ and $C_i$ in the circuit of Prob. 13.3.6. Neglect the effects of biasing resistors and feedback components.

13.3.11. Determine a suitable value of neutralizing inductance in the circuit of Fig. 13.22 if $C_{ob} = 5$ pF and $f_o = 10$ MHz.

13.3.12. Determine suitable values of $C_A$ and $C_n$ in the circuit of Fig. 13.23 if $C_{ob} = 5$ pF, $f_o = 10$ MHz, and $L = 20$ μH.

13.3.13. What advantages does mismatching have over unilateralization for achieving amplifier stability?

### 13.1e  Analysis Using Admittance Parameters

The performance of active devices (bipolar transistors, FETs) operating linearly may be predicted using a wide variety of models. At high frequencies, the hybrid-pi model for bipolar transistors has been described and used to predict circuit performance. The advantage of the hybrid-pi model is that its parameters can be identified with the physical structure of the device; the disadvantage is that it is fairly complex and time-consuming transformations are usually required.

The $h$-parameter model of Chap. 6 could be used at high frequencies provided the reactive components of each parameter are also included. The measurement of $h$ parameters, though, requires that the input and output be open-circuited (for $h_r$ and $h_o$), a condition difficult to achieve at high frequencies because of unavoidable shunt capacitances. For these reasons, a more suitable model, involving admittance ($y$) parameters, is commonly used. The $y$ parameters can be obtained experimentally at any specific frequency; they all involve short-circuiting the input or output, a condition easily achieved by using shunt capacitors or coaxial lines of the proper length. $y$ parameters can be used to analyze any active device, bipolar or field effect.

Consider the equivalent circuit of Fig. 13.24. The input voltage and current are $V_1$ and $I_1$, respectively; similarly $V_2$ and $I_2$ are the device's output voltage and current. There are two admittances, $y_i$ and $y_o$. There are also two controlled current generators. The $y_f V_1$ generator produces a current proportional to the input voltage $V_1$. The parameter $y_f$ is the forward transfer admittance and is a measure of the device's gain. The $y_r V_2$ generator produces a current at the input proportional to the output voltage $V_2$. The parameter $y_r$ is the reverse transfer admittance. It is a measure of the internal feedback and is due to $r_{b'c}$ and $C_{b'c}$ in common emitter bipolar transistors; in common source FETs, $y_r$ is due to $C_{gd}$. In a unilateralized device, $y_r$ is obviously zero. Note that including $y_r$ in the input of the equivalent circuit in Fig. 13.24 has the effect of making the input depend on the output, which, of course, is the case when the device is nonunilateral.

The analyses to follow involve work with admittances. Therefore it would be wise to bear in mind the following facts:

a. Admittance is the ratio of current to voltage
b. Admittance is the inverse of impedance
c. Parallel admittances can be combined by simple addition
d. Admittances are generally complex quantities of the following general form:

$$y = g + jb \tag{13.29}$$

where the real part $g$ is the conductance and the imaginary part $b$ the susceptance, both in mhos (throughout our discussion, lowercase letters for

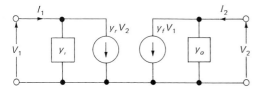

**Fig. 13.24** *Model of active device using admittance parameters.*

$y$, $g$, and $b$ will be used to denote parameters of the active device, while capital letters for $Y$, $G$, and $B$ apply to the source and load)

e. Conductance is the inverse of resistance
f. Susceptance is the inverse of reactance

For the equivalent circuit of Fig. 13.24 we can write

Input circuit: $\quad I_1 = V_1 y_i + V_2 y_r \quad$ (13.30)
Output circuit: $\quad I_2 = V_1 y_f + V_2 y_o \quad$ (13.31)

If the output is short-circuited, $V_2 = 0$ and Eq. (13.30) becomes

$$I_1 = V_1 y_i$$

from which

$$y_i = \frac{I_1}{V_1}\bigg|_{V_2=0} \quad (13.32)$$

The parameter $y_i$ is therefore the input admittance with the output short-circuited; its units are mhos. Again, with the output short-circuited, Eq. (13.31) becomes

$$I_2 = V_1 y_f$$

and

$$y_f = \frac{I_2}{V_1}\bigg|_{V_2=0} \quad (13.33)$$

The forward transfer admittance is therefore the ratio of $I_2$ and $V_1$ when the output is short-circuited; its units are also mhos.

If we now short-circuit the input, we have $V_1 = 0$, and Eq. (13.30) and (13.31) become

$$I_1 = V_2 y_r$$

and

$$I_2 = V_2 y_o$$

from which

$$y_r = \frac{I_1}{V_2}\bigg|_{V_1=0} \quad (13.34)$$

and

$$y_o = \frac{I_2}{V_2}\bigg|_{V_1=0} \quad (13.35)$$

Note that both the reverse transfer admittance $y_r$ and the output admittance $y_o$ have units of mhos.

The foregoing derivations suggest ways in which the $y$ parameters can be obtained experimentally. For example, AC-short-circuiting the output forces $V_2$ to equal 0; $V_1$, $I_1$, and $I_2$ are then measured and substituted in Eqs. (13.32) and (13.33) to yield $y_i$ and $y_f$, respectively.

Perhaps it ought to be stressed that $y$ parameters are properties of the active device, at a given frequency and at a given operating point. They are independent of the source and load that may be connected to the active device. If the operating point shifts or the operating frequency is changed, though, the $y$ parameters also change.

## Example 13.18

The following small-signal measurements are made on a field effect transistor which is biased in the linear portion of its operating region:

a.    $V_1 = 1 \text{ V } \underline{/0°}$      $I_1 = 1 \text{ mA } \underline{/88.6°}$
     $V_2 = 0$            $I_2 = 5 \text{ mA } \underline{/0°}$
b.    $V_1 = 0$            $I_1 = 0.2 \text{ mA } \underline{/90°}$
     $V_2 = 1 \text{ V } \underline{/0°}$      $I_2 = 1 \text{ mA } \underline{/84.3°}$

Determine all four $y$ parameters and identify the conductance $g$ and susceptance $b$ associated with each.

## Solution

The data in (a) involve a shorted output ($V_2 = 0$); therefore they can be used to compute $y_i$ and $y_f$:

$$y_i = \left.\frac{I_1}{V_1}\right|_{V_2=0} \tag{13.32}$$

$$y_i = \frac{10^{-3}\;\underline{/88.6°}}{1\;\underline{/0°}} = 10^{-3}\;\underline{/88.6°}$$

$$y_f = \left.\frac{I_2}{V_1}\right|_{V_2=0} \tag{13.33}$$

$$y_f = \frac{(5 \times 10^{-3})\;\underline{/0°}}{1\;\underline{/0°}} = (5 \times 10^{-3})\;\underline{/0°}$$

The data in (b) involve a shorted input ($V_1 = 0$); therefore they can be used to compute $y_o$ and $y_r$:

$$y_o = \left.\frac{I_2}{V_2}\right|_{V_1=0} \tag{13.35}$$

$$y_o = \frac{10^{-3}\,\underline{/84.3°}}{1\,\underline{/0°}} = 10^{-3}\,\underline{/84.3°}$$

$$y_r = \frac{I_1}{V_2}\bigg|_{V_1=0} \tag{13.34}$$

$$y_r = \frac{(0.2 \times 10^{-3})\,\underline{/90°}}{1\,\underline{/0°}} = (0.2 \times 10^{-3})\,\underline{/90°}$$

To separate each $y$ parameter into a conductance and susceptance, they must be expressed in rectangular, as opposed to polar, form. The conversion is accomplished as follows:

$$y_i = g_i + jb_i$$

$$y_i = 10^{-3}(\cos 88.6° + j\sin 88.6°)$$
$$y_i = (25 \times 10^{-6}) + j10^{-3} \qquad g_i = 25 \times 10^{-6}\text{ mho} \qquad b_i = 10^{-3}\text{ mho}$$

$$y_o = g_o + jb_o$$

$$y_o = 10^{-3}(\cos 84.3° + j\sin 84.3°)$$
$$y_o = 10^{-4} + j10^{-3} \qquad g_o = 10^{-4}\text{ mho} \qquad b_o = 10^{-3}\text{ mho}$$

$$y_f = g_f + jb_f$$

$$y_f = (5 \times 10^{-3}) + j0 \qquad g_f = 5 \times 10^{-3}\text{ mho} \qquad b_f = 0$$

$$y_r = g_r + jb_r$$

$$y_r = 0 + j(0.2 \times 10^{-3}) \qquad g_r = 0 \qquad b_r = 0.2 \times 10^{-3}\text{ mho}$$

Further subscripts are usually added to $y$ parameters in order to denote the particular configuration. For example, $y_{fe}$ is the forward transfer admittance for a common emitter (bipolar) transistor configuration. Similarly $y_{rs}$ is the reverse transfer admittance for a common source (field effect) transistor configuration. It is also possible to convert the $y$ parameters from one configuration to another; similarly, $y$ parameters can be converted to any other set of parameters ($z$, $h$), and vice versa, through the use of appropriate equations.

### Example 13.19

Determine the common source $y$ parameters for the 2N3823 FET at a frequency of 100 MHz using the data sheets in Appendix 10.

### Solution

The real and imaginary parts of the common source $y$ parameters are given on the fourth page of the data sheet:

$$y_{is} = 0 + j3 \text{ mmho}$$
$$y_{fs} = 4.5 - j1 \text{ mmho}$$
$$y_{rs} = 0 - j0.9 \text{ mmho}$$
$$y_{os} = 0 + j1 \text{ mmho}$$

Once the $y$ parameters for the active device are known, we may add a source and load, as shown in Fig. 13.25. Both the source and load are assumed, in general, to involve resistive as well as reactive components. We are now ready to determine a number of performance measures, such as input and output admittances and power gain.

The quantity $Y_i$ is the amplifier input admittance—that is, the ratio of $I_1$ to $V_1$ with the output properly loaded. The input admittance for the amplifier should not be confused with $y_i$, which is the input admittance for a short-circuited output. $Y_i$ is a property of the amplifier *circuit*, while $y_i$ is a property of the amplifying *device*. A similar observation applies to $y_o$ and $Y_o$; the former is a property of the active device and is measured with a short-circuited input. The latter ($Y_o$) is a property of the circuit and is given by the ratio $I_2/V_2$ with the source admittance properly connected and the current source removed ($I_S = 0$).

In the circuit of Fig. 13.25, we can write

$$y_f V_1 = -V_2(y_o + y_L)$$

$$V_2 = -\frac{y_f V_1}{y_o + Y_L} \quad ^1 \tag{13.36}$$

Substitution of the above into Eq. (13.30) yields

$$I_1 = V_1 y_i + \frac{-y_f V_1}{y_o + Y_L} y_r$$

$$Y_i = \frac{I_1}{V_1} = y_i - \frac{y_f y_r}{y_o + Y_L} \tag{13.37}$$

Equation (13.37) gives the input admittance for any load admittance $Y_L$.

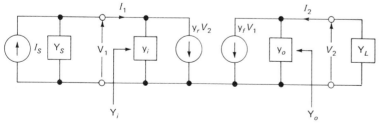

Fig. 13.25   y-parameter model with source and load added.

---

[1] The negative sign is due to the fact that the direction of current supplied by the $y_f V_1$ generator produces a polarity for $V_2$ opposite to that shown in the figure.

To obtain an expression for the output admittance, the ratio $I_2/V_2$ is determined while $Y_S$ is in place but $I_S = 0$. On the input side we can then write

$$y_r V_2 = -V_1(y_i + Y_S)$$

$$V_1 = \frac{-V_2 y_r}{y_i + Y_S} \qquad (13.38)$$

Substitution of the above into Eq. (13.31) yields

$$I_2 = \frac{-y_f V_2 y_r}{y_i + Y_S} + y_o V_2$$

$$Y_o = \frac{I_2}{V_2} = y_o - \frac{y_f y_r}{y_i + Y_S} \qquad (13.39)$$

Equation (13.39) gives the output admittance for any source admittance $Y_S$.

**Example 13.20**

For the data given below, determine

a. The input admittance, in rectangular form
b. The input impedance, in polar form
c. The equivalent parallel input resistance and capacitance

### y parameters at f = 10 MHz

$y_i = 10^{-4} + j10^{-3}$ mho   $Y_L = 0.02 + j0$ mho
$y_o = 10^{-6} + j10^{-3}$ mho
$y_r = 0 - j10^{-4}$ mho   $Y_S = 0.02 + j0$ mho
$y_f = 10^{-2} + j0$ mho

**Solution**

a. The input admittance is given by

$$Y_i = y_i - \frac{y_f y_r}{y_o + Y_L} \qquad (13.37)$$

$$Y_i = 10^{-4} + j10^{-3} - \frac{-j10^{-6}}{10^{-6} + j10^{-3} + 0.02}$$

$$Y_i = 0.102 + j1.05 \text{ mmho}$$

b. The input impedance is the reciprocal of $Y_i$:

$$Z_i = \frac{1}{Y_i}$$

$$Z_i = \frac{1}{(0.102 + j1.05)(10^{-3})}$$

$$Z_i = \frac{1,000}{1.055 \,/84.4°}$$

$$Z_i = 950 \,/\!-\!84.4°\ \Omega$$

c. The total input admittance may be separated into its real and imaginary parts, yielding an equivalent parallel conductance and susceptance, respectively:

$$Y_i = G_i + jB_i \tag{13.40}$$

where $G_i = 0.102$ mmho and $B_i = 1.05$ mmho. The equivalent shunt input resistance is

$$R_i = \frac{1}{G_i} = \frac{1}{0.102 \times 10^{-3}} = 9.8\ \text{k}\Omega$$

The susceptance is positive, yielding a capacitive reactance:

$$\frac{1}{2\pi f C_i} = \frac{1}{1.05 \times 10^{-3}}$$

$$C_i = \frac{1.05 \times 10^{-3}}{6.28(10 \times 10^6)}$$

$$C_i = 16.7\ \text{pF}$$

To determine power gain, several different measures may be used.[1] The actual power gain $G_p$ is simply the ratio of actual load power to actual input power:

$$G_p = \frac{P_o}{P_i} \tag{13.41}$$

In general, $P = |V|^2/R$ and $R = 1/G$; therefore we can write

$$G_p = \frac{|V_2|^2 G_L}{|V_1|^2 G_i} \tag{13.42}$$

Now $G_i$ is the real part of $Y_i$ which is given by Eq. (13.37); also $V_2$ is given in terms of $V_1$ by Eq. (13.36), allowing us to express $G_p$ as

$$G_p = \frac{|y_f|^2\, G_L}{|y_o + Y_L|^2\, \mathrm{Re}\left(y_i - \dfrac{y_f y_r}{y_o + Y_L}\right)} \tag{13.43}$$

[1] See Sec. 4.1.

To maximize $G$, a conjugate power match is required at the load, so that $Y_L = Y_o^*$. It is simpler, though, to let $Y_L = y_o^*$ instead. This yields

$$y_o + Y_L = (g_o + jb_o) + (g_o - jb_o)$$
$$y_o + Y_L = 2g_o$$

The result is almost a power match,[1] and the maximum unneutralized power gain $G_{pmu}$ is

$$G_{pmu} \simeq \frac{|y_f|^2}{4g_o g_i - 2Re(y_f y_r)} \tag{13.44}$$

If the circuit is unstable, $G_{pmu}$ becomes infinite, and the amplifier oscillates. If the device is unilateral (perfectly neutralized), $y_r = 0$, and we have

$$G_{pmn} = \frac{|y_f|^2}{4g_o g_i} \tag{13.45}$$

Note that neither $G_p$, $G_{pmu}$, nor $G_{pmn}$ includes the effects of source impedance. For this purpose, the transducer power gain $G_{pt}$ is used because, in addition to the active device parameters, it also includes the effects of source and load. Transducer gain is defined as follows:

$$G_{pt} = \frac{P_o}{P_{i(\max)}} \tag{13.46}$$

where $P_{i(\max)}$ is the maximum power available from the source. For the circuit of Fig. 13.25, we can state that maximum power is available from the source if $Y_i = Y_S^*$, in which case $G_i$, the real part of $Y_i$, is equal to $G_S$, the real part of $Y_S$. The imaginary components of $Y_i$ and $Y_S$ cancel so that $I_S$ divides equally between $G_S$ and $G_i$. This means that the current through $G_i$ is $0.5 I_S$, and the input power is

$$P_{i(\max)} = (0.5 I_S)^2 \left(\frac{1}{G_i}\right)$$

$$P_{i(\max)} = \frac{I_S^2}{4 G_i} = \frac{I_S^2}{4 G_S} \tag{13.47}$$

On the input side, $I_S$ flows through the parallel combination of $Y_S$ and $Y_i$, producing a voltage $V_1$:

$$V_1 = I_S \frac{1}{Y_S + Y_i} \tag{a}$$

---

[1] Linvill has shown that the error involved in this simplification is no greater than 3 dB unless the device is potentially unstable; see chap. 11 of Reference 15 at the end of this chapter.

Substituting $Y_i$ from Eq. (13.37) into Eq. (a), we get

$$V_1 = \frac{I_S}{Y_S + y_i - [y_f y_r/(y_o + Y_L)]}$$

$$V_1 = \frac{I_S(y_o + Y_L)}{(Y_S + y_i)(y_o + Y_L) - y_f y_r} \quad (b)$$

On the output side, the load power is

$$P_o = |V_2|^2 G_L \quad (13.48)$$

Substituting Eqs. (13.36) and (b) into Eq. (13.48), we have

$$P_o = \frac{|y_f|^2 |I_S|^2 |y_o + Y_L|^2 G_L}{|y_o + Y_L|^2 |(Y_S + y_i)(y_o + Y_L) - y_f y_r|^2}$$

and

$$G_{pt} = \frac{P_o}{P_{i(max)}} = \frac{4|y_f|^2 G_S G_L}{|(Y_S + y_i)(y_o + Y_L) - y_f y_r|^2} \quad (13.49)$$

**Example 13.21**

Using the given $y$ parameters:

$y_i = 0.025 + j1$ mmho
$y_o = 0.1 + j1$ mmho
$y_r = j0.2$ mmho
$y_f = 5$ mmho

determine:

a. The maximum unneutralized power gain
b. The maximum neutralized power gain

**Solution**

a. The maximum unneutralized power gain is

$$G_{pmu} = \frac{|y_f|^2}{4g_o g_i - 2Re(y_f y_r)} \quad (13.44)$$

$$G_{pmu} = \frac{25 \times 10^{-6}}{(4 \times 10^{-4})(25 \times 10^{-6}) - 2Re[(5 \times 10^{-3})(j0.2 \times 10^{-3})]}$$

$G_{pmu} = 2{,}500$, or 34 dB

b. The maximum neutralized power gain is

$$G_{pmn} = \frac{|y_f|^2}{4g_i g_o} \tag{13.45}$$

$G_{pmn} = 2{,}500$, or 34 dB

In general, we wish to maximize $G_{pt}$, consistent with a requirement for circuit stability. The conditions leading to maximization of $G_{pt}$ are fairly complex and will not be derived here.[1] It can be stated, though, that when mismatching is used to achieve circuit stability, $G_{pt}$ will be maximized if the input and output are equally mismatched. We can therefore define a mismatch factor $m$ as follows:

$$m = \frac{G_L}{g_o} = \frac{G_S}{g_i} \tag{13.50}$$

A quick way to determine how much power gain is lost for any given mismatch is to use the graph of Fig. 13.26. Here the mismatch factor $m$ is plotted along the horizontal axis, and the loss, in decibels, is read along the vertical axis. When $m = 1$, there is no mismatch, and the loss is 0 dB. For $m = 10$, the loss is 4.8 dB, and so on.

**Example 13.22**

Using the $y$ parameters of Example 13.21, estimate the transducer gain. The load and source conductances are $G_L = 2.36$ mmho; $G_S = 0.59$ mmho. Select $B_L = b_o^*$ and $B_S = b_i^*$.

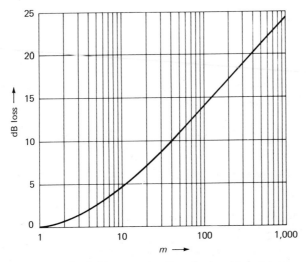

Fig. 13.26 Decibel loss as a function of mismatch factor $m$.

[1] See chap. 11 of Reference 15 at the end of this chapter. Also see Reference 13.

## Solution

$$Y_S = G_S + jB_S$$
$$Y_S = G_S - jb_i$$
$$Y_S = 0.59 \times 10^{-3} - j10^{-3}$$
$$Y_L = G_L + jB_L$$
$$Y_L = G_L - jb_o$$
$$Y_L = 2.36 \times 10^{-3} - j10^{-3}$$

The transducer gain is

$$G_{pt} = \frac{4|y_f|^2 G_S G_L}{|(Y_S + y_i)(y_o + Y_L) - y_f y_r|^2} \tag{13.49}$$

$$G_{pt} = 42.5, \text{ or } 16.3 \text{ dB}$$

### Example 13.23

The common source FET amplifier shown in Fig. 13.27 is to be designed for a center frequency of 50 MHz. Synchronous tuning (input and output circuits tuned to the same frequency) is to be used. A bandwidth of 10 MHz or less is required. The source and load are each 50 Ω, purely resistive. Both inductors are 0.5 µH with a Q of 10. Assume that the biasing resistors and DC supply have been selected to yield the proper operating point. Determine $C_2$ and the turns ratios $n_3/n_2$ and $n_2/n_1$, if the following data apply:

$$y_{is} = 0.1 + j1.2 \text{ mmho}$$
$$y_{fs} = 5 - j0.4 \text{ mmho}$$

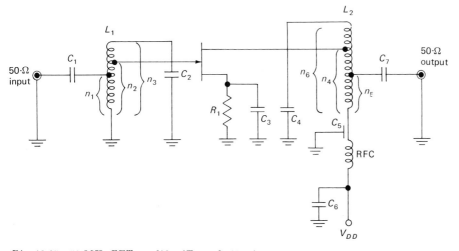

Fig. 13.27   50-MHz FET amplifier (Example 13.23).

$y_{os} = 0.01 + j0.3$ mmho
$y_{rs} = 0 - j0.15$ mmho
$Y_S$ to be seen by FET $= 3.1 - j1.2$ mmho
$Y_L$ to be fed by FET $= 0.31 - j0.3$ mmho

**Solution**

a. We will design each tuned circuit for a 3-dB bandwidth of 10 MHz. The overall response will have a narrower bandwidth; this is acceptable because the specifications require 10 MHz or less.

b. The loaded $Q$ is

$$Q_L = \frac{f_o}{B} = \frac{50 \times 10^6}{10 \times 10^6} = 5$$

c. Shown in Fig. 13.28 is the input equivalent circuit, as seen across the full ($n_3$) turns of $L_1$. The input resistance and capacitance to the FET are determined as follows:

$$Y_i = y_i - \frac{y_f y_r}{y_o + Y_L} \tag{13.37}$$

$Y_i = 10^{-4} + j(1.2 \times 10^{-3})$

$$- \frac{(5 - j0.4)(10^{-3})[-j(0.15 \times 10^{-3})]}{10^{-5} + j(3 \times 10^{-4}) + 0.31 \times 10^{-3} - j(0.3 \times 10^{-3})}$$

$Y_i = 0.287 + j3.54$ mmho

$R_i = \dfrac{1}{0.287 \times 10^{-3}} \simeq 3.5$ k$\Omega$

$B_i = 3.54$ mmho
$B_i = 2\pi f C_i$

$C_i = \dfrac{3.54 \times 10^{-3}}{6.28(50 \times 10^6)} = 11.3$ pF

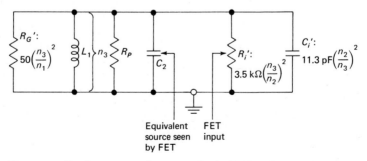

Fig. 13.28 Total input equivalent circuit for the FET amplifier of Fig. 13.27.

When reflected across $L_1$, $R_i$ is multiplied by $(n_3/n_2)^2$ while $C_i$ is divided by $(n_3/n_2)^2$.

d. $R_P$ is the equivalent parallel resistance for $L_1$. $R'_G$ is the transformed value of generator resistance. $C_2$ is the tuning capacitance for $L_1$.

e. For $L = 0.5~\mu\text{H}$ and an unloaded $Q$ of 10, $R_P$ is

$$R_P = QX_P$$
$$R_P = (10)(6.28)(50 \times 10^6)(0.5 \times 10^{-6})$$
$$R_P = 1{,}570~\Omega$$

f. For a loaded $Q$ of 5, we require an overall $R'_P = 785~\Omega$. To achieve this, the equivalent parallel combination of $R'_G$ and $R'_i$ must equal $1{,}570~\Omega$:

$$R'_G \| R'_i = 1{,}570 \qquad (a)$$

But the FET input must see an $R_S = 1/(3.1 \times 10^{-3}) = 323~\Omega$:

$$(R'_G \| R_P)\left(\frac{n_2}{n_3}\right)^2 = 323 \qquad (b)$$

We can rewrite Eq. (a) as

$$\left[50\left(\frac{n_3}{n_1}\right)^2\right] \| \left[3{,}500\left(\frac{n_3}{n_2}\right)^2\right] = 1{,}570 \qquad (c)$$

and Eq. (b) as

$$\left\{\left[50\left(\frac{n_3}{n_1}\right)^2\right] \| 1{,}570\right\}\left(\frac{n_2}{n_3}\right)^2 = 323 \qquad (d)$$

Solution of Eqs. (c) and (d) yields

$$\frac{n_3}{n_1} = 6.15 \quad \text{and} \quad \frac{n_3}{n_2} = 1.62$$

The above turns ratios ensure that the FET input is driven from an equivalent $323~\Omega$ while the total resistance across $L_1$ is $785~\Omega$.

g. The total tuning capacitance is

$$C = \frac{1}{4\pi^2 f^2 L} \qquad (13.11)$$

$$C = \frac{1}{(4\pi^2)(50)(50 \times 10^{12})(0.5 \times 10^{-6})}$$
$$C = 20~\text{pF}$$

The FET's 11.3-pF input capacitance appears as $(11.3 \times 10^{-12})(n_2/n_3)^2 = 4.3~\text{pF}$ when seen across the whole $n_3$ turns; therefore a net capacitance of $20 - 4.3 = 15.7~\text{pF}$ appears to be required. The coil, however, will have some distributed capacitance of its own; therefore a variable 1 to 8 pF plus a fixed 5 pF for $C_2$ should be satisfactory. The variable portion of $C_2$ is experimentally adjusted to yield maximum power output. The output tuned circuit design is left as an exercise for the student. The remaining components in the circuit of Fig. 13.27 are easily selected. $C_1$, $C_3$, $C_7$, and $C_6$ should all have negligible reactance at 50 MHz; $C_5$ is a feedthrough with sufficient capacitance to act as a good bypass capacitor. $C_5$, $C_6$, and the RFC (RF choke) form a power supply decoupling network. The RFC is simply a coil with a self-resonant frequency equal to 50 MHz.

## Problem Set 13.4

13.4.1. What advantages do $y$ parameters have at high frequencies over $h$ parameters?

13.4.2. Are $y$ parameters limited to field effect transistors, or can they also be used for other active devices?

13.4.3. What physical effect is responsible for the existence of $y_r$ in FETs? In bipolars?

13.4.4. What desirable characteristics does an amplifier with $y_r = 0$ have?

13.4.5. Obtain the common source $y$ parameters for the 2N3823 at 200 MHz using the data sheet reproduced in Appendix 10; the test conditions are $V_{DS} = 15$ V, $V_{GS} = 0$ V, $T_A = 25°C$.

13.4.6. The following sets of measurements are made on an active device:

  a. $V_1 = 0$; $V_2 = 0.2$ V $/0°$; $I_1 = 0.05$ mA $/90°$; $I_2 = 0.5$ mA $/75°$
  b. $V_1 = 0.2$ V $/0°$; $V_2 = 0$; $I_1 = 100\mu A/85°$; $I_2 = 1$ mA $/0°$

  Determine all four $y$ parameters in polar form.

13.4.7. Express the $y$ parameters of Prob. 13.4.6 in rectangular form. Identify the conductances and susceptances in each case.

13.4.8. Using the $y$ parameters of Prob. 13.4.6, determine the input admittance if the load is a pure 50-$\Omega$ resistance.

13.4.9. What is the equivalent input resistance in Prob. 13.4.8? The equivalent input capacitance if $f = 10$ MHz?

13.4.10. Using the $y$ parameters of Prob. 13.4.6, determine the output admittance if the source impedance is 50 $\Omega$, purely resistive.

13.4.11. What is the equivalent output resistance in Prob. 13.4.10? The equivalent output capacitance if $f = 10$ MHz?

13.4.12. Determine the maximum unneutralized gain (in decibels) for a device with the $y$ parameters of Prob. 13.4.6.

13.4.13. What is the maximum neutralized power gain (in decibels) for the device of Prob. 13.4.6?

13.4.14. Determine the maximum neutralized power gain, given the following $y$ parameters:

$$y_i = 0.03 + j0.8 \text{ mmho}$$
$$y_o = 0.12 + j0.8 \text{ mmho}$$
$$y_r = 0 + j0.1 \text{ mmho}$$
$$y_f = 4 + j0 \text{ mmho}$$

13.4.15. Estimate the transducer gain in Prob. 13.4.14 for a mismatch factor of 10. Assume $B_S = -b_i$ and $B_L = -b_o$.

13.4.16. For Example 13.23, determine $C_4$ and the turns ratios $n_6/n_4$ and $n_4/n_5$. Assume $L_2 = L_1$.

### 13.1f  Automatic Gain Control (AGC)

An additional consideration in the design of RF amplifiers is the dynamic range. If the signal is too large, for example, the circuit may be overloaded, resulting in excessive distortion. Even with minor overloads, the transistor parameters may change, thereby affecting the amplifier response. In a multistage amplifier, the problem is compounded because of the large gains involved.

A general technique for minimizing these problems is *automatic gain control* (AGC). With AGC, the gain of the initial stages is automatically reduced whenever the signal level is high. Large signals can now be handled that before would have caused overloads without AGC; the dynamic range of the amplifier is therefore extended.

Shown in Fig. 13.29 is a graph of $|h_{fe}|$ as a function of $I_C$. Note that $|h_{fe}|$ increases initially with $I_C$ until a peak is reached, then decreases as $I_C$ is increased further. A family of curves is shown, each for a different $V_{CE}$. At low collector currents, the curves are bunched, indicating that $|h_{fe}|$ is not heavily dependent on $V_{CE}$. At higher collector currents, though, $|h_{fe}|$ rises with increasing $V_{CE}$.

There are two basic AGC techniques. One is called forward AGC and the other reverse AGC. With reverse AGC, when the signal level increases, the DC collector current is automatically reduced, say from $I_C$ to $I'_C$, as shown in Fig.

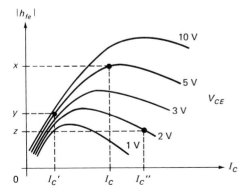

**Fig. 13.29**  *Dependence of* $|\mathrm{h_{fe}}|$ *on* $\mathrm{V_{CE}}$ *and* $\mathrm{I_C}$.

13.29 and on the output characteristics of Fig. 13.30. If the DC load is relatively low, $V_{CE}$ remains fairly constant at 5 V, and $|h_{fe}|$ drops from $x$ to $y$. The amplifier's gain is therefore reduced, and succeeding stages are not overloaded.

With forward AGC, when the signal level rises, $I_C$ is automatically increased, say from $I_C$ to $I_C''$. As $I_C$ increases, there is more of a voltage drop across the transistor's DC load (collector resistor and emitter resistor), which leaves less voltage from collector to emitter. This effect can be increased if the DC resistance is high, as indicated by the high resistance DC load line in Fig. 13.30$b$. Thus increasing $I_C$ results in a decreasing $V_{CE}$, from 5 to possibly 2 V (depending on the actual value of DC load). Both of these factors contribute to a reduction in $|h_{fe}|$, from $x$ to $y$ in Fig. 13.29.

It should be noted that with reverse AGC the quiescent current is reduced, which limits the transistor's maximum signal-handling capability at a time when it is needed most. With forward AGC, the gain reduction is achieved by an increase in the quiescent collector current, which also increases the maximum possible signal swing. A well-designed AGC transistor can automatically vary its own gain by as much as 25 dB in response to signal-level changes.

A typical RF amplifier involving AGC is shown in Fig. 13.31. The signal is received by the antenna and coupled, through transformer $T_1$, to the transistor base. $C_2$ is a coupling capacitor with negligible reactance at the frequency of operation; it prevents the base from being DC-grounded through the low DC resistance of $T_1$'s secondary winding. $C_1$ and $T_1$ form a parallel resonant circuit to provide selectivity.

The DC operating point is set by $V_{CC}$, $R_{B2}$, $R_E$, and $R_{B1}$. In a circuit without AGC, $R_{B1}$ would be tied directly to the $V_{CC}$ supply. Now $R_{B1}$ is tied to the AGC input, also a DC supply, but whose value depends on the signal strength.

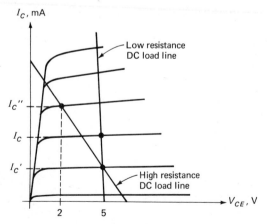

Fig. 13.30 Typical load lines for comparison of forward and reverse AGC.

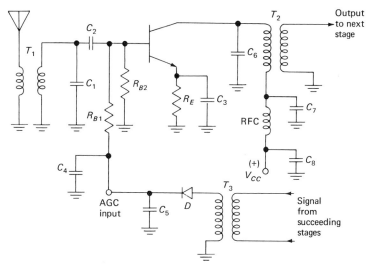

Fig. 13.31  RF amplifier with AGC.

A portion of the signal from a succeeding stage is coupled through $T_3$ to the diode-$C_5$ combination. The diode half-wave rectifies the signal, and $C_5$ filters out the ripple, yielding a DC voltage across $C_5$ whose value is proportional to the signal strength. As long as the signal is constant, the AGC voltage across $C_5$ is constant, and the transistor's operating point remains the same. If the signal level increases, though, the AGC input becomes more positive; this results in more base current and hence more collector current. Now the transistor's gain decreases because of forward AGC action. Note that this process involves negative feedback, allowing us to achieve control over the signal level.

Capacitors $C_4$, $C_3$, $C_7$, and $C_8$ are all bypass capacitors. The $C_7$, $C_8$, and RFC (RF choke) combination form a low-pass filter (also called decoupling network) to prevent stray signal coupling through the power supply. The RFC is self-resonant at the frequency of operation. Capacitor $C_6$ and transformer $T_2$ make up the output tuned circuit. No neutralizing components are shown, indicating that stability in this circuit is probably being achieved by mismatching.

AGC for FETs can be illustrated using the output characteristics of Fig. 13.32. Assume the $Q$ point is as shown. With a low DC resistance, oad line $A$ applies, and reverse AGC can be used. Thus as the signal level increases, the operating point is shifted from $Q$ to $Q'$. At $Q'$ the forward transadmittance (or transconductance) $y_f$ is reduced, yielding a lower gain. With a high DC resistance, load line $B$ applies, and forward AGC can be used. As the signal level increases, the operating point is shifted from $Q$ to $Q''$. At $Q''$, the output impedance $r_d$ is reduced ($y_o$ increases), causing a gain reduction. If the $Q$ point is

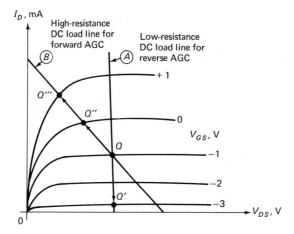

Fig. 13.32  Typical load lines for FET AGC.

shifted all the way to $Q'''$, enhancement-mode operation results. The gate is forward-biased; therefore the input impedance drops ($y_i$ increases) and the gain is further reduced.

### Problem Set 13.5

13.5.1. What is AGC? Why is it used?
13.5.2. What is the difference between forward and reverse AGC? What advantages does forward AGC have? Can you think of disadvantages?
13.5.3. Why should the DC load resistance be relatively high when using forward AGC in a bipolar transistor?
13.5.4. How does $|h_{fe}|$ depend on $V_{CE}$ at low collector currents? At high collector currents?
13.5.5. How does the gain of FETs depend on $I_D$ at near constant $V_{DS}$?
13.5.6. What causes the gain of FETs to decrease when the $Q$ point is shifted from the pinch-off region ($Q$) to the triode region ($Q''$)? To the enhancement-mode region ($Q'''$)?
13.5.7. In the circuit of Fig. 13.31, what would happen if the diode were reversed? What would happen if the value of $C_5$ were insufficiently large to smooth out the ripple?

## 13.2  NONLINEAR CIRCUITS

A nonlinear circuit may be defined as any circuit that produces frequencies not present in the input signal. These additional frequency components may be desirable or undesirable. For example, if the circuit generates a new frequency component $f_o$ which is desired, then the output circuit (usually a parallel $LC$ network) is designed to resonate at $f_o$ so as to maximize the gain.

The circuit will usually generate undesirable frequency components too. If these frequency components are far removed from $f_o$, then the parallel tuned circuit may provide sufficient attenuation to make them insignificant compared with the desired component at $f_o$.

Consider the diagram of Fig. 13.33. Here the frequency spectrum for $I_o$

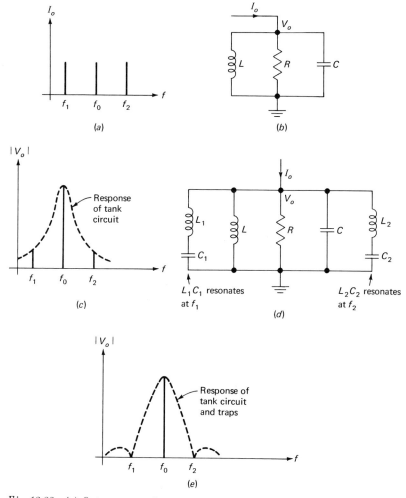

*Fig. 13.33* (a) *Output current frequency spectrum;* (b) *parallel RLC circuit tuned at* $f_o$; (c) *output voltage spectrum for the current of* (a) *and the circuit of* (b); (d) *tuned circuit with traps;* (e) *output voltage spectrum for the current of* (a) *and circuit of* (d).

consists of three components, all equal in magnitude. The desired frequency component is $f_o$; the other components $f_1$ and $f_2$ are to be attenuated as much as possible. The output voltage is taken across the $RLC$ tuned circuit of Fig. 13.33b, so that the magnitude of $V_o$ for any frequency component of $I_o$ is proportional to the impedance of the $RLC$ network at that frequency. The impedance is the dashed curve superimposed on the plot of $|V_o|$ versus $f$ shown in Fig. 13.33c. Note that at $f_o$, the impedance is highest; therefore this frequency component receives the highest gain. At $f_1$ and $f_2$ the impedance, and hence the gain, drops, and $|V_o|$ at these frequencies is relatively low.

If the $f_1$ and $f_2$ components are not sufficiently attenuated relative to the $f_o$ component, additional attenuation at $f_1$ and $f_2$ must be provided. This can be accomplished by using another stage of amplification so that the $f_o$ component is amplified again while the $f_1$ and $f_2$ components are further attenuated. It is also possible to use two coupled $RLC$ networks to achieve a steeper roll-off on each side of $f_o$, and hence greater attenuation at $f_1$ and $f_2$. A third method, illustrated in Fig. 13.33d, may also be used. Here the $L_1C_1$ and $L_2C_2$ networks, called wave traps, are series resonant at $f_1$ and $f_2$, respectively. Thus at $f_1$, the very low impedance of the $L_1C_1$ network shunts the rest of the circuit, and $V_o$ drops to a very low value. A similar effect occurs at $f_2$ due to the $L_2C_2$ trap.

The $L_1C_1$ network, which is resonant at $f_1$, behaves inductively for frequencies above $f_1$. The $L_2C_2$ network, which is resonant at $f_2$, is capacitive for frequencies below $f_2$. Therefore at $f_o$, the $L_1C_1$ trap is inductive, and its equivalent inductance can be lumped with $L$. Also at $f_o$, the $L_2C_2$ trap is capacitive, and its equivalent capacitance adds to $C$. The net result is that $f_o$ is not only determined by $L$ and $C$ but also depends on the values of trap components.

The frequency response for the total network in Fig. 13.33d is shown on the graph in Fig. 13.33e. Note the attenuation at $f_1$ and $f_2$. How much these frequency components can be attenuated is basically a function of the trap $Q$'s. A high $Q$ means low series resistance so that when either trap resonates ($f_1$ or $f_2$), it appears as a very low resistance, and $|V_o|$ at that frequency is significantly attenuated.

### 13.2a   Class C Amplification

A class C amplifier is one in which the active device is normally cut off; hence no output current flows. If a strong input signal is applied, the active device can be made to conduct for a portion of the cycle, called the conduction angle. It will be recalled that for class A, the conduction angle is 360°, and for class B, 180°; for class C it is less than 180°. A class C amplifier block diagram, with input signal waveform and frequency spectrum, is shown in Fig. 13.34.

A typical waveform of output current is shown in Fig. 13.35a. Note that $i_o$ is a rather distorted version of the input signal. The distortion arises because

RF Circuits 577

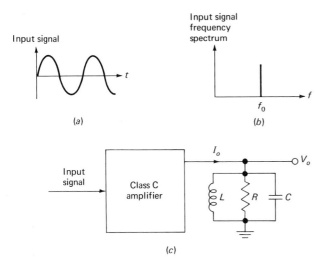

Fig. 13.34 (a) Time graph of input signal; (b) frequency spectrum for (a); (c) diagram of class C amplifier and output tuned circuit.

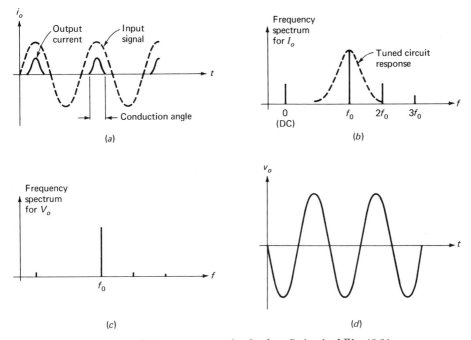

Fig. 13.35 Waveforms and frequency spectra for the class C circuit of Fig. 13.34.

current flows only during that portion of the cycle (conduction angle) for which the active device conducts.

If a class C amplifier produces so much distortion, why is it then used? The answer lies in the fact that average power dissipation can be very low, and hence efficiency high, when the device conducts for less than a full cycle. This is of real importance when a lot of power must be delivered to the load, as in the last stage of a radio transmitter. Practical efficiencies of around 70 percent can be achieved. The distortion can be eliminated by noting that the frequency spectrum for $I_o$, which is shown in Fig. 13.35b, includes the fundamental ($f_o$ = signal frequency) in addition to other harmonics and a DC component. The filtering action of the output tuned circuit attenuates frequencies other than $f_o$ so that the spectrum for $V_o$, shown in Fig. 13.35c, is mainly due to the fundamental $f_o$.

The time graph for $v_o$, in Fig. 13.35d, is very similar to the input signal. There may still be some distortion because the filtering action is not perfect, but such distortion can be further eliminated by using a double-tuned circuit or wave traps at the offending frequencies.

### 13.2b  Frequency Multiplication

Very often in communication systems there is a need for a signal whose frequency is a given multiple of some other frequency. For example, an 18-MHz signal may be required for a specific application, while a 1-MHz signal may already be available. If the 1-MHz signal is multiplied (in frequency) 18 times, the desired 18-MHz signal is produced. This can be done using two triplers: 1 × (3 × 3) = 9 followed by a doubler (9 × 2 = 18).

Frequency multiplication is a consequence of harmonic generation. Harmonics were discussed in Chap. 4 and generally represent undesirable distortion. Whenever an amplifier is overdriven, operation throughout a large range of its operating characteristics distorts the signal, resulting in the generation of harmonics. Depending on the device, its operating point, and signal level, many different harmonics can be generated, but generally the amplitude of harmonics beyond the fifth gets progressively smaller. Fairly high efficiencies can be obtained if the active device is biased class C, as discussed in the previous section.

The diagram of Fig. 13.36a illustrates frequency multiplication. The input signal, shown in Fig. 13.36b, is purely sinusoidal; therefore it has only one frequency component, $f_o$, as indicated in the spectrum of Fig. 13.36c. If the signal level is sufficiently high, the amplifier is overdriven and produces harmonics of output current as shown in Fig. 13.36d. The desired harmonic is simply selected by tuning the output tank circuit to its frequency. For a ×3 multiplier the tank circuit is tuned to $3f_o$, and the output voltage then has a spectrum as in Fig. 13.36e and a time graph as in Fig. 13.36f.

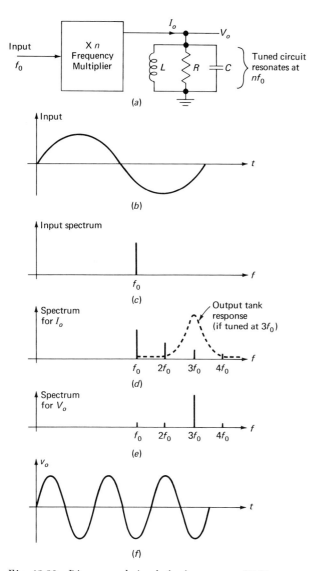

Fig. 13.36  *Diagram and signals for frequency multiplier.*

For example, assume $f_o = 1$ MHz. The frequency spectrum for $I_o$ includes the fundamental 1 MHz and a number of harmonics, 2, 3, 4, and 5 MHz, and so on. For a ×3 multiplier, the output circuit is simply tuned to 3 MHz; the remaining harmonics and the fundamental are attenuated. For a ×9 multiplier, two ×3 multipliers can be cascaded. Generally ×2 and ×3 frequency multipliers are most common, since the generation of second and third harmonics is most efficient.

Varactors are also used in frequency-multiplication applications. These are reverse-biased diodes whose capacitance is a function of the applied voltage. A changing voltage produces a time-varying capacitance that gives rise to the generation of harmonics. The desired harmonic is then selected as indicated earlier.

### 13.2c Mixing

Mixing is a process in which two signals are combined to yield an output signal whose frequency content includes both the sum and difference of the two input frequencies.

Amplitude modulation (used for the transmission of AM broadcasts) is one application that involves mixing. This will be illustrated using the diagram of Fig. 13.37. There are two inputs to the active device. One is the carrier, whose frequency $f_o$ is the transmitting frequency. The other input is the modulating signal, which represents the program that is to be transmitted. In standard broadcasts, the program is made up of audio frequencies (speech and music) that cover the range from a low of 30 to a high of about 10,000 Hz. To keep our discussion simple, we will assume that the "program" is a constant 1,000-Hz signal (this corresponds to a moderately high pitched whistle). Once the principles are understood on this basis, it is a simple matter to extend them so that the full audio spectrum is included. The carrier frequency $f_o$ is assigned to the particular station somewhere in the AM transmitting range (535 to 1,605 kHz). We will assume $f_o = 1$ MHz for our discussion.

We therefore have two inputs to the mixer. One is the carrier, whose frequency $f_o$ is 1 MHz, and the other is the modulating signal, whose frequency

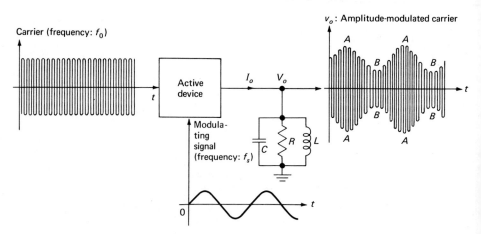

*Fig. 13.37* Block diagram of amplitude modulation system.

$f_s = 1$ kHz. Note that in this case $f_o \gg f_s$; however in other applications this is not necessarily so. Frequency spectra for $f_o$ and $f_s$ are shown in Fig. 13.38a and b, respectively.

If the modulating signal were not present, the active device would simply amplify the carrier. The output is tuned at the carrier frequency; therefore the output would contain only $f_o$. Some harmonics might be generated if the carrier amplitude were high, but this is only incidental to the process, and the output tank circuit would attenuate any such harmonics.

Once the modulating signal is applied, the gain of the active device is *modulated* to yield the output shown. The modulating signal simply shifts the operating point sinusoidally at the $f_s$ frequency. As the operating point is changed, the gain, which is a function of the $Q$ point, also varies in a manner similar to AGC. Since the gain of the active device is no longer constant, the carrier is amplified at a varying rate. At points $A$, the gain is highest, while at points $B$, the gain is lowest. The output waveform is said to be amplitude-modulated because the carrier amplitude is now related to the instantaneous signal amplitude.

When a carrier $f_o$ is amplitude-modulated by a signal $f_s$, the resulting output waveform contains the following frequency components:

Carrier: $f_o$
Sum of carrier and signal: upper side frequency: $f_o + f_s$
Difference of carrier and signal: lower side frequency: $f_o - f_s$

The sum and difference frequencies indicated above, also called side frequencies, are shown on the frequency spectrum of Fig. 13.38c. For our example, these are 1.001 MHz and 999 kHz. The information that is to be transmitted (signal) is contained in these side frequencies. The output is tuned to $f_o$ but must have a bandwidth wide enough to pass the side frequencies. When the modulating signal includes many audio-frequency components (as is usually the case), an upper and

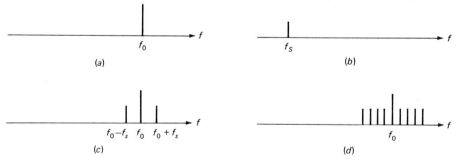

Fig. 13.38 *Frequency spectra for Fig. 13.37:* (a) *carrier* $f_o$; (b) *signal* $f_s$; (c) *modulated carrier, includes* $f_o$, *upper and lower sidebands;* (d) *modulated carrier when modulating signal has more than one frequency component.*

lower side frequency for each component is generated, as indicated in Fig. 13.38d. The side frequencies then make up upper and lower sidebands. If the highest signal frequency is 10 kHz, then the highest side frequency is 1,010 kHz and the lowest side frequency is 990 kHz. The tuned circuit must therefore have a minimum bandwidth of 20 kHz.

In conclusion, mixing is a nonlinear process that involves the generation of the sum and difference of two input frequencies. Based on the actual method used, the input frequencies and their harmonics may also appear at the output. What is more, the harmonics may also mix to yield their own sum and difference frequencies, a process earlier discussed[1] as intermodulation distortion. Often the tuned circuit can satisfactorily attenuate these additional frequency components because they are outside its passband. When this is not the case, however, wave traps may have to be used.

### Problem Set 13.6

13.6.1. Define a nonlinear process.
13.6.2. Compare the conduction angles for class A, B, and C operation.
13.6.3. Why is class C operation used? What are some of its applications?
13.6.4. In a class C amplifier the output current is a heavily distorted version of the input signal because the device is cut off for over 50 percent of the time. How is it then possible to reproduce a reasonably clean output voltage waveform?
13.6.5. What is a wave trap, and when is it used?
13.6.6. An 89.9-MHz signal is desired. ×2 and ×3 frequency multipliers are cascaded to yield the desired output at 89.9 MHz. The input signal is applied to the ×2 multiplier.

    a. What is the overall frequency-multiplication factor?
    b. At what frequency should the output of the ×3 multiplier be tuned?
    c. At what frequency should the output of the ×2 multiplier be tuned?
    d. What should be the input frequency to the ×2 multiplier?

13.6.7. A mobile ship AM station is assigned a 2,000-kHz carrier frequency. The audio-signal frequency range is 100 to 3,000 Hz. What are the highest and lowest side frequencies? What is the required bandwidth?
13.6.8. Determine the output frequencies in the block diagram of Fig. 13.39.
13.6.9. In Prob. 13.6.8, assume that second harmonics of both input signals are also produced by the mixer and that these harmonics produce sum and difference frequencies. What frequencies are present in the output now?

[1] See Sec. 4.6.

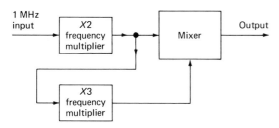

*Fig. 13.39  Diagram for Prob. 13.6.8.*

## REFERENCES

1. Kiver, M. S.: "Transistors," McGraw-Hill Book Company, New York, 1962.
2. Amos, S. W.: "Principles of Transistor Circuits," Hayden Book Company, Inc., New York, 1965.
3. Cowles, L. G.: "Transistor Circuits and Applications," Prentice-Hall, Inc., Englewood Cliffs, N.J., 1968.
4. Joyce, M. V., and K. K. Clarke: "Transistor Circuit Analysis," Addison-Wesley Publishing Company, Inc., Reading, Mass., 1961.
5. Texas Instruments Incorporated: "Transistor Circuit Design," McGraw-Hill Book Company, New York, 1963.
6. Gibbons, J. F.: "Semiconductor Electronics," McGraw-Hill Book Company, New York, 1966.
7. Fitchen, F. C.: "Transistor Circuit Analysis and Design," 2d ed., D. Van Nostrand Company, Inc., Princeton, N.J., 1966.
8. "General Electric Transistor Manual," 7th ed.
9. Hejhall, R.: RF Small-signal Design Using Admittance Parameters, *Application Note* 215, Motorola Semiconductor Products Inc., Phoenix, 1966.
10. Tatorn, C.: Whats and Whys about $Y$-Parameters, *Application Note* 158, Motorola Semiconductor Products Inc., Phoenix, 1966.
11. Leonard, D. N.: FM RF Amplifiers and Mixers Using Junction Field Effect Transistors, *Bulletin* CA-95, Texas Instruments Incorporated, Dallas, 1967.
12. Farell, C. L.: AGC Characteristics of FET Amplifiers, *Bulletin* CA-104, Texas Instruments Incorporated, Dallas, 1969.
13. Stern, A. P.: Stability and Power Gain of Tuned Transistor Amplifiers, *Proc. IRE*, March 1957, pp. 335–343.
14. Wolfendale, E.: "Transistor Circuit Design and Analysis," Heywood, London, 1966.
15. Linvill, J. C., and J. F. Gibbons: "Transistors and Active Circuits," McGraw-Hill Book Company, New York, 1961.
16. Smith, F. L.: "Radio Designer's Handbook," 4th ed., Iliffe Books Ltd., London, 1953.

# chapter 14

## oscillators

This chapter discusses basic properties of oscillators, including specific examples of practical oscillator circuits. Both the material on feedback (Chap. 10) and tuned circuits (Chap. 13) should be reviewed for the discussions to follow.

## 14.1 BASIC CONCEPTS AND DEFINITIONS

An oscillator may be described as a source of alternating voltage (or current). Such sources are widely used to provide the proper waveforms for the operation and testing of electronic circuits.

The definition of an oscillator generally excludes AC power sources, such as the 60-Hz line voltage, or any other source derived by the conversion of mechanical or heat energy into electric energy. Instead an oscillator generates its output through the conversion of electric energy at DC (zero frequency) to electric energy at a higher frequency.

This point can be further illustrated by comparing an amplifier and an oscillator as in Fig. 14.1. An amplifier delivers an output signal whose waveform corresponds to the input signal but whose power level is generally higher. The additional power content in the output signal is supplied by the DC power source used to bias the active device. The amplifier can therefore be described as an energy converter; it accepts energy from the DC power supply and converts it to

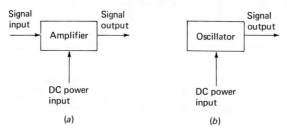

*Fig. 14.1  Comparison of amplifier* (a) *with oscillator* (b).

energy at the signal frequency.[1] The process of energy conversion is controlled by the input signal; thus if there is no input signal, no energy conversion takes place, and there is no output signal.

The oscillator differs from the amplifier in one basic respect: the oscillator requires no external signal to initiate or maintain the energy conversion process. Instead an output signal is produced as long as the source of DC power is connected.

Oscillators may be classified in terms of their output waveform, frequency range, components, or circuit configuration. The output waveform may be sinusoidal (harmonic oscillator) or nonsinusoidal (relaxation oscillator). Typical relaxation oscillator outputs include square, triangular, and sawtooth waveforms. The frequency of oscillation may be as low as in the audio range or extend to several gigahertz. Because UHF and microwave oscillators require specialized components and techniques, our discussion is limited to frequencies below 300 MHz.

Oscillator circuits employ both active and passive components. The active device provides the energy conversion mechanism. Typical active devices include transistors (bipolar, field effect, unijunction) as well as two terminal devices such as tunnel and shockley (four-layer) diodes. The bipolar and field effect transistors have been discussed in earlier chapters. The unijunction transistor and tunnel diode are introduced in this chapter while the four-layer diode is briefly described in Sec. 15.3c.

Characteristics of the active device pertinent to oscillator circuits include its ability to effect the energy conversion process at the desired frequency. Generally active devices remain useful as oscillators even at frequencies beyond the range for which they are capable of satisfactory amplification. In the case of transistors, the maximum frequency of oscillation, $f_{max}$, is an important parameter to consider. This is the frequency for which the power gain is unity (0 dB). Other factors to consider are the dependence of small-signal parameters on operating point shifts, temperature, aging, and so on.

---

[1] The efficiency of energy conversion is a matter of concern when large amounts of power are involved; this topic is discussed in Sec. 12.1d.

Passive components normally determine the frequency of oscillation. They also influence stability, which is a measure of the change in output frequency (drift) with time, temperature, or other factors. Frequency stability is an important consideration; therefore the tolerances and changes in value of all components must be carefully considered. Passive devices may include resistors, inductors, capacitors, transformers, and when a high degree of frequency stability is required, resonant crystals, which are discussed later in this chapter. Although active devices also contribute resistance and capacitance, these parameters are a function of operating point, temperature, and device replacement; therefore the frequency of a well-designed oscillator should depend mainly on the value of externally connected passive components.

Capacitors used in oscillator circuits should be of high quality. Because of low losses and excellent stability, silvered mica or ceramic capacitors are generally preferred. Frequency drift is also greater for low $Q$ circuits; therefore loaded $Q$'s should be high. Usually the loaded $Q$ is determined by coil or transformer losses and other loads introduced into the circuit.

In addition to frequency stability, performance measures for oscillators may include amplitude stability, output power, and harmonic content. If the output amplitude must be maintained at some desired level, AGC can be employed. It is generally not a good practice, however, to try achieving large outputs from an oscillator circuit. Instead, the oscillator circuit is designed to produce a stable and clean waveform; the power level is then raised through successive stages of amplification. The cleanness or purity of the output waveform refers to the absence of harmonics or other undesirable frequency components.

The active device in sinusoidal oscillator circuits may be biased class A, B, or C. Class A outputs are relatively free from harmonics; class C operation is the most efficient but also the largest producer of harmonics. By proper filtering at the output, it is possible to recover whatever harmonic is desired while attenuating the rest.

An elementary sinusoidal oscillator is shown in Fig. 14.2$a$. The inductor and capacitor are reactive elements; that is, they are capable of storing energy. The capacitor stores energy in its electric field whenever there is voltage across its plates; the inductor stores energy in its magnetic field whenever current flows. Both $C$ and $L$ are assumed here to be lossless (infinite $Q$'s). Energy can be introduced into the circuit by charging the capacitor with a voltage $V$, as shown. As long as the switch is open, $C$ cannot discharge, and so $i = 0$ and $v = 0$.

Suppose now that $S$ is closed at $t_o$; this means that $v$ rises from 0 to $V$, as shown in Fig. 14.2$b$. Note that $V$ now is simultaneously across $L$ and $C$. But, according to Faraday's law, if there is voltage across an inductor, then the current through that inductor must be changing. In our case, the current was zero just before $t_o$; immediately after $t_o$, $i$ must rise, as shown in the current graph of Fig. 14.2$c$. This current represents electrons leaving the bottom plate of $C$,

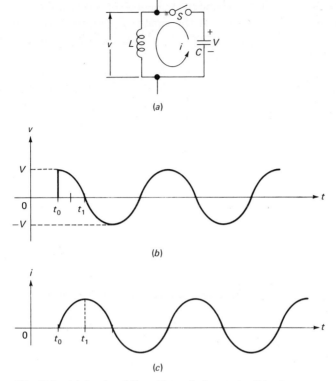

*Fig. 14.2* (a) *Lossless* LC *oscillator;* S *closes at* $t_o$; (b) *voltage waveform;* (c) *current waveform.*

moving through $L$, and entering the upper plate of $C$. Thus $C$ loses charge and energy (the less charge on a capacitor, the less energy it stores). The energy is simply transferred to the coil's magnetic field by the current. We still have the same energy as before, except that instead of being entirely stored in the capacitor's electric field, part of it is now stored in the coil's magnetic field.

As the charge on $C$ is reduced, $v$ decreases (the less charge on a capacitor, the less voltage). A lower voltage means a lower rate of change of inductor current; that is, current is still increasing because there is still voltage across $L$, but its rate of increase is tapering off. Eventually, at $t_1$, all charge has been removed from the capacitor plates, and in fact $v = 0$. With zero voltage across $L$, the current cannot change. At exactly $t_1$, the current is therefore neither increasing nor decreasing; instead it is exactly at its peak value.

Note that at $t_o$, the voltage is maximum, the current is zero, and all energy is stored by the capacitor in its electric field. At $t_1$, the voltage is zero, the current is maximum, and all the energy is stored by the inductor in its magnetic field.

The current for $t > t_1$ charges $C$ with a polarity opposite to that shown in Fig. 14.2a. The voltage across $C$ is also the voltage across $L$, and if the previous polarity $(t < t_1)$ caused the current to increase, the present polarity $(t > t_1)$ must cause the current to decrease. Thus $i$ begins decreasing, and $v$ increases in the opposite direction, as shown in Fig. 14.2b.

Note that both $v$ and $i$ are sinusoidal even though no sinusoidal input was applied; all we did was deposit some charge on the capacitor plates and let the circuit proceed on its own. The sinusoidal function is the only function that satisfies the conditions that govern the exchange of energy in the circuit. The frequency of oscillation is

$$f_o = \frac{1}{2\pi \sqrt{LC}} \qquad (14.1)$$

The circuit of Fig. 14.2a is not a practical oscillator because even if lossless components were available, one could not extract energy without introducing an equivalent resistance. This would result in *damped* oscillations, as shown in Fig. 14.3a. These oscillations decay to zero as soon as the energy in the tank is consumed. If we remove too much power from the circuit, the energy may be completely consumed before the first cycle of oscillations can take place, yielding the *overdamped* response of Fig. 14.3b.

It is possible to supply energy to the tank to make up for all losses (coil losses plus energy removed), thereby maintaining oscillations of a constant amplitude. Since energy lost may be related to a positive resistance, it follows that the circuit would gain energy if an equivalent negative resistance were available, as shown in Fig. 14.4. Here $R_{neg}$ has a negative value; it supplies whatever energy the circuit loses due to its positive resistance. Certain devices, such as tunnel diodes, exhibit, over a limited range of their characteristics, an increasing current for a decreasing voltage. This effect may be used to realize the negative resistance of Fig. 14.4. The energy supplied by the negative resistance to

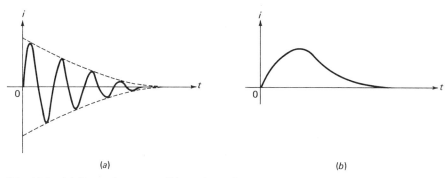

Fig. 14.3   (a) *Damped response;* (b) *overdamped response.*

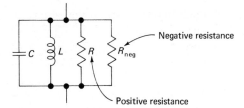

*Fig. 14.4  Oscillating* LC *circuit in which all the energy losses (positive resistance* R*) are made up by a negative resistance* $R_{neg}$.

the circuit actually comes from the DC source that is necessary to bias the device in its negative resistance region.

Another technique for producing oscillations is to use positive feedback as indicated in Fig. 14.5. Assume an amplifier with an input signal $V_i$ and an output signal $V_o$, as shown in Fig. 14.5a. $V_o$ is 180° out of phase with $V_i$ so that $A_v$ is a negative quantity. $V_o$ appears only as long as there is an input signal. Now suppose that a feedback network is added, as shown in Fig. 14.5b. The feedback network provides 180° phase shift and also some attenuation. The output from the feedback network, $V_x$, is in phase with and can be made equal in amplitude to the input signal $V_i$. If this is so, $V_x$ can be connected directly to $V_i$; the externally applied input signal can be removed, and the circuit will continue to generate an output signal! The amplifier still has an input, but the input is derived from the amplifier's own output through the feedback network. The output essentially feeds on itself and is continuously regenerated. This is positive feedback.

Note that the overall voltage amplification in Fig. 14.5b from $V_i$ to $V_x$ is 1 and that the total phase shift is zero. This may be referred to as the loop gain; it is the product of $A_v$ and $\beta_v$, where $\beta_v$ is the transfer function for the feedback network:

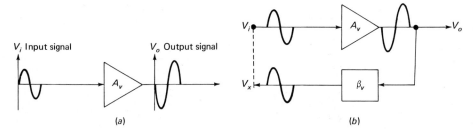

*Fig. 14.5*  (a) $A_v$ *is an amplifier that inverts and magnifies the input signal;* (b) *a portion of* $V_o$ *is fed back through the feedback network* $\beta_v$, *which inverts the signal so that* $V_x$ *is made equal in magnitude and phase to* $V_i$. $V_x$ *can now serve as the input to the amplifier, and* $V_o$ *will continue to be present even if* $V_i$ *is removed.*

$$\beta_v = \frac{V_x}{V_o} \tag{14.2}$$

Therefore

$$\beta_v A_v = 1\ \underline{/0°} \tag{14.3}$$

Equations (14.2) and (14.3) correspond to Eqs. (10.3) and (10.30b), respectively; these were developed in Chap. 10. Equation (14.3) is called the Barkhausen criterion for oscillation. When this criterion is satisfied, the closed-loop gain is infinite; that is, an output is produced without any external input.

In summarizing Eq. (14.3), we may state that, to sustain oscillations, the loop gain must be unity and the loop phase shift 0 or 360°. Since the feedback network involves passive components, with their consequent signal attenuation, it follows that the gain of the amplifier must be sufficient to make up for these losses and the power removed by the load.

The Barkhausen criterion should be satisfied at only one frequency, the frequency at which we want oscillations. Therefore the oscillator circuit must include frequency-selective elements ($L$ or $C$) that determine the oscillator frequency.

In a practical oscillator, it is not necessary to supply a signal to start the oscillations, as indicated in Fig. 14.5a. Instead, oscillations are self-starting and begin as soon as power is applied. What makes this possible is electric noise. This ever-present, otherwise undesirable element is made up of all frequency components. Therefore as soon as power is applied, there is already some energy in the circuit at $f_o$, the frequency for which the circuit is designed to oscillate. This energy is very small and is buried with all the other frequency components also present. Nonetheless it is there. Only at this frequency $f_o$ is the loop gain slightly greater than 1 and the loop phase shift zero. At all other frequencies the criterion $\beta_v A_v = 1\ \underline{/0°}$ is *not* satisfied. The magnitude of the frequency component $f_o$ is made slightly greater each time it goes around the loop. Soon the $f_o$ component is much larger than all the rest; ultimately, its amplitude is limited by the circuit's own nonlinearities (reduction of gain at high-current levels, saturation, or cutoff). This amplitude limitation does not appreciably affect the output waveform as long as the loop gain is just slightly greater than 1 (enough to start and maintain oscillations). If, however, the loop gain is too high, the active device is overdriven, and the output may be clipped on both sides, yielding a series of square waves. In any case, the sinusoidal output can always be recovered by filtering out the unwanted harmonics.

In conclusion, all practical oscillators involve

a. An active device to supply loop gain or negative resistance

b. A frequency-selective network to determine the frequency of oscillation
c. Some type of nonlinearity to limit amplitude of oscillations

### Problem Set 14.1

14.1.1. What is the basic difference between an amplifier and an oscillator?

14.1.2. What supplies the energy that an oscillator converts to generate the output waveform?

14.1.3. What is the difference between a relaxation and harmonic oscillator?

14.1.4. What active devices can be used in oscillator circuits? What property of the active device is important when considering its suitability as an oscillator?

14.1.5. What basic function do passive devices perform in oscillator circuits?

14.1.6. What is meant by frequency stability? Drift? How can frequency stability be improved?

14.1.7. What technique can be used to maintain a constant oscillator output?

14.1.8. What type of biasing (class A, B, or C) yields the cleanest sinusoidal waveform? The highest efficiency? The highest harmonic content?

14.1.9. When unwanted harmonics are generated in an oscillator, how can they be removed? What type of waveform results when all but one harmonic have been filtered out?

14.1.10. Does the energy stored by a capacitor depend on voltage or current? What is the physical energy-storage mechanism?

14.1.11. Repeat Prob. 14.1.10 for an inductor.

14.1.12. Sketch a sinusoidal signal that is

a. Damped
b. Overdamped

14.1.13. Do you think it is possible to build a circuit such as that in Fig. 14.2a?

14.1.14. Explain the operation of the circuit in Fig. 14.2a. Can you think of some way of supplying the initial energy other than charging the capacitor?

14.1.15. What is negative resistance? When is it used? What device exhibits a negative resistance characteristic?

14.1.16. What is the basic criterion for oscillations in a feedback-type circuit?

14.1.17. How can an oscillator be self-starting?

14.1.18. What is the function of nonlinearities in the active device of an oscillator circuit?

## 14.2 HARMONIC OSCILLATORS

Some common harmonic oscillators are discussed in this section. It will be recalled that a harmonic oscillator generates a sinusoidal output.

Shown in Fig. 14.6a is the diagram of a feedback-type harmonic oscillator. The DC equivalent circuit is shown in Fig. 14.6b. The DC operating point is set by $V_{CC}$, $R_B$, and $R_E$.

The AC equivalent circuit is shown in Fig. 14.6c. The transformer provides

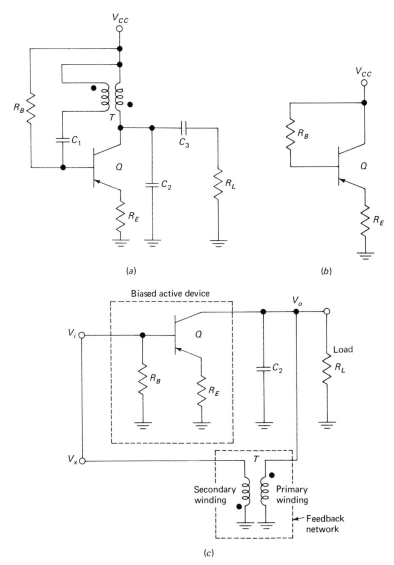

Fig. 14.6  (a) Feedback-type oscillator circuit; (b) DC equivalent circuit for (a); (c) AC equivalent circuit for (a) redrawn to emphasize positive feedback connections.

180° phase shift to ensure positive feedback so that, at the desired frequency of oscillation, the total phase shift from $V_i$ (transistor base) to $V_x$ is made equal to unity by proper choice of the transformer turns ratio. $R_E$ also controls and stabilizes the gain through negative feedback. $C_2$ and the transformer's equivalent inductance make up a resonant circuit that determines the frequency of oscillation. $C_1$ is used to block DC (otherwise the base would be directly tied to $V_{CC}$ through the low DC resistance of the transformer primary). $C_1$ has negligible reactance at the frequency of oscillation; therefore it is not a part of the frequency-determining network; the same applies to $C_3$.

Note the essential ingredients for an oscillator. There is an active device, suitably biased to provide necessary gain. Since the active device produces 180° phase shift (from base to collector), a transformer in the feedback loop provides an additional 180° to yield a loop phase shift of 0°. The feedback factor $\beta_v$ is equivalent to the transformer's turns ratio. There is also a tuned circuit, to determine the frequency of oscillation.

The load is in parallel with $C_2$ and the transformer. If the load is resistive, which is usually the case, the $Q$ of our tuned circuit and the loop gain are both affected; this must be taken into account when determining the minimum gain required for oscillation. If the load has a capacitive component, then the value of $C_2$ should be reduced accordingly. $C_2$ or a portion of it can be variable for accurately setting the oscillator frequency.

Sometimes when an oscillator is first built, it may fail to oscillate, or it may oscillate only when an input signal is initially applied. If this is the case, the circuit should be tested for loop gain. This is done by disconnecting $V_i$ from $V_x$, as in Fig. 14.6c, and testing the circuit as an amplifier whose input is $V_i$ and whose output is $V_x$, as shown in Fig. 14.7. Note that the load must be connected and the transformer secondary must be terminated with an impedance $Z_i$ which is equivalent to the impedance that it normally sees when connected to the transistor base. For self-starting oscillations, $V_x$ must be slightly greater than, and in

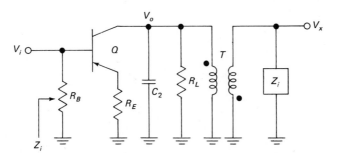

Fig. 14.7 Circuit of Fig. 14.6 redrawn to illustrate the concept of loop gain. Loop gain is the ratio of $V_x$ to $V_i$ when the transformer secondary is loaded with $Z_i$. For oscillations, $V_x/V_i = 1/\underline{0°}$.

phase with, $V_i$, at the desired frequency of oscillation. If the circuit did not oscillate when first connected, it is possible that these gain and phase criteria were not met. One way to increase the gain would be to bypass part of $R_E$. If the gain is too low because of excessive loading from $R_L$, a variety of impedance transformation techniques can be used, as discussed in Sec. 13.1c. These techniques can transform $R_L$ to a higher value when seen across the full tank circuit. This also yields a higher loaded $Q$ and improved frequency stability.

### 14.2a  The RC Phase Shift Oscillator

At low frequencies (around 100 kHz or less), resistors and capacitors are usually employed to determine the frequency of oscillation. Various circuit configurations may be used, including ladder and bridge types; our discussion here is limited to the ladder type.

A block diagram of a ladder-type $RC$ phase shift oscillator is shown in Fig. 14.8a. If the phase shift through the amplifier is 180°, then oscillations may occur at the frequency where the $RC$ network produces an additional 180° phase shift.

A common emitter transistor version of the diagram in Fig. 14.8a is shown in Fig. 14.8b. The feedback network can be analyzed by redrawing it, as in Fig. 14.8c; here $R_3$ represents whatever resistance the $RC$ network sees when it is connected to the transistor base. The $RC$ sections can all be equal ($R_1 = R_2 = R_3$ and $C_1 = C_2 = C_3$) or graded ($R_3 = 10R_2 = 100R_1$; $C_3 = 0.1C_2 = 0.01C_1$). If graded sections are used, the analysis is simplified, because loading of each section on the previous one can be neglected, and the voltage transfer function for all three sections is the product of each individual transfer function.[1] With graded sections, $R_3C_3 = R_2C_2 = R_1C_1$, which for simplicity we will call $RC$. For the circuit of Fig. 14.8c we can then write

$$\frac{V_1}{V_o} = \frac{R}{R + 1/(j\omega C)} = \frac{j\omega RC}{j\omega RC + 1}$$

and $V_x/V_2 = V_2/V_1 = V_1/V_o$. The overall transfer function is

$$\frac{V_x}{V_o} = \frac{V_x}{V_2}\frac{V_2}{V_1}\frac{V_1}{V_o}$$

$$\frac{V_x}{V_o} = \left(\frac{j\omega RC}{j\omega RC + 1}\right)^3 \qquad (14.4)$$

Provided that the phase shift through the amplifier is 180°, the circuit will oscillate at the frequency for which the phase shift from $V_o$ to $V_x$ is also 180°.

[1] See Sec. 4.5c.

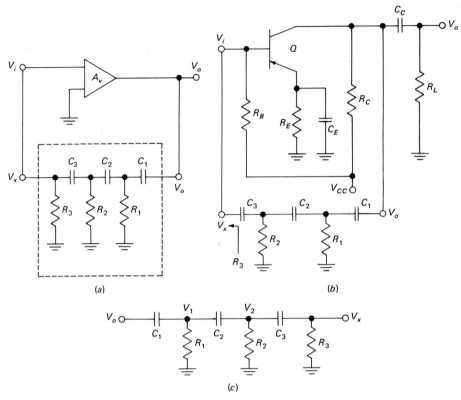

Fig. 14.8 *Amplifier with RC feedback network;* (b) *common emitter version of the circuit in* (a); (c) *feedback network.*

Since loading between stages is negligible, this is also the frequency when the phase shift through each section is $180/3 = 60°$. An expression for the phase shift through each RC section can be obtained as follows:

$$\phi\left(\frac{V_x}{V_2}\right) = \phi\left(\frac{V_2}{V_1}\right) = \phi\left(\frac{V_1}{V_o}\right) = \phi\left(\frac{j\omega RC}{j\omega RC + 1}\right)$$

$\phi$ = phase of numerator − phase of denominator

$$\phi = 90° - \tan^{-1}\frac{\omega RC}{1} \tag{14.5}$$

When $\phi = 60°$, we have

$60° = 90° - \tan^{-1}(\omega_o RC)$
$\tan^{-1}(\omega_o RC) = 30°$

$\omega_o RC = \tan 30°$
$\omega_o RC = 0.577$
$$f_o = \frac{0.577}{2\pi RC} \tag{14.6}$$

The losses through the feedback network at $f_o$ can be determined by substituting Eq. (14.6) and solving for the magnitude of $V_x/V_o$ in Eq. (14.4):

$$\left|\frac{V_x}{V_o}\right| = \left(\frac{0.577}{\sqrt{1 + 0.577^2}}\right)^3$$

$$\left|\frac{V_x}{V_o}\right| = \left(\frac{0.577}{1.58}\right)^3 = \frac{1}{8}$$

The voltage is therefore attenuated by a factor of 8, and the active device must have an open-loop voltage amplification greater than 8 to ensure sufficient loop gain for oscillations.

### Example 14.1

Assume that the amplifier of Fig. 14.8a has 180° phase shift, infinite input impedance, and zero output impedance. Determine the value of passive components required for 1-kHz oscillations using graded sections. What is the minimum open-loop voltage amplification?

### Solution

Since the amplifier has an infinite input impedance, there is no loading on the $RC$ feedback network. Similarly, the amplifier's zero output impedance means that $V_o$ is independent of any load connected at the output; therefore the impedance looking into the $RC$ feedback network does not affect the amplifier gain or phase shift.

To use graded sections, we may select $R_3 = 100$ k$\Omega$, $R_2 = 10$ k$\Omega$, and $R_1 = 1$ k$\Omega$. The $RC$ product needed for oscillations at 1 kHz is determined as follows:

$$RC = \frac{0.577}{2\pi f_o} \tag{14.6}$$

$$RC = \frac{0.577}{2\pi(10^3)}$$

$$RC = 92 \times 10^{-6}$$

From the above we can compute each capacitor value:

$$C_1 = \frac{92 \times 10^{-6}}{R_1} = \frac{92 \times 10^{-6}}{10^3}$$
$$C_1 = 0.092 \; \mu\text{F}$$

Similarly, $C_2 = 0.009 \; \mu\text{F}$ and $C_3 = 900 \; \text{pF}$.

Since the voltage in this type of feedback circuit is attenuated eight times and there is no loading at either the input or output, it follows that the amplifier's open-loop voltage amplification, at 1 kHz, must be slightly greater than 8. This will ensure sufficient loop gain for self-starting oscillations.

Graded sections are not always practical, mainly because the loading cannot always be neglected. If equal $RC$ sections are used in the feedback network, the voltage loss at $f_o$ increases to 29 (see Prob. 14.2.7), but there are advantages too. With all sections equal, it is possible to gang the capacitors (vary their value using a single control) to set or vary the frequency of oscillation.

When loading is not negligible, the complete circuit is analyzed as a unit, by writing loop or node equations. Oscillations occur when the closed-loop gain approaches infinity. The solution of a multiloop network can generally be carried out using determinants. All such solutions involve a numerator and denominator determinant, as follows:

$$A = \frac{M}{\Delta}$$

where $A$ is the amplitude of any voltage or current, $M$ is the numerator determinant, and $\Delta$, which is common to all $A$'s, is the denominator determinant. The conditions for oscillation can be determined by setting $\Delta = 0$.

We will illustrate the above procedure for the circuit of Fig. 14.8$b$. To simplify the analysis, $R_B$ is assumed large compared with the transistor's input resistance. Also, the transistor's $h_{oe}$ and $h_{re}$ are assumed to be negligible. The load $R_L$ and collector bias resistor can be lumped into a single AC load $R'_L = R_C \| R_L$, and the transistor's $h_{ie}$ can be one of the feedback-network components, which we will call $R$. The result is the simplified AC equivalent circuit of Fig. 14.9$a$. All $RC$ sections are equal, and the circuit may be redrawn as in Fig. 14.9$b$. We can further simplify the analysis by transforming the current generator shunted by $R_L$ (Norton circuit) to a voltage generator in series with $R'_L$ (Thevenin circuit), as in Fig. 14.9$c$. Loop equations can now be written as follows:

$$I_b \left( R + \frac{1}{j\omega C} + R \right) - I_1 R = 0 \quad (a)$$

$$I_1 \left( R + \frac{1}{j\omega C} + R \right) - I_b R - I_2 R = 0 \quad (b)$$

$$I_2\left(R'_L + \frac{1}{j\omega C} + R\right) + I_b(h_{fe}R'_L) - I_1 R = 0 \tag{c}$$

Equations (a), (b), and (c) may be rearranged as follows:

$$I_b\left(2R + \frac{1}{j\omega C}\right) + I_1(-R) + I_2(0) = 0 \tag{d}$$

$$I_b(-R) + I_1\left(2R + \frac{1}{j\omega C}\right) + I_2(-R) = 0 \tag{e}$$

$$I_b(h_{fe}R'_L) + I_1(-R) + I_2\left(R + R'_L + \frac{1}{j\omega C}\right) = 0 \tag{f}$$

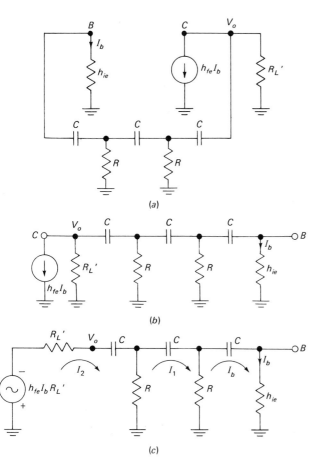

**Fig. 14.9** Equivalent circuits for the analysis of RC phase shift oscillator.

The denominator determinant[1] for all unknowns is

$$\Delta = \begin{vmatrix} 2R + \dfrac{1}{j\omega C} & -R & 0 \\ -R & 2R + \dfrac{1}{j\omega C} & -R \\ h_{fe}R'_L & -R & R + R'_L + \dfrac{1}{j\omega C} \end{vmatrix} \quad (g)$$

Oscillations occur when $\Delta = 0$:

$$\Delta = \left(2R + \dfrac{1}{j\omega C}\right)^2 \left(R + R'_L + \dfrac{1}{j\omega C}\right) - (-R)(-R)\left(2R + \dfrac{1}{j\omega C}\right)$$
$$- (-R)(-R)\left(R + R'_L + \dfrac{1}{j\omega C}\right) + (-R)(-R)(h_{fe}R'_L) = 0 \quad (h)$$

After some simplification, Eq. (h) becomes

$$\Delta = \left[R^3 + R^2 R'_L(h_{fe} + 3) - \dfrac{5R + R'_L}{\omega^2 C^2}\right] + j\left(\dfrac{1}{\omega^3 C^3} - \dfrac{6R^2 + 4RR'_L}{\omega C}\right) = 0$$

For $\Delta = 0$, both its real and imaginary parts must equal zero. Setting $\text{Im}(\Delta) = 0$ yields

$$\dfrac{1}{\omega^3 C^3} = \dfrac{6R^2 + 4RR'_L}{\omega C}$$

$$\omega = \dfrac{1}{C\sqrt{6R^2 + 4RR'_L}} \quad (14.7)$$

Setting $\text{Re}(\Delta) = 0$ yields

$$R^3 + R^2 R'_L(h_{fe} + 3) = \dfrac{5R + R'_L}{\omega^2 C^2}$$

The information in Eq. (14.7) can be used to substitute for $\omega^2 C^2$, as follows:

$$R^3 + R^2 R'_L(h_{fe} + 3) = (5R + R'_L)(6R^2 + 4RR'_L)$$

which after simplification becomes

$$h_{fe} = 29\dfrac{R}{R'_L} + 23 + 4\dfrac{R'_L}{R} \quad (14.8)$$

Note that setting $\text{Im}(\Delta) = 0$ gave us information about the frequency of oscillation, while setting $\text{Re}(\Delta) = 0$ provided the conditions (minimum $h_{fe}$) required for oscillations.

---

[1] See Appendix 4.

## Example 14.2

The following data apply to the circuit of Fig. 14.8b: $h_{ie} = 1.8$ k$\Omega$, $1/h_{oe} = 20$ k$\Omega$, $h_{re}$ is negligible; $R_E = 2$ k$\Omega$, $R_C = 4$ k$\Omega$, $R_B = 68$ k$\Omega$, $R_L = 1$ k$\Omega$. Select suitable feedback-network components for 10-kHz oscillations. Use equal $RC$ sections. What is the minimum $h_{fe}$ required?

## Solution

Since $R_B \gg h_{ie}$, we can neglect $R_B$, and the impedance looking into the base is approximately $h_{ie} = 1{,}800$ $\Omega$. This corresponds to $R_3$ in Fig. 14.8c. If ungraded sections are used, we can select $R = R_1 = R_2 = R_3 = 1{,}800$ $\Omega$. Since $R_L \ll 1/h_{oe}$, we can neglect $h_{oe}$, and the equivalent circuits of Fig. 14.9 apply. The net load seen by the collector is $R'_L = R_L \| R_C = 800$ $\Omega$.

Equations (14.7) and (14.8) can be used to determine the capacitance and $h_{fe}$ required, respectively,

$$C = \frac{1}{2\pi f_o \sqrt{6R^2 + 4RR'_L}} \tag{14.7}$$

$$C = \frac{1}{2\pi(10^4) \sqrt{6(1{,}800^2) + 4(1{,}800)(800)}}$$

$C = 3{,}170$ pF

$$h_{fe} = 29\frac{R}{R'_L} + 23 + 4\frac{R'_L}{R} \tag{14.8}$$

$$h_{fe} = 29\frac{1{,}800}{800} + 23 + 4\frac{800}{1{,}800}$$

$h_{fe} = 91$

The circuit will not oscillate unless the transistor's $h_{fe}$ is greater than 91.

## Problem Set 14.2

Problems 14.2.1 through 14.2.5 apply to the circuit of Fig. 14.6a:

14.2.1. What is the function of

    a. $R_B$          d. $C_2$
    b. $R_E$          e. $C_3$
    c. $C_1$          f. $T$

14.2.2. Do you think the circuit would oscillate if the transformer secondary winding were reversed? Explain your answer.

14.2.3. Give an example of how a low-resistance load could be coupled to this oscillator. Repeat for a high-resistance load.

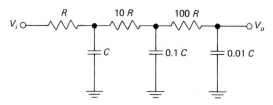

Fig. 14.10   Graded sections RC network.

14.2.4. If the circuit failed to oscillate, what would you do?

14.2.5. How would bypassing $R_E$ affect the $Q$ of the output tank circuit?

Problems 14.2.6 through 14.2.9 apply to the circuit of Fig. 14.8a.

14.2.6. Assuming equal $RC$ sections, infinite input, and zero output impedance, show that the frequency of oscillation is $f_o = 0.065/RC$.

14.2.7. In Prob. 14.2.6, show that the voltage ratio $V_x/V_o$ at $f_o$ is 1/29. What is the minimum open-loop voltage amplification required for oscillations?

14.2.8. If the amplifier has zero output and infinite input impedance, determine the value of all feedback components for oscillations at 10 kHz. Assume graded sections and $R_3 = 100$ kΩ.

14.2.9. Repeat Prob. 14.2.8 using equal $RC$ sections and the results of Prob. 14.2.6. Assume all $R = 100$ kΩ.

14.2.10. The transistor in the circuit of Fig. 14.8b has $h_{ie} = 2{,}000$ Ω, negligible $h_{oe}$, and $h_{re}$. Assume $R_B$ is large, $R_C = 10$ kΩ, and $R_L = 10$ kΩ. Use $h_{ie}$ in place of $R_3$. Select the remaining feedback-network capacitors and resistors for oscillations at 50 kHz using equal $RC$ sections. What is the minimum $h_{fe}$ required?

14.2.11. Show that the voltage loss for the graded $RC$ network in Fig. 14.10 at the frequency of 180° phase shift is 8. Can you think of any application for this network?

## 14.2b   The Colpitts Oscillator

Shown in Fig. 14.11a is the AC circuit of a Colpitts oscillator. The DC biasing components are not shown here for simplicity. The output is developed across the tank circuit; the positive feedback required for oscillations is derived by capacitive tapping of the tank. Although the circuit is shown here in the common base configuration, by grounding the emitter or the collector, the same circuit can be operated in the common emitter or common collector configuration.

A common emitter Colpitts oscillator is shown in Fig. 14.11b; its equivalent circuit, including a transistor model, is shown in Fig. 14.11c.

The feedback connection between $V_x$ and $V_i$ has been left open in Fig.

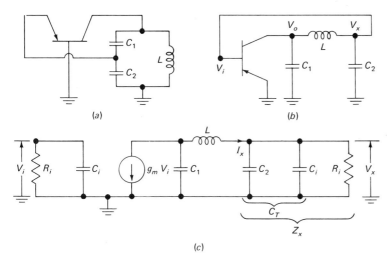

*Fig. 14.11  Colpitts oscillator circuits.*

14.11c. We will analyze the circuit to determine the conditions for which $V_x$ is identical to $V_i$, in which case the loop can be closed and sustained oscillations will result. To do this we must load $C_2$ with the same circuit it sees when $V_i$ and $V_x$ are connected; that is, we must add the input resistance $R_i$ and input capacitance $C_i$, as shown in the figure.

$R_i$ is approximately equal to $r_{b'e}$, while $C_i$ is the total input capacitance ($C_{b'e}$ plus the Miller effect equivalent of $C_{b'c}$). Neglecting $r_{bb'}$ makes the analysis a bit less accurate; however, most oscillator circuits must be experimentally adjusted to the desired frequency, and therefore accurate analyses are usually unnecessary.

Let $C_T$ equal the parallel combination of $C_2$ and $C_i$. Also $Z_x$ is the impedance of the parallel $C_T$ and $R_i$ combination. We can therefore write

$$V_x = I_x Z_x \qquad (a)$$

$$I_x = -g_m V_i \frac{1/(j\omega C_1)}{1/(j\omega C_1) + j\omega L + Z_x} \qquad (b)$$

For oscillations, $V_x$ must equal $V_i$:

$$V_x = V_i \qquad (c)$$

Substituting for $V_x$ from Eq. (a) and for $V_i$ from Eq. (b), Eq. (c) becomes

$$I_x Z_x = \frac{I_x[1/(j\omega C_1) + j\omega L + Z_x]}{-g_m[1/(j\omega C_1)]} \qquad (d)$$

which after substitution of

$$Z_x = \frac{R_i[1/(j\omega C_T)]}{R_i + 1/(j\omega C_T)} \qquad (e)$$

yields

$$1 + g_m R_i - \omega^2 L C_1 + j(\omega C_1 R_i + \omega C_T R_i - \omega^3 L C_1 C_T R_i) = 0 \qquad (f)$$

Setting the imaginary part of Eq. (f) equal to zero yields the frequency of oscillation:

$$\omega^3 L C_1 C_T R_i = \omega C_1 R_i + \omega C_T R_i$$

$$\omega^2 = \frac{C_1 + C_T}{L C_T C_1}$$

$$f_o = \frac{1}{2\pi \sqrt{LC}} \qquad (14.1)$$

where $C = C_1 C_T/(C_1 + C_T)$. Setting the real part of Eq. (f) equal to zero yields the conditions for sustained oscillations:

$$1 + g_m R_i = \omega^2 L C_1$$

$$1 + g_m R_i = \frac{C_1 + C_T L C_1}{L C_T C_1}$$

$$1 + g_m R_i = 1 + \frac{C_1}{C_T}$$

$$g_m R_i = \frac{C_1}{C_T}$$

Since $R_i \simeq r_{b'e}$, $g_m R_i \simeq g_m r_{b'e} = h_{fe}$.[1] Therefore the minimum $h_{fe}$ for oscillations is

$$h_{fe} = \frac{C_1}{C_T} \qquad (14.9)$$

If the circuit does not oscillate when first connected, $C_2$ (which is part of $C_T$) should be increased until the positive feedback is sufficient for self-starting oscillations.

A typical Colpitts oscillator circuit is shown in Fig. 14.12a. The operating point is set by $V_{CC}$, $R_1$, $R_2$, and $R_3$. $C_5$ is a bypass capacitor at $f_o$; therefore the base is AC-grounded. The RFC (RF choke) is selected to be self-resonant at $f_o$, therefore exhibiting a high impedance. This reduces the loading on $C_2$ from $R_3$. Without the RFC, $R_3$ would be directly across $C_2$, lowering the Q of the tuned circuit. $C_4$ is a power supply bypass, while $C_3$ is a DC blocking capacitor; both have negligible reactance at $f_o$. The complete AC equivalent circuit is shown in Fig. 14.13b. Note that, except for the addition of a load, this circuit is identical to

---

[1] This relationship is given by Eq. (6.15) in Sec. 6.3.

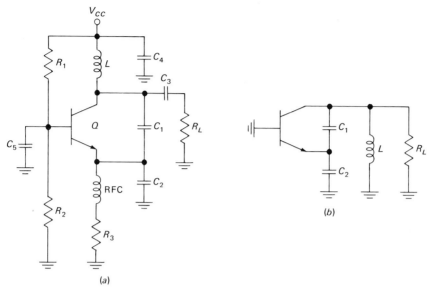

Fig. 14.12 (a) Bipolar transistor Colpitts oscillator; (b) equivalent circuit for (a).

that of Fig. 14.11a. The load does not have to be connected directly in parallel with $L$ as shown here. Depending on its value, $R_L$ could be transformed to a more suitable value by tapping $L$.

The 100-MHz FET Colpitts oscillator of Fig. 14.13 is now described. The DC operating point is set by $V_{DD}$ in conjunction with source resistor $R_1$. The polarity of DC source current through $R_1$ makes the source positive with respect to ground; since the gate is grounded, $V_{GS}$ is negative, which is the polarity required for proper biasing. The operating point determines the value of AC parameters.

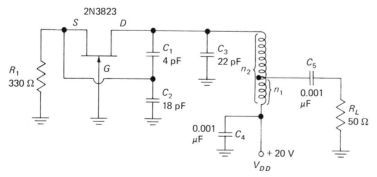

Fig. 14.13 100-MHz FET Colpitts oscillator. (Courtesy of Texas Instruments Incorporated.)

The output AC equivalent circuit is shown in Fig. 14.14. $R'_L$ is the transformed value of $R_L$ when seen across the full $n_2$ turns of $L$; therefore $R'_L = R_L(n_2/n_1)^2$. $R_o$ is the FET's output resistance. The bottom end of $L$ is AC-grounded through the low reactance of $C_4$. $L$, in conjunction with $C_3$, $C_1$, $C_2$, and $C_i$, forms a parallel resonant circuit at 100 MHz, the frequency of oscillation. The junction of $C_1$ and $C_2$ is connected to the source; therefore the input resistance ($R_i$), input capacitance ($C_i$), and source-biasing resistor ($R_1$) all shunt $C_2$. $C_i$ simply adds to $C_2$, yielding $C'_2 = C_2 + C_i$. The resistances $R_1$ and $R_i$ can also be combined to yield $R'_i = R_1 \| R_i$. The ratio of $C_1$ and $C'_2$ determines the transformed value of $R'_i$ when seen across the full tank circuit.

The relative values of $C_1$ and the total impedance from source to ground ($C'_2$ in parallel with $R'_i$) determine the feedback factor $\beta_v$. Therefore, by adjusting the relative values of $C_1$ and $C_2$, it is possible to increase or decrease the feedback factor. This control is useful to ensure self-starting oscillations.

## Example 14.3

For the 100-MHz FET oscillator of Figs. 14.13 and 14.14, assume the following values: $L = 0.1$ µH with an unloaded $Q$ of 200, $R_o = 100$ kΩ, $R_i = 300$ Ω, $C_i = 2$ pF. Determine the loaded $Q$ of the output circuit.

### Solution

a. The total capacitance from source to ground is

$$C'_2 = C_i + C_2 = 20 \text{ pF}$$

b. The total resistance from source to ground is

$$R'_i = R_i \| R_1 = 300 \| 330 = 157 \text{ Ω}$$

c. $R'_i$, when seen across the full tank circuit, is transformed to

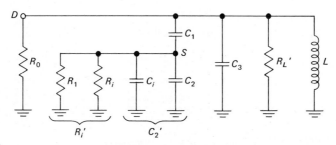

Fig. 14.14 Output equivalent circuit for the oscillator of Fig. 14.13.

$$R = R'_i\left(1 + \frac{C'_2}{C_1}\right)^2 = 157(36) \qquad (13.15a)$$

$R = 5.66 \text{ k}\Omega$

d. The value of $R'_L$ is

$$R'_L = R_L\left(\frac{n_2}{n_1}\right)^2 = 50(24^2)$$

$R'_L = 28.8 \text{ k}\Omega$

e. The coil's $R_P$ is

$$R_P = Q_U X_P \qquad (13.7)$$
$X_P = 2\pi fL = (6.28)(100 \times 10^6)(0.1 \times 10^{-6}) = 62.8 \text{ }\Omega$
$R_P = 200(62.8)$
$R_P = 12{,}600 \text{ }\Omega$

f. The total resistance shunting the output tank circuit is

$R_T = R_o \| R \| R'_L \| R_P$
$R_T \simeq 3.3 \text{ k}\Omega$

g. The loaded $Q$ is

$$Q_L = \frac{R_T}{X_P} \qquad (13.7)$$

$$Q_L = \frac{3{,}300}{62.8} = 52.5$$

A modified version of the Colpitts oscillator that results in greater frequency stability is the Clapp oscillator, whose AC equivalent circuit is shown in Fig. 14.15. The only difference between this circuit and that of Fig. 14.11a is the addition of $C_3$. Without $C_3$, the resonant frequency is determined by $L$ and $C$, where $C \simeq C_1 C_2/(C_1 + C_2)$. $C_2$, however, is shunted by the transistor's input capacitance $(C_i)$, which is a function of operating point and temperature and may vary from one device to another. If $C_i$ is significant compared with $C_2$, the oscillator frequency may not be sufficiently stable. The addition of $C_3$ makes the resonant frequency depend more on $C_3$ than on $C$. Since $C_3$ is not subject to the variations of $C_i$, the net result is an improvement in frequency stability.

To determine $f_o$, note that at frequencies above the series resonant frequency of $L$ and $C_3$, the $L$-$C_3$ branch is inductive. When the magnitude of inductive reactance for the $L$-$C_3$ branch is equal to the magnitude of capacitive reactance for the $C_1$-$C_2$ combination, parallel resonance occurs:[1]

---

[1] The losses in $L$ will influence the parallel resonant frequency to some degree; however, this effect is negligible for high $Q$ coils.

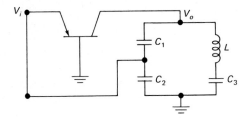

*Fig. 14.15 Clapp oscillator.*

$$\left| j\omega_o L - \frac{j}{\omega_o C_3} \right| = \left| \frac{-j}{\omega_o C} \right|$$

$$\omega_o L - \frac{1}{\omega_o C_3} = \frac{1}{\omega_o C}$$

from which

$$\omega_o^2 = \frac{1}{L}\left(\frac{1}{C_3} + \frac{1}{C}\right)$$

$$f_o = \frac{1}{2\pi\sqrt{L\dfrac{C_3 C}{C_3 + C}}} \tag{14.10}$$

If $C_3 \ll C$, the tuning capacitance is mainly $C_3$, which is independent of the transistor's input capacitance. This simplifies setting of the oscillator frequency; without $C_3$, $f_o$ could be varied with $C_1$ or $C_2$, which also affected the feedback factor. Now $f_o$ can be set by varying $C_3$ without disturbing the feedback factor, which is still determined by $C_1$ and $C_2$.

### 14.2c  The Hartley Oscillator

The Hartley oscillator differs from the Colpitts oscillator in the manner that the output tank circuit is tapped to produce the feedback signal. As shown in Fig. 14.16, $L$ is tapped instead of the tuning capacitor. The turns ratio $n_2/n_1$

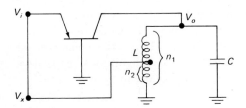

*Fig. 14.16 Hartley oscillator.*

determines the ratio of feedback voltage $V_x$ to the total voltage $V_o$ developed across the tank. The feedback factor $\beta_v$ is therefore

$$\beta_v = \frac{V_x}{V_o} \tag{14.2}$$

$$\beta_v = \frac{n_2}{n_1} \tag{14.2a}$$

To increase the feedback, for example, the tap is raised so that $n_2$ is greater. Usually $n_2/n_1$ is made somewhat larger than the minimum value to ensure self-starting oscillations under a variety of operating conditions.

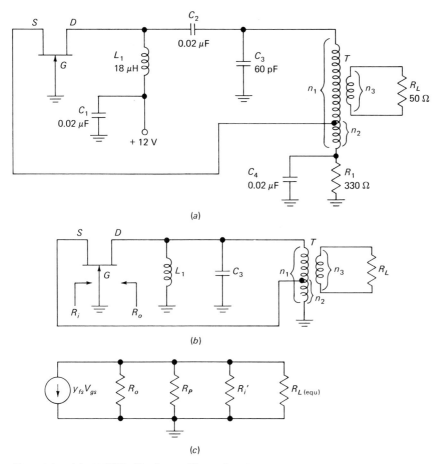

Fig. 14.17 (a) 20-MHz Hartley oscillator; (b) AC equivalent circuit for (a); (c) output equivalent circuit for (a). (Courtesy of Texas Instruments Incorporated.)

A common gate FET Hartley oscillator is shown in Fig. 14.17a. The AC equivalent circuit is shown in Fig. 14.17b. Note that $C_1$, $C_2$, and $C_4$ all behave as short circuits at $f_o$, 20 MHz in this circuit. $R_1$ is effectively between source and ground and AC-bypassed with $C_4$. The resonant frequency is mainly determined by $L_1$, $C_3$, and the transformer's equivalent inductance and capacitance. The amount of positive feedback required for self-starting oscillations determines the turns ratio $n_2/n_1$. The transformer also performs the function of transforming the 50-$\Omega$ load into a more suitable value across the full tank circuit.

The conditions for oscillation can be easily estimated using Barkhausen's criterion:

$$\beta_v A_v = 1 \, \underline{/0°} \tag{14.3}$$

The voltage amplification is

$$A_v \simeq R'_L y_{fs} \tag{9.24a}$$

where $R'_L$, the total resistance seen by the drain, is assumed to be much smaller than $r_d$. If we substitute the feedback factor $\beta_v$ from Eq. (14.2a) and $A_v$ from Eq. (9.24a), Barkhausen's criterion becomes

$$\frac{n_2 R'_L y_{fs}}{n_1} = 1 \, \underline{/0°} \tag{14.3a}$$

from which

$$\frac{n_2}{n_1} = \frac{1}{R'_L y_{fs}} \tag{14.11}$$

### Example 14.4

For the 20-MHz oscillator of Fig. 14.17a, assume FET parameters as follows: $r_d = 100$ k$\Omega$, minimum $y_{fs} = 3{,}000$ $\mu$mho. The combined $R_P$ for $L_1$ and the transformer is 15 k$\Omega$; the turns ratio $n_1/n_3 = \sqrt{40}$. Select a suitable turns ratio $n_1/n_2$ for self-starting oscillations.

### Solution

The output equivalent circuit of Fig. 14.17c will be used. The FET's output resistance $R_o \simeq r_d = 100$ k$\Omega$. The combined $R_P$ of $L_1$ and the transformer is 15 k$\Omega$. The FET's input resistance can be estimated using Eq. (9.23a):[1]

$$R_i \simeq \frac{r_d}{1 + y_{fs} r_d} \tag{9.23a}$$

$$R_i = 333 \, \Omega$$

---

[1] $R_S$ in Eq. (9.23a) corresponds to $R_1$ in the circuit of Fig. 14.17a; it can be neglected because it is bypassed with $C_4$. Also note that $y_{os} = 1/r_d$.

When seen across the full $n_1$ turns, this becomes

$$R'_i = 333\left(\frac{n_1}{n_3}\right)^2$$

The 50-Ω load is transformed to

$$R_{L(\text{equ})} = 50\left(\frac{n_1}{n_3}\right)^2 = 50(40) = 2{,}000 \ \Omega$$

The total resistance across the output circuit is

$$R'_L = R_o \| R_P \| R'_i \| R_{L(\text{equ})}$$

$$R'_L = 10^5 \| 15 \times 10^3 \| 333\left(\frac{n_1}{n_2}\right)^2 \| 2{,}000$$

Clearly $10^5$ is large compared with the rest; therefore $R'_L$ is

$$R'_L \simeq 1{,}750 \| 333\left(\frac{n_1}{n_2}\right)^2$$

When the above value of $R'_L$ is substituted into Eq. (14.11), the criterion for oscillation is

$$\frac{n_2}{n_1} = \frac{1}{[1{,}750\|333(n_1/n_2)^2](3 \times 10^{-3})} \qquad (14.11)$$

$$\frac{n_1}{n_2} = \frac{(1{,}750)(333)(n_1/n_2)^2(3 \times 10^{-3})}{1{,}750 + 333(n_1/n_2)^2}$$

Let $n_1/n_2 = x$:

$$x = \frac{1{,}750 x^2}{1{,}750 + 333 x^2}$$

$$1{,}750 + 333 x^2 = 1{,}750 x$$

which simplifies to

$$x^2 - 5.26x + 5.26 = 0$$

yielding $x = 1.33$ or $3.93$. Since $x = n_1/n_2$, a turns ratio $n_1/n_2$ greater than 1.33 should be adequate. To ensure self-starting oscillations under all expected operating conditions, a value of 3 was actually chosen.

### 14.2d  Crystal Oscillators

A crystal oscillator employs a quartz crystal as the frequency-determining element. Quartz crystals are characterized by the piezoelectric effect, a natural phenomenon in which mechanical energy is converted to electric energy and vice versa. If an electric signal is applied across a quartz crystal, the crystal will vibrate. The intensity of vibrations is greatest at the crystal's natural frequency.

For our purposes it is sufficient to consider the symbol and equivalent circuit of Fig. 14.18a and b. A quartz crystal is simply a passive, two-terminal component whose electrical behavior can be predicted using the equivalent circuit of Fig. 14.18b. $L_1$, $C_1$, and $R_S$ are not separately accessible components. $L_1$ is an inductive effect related to the mass, while $R_S$ accounts for the energy losses. The crystal is mounted between electrodes yielding parallel capacitance $C_2$.

The $Q$ of crystals can be very high, much higher than any practical inductor. $Q$'s of 10,000 are quite common, while for specially designed units, $Q$'s as high as 1 million have been reported. This means that the effect of $R_S$ on resonance and impedance can be largely ignored.

The crystal impedance, neglecting $R_S$, is shown in Fig. 14.18c. At low frequencies the crystal behaves capacitively. At $f_1$, the $L_1$-$C_1$ branch resonates, and the total impedance approaches zero (in reality it is $R_S$). The series resonant frequency for the crystal is therefore

$$f_1 = \frac{1}{2\pi \sqrt{L_1 C_1}} \tag{14.12}$$

Above $f_1$, the crystal behaves inductively. The equivalent inductance of the $L_1$-$C_1$ branch resonates with $C_2$ at the parallel resonant frequency $f_2$:

$$f_2 = \frac{1}{2\pi \sqrt{L_1 C_1 C_2 / (C_1 + C_2)}} \tag{14.13}$$

which can also be expressed as

$$f_2 = f_1 \sqrt{1 + \frac{C_1}{C_2}} \tag{14.14}$$

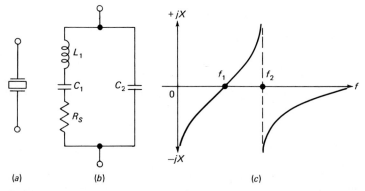

Fig. 14.18  Resonant crystal: (a) symbol; (b) equivalent circuit; (c) impedance characteristic.

Usually $C_1/C_2$ is very small and $f_2$ is very close to $f_1$. The crystal therefore exhibits the properties of series resonance (low resistance) at $f_1$ and parallel resonance (high resistance) at $f_2$. Beyond $f_2$, the crystal behaves capacitively. A complete equivalent circuit would include additional branches similar to the $L_1$-$C_1$-$R_S$ branch. These give rise to overtones, that is, series resonant frequencies that are harmonically related to $f_1$. Therefore a real crystal also resonates at $2f_1$, $3f_1$, $4f_1$, and so on.

Crystals can be manufactured with values of $f_1$ as low as 10 kHz; at these frequencies the crystal is relatively thick. On the high-frequency side, $f_1$ can be as high as 10 MHz; here the crystal is very thin. Beyond 10 MHz, overtone operation can be employed to yield higher series resonant frequencies.

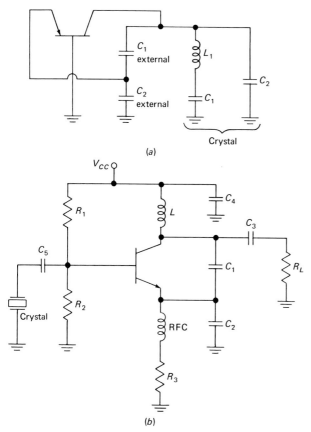

Fig. 14.19 (a) Clapp crystal oscillator; (b) Colpitts crystal oscillator.

The temperature coefficient of crystals is usually small and can be made zero. When extreme temperature stability is required, the crystal may be housed in an oven to maintain it at a constant temperature. The high $Q$ of crystals also contributes to the relatively drift-free operation of crystal oscillators.

The equivalent circuit for a typical Clapp oscillator using a crystal is shown in Fig. 14.19a. Here the crystal's $L_1$ and $C_1$ replace $L$ and $C_3$, respectively, in the circuit of Fig. 14.15. The crystal's $C_1$ can be as low as 0.001 pF! For the Clapp oscillator, this means that $f_o$, the frequency of oscillation, is determined almost entirely by the crystal. If we wish to change $f_o$, the crystal is simply replaced with one that resonates at the desired frequency.

A Colpitts crystal oscillator is shown in Fig. 14.19b. In order for the circuit to oscillate, the base must be AC-grounded. In the circuit of Fig. 14.12a this is accomplished with $C_5$. $C_5$ is still used here to keep DC out of the crystal; the crystal impedance is high on each side of $f_1$ but takes a sharp dip at $f_1$. Thus the base is AC-grounded only at $f_1$, and the circuit's frequency of oscillation $f_o$ becomes $f_1$. The output circuit is also tuned at $f_o$.

The $Q$ of the output circuit cannot approach that of the crystal. Without the crystal, $C_5$ AC-grounds the base not just at $f_o$ but for a wide range of frequencies on each side of $f_o$; therefore $f_o$ can vary within the frequency-selectivity curve of the output circuit. Even with a loaded $Q$ as high as 100, the bandwidth of the output tuned circuit is usually wide enough to allow relatively large variations in $f_o$.

A crystal oscillator can exhibit frequency drifts as low as 10 ppm (parts per million) per °C.

### 14.2e  Tunnel Diode Oscillators

The tunnel diode is an interesting semiconductor device that, when suitably biased, exhibits the property of negative resistance. The symbol and voltage-current characteristics for a tunnel diode are shown in Fig. 14.20a and b, respectively.

Consider the plot of current versus voltage shown in Fig. 14.20b. This applies to forward voltage and current, which is the polarity shown in Fig. 14.20a. The dashed curve superimposed on the graph is the characteristic for a conventional diode. When $V_F$ is increased from zero, $I_F$ increases more rapidly than for a conventional diode. This steep rise in current is due to tunneling, a phenomenon in which carriers cross a PN junction even though their energy level is insufficient to overcome the potential barrier that exists across the junction. For this effect to occur, a very narrow depletion region is necessary; this is obtained by heavy doping of both P and N regions.

The current and voltage corresponding to point $x$ in Fig. 14.20b are the peak point current ($I_P$) and peak point voltage ($V_P$). As $V_F$ is increased beyond

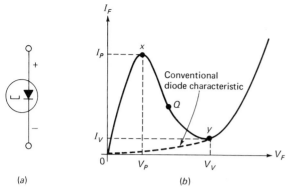

Fig. 14.20  Tunnel diode: (a) symbol; (b) forward characteristic.

$V_P$, $I_F$ decreases. The decrease in current with an increasing voltage is a negative resistance characteristic. The current and voltage corresponding to point $y$ are the valley point current ($I_V$) and valley point voltage ($V_V$), respectively. Note that the valley point voltage is greater than the peak point voltage. Beyond the valley point voltage, the tunnel diode behaves as a conventional diode, and $I_F$ increases with increasing $V_F$. Typical specifications for a germanium tunnel diode are

Peak point voltage:  $V_P = 50$ mV
Peak point current:  $I_P = 1$ mA
Valley point current:  $I_V = 0.1$ mA
Valley point voltage:  $V_V = 350$ mV

At any operating point $Q$ between $x$ and $y$ on the characteristic of Fig. 14.20b, the dynamic resistance is negative. The dynamic resistance is the ratio of voltage to current along a tangent to the curve at a specific operating point:

$$\text{Dynamic resistance:} \quad r = \frac{\Delta V_F}{\Delta I_F} \tag{14.15}$$

To allow for maximum signal swing, the tunnel diode should be biased somewhere in the middle of the negative resistance region. Here values of $r$ around $-250\ \Omega$ are typical.

A tunnel diode DC biasing circuit is shown in Fig. 14.21a. The DC biasing resistors and battery can be replaced with their Thevenin equivalent, as shown in Fig. 14.21b. Depending on the choice of $R_T$ and $E_T$, a number of possibilities arise. Two different modes of operation are illustrated in Fig. 14.22.

If load line $R_{T1}$ applies, there is only one intersection with the diode characteristic, at $Q$. Note that this load line has a steeper slope than the diode

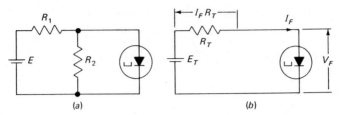

Fig. 14.21  (a) DC biasing circuit for tunnel diode; (b) Thevenin equivalent circuit for (a); $E_T = ER_2/(R_1 + R_2)$, $R_T = R_1R_2/(R_1 + R_2)$.

characteristic in this region; therefore $R_{T1} < |r|$. The $R_{T1}$ load line results in a stable operating point, and the diode is useful as an amplifier or oscillator.

If load line $R_{T2}$ applies, there are three intersections with the diode characteristic, at $Q$, $Q_1$, and $Q_2$. Of these, only $Q_1$ and $Q_2$ are stable, and the diode may be at either of these two points. $Q$ is not stable; if the diode happens to be at $Q$, it will quickly switch to either of its two stable points, $Q_1$ or $Q_2$. This can be illustrated as follows. Suppose that the diode is at $Q$. Because of thermal agitation, there are always small noise currents. At any instant, the noise will either add to or subtract from the quiescent current. If the noise is additive, $I_F$ momentarily increases; this produces a larger voltage drop across $R_T$ (Fig. 14.21b), therefore reducing $V_F$. But we are in the negative resistance region, and reducing $V_F$ increases $I_F$. This additional increase in $I_F$ further reduces $V_F$, causing more $I_F$, and so on. Very quickly the diode switches from $Q$ to $Q_1$. Once at $Q_1$, the diode can remain there indefinitely, because the diode resistance is positive. Had the noise initially reduced $I_F$, the diode would have switched to $Q_2$. Because there are only two stable points, this diode circuit is said to be bistable and is useful in switching applications.

The selection of $R_T$ is therefore limited by the constraint that, for a stable operating point in the negative resistance region, $R_T < |r|$. In practice $R_T$ cannot

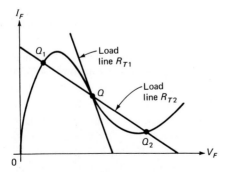

Fig. 14.22  Tunnel diode characteristic with stable load line ($R_{T1}$) and unstable load line ($R_{T2}$).

be made too small; otherwise excessive power is required from the DC supply. A compromise often used is to make $R_T \simeq \frac{1}{3}|r|$.

### Example 14.5

Determine the dynamic resistance of the tunnel diode whose characteristic is given in Fig. 14.23. The operating point is at $I_F = 0.5$ mA and $V_F = 120$ mV.

### Solution

A tangent is drawn to the curve at $Q$. The voltage-current ratio along this tangent is the diode's dynamic resistance:

$$r = \frac{\Delta V_F}{\Delta I_F} = \frac{(180 - 0)(10^{-3})}{(0 - 1.5)(10^{-3})} \tag{14.15}$$
$$r = -120 \ \Omega$$

### Example 14.6

Select a suitable load line to bias the circuit of Fig. 14.21b at the $Q$ point shown in Fig. 14.23. What are $R_T$ and $E_T$? If a 1.5-V battery is available, determine $R_1$ and $R_2$.

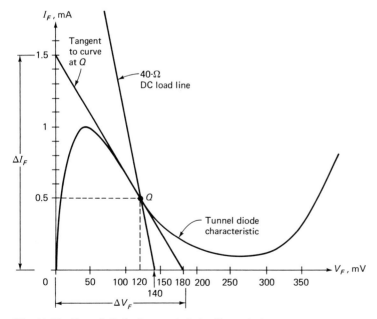

Fig. 14.23  Tunnel diode characteristic for Example 14.5.

## Solution

Since $|r| = 120 \, \Omega$ (Example 14.5), $R_T$ is chosen as $120/3 = 40 \, \Omega$. The $40\text{-}\Omega$ DC load line is shown on the characteristic of Fig. 14.23. At $Q$, $I_F = 0.5$ mA and $V_F = 120$ mV; therefore $E_T$ is

$$E_T = I_F R_T + V_F$$
$$E_T = (0.5 \times 10^{-3})(40) + (120 \times 10^{-3})$$
$$E_T = 140 \text{ mV}$$

If a 1.5-V battery is available, we can write

$$R_T = \frac{R_1 R_2}{R_1 + R_2} = 40 \qquad (a)$$

and

$$E_T = E \frac{R_2}{R_1 + R_2}$$
$$0.14 = 1.5 \frac{R_2}{R_1 + R_2} \qquad (b)$$

Simultaneous solution of Eqs. (a) and (b) yields

$$R_1 = 430 \, \Omega \qquad R_2 = 70 \, \Omega$$

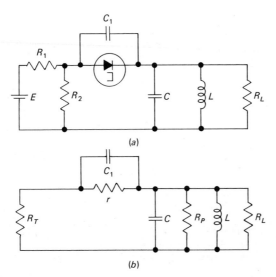

Fig. 14.24 (a) Tunnel diode oscillator; (b) AC equivalent circuit for (a).

The circuit diagram for a tunnel diode oscillator is shown in Fig. 14.24a. The DC equivalent circuit is the same as in Fig. 14.21b, while the AC equivalent circuit is shown in Fig. 14.24b. Note that $R_T$ is the parallel combination of $R_1$ and $R_2$.

In this circuit energy is absorbed by the biasing resistance $R_T$, the coil's resistance $R_P$, and the load $R_L$. All these resistances are positive. The tunnel diode's AC equivalent circuit may be approximated with a single resistance $r$ (whose value is negative). A complete AC equivalent circuit would include the diode's diffusion capacitance in parallel with $r$ as well as lead resistance and self-inductance. These additional components, however, become significant only at very high frequencies. The externally added capacitance $C_1$ swamps the diode's diffusion capacitance.

For sustained oscillations, the diode's negative resistance must supply whatever energy is lost in all the positive resistances, including the load. The energy, of course, comes from the battery; the diode simply converts it to energy at the frequency of oscillation. Since the negative resistance region is limited, the voltage and current swings that are possible are also limited; output power levels of a few hundred microwatts are typical.

The equivalent circuit of Fig. 14.24b is redrawn in Fig. 14.25. Here $R'_L = R_P \| R_L$. The conditions for oscillation may be determined by writing an expression for the total admittance between points $j$ and $k$:

$$Y_{jk} = Y_1 + Y_2$$
$$Y_1 = \frac{1}{Z_1}$$

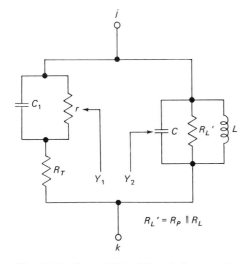

Fig. 14.25  Tunnel diode AC equivalent circuit useful for deriving oscillating criteria.

$$Z_1 = R_T + r \left\| \frac{1}{j\omega C_1} \right.$$

$$Y_1 = \frac{r + R_T + \omega^2 r^2 C_1^2 R_T}{(r + R_T)^2 + (r\omega C_1 R_T)^2} + j \frac{\omega r C_1 (r + R_T) - r\omega C_1 R_T}{(r + R_T)^2 + (r\omega C_1 R_T)^2}$$

$$Y_2 = j\omega C + \frac{1}{R_L'} + \frac{1}{j\omega L}$$

$$Y_{jk} = \frac{1}{R_L'} + \frac{r + R_T + \omega^2 r^2 C_1^2 R_T}{(r + R_T)^2 + (r\omega C_1 R_T)^2} + j \left[ \frac{\omega r C_1 (r + R_T) - r\omega C_1 R_T}{(r + R_T)^2 + (r\omega C_1 R_T)^2} + \omega C - \frac{1}{\omega L} \right]$$

Setting $\text{Re}(Y_{jk}) = 0$ yields

$$(r + R_T)^2 + (r\omega C_1 R_T)^2 + R_L'(r + R_T) + R_L' \omega^2 r^2 C_1^2 R_T = 0$$
$$(r + R_T)(r + R_T + R_L') + \omega^2 r^2 C_1^2 R_T (R_T + R_L') = 0$$

Usually $R_L' \gg R_T$ and $R_L' \gg (R_T + r)$; therefore the above equation simplifies to

$$(r + R_T)(R_L') + \omega^2 r^2 C_1^2 R_T R_L' = 0$$

from which

$$R_T = \frac{-r}{1 + \omega^2 r^2 C_1^2} \qquad (14.16)$$

Equation (14.16) yields the conditions required for oscillation at any frequency $\omega$. It can also be solved for $\omega$, which then represents the highest possible frequency of oscillation for any given $R_T$:

$$f_{\max}{}^2 = \frac{-(r + R_T)}{r^2 R_T C_1^2 (2\pi)^2} \qquad (14.17)$$

Beyond $f_{\max}$, the negative resistance is no longer sufficient to cancel the total positive resistance, and oscillations cannot be sustained. The actual frequency of oscillation is determined by setting $\text{Im}(Y_{jk}) = 0$:

$$\frac{\omega r^2 C_1}{(r + R_T)^2 + \omega^2 r^2 C_1^2 R_T^2} + \omega C - \frac{1}{\omega L} = 0$$

$$\omega^2 L r^2 C_1 + (\omega^2 LC - 1)[(r + R_T)^2 + \omega^2 r^2 C_1^2 R_T^2] = 0$$

Substituting for $\omega^2$ from Eq. (14.17) yields

$$\omega^2 L r^2 C_1 + (\omega^2 LC - 1)[(r + R_T)^2 - (r + R_T) R_T] = 0$$

which simplifies to

$$\omega^2 L r^2 C_1 + (\omega^2 LC - 1)(r + R_T)(r) = 0$$

$$\omega^2 = \frac{1}{LC\{1 + [rC_1/C(r + R_T)]\}}$$

$$f_o = \frac{1}{2\pi \sqrt{LC} \sqrt{1 + \frac{rC_1}{C(r + R_T)}}} \tag{14.18}$$

### Example 14.7

Using the data from Examples 14.5 and 14.6, determine the maximum frequency of oscillation for the circuit of Fig. 14.24 if $C_1 = 10$ pF.

**Solution**

The pertinent data are $r = -120$ Ω, $R_T = 40$ Ω, $C_1 = 10$ pF. The maximum frequency of oscillation is

$$f_{max}^2 = \frac{-(r + R_T)}{r^2 R_T C_1^2 (2\pi)^2} \tag{14.17}$$

$$f_{max}^2 = \frac{-(-120 + 40)}{(-120)^2 (40 \times 10^{-22})(2\pi)^2}$$

$f_{max}^2 = 3.47 \times 10^{16}$

$f_{max} = 186$ MHz

### Example 14.8

For the circuit of Example 14.7, assume $L = 0.1$ μH and determine $C$ for oscillations at 100 MHz. The equivalent resistance of the tank circuit ($R_P$ of coil in parallel with $R_L$) is high compared with $R_T$.

**Solution**

Equation (14.18) may be used to solve for $C$:

$$10^8 = \frac{1}{6.28 \sqrt{10^{-7} C} \sqrt{1 + \frac{(-120 \times 10^{-11})}{C(-120 + 40)}}} \tag{14.18}$$

$C = 10$ pF

### Problem Set 14.3

Problems 14.3.1 through 14.3.7 apply to the circuit of Fig. 14.12a.

14.3.1. What type of oscillator circuit is this?
14.3.2. What is the function of $C_5$? $C_3$? RFC?

14.3.3. Draw a DC equivalent circuit.

14.3.4. If the circuit fails to oscillate, how would you test it for loop gain?

14.3.5. If it were necessary to increase the amount of positive feedback, what component value would you change and in what way?

14.3.6. If $L = 1$ $\mu$H, determine $C$ for oscillations at 10.7 MHz.

14.3.7. Using the data of Prob. 14.3.6, determine $C_1$ and $C_2$ if $C_i = 10$ pF and $h_{fe} = 10$ at 10.7 MHz.

Problems 14.3.8 and 14.3.9 apply to the circuit of Fig. 14.13.

14.3.8. If $L = 0.1$ $\mu$H with an unloaded $Q$ of 200, $R_o = 100$ k$\Omega$, $R_i = 300$ $\Omega$, and $C_i = 2$ pF, determine the turns ratio $n_2/n_1$ for a loaded $Q$ of 50 when $R_L = 75$ $\Omega$.

14.3.9. How would you eliminate the loading on the output tuned circuit due to $R_1$?

14.3.10. What advantages does a Clapp oscillator have over a Colpitts type?

14.3.11. What is the basic difference between a Colpitts and Hartley oscillator?

14.3.12. How could the Hartley oscillator of Fig. 14.17a be modified to a Colpitts oscillator?

14.3.13. What advantages does a crystal have over a conventional tuned circuit?

14.3.14. A given crystal series resonates at 1 MHz. Its $Q$ is 20,000, and $C_1 = 0.005$ pF. Determine $L_1$ and $R_S$.

14.3.15. The crystal of Prob. 14.3.14 has an equivalent $C_2$ of 0.5 pF. Determine the parallel resonant frequency $f_2$. What is the equivalent $R_P$ at $f_2$?

14.3.16. What is the basic difference between the oscillator of Fig. 14.19b and that of Fig. 14.12a?

14.3.17. In what way is a tunnel diode different from a conventional diode?

14.3.18. Draw the symbol for a tunnel diode and indicate the correct bias polarity for negative resistance operation.

14.3.19. For the DC circuit of Fig. 14.21a, assume $E = 1.5$ V and $R_T = 50$ $\Omega$. The diode is to be biased at 100 mV and 0.65 mA. Determine $R_1$ and $R_2$.

14.3.20. What is meant by an unstable operating point? For stable operation in the circuit of Fig. 14.21b, what relationship must hold between $R_T$ and the magnitude of diode negative resistance?

14.3.21. Estimate the dynamic resistance on the characteristic of Fig. 14.23 at $V_F =$

    a. 20 mV      d. 200 mV
    b. 40 mV      e. 260 mV
    c. 60 mV      f. 360 mV

Problems 14.3.22 and 14.3.23 apply to the circuit of Fig. 14.24a.

14.3.22. What is the maximum frequency of oscillation if $R_T = 60$ $\Omega$, $r = -180$ $\Omega$, and $C_1 = 15$ pF?

**14.3.23.** Given $L = 0.2\ \mu\text{H}$, determine $C$ for oscillations at 40 MHz. The data of Prob. 14.3.22 apply.

## 14.3 RELAXATION OSCILLATORS

Two types of relaxation oscillators are discussed in this section. One is the unijunction transistor oscillator, which can generate an approximately sawtooth waveform; the other is the astable multivibrator, which produces a square-wave output.

### 14.3a The Unijunction Transistor Oscillator

The unijunction transistor, as the name implies, is characterized by a single PN junction. It exhibits a negative resistance characteristic that makes it useful in oscillator circuits.

The symbol for a UJT (unijunction transistor) is given in Fig. 14.26a. Note that there are three terminals: base 1 ($B1$), base 2 ($B2$), and emitter ($E$). The physical structure can be explained using the diagram in Fig. 14.26b. Here a lightly doped $N$-type silicon bar acts as the base. A P-type impurity is introduced into the base, producing the single PN junction. This PN junction exhibits the properties of a conventional diode; therefore the emitter is a *rectifying* contact. $B2$ and $B1$, on the other hand, are *ohmic* contacts.

A simplified equivalent circuit for the UJT is shown in Fig. 14.27. $V_{BB}$ is a source of biasing voltage connected between $B2$ and $B1$. If the emitter is open, the total resistance from $B2$ to $B1$ is simply the resistance of the silicon bar; this is referred to as the interbase resistance $R_{BB}$. $R_{B2}$ is the resistance between $B2$ and point $a$, while $R_{B1}$ is the resistance from point $a$ to $B1$; therefore the interbase resistance is

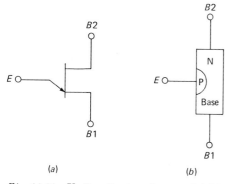

Fig. 14.26 *Unijunction transistor symbol* (a) *and structure* (b).

$$R_{BB} = R_{B1} + R_{B2} \tag{14.19}$$

The diode accounts for the rectifying properties of the PN junction. $V_T$ is the diode's threshold voltage. With the emitter open, $I_E = 0$ and $I_1 = I_2$. The interbase current is given by Ohm's law:

$$I_1 = I_2 = \frac{V_{BB}}{R_{BB}}$$

For example, if $V_{BB} = 20$ V and $R_{BB} = 10$ kΩ, then $I_1 = I_2 = 2$ mA.

Note that part of $V_{BB}$ is dropped across $R_{B2}$ while the rest is dropped across $R_{B1}$. We may refer to that portion of $V_{BB}$ which is dropped across $R_{B1}$ as $V_a$. Using simple voltage-divider relationships,

$$V_a = V_{BB} \frac{R_{B1}}{R_{B1} + R_{B2}} \tag{14.20}$$

The voltage division factor is given a special symbol and name: $\eta$ = intrinsic standoff ratio:

$$\eta = \frac{R_{B1}}{R_{B1} + R_{B2}} \tag{14.21}$$

so that

$$V_a = \eta V_{BB} \tag{14.22}$$

The intrinsic standoff ratio is a property of the UJT; it is always less than 1 and usually ranges between 0.4 and 0.85. If $V_{BB} = 20$ V and $\eta = 0.6$, then the voltage developed at point $a$ relative to $B_1$ is $V_a = \eta V_{BB} = 12$ V. The remaining 8 V is obviously dropped across $R_{B2}$.

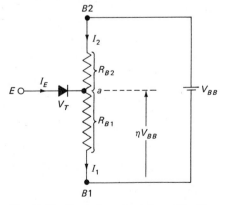

Fig. 14.27 Equivalent circuit for unijunction transistor.

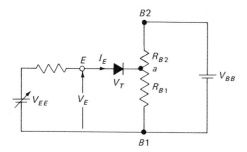

Fig. 14.28  Circuit for experimental determination of UJT characteristics.

As long as $I_E = 0$, the circuit of Fig. 14.27 behaves as a voltage divider. Assume now that $V_E$ is gradually increased from zero, using an emitter supply $V_{EE}$, as indicated in Fig. 14.28. The diode remains reverse-biased, and except for leakage, no emitter current flows until $V_E$ is sufficiently positive to forward-bias the diode. Appreciable emitter current begins to flow when

$$V_E = V_T + V_a \tag{14.23}$$

where $V_T$ is the diode's threshold voltage and $V_a = \eta V_{BB}$. The value of $V_E$ that causes the diode to start conducting is called the peak point voltage:

$$V_P = V_T + \eta V_{BB} \tag{14.24}$$

The graph of Fig. 14.29 displays the relationship between emitter voltage and current. Note that $V_E$ is plotted on the vertical axis, while $I_E$ is plotted on the

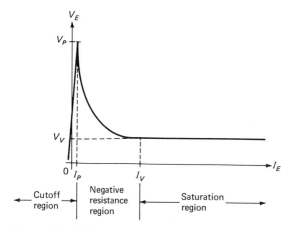

Fig. 14.29  UJT characteristic.

horizontal axis. The region from $V_E = 0$ to $V_E = V_P$ is called the cutoff region, because no emitter current (except for some leakage) flows. Once $V_E$ exceeds the peak point voltage, $I_E$ increases, but $V_E$ decreases. We are now in the negative resistance region, which extends to the valley point ($V_V$ and $I_V$). Beyond the valley point, $I_E$ increases with $V_E$; this is the saturation region, which exhibits a positive resistance characteristic.

The physical process responsible for the negative resistance characteristic is called conductivity modulation. When the peak point voltage is exceeded, holes from the P-type emitter are injected into the N-type base. Here they move toward $B1$ to complete their path by reentering at the negative terminal of $V_{EE}$. The base region between point $a$ and $B1$ now has excess holes with their inherent positive charge. In an effort to keep this region electrically neutral, free electrons diffuse toward $B1$. This means that the number of charge carriers in this area (holes and electrons) has increased. The increase in charge carriers represents a decrease in resistance. $R_{B1}$ can therefore be considered as a variable resistance whose conductivity is modulated (changed) by $I_E$. As $I_E$ increases, $R_{B1}$ decreases. Typical values of $R_{B1}$ may range from 5 kΩ (when $I_E = 0$) to 50 Ω (when $I_E$ is maximum).

Since $\eta$ is a function of $R_{B1}$ [Eq. (14.21)], it follows that the reduction of $R_{B1}$ causes a corresponding reduction in intrinsic standoff ratio. Thus as $I_E$ increases, $R_{B1}$ decreases, $\eta$ decreases, and $V_a$ decreases. The decrease in $V_a$ causes more emitter current to flow, which causes a further reduction of $R_{B1}$, $\eta$, and $V_a$. The process is evidently regenerative, and $V_a$, as well as $V_E$, quickly drops while $I_E$ increases. Note that, although it decreases in value, $R_{B1}$ is always a positive resistance. It is only the dynamic resistance in the region between peak and valley points that is negative.

The curve of Fig. 14.29 applies to a specific value of $V_{BB}$. A family of such curves results when different values of $V_{BB}$ are used.

There are two UJT parameters whose temperature dependence should be considered here. The diode threshold voltage decreases with temperature. The interbase resistance $R_{BB}$, because of silicon's positive temperature coefficient, increases with temperature. It is possible to compensate for these effects using external resistance, as will be seen shortly.

**Example 14.9**

A given UJT has an interbase resistance of 8 kΩ; also $R_{B1} = 6$ kΩ when $I_E = 0$. Determine

a. The UJT current if $V_{BB} = 16$ V and $V_E$ is less than the peak point voltage
b. The intrinsic standoff ratio and $V_a$, using the data from (a)
c. The peak point voltage

## Solution

The available information is $R_{BB} = 8$ kΩ, $R_{B1} = 6$ kΩ; since $R_{BB} = R_{B1} + R_{B2}$, it follows that $R_{B2} = 2$ kΩ.

a. As long as $V_E < V_P$, $I_E = 0$ and

$$I_1 = I_2 = \frac{V_{BB}}{R_{BB}} = \frac{16}{8 \times 10^3} = 2 \text{ mA}$$

b.

$$\eta = \frac{R_{B1}}{R_{B1} + R_{B2}} \qquad (14.21)$$

$\eta = 6/8 = 0.75$

$$V_a = \eta V_{BB} \qquad (14.22)$$

$V_a = 0.75 \times 16 = 12$ V

c. The peak point voltage is

$$V_P = V_T + \eta V_{BB} \qquad (14.24)$$

The diode's threshold voltage may be assumed as 0.5 V, yielding

$V_P = 0.5 + 12 = 12.5$ V

We will now consider a UJT oscillator circuit as shown in Fig. 14.30a. Externally connected resistors $R_2$ and $R_1$ are used for reasons to be discussed shortly. $R_2$ may be a few hundred ohms, while $R_1$ should be less than 50 Ω; $R_1$ is often omitted. The DC source $E$ supplies the necessary bias. The interbase voltage $V_{BB}$ is the difference between $E$ and the voltage drops across $R_2$ and $R_1$. Usually the interbase resistance $R_{BB}$ is much larger than $R_2$ and $R_1$ so that $V_{BB} \simeq E$. Before proceeding further, however, one should not confuse $R_{B1}$ and $R_{B2}$ with $R_1$ and $R_2$. $R_{B1}$ and $R_{B2}$ are the interbase resistance components; they are *internal* resistances of the UJT and do not appear in the circuit diagram of Fig. 14.30a. $R_2$ and $R_1$ are actual resistors *external* to the UJT. In the complete equivalent circuit of Fig. 14.30b, $R_2$ is in series with $R_{B2}$, while $R_1$ is in series with $R_{B1}$.

As soon as power is applied to the circuit of Fig. 14.30a, $C$ begins to charge toward $E$. The voltage across $C$, which is also $v_E$, rises exponentially with a time constant

$$\tau = RC \qquad (14.25)$$

As long as $v_E < V_P$, $I_E = 0$. As indicated in the circuit of Fig. 14.30b, there are simply two parallel branches across $E$; one is the series $RC$ circuit, while the other is the series combination of $R_1$, $R_{BB}$, and $R_2$. The diode remains reverse-biased as long as $v_E < V_P$.

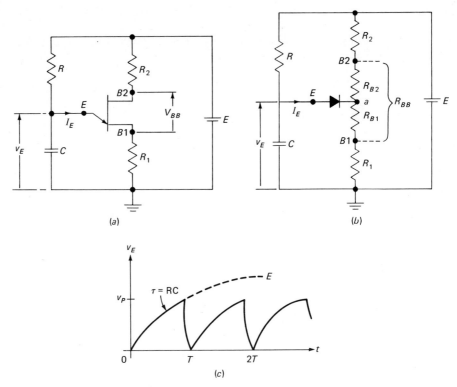

**Fig. 14.30** (a) *Circuit of UJT oscillator;* (b) *AC equivalent circuit for* (a); (c) *emitter voltage waveform for the circuit of* (a).

When the capacitor voltage has risen to the firing potential, the diode conducts and $R_{B1}$ decreases. We have in effect connected a low resistance across $C$, through which it discharges. The voltage across the capacitor, $v_E$, drops very quickly to the valley voltage $V_V$ because of the fast time constant due to the low value of $R_{B1} + R_1$. As soon as $v_E$ drops below $V_a + V_T$, though, the diode is no longer forward-biased, and it stops conducting. We now revert to the previous state, and $C$ begins to charge once more toward $E$.

The emitter voltage waveform is shown in Fig. 14.30c. Note that at $t = 0$, $v_E$ rises exponentially toward $E$ but drops to a very low value after it reaches the peak point voltage $V_P$. The time for $v_E$ to drop from $V_P$ to its minimum value is relatively small and usually neglected. The period $T$ can therefore be approximated as

T:   Time required for $v_E$ to rise from 0 to $V_P$ (a)

As long as $R_{BB} \gg R_1 + R_2$, $V_{BB} \simeq E$ and

$$V_P \simeq \eta E \tag{b}$$

The equation that describes the exponential rise of voltage across a capacitor is

$$v_E = E(1 - e^{-t/(RC)}) \tag{c}$$

where $v_E$ is the instantaneous emitter voltage, $E$ the "target" voltage that $C$ aims to reach, $e$ the base of natural logarithms, $t$ the time in seconds, and $RC$ the time constant in seconds. Now at $t = T$, $v_E = V_P = \eta E$; substituting these data in Eq. (c) yields

$$\eta E = E(1 - e^{-T/(RC)})$$
$$\eta = 1 - e^{-T/(RC)}$$
$$e^{-T/(RC)} = 1 - \eta$$
$$e^{T/(RC)} = \frac{1}{1 - \eta}$$
$$\frac{T}{RC} = \ln\left|\frac{1}{1 - \eta}\right|$$
$$T = RC \ln\left|\frac{1}{1 - \eta}\right|$$

Now let

$$K = \ln\left|\frac{1}{1 - \eta}\right| \tag{14.26}$$

which yields

$$T = RCK$$

The frequency of oscillation is $1/T$; therefore

$$f_o = \frac{1}{RCK} \tag{14.27}$$

The parameter $K$ is plotted, as a function of $\eta$, in Fig. 14.31. As an example, for $\eta = 0.75$, $K \simeq 1.4$.

### Example 14.10

The UJT circuit of Fig. 14.30a is to oscillate at 1 kHz. If $R = 10$ k$\Omega$, determine $C$. Assume $\eta = 0.65$.

### Solution

The parameter $K$ is read from the graph of Fig. 14.31; for $\eta = 0.65$, $K \simeq 1.05$. The required capacitance is determined using Eq. (14.27):

$$C = \frac{1}{f_o R K} \tag{14.27}$$

$$C = \frac{1}{(10^3)(10^4)(1.05)} \simeq 0.095 \ \mu\text{F}$$

There are two additional outputs possible for the UJT oscillator just discussed. One of these is the voltage developed at $B1$ (across $R_1$) due to the capacitor discharge, while the other is the voltage developed at $B2$. A complete set of all three output waveforms is shown in Fig. 14.32. When the UJT fires (at $t = T$), $V_a$ drops, causing a corresponding voltage drop at $B2$. The duration of pulses at $B1$ and $B2$ is determined by the discharge time of $C$. If $R_1$ is very small, $C$ discharges very quickly, and the pulses generated at each base are quite narrow. If $R_1$ is zero, obviously no pulses appear at $B1$; similarly if $R_2 = 0$, no pulses can be generated at $B2$. $R_1$ therefore serves no purpose other than to develop a voltage proportional to the current. If these voltage pulses are not needed, $R_1$ can be omitted. If $R_1$ is too large, its positive resistance may swamp

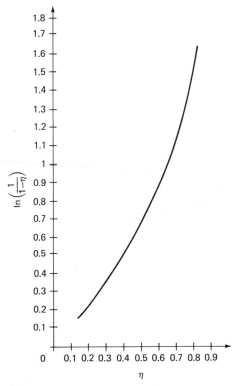

Fig. 14.31 Plot of $\ln\left(\frac{1}{1-\eta}\right)$ as a function of intrinsic standoff ratio $\eta$.

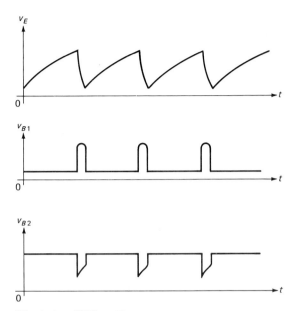

Fig. 14.32   UJT oscillator waveforms.

the unijunction's negative resistance and prevent the UJT from switching back after it has fired.

$R_2$, in addition to providing a source of pulses at $B2$, is useful for temperature stabilization of the UJT's peak point voltage. This is given by

$$V_P = V_T + \eta V_{BB} \qquad (14.24)$$

As the temperature rises, $V_T$ decreases. The temperature coefficient of $R_{BB}$ is positive; $R_2$ is essentially independent of temperature. It is therefore possible to select $R_2$ so that $\eta V_{BB}$ increases with temperature by the same amount as $V_T$ decreases. This yields a constant peak point voltage, which in turn stabilizes the frequency of oscillation.

The selection of the timing resistor and capacitor must also be considered. In the circuit of Fig. 14.30a, prior to firing, $R$ is required to pass only the capacitor-charging current. At the instant that the peak point voltage is reached, $R$ must supply the peak point current. It is therefore necessary that the current through $R$ be slightly greater than the peak point current:

Current through $R > I_P$

$$\frac{E - V_P}{R} > I_P$$

$$R < \frac{E - V_P}{I_P} \qquad (14.28)$$

Once the UJT fires, $v_E$ drops to the valley voltage $V_V$. $I_E$ should not be allowed to increase beyond the valley current $I_V$; otherwise the UJT is taken into the saturation region and does not switch back. $R$ must therefore be selected large enough to ensure that

$$I_E < I_V$$
$$\frac{E - V_V}{R} < I_V$$
$$R > \frac{E - V_V}{I_V} \tag{14.29}$$

Note that Eqs. (14.28) and (14.29) place lower and upper bounds on the value of $R$. As long as $R$ is chosen between these extremes, reliable oscillator operation should result. It should be remembered, though, that $R_1$ and $R_2$ must be small compared to $R_{BB}$.

**Example 14.11**

Using the static characteristics for the 2N489 UJT in Fig. 14.33, estimate $V_P$, $V_V$, and $I_V$. $V_{BB}$ is 20 V. Determine a suitable range for $R$ if the peak point current is 20 μA.

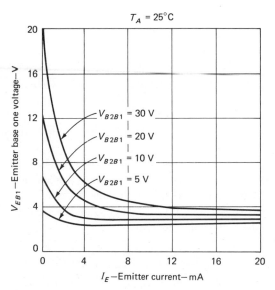

Fig. 14.33  Static emitter characteristics for the 2N489 UJT. (Courtesy of Texas Instruments Inc.)

## Solution

The peak point voltage is read from the characteristic as $V_P \simeq 12.5$ V. The peak point current is given as 20 μA. The valley point voltage is read as $V_V \simeq 3$ V. The valley point current is $I_V \simeq 10$ mA. A suitable range for $R$ is now determined:

$$R < \frac{E - V_P}{I_P} \tag{14.28}$$

As long as $V_{BB} \simeq E$,

$$R < \frac{20 - 12.5}{20 \times 10^{-6}}$$
$$R < 373 \text{ k}\Omega$$

The lower limit is

$$R > \frac{E - V_V}{I_V} \tag{14.29}$$

$$R > \frac{20 - 3}{10 \times 10^{-3}}$$
$$R > 1{,}700 \ \Omega$$

There are many other applications of UJTs in timing, voltage sensing, and triggering circuits. The student is referred to the literature for exploring some of these other applications.

### 14.3b  The Astable Multivibrator

The astable multivibrator (multi, for short) is a circuit without any stable states. This means that voltage and current levels constantly switch between their allowable extremes; that is, the circuit oscillates. Other types of multivibrators include the monostable (one stable state) and bistable (two stable states). Some of these circuits are discussed in Chap. 15.

A circuit diagram to illustrate operation of the astable multi is shown in Fig. 14.34. Since no two transistors are exactly alike, assume that when the circuit is first connected, $Q_1$ conducts slightly more than $Q_2$. As the current through $Q_1$ increases from zero, $v_{C1}$ drops. This drop is coupled by $C_1$ to the base of $Q_2$, therefore reducing its forward base-emitter voltage and causing $Q_2$ to conduct less. As the current through $Q_2$ decreases, $v_{C2}$ rises toward $V_{CC}$. This rise is coupled by $C_2$ to the base of $Q_1$, thereby increasing this transistor's forward base-emitter voltage. Thus $Q_1$ conducts more and $Q_2$ conducts less, each action reinforcing the other. Soon $Q_1$ is saturated, and $Q_2$ is cut off. Assuming low-leakage silicon transistors and $V_{CC} = 12$ V, we have $v_{C1} \simeq 0.3$ V, $v_{B1} \simeq 0.7$ V, and $v_{C2} \simeq 12$ V. These levels are shown, for $t < 0$, on the appropriate graphs of Fig. 14.35. During this

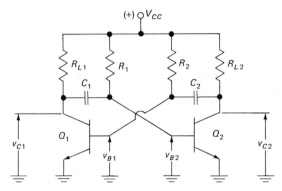

Fig. 14.34  Astable multivibrator circuit.

time, $C_1$ charges toward $V_{CC}$, as indicated in the equivalent circuit of Fig. 14.36. The time constant is $\tau_1 = R_1C_1$. Note that the voltage at the junction of $C_1$ and $R_1$ is also $v_{B2}$. Thus $v_{B2}$ now follows an exponential curve rising toward 12 V with a time constant $\tau_1$, as shown, for $t < 0$, on the appropriate graph in Fig. 14.35.

The circuit will make a transition (change state) as soon as $v_{B2}$ is sufficiently positive to cut in $Q_2$. This occurs at $t = 0$, when $v_{B2} \simeq 0.5$ V. $Q_2$, which had been cut off, now conducts; therefore $v_{C2}$ drops from 12 to 0.3 V. This 11.7-V drop is coupled, through $C_2$ to the base of $Q_1$, so that $v_{B1}$ also drops by 11.7, to $-11$ V. This voltage cuts off $Q_1$, and $v_{C1}$ rises toward 12 V with a time constant $\tau_2 = R_{L1}C_1$. The rise in $v_{C1}$ is coupled, through $C_1$ to the base of $Q_2$, causing a small overshoot. Soon $v_{B2}$ settles at 0.7 V, which is the voltage required to saturate $Q_2$. With $Q_2$ ON and $Q_1$ OFF, our levels are $v_{C2} \simeq 0.3$ V, $v_{B2} \simeq 0.7$ V, and $v_{C1} \simeq 12$ V; $v_{B1}$ is rising exponentially from $-11$ to $+12$ V with a time constant $\tau_3 = R_2C_2$. These levels are shown, on the appropriate graphs of Fig. 14.35, for the period between $t = 0$ and $t = t_1$.

At $t_1$, $v_{B1}$ reaches the cut-in voltage for $Q_1$, and another circuit transition takes place. Note that the cut-in value for $Q_1$ is around 0.5 V; this is halfway between $v_{B1}$'s starting voltage $(-11$ V) and target voltage $(+12$ V). Now it takes $0.69\tau$ to reach the halfway point on any exponential rising curve; therefore the period from $t = 0$ to $t = t_1$, which we will call $T_A$, is

$$T_A = 0.69\tau_3$$
$$T_A = 0.69R_2C_2 \tag{14.30}$$

In a similar fashion, the period from $t_1$ to $t_2$ is governed by $\tau_1 = R_1C_1$ and

$$T_B = 0.69R_1C_1 \tag{14.31}$$

It is, of course, possible to make $T_A = T_B$; in general, though, this is not necessarily true. The total period is $T_A + T_B$, and the frequency of oscillation is

$$f = \frac{1}{T_A + T_B} \tag{14.32}$$

The minimum $h_{FE}$ required to saturate the ON transistor is now determined. We will assume $Q_1$ in Fig. 14.34 to be ON and $v_{B1} \simeq v_{C1} \simeq 0$ V, yielding

$$I_{C1(ON)} \simeq \frac{V_{CC}}{R_{L1}}$$

$$I_{B1(ON)} \simeq \frac{V_{CC}}{R_2}$$

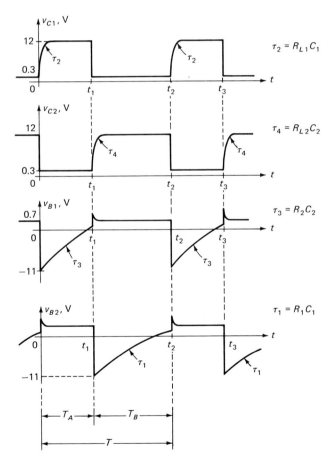

Fig. 14.35 Waveforms for the multivibrator of Fig. 14.34. Note that during $T_A$, $Q_1$ is OFF and $Q_1$ is ON while during $T_B$, $Q_1$ is ON and $Q_2$ is OFF.

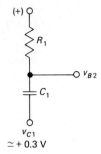

Fig. 14.36 While $Q_1$ is ON in the circuit of Fig. 14.34, $v_{B2}$ rises exponentially toward $V_{CC}$ with a time constant $\tau_1 = R_1 C_1$.

Minimum $h_{FE1} = \dfrac{I_{C1(ON)}}{I_{B1(ON)}}$

$$\underline{h_{FE1} = \dfrac{R_2}{R_{L1}}} \qquad (14.33)$$

Similarly

$$\underline{h_{FE2} = \dfrac{R_1}{R_{L2}}} \qquad (14.34)$$

In symmetric circuits, $R_{L1} = R_{L2} = R_L$ and $R_1 = R_2 = R$, yielding

$$\underline{h_{FE} = \dfrac{R}{R_L}} \qquad (14.35)$$

### Example 14.12

Estimate the frequency of oscillation in the circuit of Fig. 14.34 if $R_1 = R_2 = 10\ \text{k}\Omega$ and $C_1 = C_2 = 0.01\ \mu\text{F}$. What is the minimum $h_{FE}$ if $R_{L1} = R_{L2} = 1\ \text{k}\Omega$?

### Solution

Both time constants are equal; therefore

$T_A = T_B = (0.69 \times 10^4)(10^{-8}) = 69\ \mu\text{s}$

The total period is

$T_A + T_B = 2(69 \times 10^{-6}) = 138\ \mu\text{s}$

The frequency of oscillation is

$$f_o = \frac{1}{138 \times 10^{-3}}$$
$$f_o = 7{,}250 \text{ Hz}$$

The minimum $h_{FE}$ is

$$\underline{h_{FE}} = \frac{R}{R_L} = \frac{10^4}{10^3} = 10 \tag{14.35}$$

### Problem Set 14.4

14.4.1. Describe the following terms:
   a. Interbase resistance
   b. Intrinsic standoff ratio
   c. Rectifying contact
   d. Peak point voltage
   e. Conductivity modulation

14.4.2. What are the three regions of operation for a unijunction transistor?

Problems 14.4.3 through 14.4.7 apply to the circuit of Fig. 14.30a.

14.4.3. Estimate the frequency of oscillation if $R_1 = R_2 = 0$, $\eta = 0.5$, $R = 100$ k$\Omega$, $C = 0.1$ $\mu$F.

14.4.4. What voltage appears at B2 and B1 in Prob. 14.4.3?

14.4.5. What is the function of $R_1$? What happens if $R_1$ is too large?

14.4.6. What is the function of $R_2$? What is a typical value?

14.4.7. Given $V_P = 10$ V, $V_V = 2$ V, $I_P = 10$ $\mu$A, $I_V = 15$ mA, $E = 20$ V, $R_2 = R_1 = 0$, determine a range of permissible values for $R$.

14.4.8. Explain qualitatively how the circuit of Fig. 14.34 oscillates.

14.4.9. Given an astable multi of the type in Fig. 14.34, with $V_{CC} = 15$ V and low-leakage silicon transistors, estimate the base and collector voltages for each transistor when $Q_1$ is ON and $Q_2$ is OFF.

14.4.10. In the circuit of Fig. 14.34, an output is desired in which $T_A = 2T_B$. The frequency of oscillation must be 150 kHz. If $R_1 = R_2 = 2$ k$\Omega$, determine $C_1$ and $C_2$.

14.4.11. If $R_L = 100$ $\Omega$ for both transistors in Prob. 14.4.10, determine the minimum $h_{FE}$ required.

## REFERENCES

1. Mulvey, J.: "Semiconductor Device Measurements," Tektronix, Inc., Beaverton, Ore., 1968.

2. Kiver, M. S.: "Transistors," McGraw-Hill Book Company, New York, 1962.
3. Amos, S. W.: "Principles of Transistor Circuits," Hayden Book Company, Inc., New York, 1965.
4. Cowles, L. G.: "Transistor Circuits and Applications," Prentice-Hall, Inc., Englewood Cliffs, N.J., 1968.
5. Joyce, M. V., and K. K. Clarke: "Transistor Circuit Analysis," Addison-Wesley Publishing Company, Inc., Reading, Mass., 1961.
6. Texas Instruments Incorporated: "Transistor Circuit Design," McGraw-Hill Book Company, New York, 1963.
7. Ristenbatt, M. P., and R. L. Riddle: "Transistor Physics and Circuits," Prentice-Hall, Inc., Englewood Cliffs, N.J., 1966.
8. Gibbons, J. F.: "Semiconductor Electronics," McGraw-Hill Book Company, New York, 1966.
9. "General Electric Transistor Manual," 7th ed., Syracuse, 1964.
10. Farell, C. L.: Field Effect Transistor Oscillators, *Bulletin* CA-99, Texas Instruments Incorporated, Dallas, 1969.
11. Butler, F.: Transistor R-C Oscillators and Selective Amplifiers, *Wireless World*, December 1962, pp. 583–589.
12. Chow, W. F.: Crystal Controlled High Frequency Transistor Oscillators, *Semiconductor Products*, September 1959, pp. 21–27.
13. Crawford, R. H., and R. T. Dean: The How and Why of Unijunction Transistors, Theory, Operation and Circuits, *Bulletin* CA-68, Texas Instruments Incorporated, Dallas, 1969.

# chapter 15

## pulse circuits

This chapter treats circuits involving abruptly changing signals, as opposed to sinusoidal signals whose changes are relatively gradual. These abrupt changes in current or voltage levels are referred to as *pulses*, and we will study the manner in which various devices (bipolar transistors, silicon-controlled rectifiers, FETs) respond to such signals.

Pulse circuits find applications in virtually all electronics-based industries. In communications, various types of pulse code modulation are employed; radar utilizes pulses to track targets. Digital computers require circuits that can be switched very rapidly between two "states," using appropriate pulses.

Generally, a study of pulse circuits involves their generation, processing, and applications. We shall concentrate mainly on generation and processing.

## 15.1 PROPERTIES OF PULSES

A pulse may be defined as an abrupt discontinuity in voltage or current. It is usually of brief duration and can either be one shot (single pulse) or repetitive. Shown in Fig. 15.1 is the time graph for a typical series of voltage pulses. This is often referred to as a pulse train. The level from which the pulses rise, usually called the base line, is 0 V in this case, but in practice it can take on any value, positive or negative. Pulse polarity can also be positive or negative; here the pulses are positive-going.

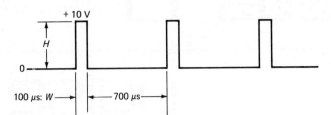

Fig. 15.1 Pulse train.

For the pulse train of Fig. 15.1, the following parameters can be defined:

a. Pulse height (difference between peak and baseline levels): $H = 10$ V
b. Pulse width (also called pulse duration): $W = 100$ μs
c. Period: $T = 800$ μs
d. Pulse repetition frequency: $PRF = \dfrac{1}{T} = \dfrac{1}{800 \times 10^{-6}} = 1{,}250$ Hz;
e. Duty cycle $= \dfrac{\text{pulse ON time}}{\text{period}} = \dfrac{100}{800} = 12.5$ percent

In practice, all pulses depart from the ideal of Fig. 15.1. Figure 15.2 illustrates a practical pulse; some of the details have been exaggerated to define them more easily. The base line is denoted as 0 percent, while the ultimate level at which the pulse height settles is 100 percent. The distance between these extremes is the pulse height. Other pertinent parameters are

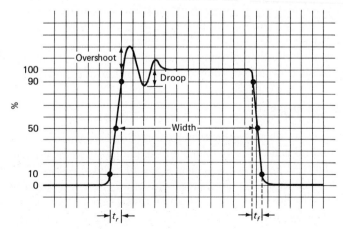

Fig. 15.2 Typical pulse waveform: horizontal sensitivity, 50-ns/division; vertical sensitivity, 1-V/division.

Rise time $t_r$: time required for pulse to rise from the 10 to the 90 percent level

Fall time $t_f$: time for the pulse to fall from the 90 to the 10 percent level

Pulse width $W$: time between 50 percent levels

Overshoot $O$: amount by which ultimate pulse height level is exceeded; usually expressed as a percent of pulse height

Droop $D$: amount by which pulse drops below ultimate pulse height; may also be expressed as a percent of pulse height

For the pulse of Fig. 15.2, the following parameters may be determined:

a. Rise time: $t_r = 50$ ns
b. Pulse width: $W = 625$ ns
c. Fall time: $t_f = 50$ ns
d. Pulse height: $H = 10$ V
e. Overshoot: $O = \dfrac{2}{10} = 20$ percent
f. Droop: $D = \dfrac{1.3}{10} = 13$ percent

It has been indicated earlier that the *time* and *frequency* properties of waveforms are related. The pulse train of Fig. 15.1, as an example, may be broken down into an infinite number of sinusoidal components, each with its own amplitude, frequency, and phase. The rise time of the pulses is a measure of the high-frequency content, while the *sag* or *tilt* (departure from a perfectly flat top) gives an indication of the low-frequency content. A pulse with zero rise time must contain a sinusoidal component of infinite frequency; in practice, this can only be approached.

To explore further this *time-frequency* relationship, consider the circuit of Fig. 15.3. If we assume a voltage step input (zero rise time), the output rises exponentially from zero toward its ultimate value (100 percent). The time constant is $\tau = RC$ s; it takes approximately $5\tau$ to reach 100 percent. The rise time is the time interval between the 10 and 90 percent levels; using the rising exponential curve of Fig. 15.4, we have

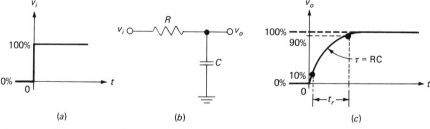

*Fig. 15.3* Response of RC circuit to step input; (a) step input; (b) RC circuit; (c) exponential output.

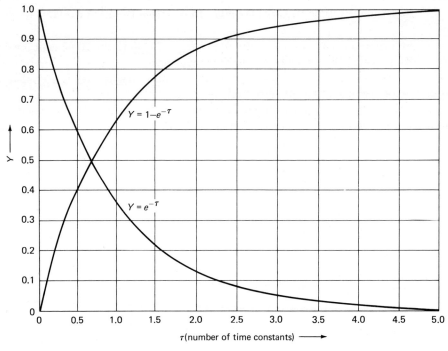

Fig. 15.4 *Exponential graphs.*

Time required to reach 90 percent: $2.3\tau$
Time required to reach 10 percent: $0.1\tau$
Rise time: $t_r = 2.3\tau - 0.1\tau = 2.2\tau$
$$t_r = 2.2\tau = 2.2RC \tag{15.1}$$

The above is what we refer to as *time domain* analysis. To relate the rise time to the high-frequency content of the pulse, we must compute the high-frequency cutoff for the circuit:

$$\frac{V_o}{V_i} = \frac{1/(j\omega C)}{R + 1/(j\omega C)} = \frac{1}{1 + j\omega/(1/RC)}$$

which indicates a 3-dB high-frequency cutoff of

$$\omega_H = \frac{1}{RC}$$

$$RC = \frac{1}{\omega_H} = \frac{1}{2\pi f_H}$$

By substituting the above results for $RC$ in Eq. (15.1), the relationship between $t_r$ and $f_H$ is

$$t_r = 2.2RC = 2.2\,\frac{1}{2\pi f_H}$$

$$t_r = \frac{0.35}{f_H} \tag{15.2}$$

The above expression applies whenever the circuit involves a single $RC$ or $RL$ section; when more than one such section is involved, the results are still useful, although less accurate.

### Example 15.1

A given amplifier's high-frequency response is down 3 dB at 7 MHz and drops at the rate of 6 dB per octave from then on. Determine the rise time of an ideal pulse after it has passed through such an amplifier.

### Solution

$$t_r = \frac{0.35}{f_H} \tag{15.2}$$

$$t_r = \frac{0.35}{7 \times 10^6} = 50 \text{ ns}$$

## Problem Set 15.1

15.1.1. For the pulse train shown in Fig. 15.5, determine

    a. Base-line voltage
    b. Pulse height
    c. Pulse width
    d. Period
    e. PRF
    f. Duty cycle
    g. Rise time
    h. Fall time
    i. $f_H$

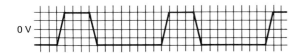

Fig. 15.5 Pulse train for Prob. 15.1.1: vertical sensitivity, 5-V/division; horizontal sensitivity, 0.1-µs/division.

**15.1.2.** A particular oscilloscope has a frequency response that is down 3 dB at 450 kHz and drops off at 6 dB per octave from there on. What is the minimum pulse rise time that can be displayed?

## 15.2 LINEAR WAVE SHAPING

A pulse, or any other waveform, may be reshaped, intentionally or unintentionally, by various circuit elements, active or passive.

Linear wave shaping involves a process in which the "output" contains no frequencies other than those in the "input." Some of the input frequencies may be removed, amplified, or attenuated, but no new frequencies are added in the process. Normally, linear wave shaping results when a signal is passed through a linear amplifier (class A) or a passive network, such as a resistive attenuator, an $RC$ network, a pulse transformer, or a delay line.

Some examples of linear wave shaping are attenuation, amplification, phase inversion, time delay, differentiation, and integration. Techniques used to perform some of these functions are discussed in the sections to follow.

### 15.2a Attenuation Networks

The resistive voltage divider of Fig. 15.6 provides signal attenuation without altering any other signal property. The output voltage $v_o$ is simply four-fifths of the input voltage $v_i$; in this case: $10(4/5) = 8$ V. It would be an easy task to analyze circuits of this type if it were not for the problems of shunt capacitance, which is always present.

As an example, consider the circuit of Fig. 15.7a. This is a voltage divider shunted by a 20-pF capacitor. To simplify the analysis, a Thevenin equivalent circuit is developed in Fig. 15.7b. The output voltage aims for

$$v_o = v_i \frac{R_2}{R_1 + R_2} = v_i'$$

$$v_i' = 10 \frac{1}{5} = 2 \text{ V}$$

The resulting pulse, shown in Fig. 15.7c, has a rise time $t_r = 2.2RC = (2.2)(800 \times$

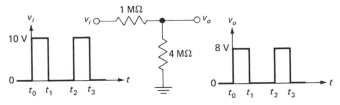

*Fig. 15.6  Resistive voltage divider.*

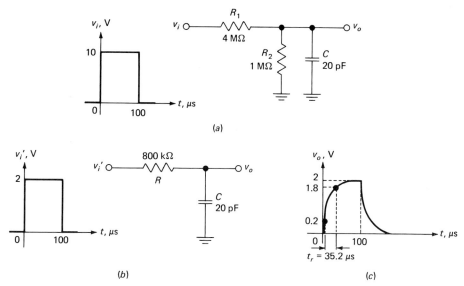

Fig. 15.7 (a) Voltage divider with shunt capacitance; (b) Thevenin equivalent circuit for (a); (c) output voltage.

$10^3)(20 \times 10^{-12}) = 35.2$ μs. Note that the output pulse would be further distorted if the input pulse were of shorter duration. For example, a 10-μs input pulse would result in a highly attenuated triangular wave; any reference to rise time would then be rather pointless.

To minimize the effects of shunt capacitance, voltage dividers are usually "compensated," as in Fig. 15.8a. We will assume the circuit to be driven from a constant voltage generator (zero internal resistance), producing a voltage step whose height is $V$ and $t_r = 0$.

Prior to $t = 0$, both input and output are 0 V. At $t = 0$, the voltage step is applied across the input. The output is determined by the voltage-divider action of the $R_2$-$C_2$ and $R_1$-$C_1$ networks. Capacitors behave as effective short circuits to sudden voltage changes; hence both resistors are out of the picture at $t = 0$, and the voltage division is determined entirely by the capacitors. Using Fig. 15.8b, the output voltage at $t = 0$ is

$$v_{oo} = V \frac{1/C_2}{1/C_1 + 1/C_2}$$

$$v_{oo} = V \frac{C_1}{C_1 + C_2} \qquad (15.3)$$

What happens in the steady state? Eventually both capacitors are fully charged at some final level and appear as open circuits. The only components to have current then are the resistors, and the output voltage is controlled by the $R_1$-$R_2$ voltage divider. This yields

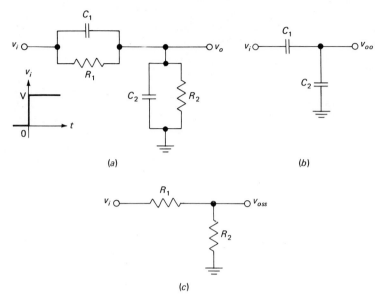

Fig. 15.8 (a) *Compensated voltage divider;* (b) *equivalent circuit for* t = 0; $v_{oo} = VC_1/(C_1 + C_2)$; (c) *steady state equivalent circuit;* $v_{oss} = VR_2/(R_1 + R_2)$.

$$v_{oss} = V \frac{R_2}{R_1 + R_2} \tag{15.4}$$

as indicated in Fig. 15.8c. Thus $v_o$ has an initial value ($v_{oo}$) determined by the capacitors and a steady state value ($v_{oss}$) determined by the resistors. The output reaches its final value by following an exponential curve with a time constant $\tau = RC$, where $R$ and $C$ represent the Thevenin resistance and capacitance, respectively,

$$\tau = RC = \frac{R_1 R_2}{R_1 + R_2} (C_1 + C_2) \tag{15.5}$$

In general, there are three possibilities to be considered when determining the output of the circuit just discussed:

a. $v_{oss} > v_{oo}$: Depending on the relative values of resistance and capacitance, it is possible that the steady state output is greater than the initial output. This is illustrated in Fig. 15.9a. The time constant in the exponential rise from $v_{oo}$ to $v_{oss}$ is given by Eq. (15.5). The curvature of the output waveform is referred to as *roll-off*, and the circuit is said to be undercompensated.

b. $v_{oss} < v_{oo}$: When the steady state output is less than the initial output, the

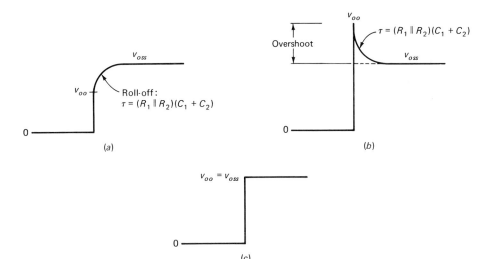

Fig. 15.9 Output waveforms for the circuit of Fig. 15.8a when it is (a) undercompensated: $v_{oss} > v_{oo}$; (b) overcompensated: $v_{oss} < v_{oo}$; (c) perfectly compensated: $v_{oss} = v_{oo}$.

waveform of Fig. 15.9b results. The time constant is again given by Eq. (15.5). There is overshoot, and the circuit is said to be overcompensated.

c. $v_{oss} = v_{oo}$: This is perfect compensation. The initial and steady state outputs are equal, and no exponential rise or decay is required to get from one to the other. This condition allows us to preserve the input rise time and is widely used in the construction of *compensated probes*. The output waveform is shown in Fig. 15.9c.

The conditions resulting in perfect compensation can be determined by equating $v_{oo}$ and $v_{oss}$ from Eqs. (15.3) and (15.4):

$$V \frac{C_1}{C_1 + C_2} = V \frac{R_2}{R_1 + R_2}$$
$$C_1 R_1 + C_1 R_2 = C_1 R_2 + C_2 R_2$$
$$C_1 R_1 = C_2 R_2 \tag{15.6}$$

### Example 15.2

Determine the voltage-division factor and rise time for the circuit of Fig. 15.8a, using the following data: $V = 30$ V, $R_1 = 6$ MΩ, $R_2 = 3$ MΩ, $C_2 = 40$ pF, and

a. $C_1 = 0$
b. $C_1 = 10$ pF
c. $C_1 = 20$ pF
d. $C_1 = 40$ pF

## Solution

The voltage-division factor is the ratio of input to steady state output voltage:

$$\frac{V}{V[R_2/(R_1+R_2)]} = \frac{R_1+R_2}{R_2} = \frac{30}{10} = \frac{3}{1}$$

This means that a voltage reduction of 3 is provided by this voltage divider. The steady state output therefore is $30/3 = 10$ V. The rise time for each value of $C_1$ is now calculated:

a. $C_1 = 0$: This represents no compensation at all. The output rises from 0 (at $t = 0$) toward 10 V with the following time constant:

$$\tau = RC = \frac{R_1 R_2}{R_1 + R_2} C_2$$

$$\tau = \left(\frac{18}{9}\right)(10^6)(40 \times 10^{-12}) = 80 \ \mu s$$

The rise time is given by Eq. (15.1):

$$t_r = 2.2\tau = 2.2(80) = 176 \ \mu s$$

The waveform is shown in Fig. 15.10a.

b. $C_1 = 10$ pF: The initial output is given by Eq. (15.3):

$$v_{oo} = V \frac{C_1}{C_1 + C_2} = 30 \frac{10}{50} = 6 \ V$$

This is clearly less than the final 10 V; hence the divider is undercompensated. The time constant is

$$\tau = RC = \frac{R_1 R_2}{R_1 + R_2}(C_1 + C_2) \qquad (15.5)$$

$$\tau = (2 \times 10^6)(50 \times 10^{-12}) = 100 \ \mu s$$

The rise time is the time interval between 10 and 90 percent levels. In our case the time required to reach 10 percent (1 V) is 0 s. The time required to reach 90 percent (9 V) is determined by noting that the exponential rise begins at 6 and ends at 10 V, and hence 100 percent of the exponential corresponds to $10 - 6 = 4$ V. It follows that 9 V is 3 V beyond the start, or 75 percent of the way toward the final level. Using the curve of Fig. 15.4, it takes $1.4\tau$ to reach the 75 percent level; hence $t_r = 1.4(100) = 140 \ \mu s$. The output waveform is shown in Fig. 15.10b.

c. $C_1 = 20$ pF: The initial output is

$$v_{oo} = 30 \frac{20}{60} = 10 \ V$$

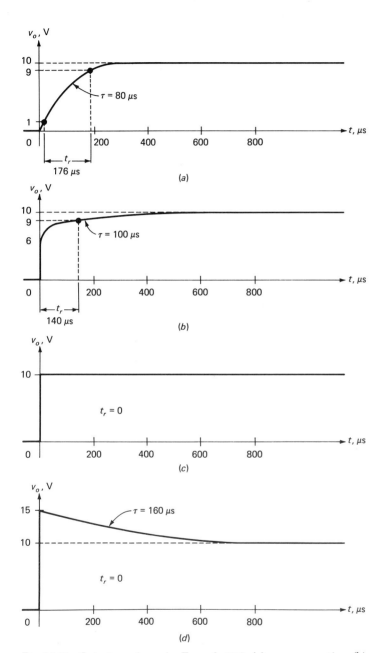

Fig. 15.10 Output waveforms for Example 15.2: (a) no compensation; (b) undercompensated; (c) perfect compensation; (d) overcompensated.

which is identical to the final output; hence there is perfect compensation. The waveform is shown in Fig. 15.10c. It is unnecessary to compute $\tau$, since initial and final values are the same.

d. $C_1 = 40$ pF: The initial output is

$$v_{oo} = 30 \frac{40}{80} = 15 \text{ V}$$

which is higher than the final output; hence there is overcompensation. The rise time is obviously 0, since the output immediately rises from 0 to 150 percent. It does take $5\tau$, however, for the output to settle at its ultimate level.

$$\tau = \frac{R_1 R_2}{R_1 + R_2} (C_1 + C_2) \tag{15.5}$$
$$\tau = (2 \times 10^6)(80 \times 10^{-12}) = 160 \text{ }\mu\text{s}$$
$$5\tau = 800 \text{ }\mu\text{s}$$

The complete waveform is shown in Fig. 15.10d.

The previous example assumed that the voltage divider was driven from a source with zero internal resistance. In practice, all sources exhibit some internal resistance which prevents us from achieving the ideal zero rise time. Nevertheless, the ideal can be approached quite closely, and practical *compensated probes* do just that. For example, when a compensated probe is used to couple pulse waveforms to an oscilloscope, $C_1$ (which is variable) is adjusted until the display exhibits the best possible rise time.

An additional benefit provided by compensated dividers is an increase in the impedance seen by the driving source. For Example 15.2, if the source is connected directly across the $R_2$-$C_2$ combination, it sees 3 M$\Omega$ shunted by 40 pF. When the compensated divider is used, however, the total load seen by the generator is the series combination of the $R_1$-$C_1$ and $R_2$-$C_2$ networks, as shown in Fig. 15.11a. The circuit has been redrawn in a slightly different orientation to emphasize that it may be considered as a balanced bridge. If the connection between $A$ and $B$ is broken, as in Fig. 15.11b, the voltages at $A$ and $B$ relative to ground are

$$V_A = V_i \frac{R_2}{R_1 + R_2} \tag{a}$$

$$V_B = V_i \frac{C_1}{C_1 + C_2} \tag{b}$$

It was shown earlier that the above equations are equal if $R_1 C_1 = R_2 C_2$, the condition for perfect compensation. Hence $V_A = V_B$, the bridge is balanced, and the potential difference between $A$ and $B$ is zero. In this case it does not

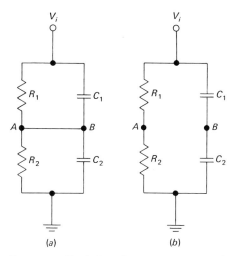

*Fig. 15.11 Equivalent circuit of a compensated voltage divider; the circuit in (a) is equivalent to that in (b) as long as $R_1C_1 = R_2C_2$.*

matter whether the connection between $A$ and $B$ is broken or not. By breaking the connection, in fact, the circuit simplifies to that of Fig. 15.11b, which represents a total load of $R_1 + R_2$ shunted by $C_1C_2/(C_1 + C_2)$. For the values of Example 15.2, the load is 9 MΩ shunted by 13.3 pF. This represents an improvement over the uncompensated $R_2$-$C_2$ combination by a factor of 3; note that the voltage is also reduced by a factor of 3 (30 to 10 V).

## Example 15.3

A given oscilloscope has an input resistance of 10 MΩ shunted by 25 pF. Design a 10 × (10 times) voltage divider that utilizes the scope input as the $R_2$-$C_2$ combination. What are the equivalent input resistance and capacitance for the compensated divider?

## Solution

The voltage division ratio is

$$\frac{R_2}{R_1 + R_2} = \frac{1}{10}$$
$$10R_2 = R_1 + R_2$$
$$9R_2 = R_1$$
$$9 \times 10^7 = R_1 = 90 \text{ MΩ}$$

For perfect compensation,

$$R_1 C_1 = R_2 C_2 \tag{15.6}$$

$$C_1 = \frac{R_2 C_2}{R_1} = \frac{10^7 (25 \times 10^{-12})}{9 \times 10^7}$$

$$C_1 = 2.78 \text{ pF}$$

The input resistance is

$$R_i = R_1 + R_2 = 100 \text{ M}\Omega$$

The input capacitance is

$$C_i = \frac{C_1 C_2}{C_1 + C_2} = \frac{2.78(25 \times 10^{-12})}{27.78}$$

$$C_i = 2.5 \text{ pF}$$

## Problem Set 15.2

**15.2.1.** A 50-$\Omega$ pulse generator produces 500-$\mu$s pulses with negligible rise time and 5-V positive amplitude into an open circuit. The base line is zero,

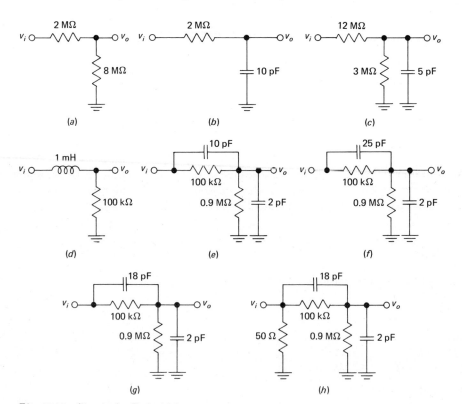

Fig. 15.12 Circuits for Prob. 15.2.

and the duty cycle is 25 percent. Determine the pulse shape (amplitude, rise time, fall time) after it has passed through each of the circuits in Fig. 15.12.

15.2.2. If the pulse of Prob. 15.2.1 is fed to an oscilloscope with an input resistance of 20 MΩ shunted by 50 pF, what is the rise time?

15.2.3. Determine $C_1$ for perfect compensation of the voltage divider in Fig. 15.13. Also determine the voltage attenuation ratio, the input resistance, and the input capacitance for the compensated circuit.

15.2.4. The input circuit to an oscilloscope involves a 20-MΩ resistance shunted by 18 pF. Design a compensated 5× probe to work into the scope input. What are the input resistance and capacitance of the probe when connected to the oscilloscope?

15.2.5. Repeat Prob. 15.2.4 with the added specification that the input resistance must be 50 MΩ. What is the input capacitance?

### 15.2b  Amplification and Phase Inversion

These two topics will be treated simultaneously since they are related. An inverter is an amplifier with 180° phase shift. In the simplest case the voltage amplification is unity ($A_v = -1$), but as long as phase inversion occurs, the circuit may be referred to as an inverter, regardless of the actual gain. Any of the circuit configurations already discussed (common emitter stage, unity gain operational amplifier) may be used to provide phase inversion if it has the necessary frequency response to preserve the pulse shape.

Pulse amplifiers (also called video amplifiers because of their application in wideband television circuits) must have a wide bandwidth, usually extending into the megahertz range. This is necessary to faithfully reproduce the rather large frequency content of pulse signals. Negative feedback is often employed to increase an amplifier's bandwidth, although it reduces the gain.

As has been pointed out earlier, an amplifier with poor low-frequency response introduces tilt (sag) in the output pulse, while poor high-frequency

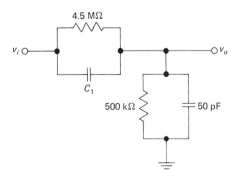

Fig. 15.13  Circuit for Prob. 15.2.3.

response increases the rise time. Further distortion of the signal may result from unequal phase shift for each frequency component.

To obtain good low-frequency response, direct coupling may be used; the absence of coupling capacitors or transformers results in the frequency response extending all the way down to DC. Direct coupling, however, may present drift problems; hence $RC$ coupling is often preferred. Although $RC$ coupling prevents amplification of DC signals, it can be adequate for pulse durations that are short compared to the time constants involved. To illustrate this point, consider the $RC$-coupled circuit of Fig. 15.14. $R_g$ is the internal resistance of the driving circuit, while $R_i$ is the input resistance to the next stage. Here $R_i = 1{,}000 \ \Omega$, $R_g = 1{,}000 \ \Omega$, and $C_c$ is to be chosen for a tilt of less than 1 percent.

As indicated in Fig. 15.14a, excessive tilt results if $C_c$ is not large enough. The initial value of $v_i = 1 R_i/(R_i + R_g) = 0.5$ V. A 1 percent tilt means that during the 100-µs pulse duration $v_i$ is allowed to decay by an amount equal to $0.01(0.5) = 0.005$ V. Since the initial portion of an exponential decay is linear, the time constant is chosen as follows:

$$\frac{\tau}{100 \ \mu s} = \frac{0.5}{0.005}$$

$$\tau = 10^{-4}(10^2) = 10^{-2}$$

But

$$\tau = (R_i + R_g)(C_c)$$

therefore

$$10^{-2} = 2{,}000 C_c$$

$$C_c = \frac{10^{-2}}{2{,}000} = 5 \ \mu F$$

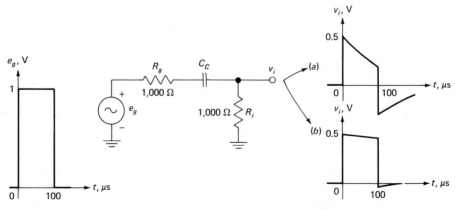

Fig. 15.14  Effect of coupling capacitor on waveform: (a) $C_c$ too small for good low-frequency coupling; (b) $C_c$ adequate.

This value of $C_c$ is not excessively large. DC voltages are also usually low in semiconductor circuits; therefore a large voltage rating for the capacitor is not required. If instead of 100 μs we had a 1,000-μs pulse, 5 μF would no longer be adequate; 50 μF would be required to maintain the same tilt.

The circuit of Fig. 15.15a illustrates one technique for extending the low-frequency response. As long as $R_1$ is large compared to the reactance of $C_1$, the equivalent circuit of Fig. 15.15b is valid and may be used to establish compensation criteria. For perfect compensation, the following must hold:

$$v_{oo} = v_{oss}$$

But

$$v_o = i_2 R_i$$

hence

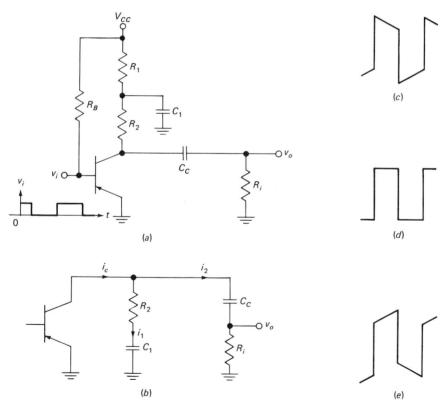

Fig. 15.15 (a) Amplifier with low-frequency compensation; (b) equivalent circuit for (a); (c) undercompensated output voltage waveform; (d) perfectly compensated output voltage waveform ($R_2 C_1 = R_i C_c$); (e) overcompensated output voltage waveform.

$$v_o = i_c \frac{R_2 + 1/(j\omega C_1)}{R_2 + 1/(j\omega C_1) + R_i + 1/(j\omega C_c)} R_i \qquad (a)$$

Now $v_{oo}$ is governed by the high-frequency response of the network; its value can be obtained by letting $\omega$ approach infinity in Eq. (a):

$$v_{oo} = i_c \frac{R_2 R_i}{R_2 + R_i} \qquad (b)$$

Similarly, $v_{oss}$ is governed by the low-frequency response and can be obtained by letting $\omega$ approach zero in Eq. (a); we first rewrite Eq. (a) as follows:

$$v_o = \frac{[i_c(j\omega C_1 R_2 + 1)/(j\omega C_1)](R_i)}{[(R_2 + R_i)(j\omega C_1 C_c) + C_c + C_1]/(j\omega C_1 C_c)} \qquad (a)$$

$$v_o = \frac{i_c(j\omega C_1 R_2 + 1)(R_i C_c)}{(R_2 + R_i)(j\omega C_1 C_c) + C_c + C_1} \qquad (a)$$

Now, as $\omega \to 0$,

$$v_{oss} = \frac{i_c R_i C_c}{C_c + C_1} \qquad (c)$$

Equating Eqs. (b) and (c) yields

$$\frac{i_c R_2 R_i}{R_2 + R_i} = \frac{i_c R_i C_c}{C_c + C_1}$$
$$R_2 C_c + R_2 C_1 = R_2 C_c + R_i C_c$$
$$R_2 C_1 = R_i C_c$$
$$C_1 = \frac{R_i C_c}{R_2} \qquad (15.7)$$

which is the condition for perfect compensation. Shown in Fig. 15.15c, d, and e are output voltage waveforms for undercompensation, perfect compensation, and overcompensation.

The high-frequency response can also be extended through appropriate compensation, as in the circuit of Fig. 15.16a. Here $C_c$ behaves as a short circuit; $R_i$ and $C_i$ are the input resistance and capacitance for the next stage.

An examination of the equivalent circuit shown in Fig. 15.16b indicates that without $L$ the high-frequency response would suffer, because of $C_i$. The inductance provides a canceling effect that extends the high-frequency response, allowing us to better preserve the rise time. This technique is often referred to as *shunt peaking*.

If the circuit is overcompensated, there will be some overshoot and possibly ringing, as shown in Fig. 15.16e. Depending on the application, some overshoot may actually be desirable. A detailed derivation of the conditions

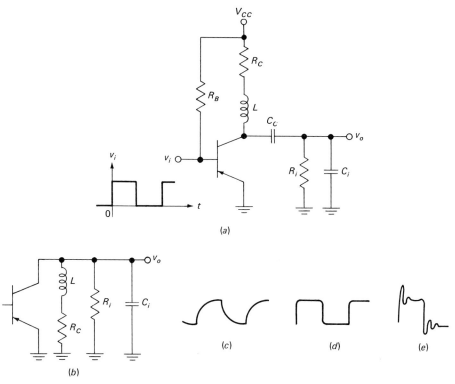

*Fig. 15.16* (a) *Amplifier with high-frequency compensation (shunt peaking);* (b) *equivalent circuit for* (a); *output voltage waveforms when circuit is undercompensated* (c), *perfectly compensated* (d), *and overcompensated* (e).

for optimum compensation is not carried out here. In practice, the circuit response to a square wave may be observed on an oscilloscope, and $L$ is adjusted to produce the desired rise time and maximum permissible overshoot.

## Problem Set 15.3

15.3.1. Why is a linear pulse amplifier required to have good low-frequency response? Good high-frequency response?

15.3.2. What advantages does direct coupling have over transformer or capacitive coupling? What are the disadvantages?

15.3.3. Estimate the percent tilt of a $\frac{1}{2}$-ms voltage pulse coupled from a 600-$\Omega$ generator through a 10-$\mu$F capacitor to an amplifier whose input resistance is 1,900 $\Omega$.

15.3.4. For the circuit of Fig. 15.15, the following information is provided: $R_1 = 10\text{k}\Omega$, $R_2 = 2\text{ k}\Omega$, $C_c = 20\ \mu\text{F}$, $R_i = 2\text{ k}\Omega$.

a. Determine the value of $C_1$ required for perfect compensation.
b. Estimate the lowest $PRF$ for which the compensation of (a) will work. *Hint:* The compensation criterion developed in the text is valid only as long as $R_1 \gg X_{C_1}$.

15.3.5. Briefly describe how shunt peaking works.
15.3.6. How can negative feedback help retain a pulse's wave shape?

### 15.2c  Differentiation and Integration

A differentiating circuit produces an output which is proportional to the time rate of change of the input. The output is said to be the derivative, with respect to time, of the input:

$$v_o = \frac{d}{dt}(v_i) \tag{15.8}$$

When a signal is displayed as a function of time, the waveform for its derivative may be obtained by plotting the slope of the curve, as indicated in Fig. 15.17a. Note that whenever the slope of $v_i$ is zero (flat portions of curve), the output is also zero. When $v_i$ changes, however, infinitely high spikes are generated. The slope of $v_i$ is $+\infty$ at $t = 0$ and $t_2$; it is $-\infty$ at $t_1$ and $t_3$. Although it is impossible to generate infinite voltages, practical circuits can and do approach the output shape shown in the figure.

It may be difficult at first to realize that this is *linear wave shaping*, since the output and input waveforms hardly resemble each other, but in fact there are no frequency components in the output other than those already present in the input. The input of Fig. 15.17a contains frequency components as low as the fundamental ($f_1 = 1/T$) and as high as infinity; evidence of an "infinite" frequency is the zero rise and fall times. The output, on the other hand, contains only one frequency component, namely $f = \infty$ (each spike lasts zero time, theoretically; therefore it corresponds to an infinite frequency component). The circuit is therefore nothing more than a *high-pass* filter; it filters out all frequency components except $f = \infty$!

The $RC$ circuit shown in Fig. 15.17b may be used to achieve an approximate differentiation provided that its time constant is short compared to the signal period ($\tau \ll T$). The circuit will be recognized as a high-pass filter whose 3-dB low-frequency cutoff is $f_L = 1/(2\pi RC)$. Hence as $\tau$ is reduced ($\tau = RC$), $f_L$ increases, and the ideal differentiator described above is more closely approached.

To improve the performance of the passive differentiating circuit just described, active circuitry involving an operational amplifier (OA) with negative feedback is often employed. The OA, it will be recalled, has very high voltage amplification and 180° phase shift. For the circuit of Fig. 15.17c, assume that the

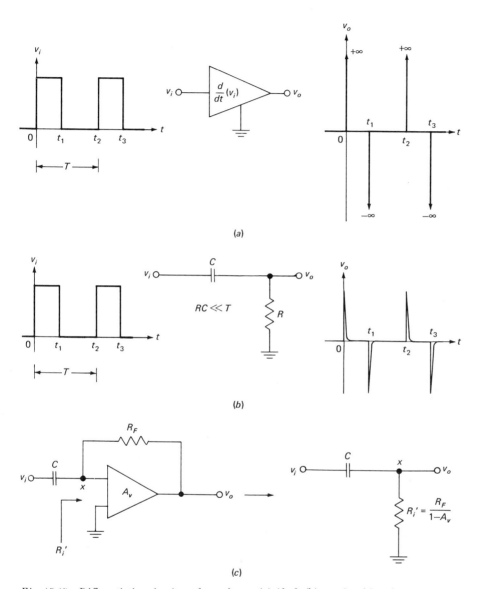

Fig. 15.17 Differentiating circuits and waveforms: (a) ideal; (b) passive; (c) active.

amplifier has an infinite input resistance and $A_v = -100{,}000$. Due to the Miller effect, the resistance looking into the amplifier from point $x$ is

$$R_i' = \frac{R_F}{1 - A_v} \simeq \frac{R_F}{10^5}$$

The input circuit is therefore equivalent to the $RC$ network shown, and $\tau = (R_F/10^5)C$; this is $10^5$ times shorter than $C$ and $R_F$ alone could produce. Hence the OA may be considered to provide an equivalent resistance which is extremely small, resulting in a very short time constant.

The circuit of Fig. 15.17c can be an excellent differentiator, but due to its

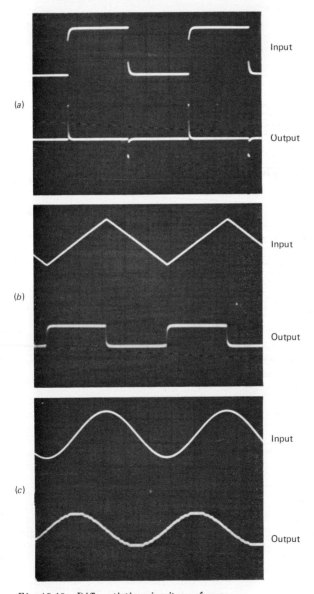

Fig. 15.18 *Differentiating circuit waveforms.*

increasing gain with frequency (it is a high-pass filter with a frequency response which rises at 6 dB per octave), high-frequency noise may also appear in the output. For this reason a small capacitor to reduce the high-frequency gain is often connected across $R_F$. As long as the capacitor is small, the differentiating characteristics of the circuit do not suffer.

It should be noted that the OA also introduces 180° phase shift; hence the output is phase-inverted. Various input and output waveforms for an active differentiator are reproduced in Fig. 15.18. The oscilloscope display for each output was inverted to cancel the 180° phase shift of the OA; hence the output waveforms shown correspond to the time rate of change of the input.

We now come to integration, which is the inverse of differentiation. Here the output, at any instant in time, is proportional to the area accumulated under the input waveform up to that time. This relationship can be expressed mathematically as follows:

$$v_o = \int v_i \, dt \tag{15.9}$$

In our case, instead of a mathematical approach, a graphic technique will be used, as we did for the differentiating circuit. This allows us to determine the shape of the output waveform.

Consider the waveforms and circuit of Fig. 15.19a. Starting at $t = 0$, the area under the input curve increases linearly until $t_1$; this produces the *ramp* output during this time. From $t_1$ to $t_2$, $v_o$ drops linearly in response to the negative area generated under the input curve. The output should therefore stay constant when there is no input (since no area is generated), rise linearly for a constant positive input, and drop linearly for a constant negative input.

Since integration is the opposite of differentiation, it should be possible to differentiate the output and recover the original input waveform. The student should do this on his own to develop a feel for linear signal processing.

If we examine the input and output waveforms just discussed, it should be evident that the high-frequency content of the input (zero rise and fall times) has been lost; thus we can think of an integrator as performing a *low-pass* filtering function, the opposite of the high-pass function performed by a differentiator.

A passive network that can realize an approximate integration is shown in Fig. 15.19b. This is a low-pass $RC$ filter whose time constant is much longer than the square-wave period. In effect, the larger the time constant, the more linear the ramp output, although the amplitude is also reduced. A close examination of Fig. 15.4 reveals, in fact, that reasonable linearity for any exponential curve is maintained up to approximately $0.1\tau$. We should therefore expect good integration if each half of the square-wave input lasts less than $0.1\tau$:

$$\frac{T}{2} < 0.1\tau$$

$$\tau > 5T \tag{15.10}$$

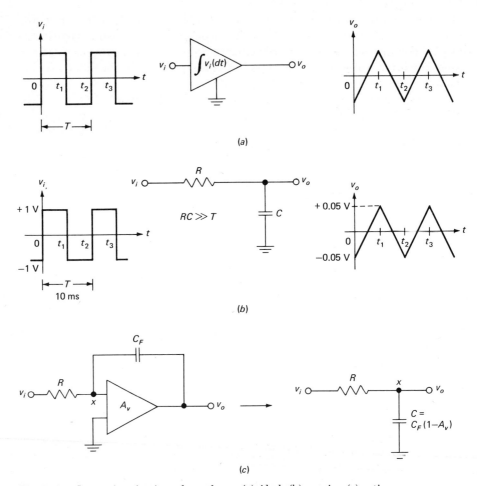

Fig. 15.19  Integrating circuits and waveforms: (a) ideal; (b) passive; (c) active.

In the circuit of Fig. 15.19b, $T = 10$ ms. Using Eq. (15.10) as a rule of thumb, the time constant should be at least 50 ms. This means that for each half-period of the input square wave (half-period $= 5$ ms $= 0.1\tau$), the voltage across the capacitor must change by 10 percent of the input voltage, that is, 0.1 V. The output waveform shown satisfies this condition. Note also that the output alternates equally between positive and negative values; in other words, there is no DC component in the output. This is because a linear circuit such as this cannot generate any new frequency components; hence there can be no DC in the output because there is none in the input.

To improve the integration process, it is clear that a larger $\tau$, involving a larger effective capacitance or resistance, is required. A large $\tau$ also results in a smaller output. The active circuit of Fig. 15.19c overcomes these problems to a large degree. Again, if the input resistance is very large, the equivalent circuit seen when looking into the amplifier from point $x$ is $C = C_F(1 - A_v)$. If we

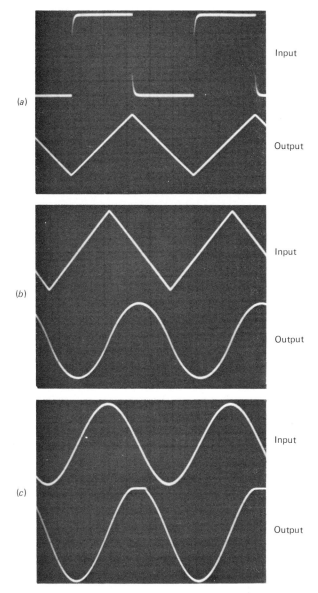

Fig. 15.20  Integrating circuit waveforms.

use $A_v = -10^5$, the Miller effect capacitance $C \simeq 10^5 C_F$. The OA therefore multiplies the available capacitance by a factor approximately equal to its open-loop voltage amplification. The resultant circuit, shown in the same figure, is equivalent to the passive circuit previously discussed; the only difference is that now we have achieved a relatively large $\tau$ without the requirement for a large capacitor.

Shown in Fig. 15.20 are several input and output waveforms for an active integrator. Again, the 180° phase shift due to the OA does not appear, because of a compensating phase inversion introduced by the oscilloscope used to display the signals.

## Problem Set 15.4

15.4.1. Indicate which of the circuits in Fig. 15.21 is performing an approximate

    a. Differentiation
    b. Integration

15.4.2. Sketch the output waveform for each of the circuits of Fig. 15.21.
15.4.3. How does a high-pass filter perform differentiation?
15.4.4. How does a low-pass filter perform integration?
15.4.5. Can an OA with zero phase shift be used to achieve differentiation or integration, as illustrated in Figs. 15.17c and 15.19c?
15.4.6. What problems would you expect if the input to an integrating circuit never changes polarity?
15.4.7. The output of a differentiating circuit is displayed on an oscilloscope. Since the input is a square wave, the output appears as a series of positive and negative spikes; however, it is noted that the negative spikes have a lower amplitude than the positive ones. What factor could account for this?

## 15.3 NONLINEAR WAVE SHAPING

Some very useful nonlinear wave shaping functions can be accomplished using electronic devices in switching configurations. A circuit element simulates a switch if, upon the application of appropriate signals, it switches from one "state" to another, quickly and reliably. The "state" should be a well-defined set of conditions, such as zero voltage, maximum current, minimum voltage, and so on. Depending on the application and the device, many possible combinations of well-defined states may exist.

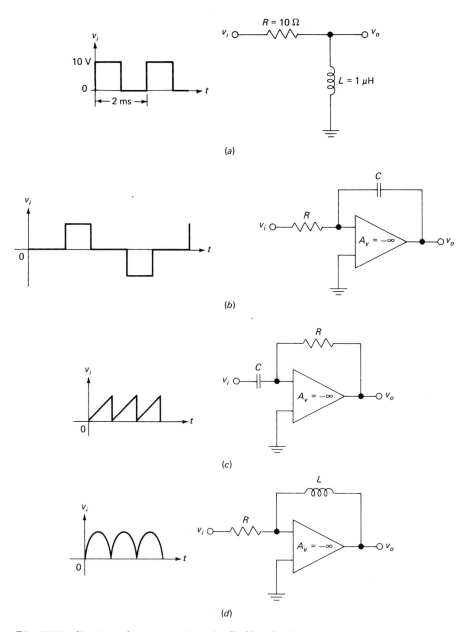

Fig. 15.21  Circuits and input waveforms for Problem Set 15.4.

### 15.3a  The Bipolar Transistor

It was shown earlier[1] that a transistor may be used to simulate a switch if the proper input signals are applied. This earlier discussion considered mainly the static behavior of a common emitter transistor switch. Static behavior has to do with circuit conditions during the time when no changes are taking place (transistor has been in its present state for a relatively long time). Although the static (OFF and ON) characteristics of a transistor switch are important, equally or more important are the transient characteristics that determine switching speed, that is, the time required to switch from one state to another. Switching speed determines to a large degree the rate at which information can be processed and the average power dissipation. While the transistor is OFF, there is negligible power dissipation because the current is small; similarly, when the transistor is ON, there is no appreciable power dissipation because the voltage is small. As long as the transistor is OFF or ON, power dissipation is not a problem. During the transient period (time that the transistor is being switched between the OFF and ON states), however, the active region must be crossed. Because power dissipation is highest in the active region, the longer the time required to cross it, the higher the average power dissipation. This is especially important when large voltage and current levels are being switched.

The transient response for the common emitter switching circuit of Fig. 15.22a is now discussed. The input voltage is the control signal, while the collector current represents the output. Input and output waveforms are shown in Fig. 15.22b and d, respectively.

Prior to $t_o$, $v_i$ is negative. Except for the cutoff current, both $i_b$ and $i_c$ are essentially zero, as indicated in Fig. 15.22c and d. During this time, the transistor is cut off.

At $t_o$, $v_i$ becomes positive, and base current flows. The collector current, however, does not respond immediately but requires a brief period of time to rise. The time required for $i_c$ to rise from 0 to the 10 percent level is called the delay time $t_d$. This is due mainly to the depletion-region capacitances that must be discharged from their OFF voltage levels before the transistor can be made to conduct. Obviously, the lower the reverse bias just before $t_o$, the less charge must be removed and the shorter the delay time. Similarly, the higher the turn-on base current $I_{B1}$, the faster these capacitances can be discharged, therefore reducing $t_d$. Other factors influencing the delay time are the transistor's $h_{FE}$, which is low at low current levels, and the time required for carriers injected by the emitter to diffuse into the base.

Following $t_d$, there is the rise time, which has been defined earlier as the time between the 10 and 90 percent levels. All transistor capacitances are some-

---

[1] See Sec. 5.7.

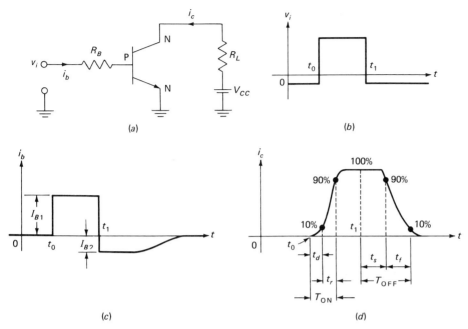

Fig. 15.22  Common emitter switching circuit and waveforms.

what responsible for $t_r$, but the main culprit is the diffusion capacitance $C_D$, which limits the high-frequency response of the transistor, therefore introducing the nonzero rise time. An improvement in $t_r$ results if the base turn-on current $I_{B1}$ is relatively large; this quickly saturates the transistor. During the rise time, the transistor is in the active region, since it is switching from the OFF to the ON state. Therefore a poor rise time generally means higher average power dissipation.

Once the collector current has reached its ultimate level, the turn-on transient is over, and a steady state, or static, situation prevails. The total turn-on time is defined as the sum of delay and rise times:

$$T_{ON} = t_d + t_r \tag{15.11}$$

At $t_1$, the input drive again becomes negative, thus initiating the process of turning the transistor off. It will be noted, however, that $i_c$ does not respond until a storage time $t_s$ has elapsed. The storage time is the time between the 100 and 90 percent levels. When a transistor is driven into saturation, both PN junctions are forward-biased. This means that there is diffusion of charge carriers into the base not only from the emitter but also from the collector. A number of such carriers are in the base at any instant and constitute *stored charge*. Before the transistor can be turned OFF, this charge must be removed. The harder a transistor is driven into saturation, the more stored charge must be removed and hence the longer the storage time. It is for this reason that nonsaturated switching is often used.

The reverse base current at $t_1$ exists as long as it is required to remove the stored charge. This current is referred to as the turn-off current $I_{B2}$ and depends on the reverse bias across the base emitter junction. To reduce storage time, $I_{B2}$ should be as large as possible.

The same factors that cause the rise time are also responsible for the fall time $t_f$. The total turn-off time is the sum of storage and fall times:

$$T_{\text{OFF}} = t_s + t_f \qquad (15.12)$$

while the total switching time is defined as

$$T_T = T_{\text{OFF}} + T_{\text{ON}} \qquad (15.13)$$

When a transistor is saturated, $t_r$ is reduced but $t_s$ increases. Generally, saturated transistors have a longer switching time than nonsaturated transistors, but their average power dissipation is lower.

### 15.3b  The Field Effect Transistor

Before proceeding, the student should review static parameters of FETs, as discussed in Chap. 9. A common source JFET switching circuit is shown in Fig. 15.23a. Output characteristics for the circuit are also given in Fig. 15.23b. The static behavior of the FET switch can be analyzed by assuming an input pulse as shown in Fig. 15.23a. The FET is kept OFF with a zero input while a positive input turns it ON.

When $v_i = 0$, $V_{GS}$ is determined by the negative supply $V_{GG}$ and the voltage-divider action of $R_1$ and $R_2$. Thus $V_{GS}$ is negative, and $I_D$ is very low. The specific value of $V_{GS}$ required to reduce $I_D$ to $I_{D(\text{OFF})}$ (usually $0.001 I_{DSS}$) is called $V_{GS(\text{OFF})}$. The device is now in the cutoff region where $V_{DS} \simeq V_{DD}$ and $I_D = I_{D(\text{OFF})} \simeq 0$. Most of the supply voltage is dropped across the FET, and negligible current flows through $R_D$. When $v_i$ is positive, $V_{GS}$ is determined by the combined effects of the negative $V_{GG}$ supply and positive input pulse. By an appropriate selection of $R_1$ and $R_2$, $V_{GS}$ can be made equal to zero, which is the condition of maximum drain current ($I_{DSS}$). Now the device is in the ON region. The voltage across the FET is very low, $V_{DS} = V_{DS(\text{ON})} \simeq 0$, and the current $I_D = I_{DSS} \simeq V_{DD}/R_D$.

The static parameters of interest are therefore $I_{DSS}$ and $V_{DS(\text{ON})}$ when the FET is ON and $V_{GS(\text{OFF})}$ and $I_{D(\text{OFF})}$ when the FET is OFF. $I_{DSS}$ is a function of temperature; as temperature rises, $I_{DSS}$ decreases because of increasing channel resistance. $V_{GS(\text{OFF})}$ is also temperature-dependent; as the temperature rises, a more negative $V_{GS}$ is required to maintain $I_{D(\text{OFF})}$ at the same level.

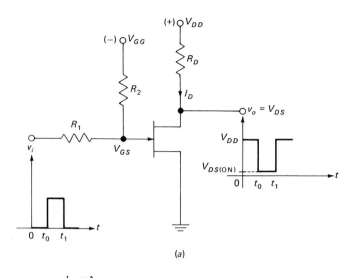

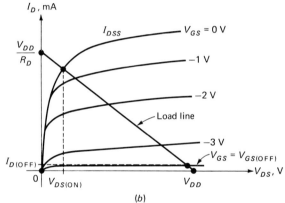

Fig. 15.23 (a) *JFET switching circuit*; (b) *output characteristics for* (a).

MOSFET switching proceeds in a similar fashion. The circuit and characteristics of Fig. 15.24 are used to illustrate static performance. No $V_{GG}$ supply is required here, because a zero gate-source voltage keeps the MOSFET OFF. Thus when $v_i = 0$, the device is OFF, $v_o \simeq V_{DD}$, and $I_D = I_{DSS}$. Note that $I_{DSS}$ here represents minimum as opposed to maximum current for JFETs. When $v_i$ is positive, the MOSFET conducts. The output voltage $v_o = V_{DS(ON)}$, and the drain current $I_D \simeq V_{DD}/R_D$.

The two preceding circuits perform the very basic function of phase inversion. Although not operating linearly, the output is a phase-inverted version of the input.

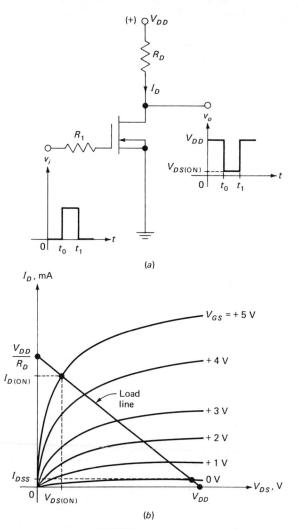

Fig. 15.24 (a) MOSFET switching circuit; (b) drain characteristics for (a).

Transient performance is now considered. The equivalent circuit of Fig. 15.25a represents the OFF state for the JFET inverter of Fig. 15.23a. The gate voltage is negative, and $C_{gs}$ is charged to $V_{GS(OFF)}$. Drain resistance $r_d$ is omitted from the diagram, since it is almost infinite when the device is OFF. $C_L$ represents load capacitance. The output $v_o \simeq V_{DD}$. $C_{gd}$ is charged to the difference in potential between its two plates, $V_{DD} - V_{GS(OFF)}$.

As soon as $v_i$ becomes positive, the gate voltage is influenced by both

Pulse Circuits 671

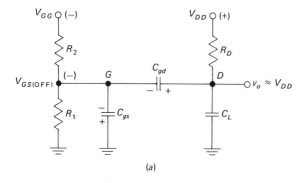

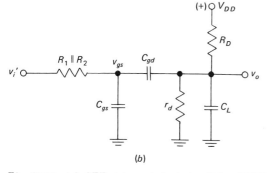

Fig. 15.25 (a) OFF-state equivalent circuit for JFET inverter of Fig. 15.23a; (b) ON-state equivalent circuit for JFET inverter of Fig. 15.23a.

$V_{GG}$ and $v_i$, yielding the Thevenin equivalent circuit shown in Fig. 15.25b. Here $v_i' \simeq 0$ V. $C_{gs}$, however, cannot discharge instantaneously, and $v_o$ cannot respond immediately. Instead, as $v_{gs}$ rises toward zero, $r_d$ drops from infinity toward its ON value. $v_o$ aims for

$$V_{DS(ON)} = \frac{V_{DD}r_d}{r_d + R_D} \tag{15.14}$$

but is again limited from reaching the low ON voltage until some time has elapsed. The complete picture is somewhat complicated, because both $C_{gs}$ and $C_{gd}$ are voltage-dependent; also $r_d$ is not constant during the turn-on time, yielding a continuously changing time constant!

In any case, both turn-on and turn-off time can be reduced with a larger drain current; the turn-off time tends to be somewhat longer than turn-on because both $C_{gs}$ and $C_{gd}$ are highest during the ON time (low voltage, high capacitance).

The transient performance of MOSFETs is similar to that of JFETs except that a voltage-dependent junction capacitance $C_{d(sub)}$ is now present between drain and substrate; this capacitance is in parallel with $C_L$. $C_{gs}$ and $C_{gd}$, however, are not junction capacitances; therefore they do not vary with voltage.

## Problem Set 15.5

15.5.1. What is the difference between static and transient conditions?
15.5.2. What factors prevent a semiconductor device (SCR, bipolar transistor, or field effect transistor) from simulating an ideal switch?
15.5.3. Define the cutoff, active, and ON regions of bipolar transistor operation in terms of the bias (forward or reverse) across each junction.
15.5.4. What are the advantages of driving a bipolar transistor hard into saturation? What are the disadvantages?
15.5.5. How does switching speed affect the average power dissipation of a bipolar transistor?
15.5.6. Indicate what factors are mainly responsible for each of the following effects in bipolar transistors:

    a. Delay time
    b. Rise time
    c. Storage time
    d. Fall time

15.5.7. What are the static parameters of interest for FET switches?
15.5.8. Does $I_{DSS}$ always represent maximum current through a field effect device?
15.5.9. What capacitances are voltage-dependent in JFETs? Are these capacitances also voltage-dependent in MOSFETs?

## 15.3c  The SCR

The silicon-controlled rectifier (SCR) is a switching device widely used in power control applications. Shown in Fig. 15.26a is a four-layer PNPN device across which an external voltage $V_1$ is applied. The polarity of $V_1$ forward-biases junctions $J_1$ and $J_3$ while reverse-biasing $J_2$. There is no net current, except for a small leakage, through the device. If the polarity of $V_1$ is reversed, $J_1$ and $J_3$ are reverse-biased, and again there is no appreciable current.

We now redraw our diagram as shown in Fig. 15.26b. This is simply an analogy to help us understand how the device operates. Note that the four-layer structure can be represented in terms of two interconnected, three-layer structures, one PNP and the other NPN. This is further developed in Fig. 15.26c where a PNP junction transistor ($Q_2$) is interconnected with an NPN junction

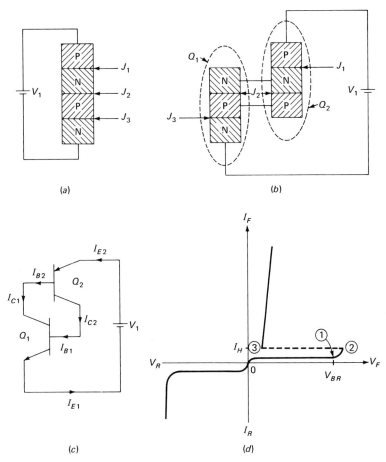

Fig. 15.26 (a) Four-layer structure; (b) the structure of (a) separated into interconnected $PNP$ and $NPN$ structures; (c) two-transistor analogy of (b); (d) voltage-current characteristic for four-layer diode.

transistor ($Q_1$) such that the collector current of $Q_1$ is also the base current of $Q_2$ and the base current of $Q_1$ is the collector current of $Q_2$. Also note that the polarity of $V_1$ provides proper bias for normal transistor operation of both $Q_1$ and $Q_2$.

Suppose now that the magnitude of $V_1$ is increased to the point where $J_2$ starts breaking down. Current through the device begins to rise. In the diagram of Fig. 15.26c, $I_{E1}$, and hence $I_{C1}$, increases. Since $I_{C1} = I_{B2}$, $I_{C2}$ also rises. But $I_{C2} = I_{B1}$, which causes $I_{C1}$ and $I_{E1}$ to rise even more. There is thus regenerative action taking place whereby an initial rise in current produces further increases of the same current. Soon the maximum current is reached, limited only by bulk and externally connected resistances. The two transistors are fully saturated (turned ON), and the voltage across the PNPN device drops to a very low value.

The voltage-current characteristic of Fig. 15.26d expresses the above behavior graphically. The reverse characteristic includes a reverse current and breakdown voltage; this is of no interest to us except to ensure that the reverse breakdown voltage is never exceeded. For $V_F$ (forward voltage) less than $V_{BR}$ (breakover voltage) the device conducts very little. As soon as the breakover voltage is reached (point 1 on the curve), the current begins to rise. At point 2 the device fires. In the two-transistor analogy circuit, this is where regeneration is sufficient to turn both $Q_1$ and $Q_2$ fully ON. Once the device has fired, the voltage drops to a very low value. The device has a very low ON resistance and can pass substantial current with a relatively low voltage drop. The resulting power dissipation is extremely low; this is a very definite advantage. The voltage remains at its low value as long as the current exceeds the "holding" current $I_H$.

The device illustrated in Fig. 15.26 is called a four-layer diode; its applications are limited, and it was introduced only as the basis for the more common SCR. The SCR is obtained by adding a third element, called *gate*, to the four-layer diode, as shown in Fig. 15.27a. The circuit, using the SCRs schematic symbol, is shown in Fig. 15.27b. $R_L$ is the load whose power we wish to control, while $v_r$ is a source of gate pulses. $v_1$ is an AC source instead of DC for reasons soon to be discussed.

The function of the gate is to control the firing of the SCR. The characteristics of Fig. 15.28 indicate that, as the gate current $I_G$ increases, the breakover voltage decreases. The range of gate control, however, is small; therefore control

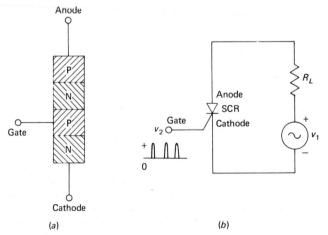

Fig. 15.27  (a) SCR structure; (b) SCR symbol and circuit.

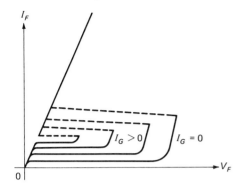

*Fig. 15.28  SCR characteristics.*

is not accomplished by changing the magnitude of gate current. Instead, enough gate current is applied each time to ensure firing of the SCR. Control of load power is achieved by proper timing of the gate pulses.

Once a gate pulse has fired the SCR, the gate loses control. The SCR remains ON even if the gate current is reduced to zero. To turn the SCR OFF, a number of techniques may be employed. One is to reverse the polarity of anode-cathode voltage; this is done each half-cycle by $v_1$ in Fig. 15.27b. Another possibility is to reduce the current through the SCR below the holding current $I_H$, thereby reverting to the high-voltage, low-current state.

Consider the waveforms of Fig. 15.29. $v_1$ is the AC supply; $R_L$ could be a light bulb whose light intensity we wish to control. If the gate pulses are as indicated in Fig. 15.29a, the time during which power is applied to $R_L$ (the conduction angle) is a small portion of the whole cycle. The average power to $R_L$, and hence the light intensity, is relatively low. If the gate pulses of Fig. 15.29b are applied instead, the conduction angle, and hence the average power to $R_L$, increases, producing a greater light intensity. Note that the SCR is turned OFF during each half-cycle when the polarity of $v_1$ makes its anode negative with respect to the cathode. Also note that when the SCR is OFF, its current is negligible, while when it is ON, its voltage is negligible. Consequently the SCR never has to dissipate appreciable power, even while controlling substantial power to $R_L$.

SCRs are normally rated in terms of the maximum reverse voltage that can be applied between anode and cathode (anode negative); this ranges from a few volts to several hundred volts. In the forward direction (anode positive) the maximum anode voltage and the maximum anode current are specified. These range to several hundred volts peak and several amperes RMS, respectively. Maximum gate-voltage, current, and power-dissipation ratings are also usually given.

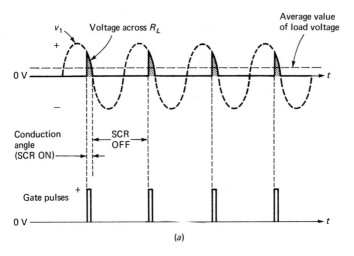

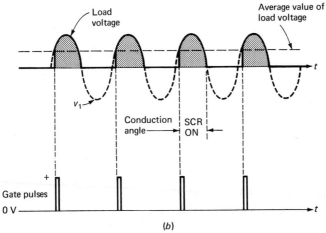

*Fig. 15.29* (a) *Timing of gate pulses produces short SCR conduction angle, and hence low average load voltage;* (b) *conduction angle is increased, yielding higher average load voltage.*

Time delays due to junction capacitances are present in the SCR; these limit the frequency at which the device can be operated, but present no major problem at low switching speeds.

Other four-layer devices operating on similar principles include the SCS (silicon-controlled switch), which has two gates; the triac, which has a single gate but equal forward and reverse characteristics; and the light-activated SCR (LASCR), whose gate is photosensitive.

An SCR-controlled circuit is shown in Fig. 15.30a. $R_L$ represents any

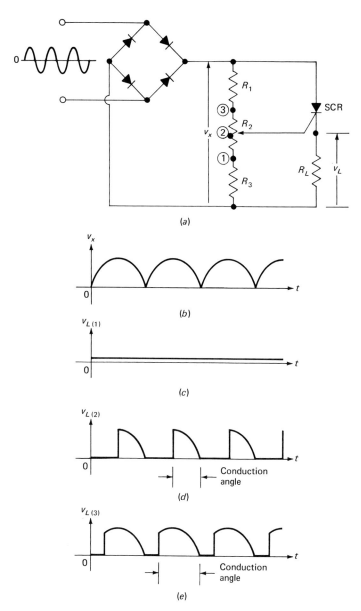

Fig. 15.30 (a) Full-wave SCR power control circuit; (b) rectified line voltage waveform; (c) load voltage for potentiometer set at position 1; (d) load voltage for potentiometer set at position 2; (e) load voltage for potentiometer set at position 3.

load whose average power we wish to control, such as a light or an electric appliance. The diode bridge rectifies the AC line voltage, yielding a full-wave rectified output $v_x$ as shown in Fig. 15.30b. Because of this full-wave rectified voltage, control over both halves of the AC cycle is possible.

With $R_2$ set at 1, the SCR cannot be fired and remains OFF throughout each half-cycle. There is no current, and hence no voltage across $R_L$, as shown in Fig. 15.30c. As the potentiometer is varied toward points 2 and 3, the SCR's anode-firing potential decreases, yielding load voltage waveforms as in Fig. 15.30d and e. The setting of $R_2$ therefore controls the conduction angle. As $R_2$ is varied, the conduction angle changes, and hence the average power to $R_L$ is controlled.

The circuit of Fig. 15.31a is now considered. Here the bridge rectifier develops a full-wave rectified voltage between $C$ and $F$ when the SCR does not conduct. This voltage is clipped at a convenient level by the Zener $D$, thus providing the necessary DC voltage to operate the unijunction transistor oscillator. The frequency of oscillation can be controlled with the variable resistor $R_1$.

The pulses generated at $B_1$ are fed to the SCR gate. Every time one of these pulses is generated, the SCR fires, and the voltage between $C$ and $F$ drops to almost zero; $R_L$ then receives the full supply voltage.

The simplified equivalent circuit for a positive-going input is shown in Fig. 15.31b. When the input changes polarity, opposite diode pairs conduct, but $R_L$ still receives the full supply (except for the small drop across the SCR and conducting diodes). Relevant waveforms are shown in Fig. 15.31c. It should be noted that varying $R_1$ varies the frequency of gate pulses, thus controlling the conduction angle. The average power delivered to $R_L$ is therefore a function of the $R_1$ setting.

## Problem Set 15.6

15.6.1. In what way does a four-layer diode differ from a two-layer diode?
15.6.2. How is an SCR turned ON?
15.6.3. What is breakover voltage?
15.6.4. What is holding current?
15.6.5. How is an SCR turned OFF?
15.6.6. Why do you think SCR power control is preferable over a simple potentiometer type of control?
15.6.7. What effect does varying $R_1$ have in the circuit of Fig. 15.31? Could the same effect be achieved by varying some other component?
15.6.8. The circuit of Fig. 15.30 can be driven directly from the 60-Hz line provided a diode is connected in series with the SCR gate to protect it against negative gate voltage. Show how such a diode should be connected.

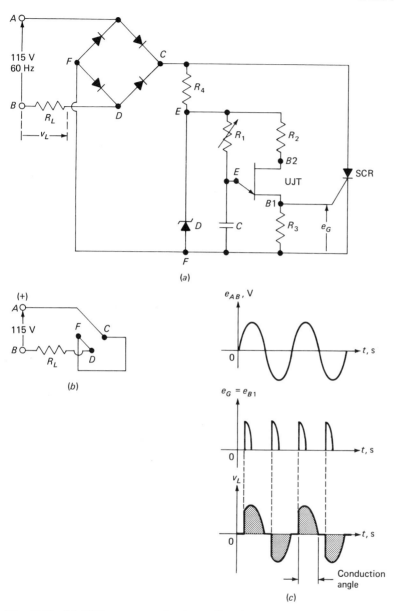

Fig. 15.31 (a) SCR power control using unijunction transistor oscillator for SCR firing; (b) equivalent circuit when SCR is conducting; (c) pertinent waveforms.

### 15.3d  Sweep Generation

A *sawtooth* or *sweep* waveform is shown in Fig. 15.32. Ideally such a signal rises at a constant rate for a period of time and then drops to the base line in zero time and resumes its sequence. A sweep may be positive-going, as in the figure, or negative-going. The base line may be zero or some other convenient level. Some generators are designed to produce voltage sweeps; others produce current sweeps.

One of the most common applications of voltage sweeps is for the horizontal deflection of the electron beam in an oscilloscope. Here the sweep is applied to the horizontal deflection plates; the rising voltage linearly deflects the electron beam from left to right across the oscilloscope screen at a rate determined by the time base setting. In television receivers, electromagnetic deflection is employed so that a current sweep applied through a deflection coil is required. In either case, linearity is important as well as timing, that is, the sweep starting point and duration.

A sweep may be free-running or triggered. When free-running, there is no control over the starting time. A triggered sweep, instead, is initiated only when an appropriate signal (trigger) is applied. Here the sweep is said to be synchronized with whatever source provides the triggering signal. As an example, when an oscilloscope display is viewed, it is important that each successive sweep start at the same level on the waveform to be displayed; otherwise a jumbled display results. Therefore the sweep is usually triggered to start whenever the AC signal to be displayed passes through a predetermined "level."

A simple technique for generating a voltage sweep is shown in Fig. 15.33. The neon bulb does not conduct until the voltage across it exceeds a given value, say $V_2$. Once this happens, the bulb conducts with a relatively low resistance; it stops conducting whenever the voltage drops below $V_1$. At first the capacitor starts charging toward $V_i$. As long as $v_o < V_2$, the neon bulb appears as an open circuit, and the voltage across the capacitor follows the usual exponential rise, as indicated by the initial portion of the output curve in the figure.

As soon as $v_o$ exceeds $V_2$, the neon bulb conducts. The capacitor starts losing its charge very quickly through the neon bulb's low-resistance path. The capacitor's voltage $v_o$ quickly drops toward zero. But once $v_o$ drops below $V_1$, the

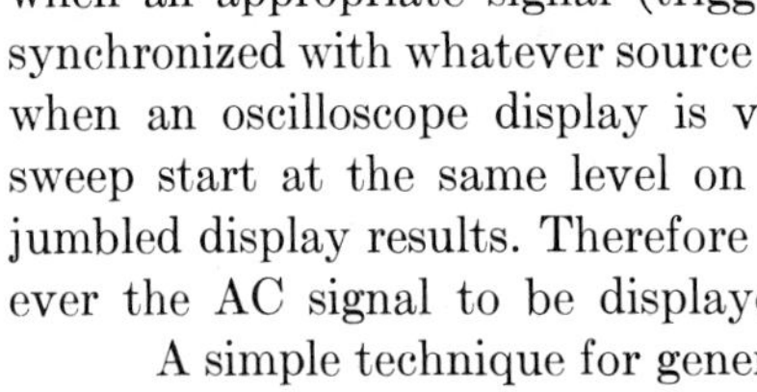

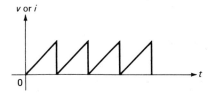

*Fig. 15.32* Typical sawtooth or sweep waveform.

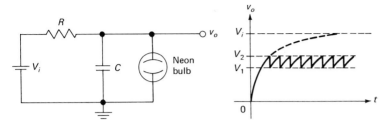

Fig. 15.33  Simple sweep generator.

neon bulb stops conducting and behaves as a large resistance, allowing $C$ to continue charging once more.

To obtain a reasonably linear sweep, only the initial portion of the exponential rise can be used ($V_2 \ll V_i$). The reset time (time for $v_o$ to drop from $V_2$ to $V_1$) is a function of the neon bulb's ON resistance, which can never be zero. For these reasons and also because the sweep cannot be triggered, this circuit finds very little application.

To improve linearity, a bootstrapping arrangement can be used. Shown in Fig. 15.34 is the basic principle from which a practical circuit can be developed. Here a fictitious variable source $v_o$ is added in series with $V_i$. Initially, $v_o$ is zero, and the voltage across the capacitor starts rising toward $V_i$. This is the same as in the circuit of Fig. 15.33. As soon as the capacitor's voltage starts to rise, though, $v_o$ is no longer zero; now the total voltage across the $RC$ combination is $V_i + v_o$, and the capacitor aims toward this modified target value. Since the target value is constantly increasing, we are always operating in the initial portion of the exponential rise, which is linear. We never get closer to the target value; hence the exponential rise never levels off! In a practical circuit there must obviously be some limit to the ultimate voltage, but this presents no great problem, because the sweep amplitude must be limited too.

Consider the circuit of Fig. 15.35a. Here the sweep voltage is developed across $C$. With a positive voltage at the base of $Q_1$, $Q_1$ is saturated and $C$ dis-

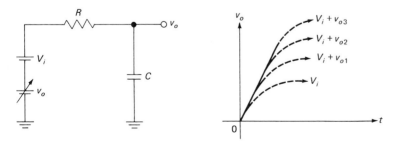

Fig. 15.34  Use of bootstrapping to obtain sweep linearity.

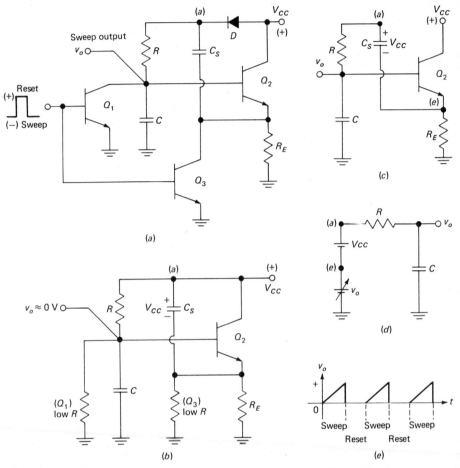

Fig. 15.35 (a) Bootstrap-type sweep generator; (b) equivalent circuit for (a) when reset; (c) equivalent circuit for (a) during sweep period; (d) equivalent circuit for (c); (e) output.

charges; therefore $v_o = 0$ (sweep is reset). Driving the base of $Q_1$ negative triggers the sweep by turning $Q_1$ OFF and allowing $C$ to charge through $R$ from $V_{CC}$.

When the sweep is reset, the equivalent circuit of Fig. 15.35b applies. Both $Q_1$ and $Q_3$ are fully ON and behave as very low resistances to ground. Whatever charge $C$ had before resetting it quickly loses through $Q_1$'s low resistance, and $v_o \simeq 0$ V. The emitter of $Q_2$ and bottom end of $C_S$ are similarly grounded through $Q_3$'s low ON resistance. $C_S$ quickly charges to $V_{CC}$ as shown.

When a negative input voltage is applied at the base of $Q_1$, the sweep is triggered. Assuming no leakage, both $Q_1$ and $Q_3$ behave as open circuits, and the equivalent circuit of Fig. 15.35c applies. $C_S$ is large; therefore it retains its previously acquired charge at $V_{CC}$ and in fact behaves as a battery during the sweep period. $D$ is now reverse-biased, effectively disconnecting point $a$ from the external supply. The emitter voltage of $Q_2$ "follows" the base voltage, which is simply $v_o$. Thus point $e$ is at a potential very close to $v_o$. This is further illustrated in the equivalent circuit of Fig. 15.35d in which a variable source $v_o$ replaces the voltage developed across $R_E$ and a fixed source $V_{CC}$ replaces $C_S$. Note the similarity between this circuit and that in Fig. 15.34.

The sweep voltage waveform is shown in Fig. 15.35e. $v_o$ rises linearly until a reset signal is applied. Thus not only is the sweep linear, but it can also be triggered. Note that during reset, $C_S$ gets back whatever charge it lost during the sweep period through the low-resistance path of $Q_3$ and $D$. To protect $Q_1$ from excessive reverse voltage, a clamping diode is sometimes connected between the collector of $Q_1$ and a positive supply $V_P$. The diode is normally reverse-biased; therefore it does not affect circuit operation until $v_o$ exceeds $V_P$. Then the diode conducts and clamps the collector of $Q_1$ to $V_P$, thus preventing $v_o$ from rising further.

## Example 15.4

Determine the peak value of the sweep voltage in the circuit of Fig. 15.35a if $R = 10$ k$\Omega$, $C = 0.01$ $\mu$F, $V_{CC} = 45$ V, and the sweep duration is 50 $\mu$s. What happens if the negative level at the base of $Q_1$ remains ON for very long?

## Solution

With a linear sweep, the rate of change is the same as in the initial portion of any exponential curve. The initial target value—$V_{CC}$ in our case—would be achieved in one time constant; that is, $v_o$ would reach 45 V in

$\tau = RC$
$\tau = 10^4(0.01 \times 10^{-6})$
$\tau = 100$ $\mu$s

Our sweep lasts 50 $\mu$s; therefore $v_o$ reaches 22.5 V.

If the base is held negative too long, $C$ keeps on charging. If a clamping diode is not connected at the collector of $Q_1$ to prevent $v_o$ from rising beyond a specified voltage, $Q_1$ could be damaged by excessive reverse voltage.

### Problem Set 15.7

15.7.1. What is meant by sweep linearity?
15.7.2. What does sweep resetting mean?
15.7.3. What is a triggered sweep? Where is it utilized?

15.7.4. What are some of the disadvantages of the sweep circuit in Fig. 15.33?

15.7.5. Explain how bootstrapping operates to yield a linear sweep in the circuit of Fig. 15.34.

15.7.6. In the circuit of Fig. 15.35a, $V_{CC} = 30$ V, $R = 10$ k$\Omega$, and $C = 0.05$ $\mu$F. For how long must the base of $Q_1$ be held negative for a peak sweep output of 10 V?

15.7.7. Using the sweep duration obtained in Prob. 15.7.6, determine what happens to the sweep amplitude when $R$ is doubled.

15.7.8. Show how a clamping diode could be connected in the circuit of Fig. 15.35a to prevent $v_o$ from rising beyond $+6$ V.

15.7.9. What do you think would happen in the circuit of Fig. 15.35a if $C_S$ were not sufficiently large?

15.7.10. Determine the reset time (time required for sweep to drop from its maximum amplitude to zero) in the circuit of Fig. 15.35a if $C = 0.02$ $\mu$F and $Q_1$'s ON resistance is 1 $\Omega$.

### 15.3e  Staircase Generation

Very often a series of *step voltages* similar to the diagram in Fig. 15.36 must be generated. One application of this "staircase" type of signal is in automatic curve tracers where a family of curves is generated on an oscilloscope screen. As an example, in tracing the output characteristics of a common emitter transistor configuration (see Fig. 5.8), each curve requires a base current drive which is one "step" higher than the previous level. We therefore require a "current step generator" to provide the necessary base drive. In other cases a "voltage step generator" may be indicated. The internal resistance of current generators is high, while for voltage generators it is low; otherwise the basic principles are the same.

Consider the circuit of Fig. 15.37a. This is not a practical circuit for generating steps, but it can be used to illustrate some basic concepts. If the switch is alternately closed and opened, the waveform in Fig. 15.37b results. Note that whenever the switch is open, $e_c$ remains the same, since the capacitor neither acquires nor loses any charge. Obviously, if a resistive load is connected

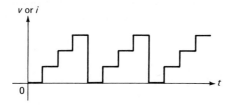

*Fig. 15.36  Step or staircase waveform.*

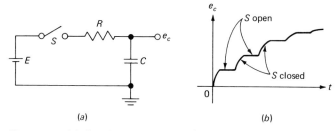

Fig. 15.37 (a) *Simple step generator;* (b) *output of* (a).

across $C$, charge is lost and $e_c$ sags. When the switch is closed, the charge on $C$ resumes its buildup, and the $e_c$ waveform begins to rise again from where it left off when the switch was open. During the intervals that the switch is closed, $e_c$ follows the usual exponential rise, and therefore the steps are not uniform. In fact, the steps get progressively smaller as the capacitor voltage approaches its final value. Also note that a nonzero rise time is required to go from one step to another. In the ideal waveform of Fig. 15.36 there is no time delay between the end of one step and the beginning of the next; all steps are also of uniform height. It is obvious that the circuit of Fig. 15.37a comes far short of meeting the requirements of an ideal step generator.

The circuit of Fig. 15.38a can produce an improved voltage staircase waveform. A pulse generator is required at the input, in this case producing negative-going pulses from a zero base line. The neon bulb fires at 28 V; this means that as long as the voltage across the bulb is less than 28 V, it behaves as a large resistance, and $C_2$ cannot discharge into it. As soon as 28 V is exceeded, however, the neon bulb fires and discharges $C_2$. The 330-$\Omega$ resistor limits the discharge current to a safe value.

To further illustrate the circuit's operation, let us assume that $R_g = 50\ \Omega$, all diodes are ideal, and all capacitors are without leakage. The input pulses are 10 ms wide, and the period is 20 ms. $C_1 = 0.1\ \mu\text{F}$ and $C_2 = 0.2\ \mu\text{F}$. For $t < 0$, the pulse generator is at $-36$ V; $C_1$ therefore charges to 36 V through $R_g$, $C_1$, $D_1$, and the generator, with a time constant $\tau = R_g C_1 = 5\ \mu\text{s}$. This is fast enough that it may be neglected compared with the input pulse duration. The charge polarity on $C_1$ is as indicated in Fig. 15.38a. During this time $D_2$ does not conduct; hence the circuit to the right of point $a$ is undisturbed. Assuming no initial charge on $C_2$, the output at point $b$ is still zero, as shown in Fig. 15.38b for $t < 0$.

At $t = 0$ the pulse generator switches from $-36$ to 0 V, and the equivalent circuit in Fig. 15.38c applies; here $D_1$ is off, but $D_2$ conducts. Current flows through the circuit in such a manner as to remove charge from $C_1$ and deposit it on $C_2$. As the voltage across $C_1$ decreases, the voltage across $C_2$ increases. The process continues until the voltage across $C_1$ is equal in magnitude to the voltage

686  Solid State Electronic Circuits

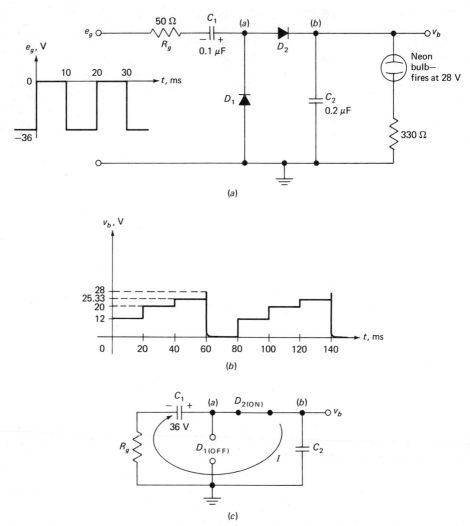

Fig. 15.38 (a) Staircase generator; (b) output of (a); (c) equivalent circuit of (a) when $e_g = 0$ V.

across $C_2$, at which time there is no potential difference across $R_g$, the current becomes zero, and there is no further exchange of charge between capacitors. The time required for this process is $5\tau = 5R_g C_1 C_2/(C_1 + C_2) = (5)(50)(0.066 \times 10^{-6}) \simeq 15$ μs. This is short compared to the duration of the input pulse, and it can be neglected. Therefore the loss of charge by $C_1$ and gain of charge by $C_2$ at $t = 0$ are almost instantaneous. The final voltage on $C_1$ and $C_2$ may be determined by using the fact that whatever charge is lost by $C_1$ is gained by $C_2$ so that there is no net loss or gain of charge in the circuit. The formula for charge on a capacitor is $Q = CV$; in our case we have

*Initial Conditions (Slightly before* t = 0*)*

Charge on $C_1$: $Q = 10^{-7}(36) = 3.6\ \mu C$
Charge on $C_2$: 0
Total charge: $3.6\ \mu C$

*First Levels (Slightly after* t = 0*)*

Charge on $C_1$: $Q = 10^{-7}V_1$
Charge on $C_2$: $Q = 2 \times 10^{-7}V_1$
Total charge: $3 \times 10^{-7}V_1$

The total charge slightly before $t = 0$ must be the same as the total charge slightly after $t = 0$, yielding

$3.6 \times 10^{-6} = 3 \times 10^{-7}V_1$
$V_1 = 12\ V$

The output at point $b$ is now at 12 V, and the neon bulb appears as a high resistance; hence $C_2$ retains its charge. The voltage across $C_1$ is also 12 V.
 The input generator switches back to $-36$ V at $t = 10$ ms. $D_1$ conducts, and $C_1$ is recharged to 36 V. Point $a$ in Fig. 15.38a is at ground potential because of $D_1$; hence $D_2$ is kept OFF, and the voltage across $C_2$ remains undisturbed at 12 V for the whole duration of the $-36$ V pulse, that is, until $t = 20$ ms. This is shown on the graph of Fig. 15.38b. At $t = 20$ ms, the input becomes 0 V once more. The equivalent circuit in Fig. 15.38c again applies. Here $C_1$ loses and $C_2$ gains charge, as follows:

*Initial Conditions (Slightly before* t = 20 ms*)*

Charge on $C_1$: $Q = 10^{-7}(36) = 3.6\ \mu C$
Charge on $C_2$: $Q = (2 \times 10^{-7})(12) = 2.4\ \mu C$
Total charge: $6\ \mu C$

*Second Levels (Slightly after* t = 20 ms*)*

Charge on $C_1$: $Q = 10^{-7}V_2$
Charge on $C_2$: $Q = 2 \times 10^{-7}V_2$
Total charge: $Q = 3 \times 10^{-7}V_2$

Equating the above, we have

$3 \times 10^{-7}V_2 = 6 \times 10^{-6}$
$V_2 = 20\ V$

The process continues and a new step is generated for every cycle of the input pulse. The third step is

$$V_3 = \frac{36 \times 10^{-7} + 20(2 \times 10^{-7})}{3 \times 10^{-7}}$$

$$V_3 = 25.33 \text{ V}$$

In general, the $n$th step is

$$V_n = \frac{E_{gm}C_1 + V_{n-1}C_2}{C_1 + C_2} \tag{15.15}$$

where $E_{gm}$ is the peak value of $E_g$—36 V in our case; $V_{n-1}$ is the step just before $V_n$. Thus for the fourth step, $n = 4$, and Eq. (15.15) becomes

$$V_4 = \frac{36C_1 + V_3C_3}{C_1 + C_2}$$

$$V_4 = \frac{36 \times 10^{-7} + 25.33(2 \times 10^{-7})}{3 \times 10^{-7}}$$

$$V_4 = 28.9 \text{ V}$$

Since the neon bulb discharges $C_2$ as soon as $V_b$ reaches 28 V, this last step is limited to 28 V. After this, the generation of steps resumes, as indicated in the output waveform of Fig. 15.38b. Note that the steps are not of uniform size; the rise time, however, is negligible compared to the step duration. As with sweep generators, it is possible to improve linearity (make the step sizes uniform) through the use of bootstrapping techniques.

### Problem Set 15.8

15.8.1. Determine each voltage step for the circuit of Fig. 15.38a if everything stays the same except $C_2 = 0.5$ μF. How many steps are generated?

15.8.2. Repeat Prob. 15.8.1 if $C_2 = 0.05$ μF.

### 15.3f  The One Shot (Monostable Multivibrator)

The *one shot* is a monostable circuit—that is, only one stable state is possible. This state is called standby and is characterized by $Q_1$ being OFF and $Q_2$ ON in the circuit of Fig. 15.39a. The circuit cannot remain indefinitely in any other state and always returns to standby.

With the circuit in standby, the collector of $Q_1$ is at a high voltage, while the collector of $Q_2$ is close to ground. If an appropriate trigger pulse is applied through $C_T$, the one shot changes state; that is, $Q_2$ goes OFF while $Q_1$ comes ON.

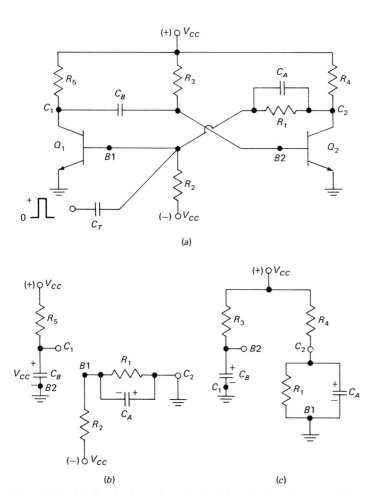

**Fig. 15.39** (a) Standard one-shot circuit; (b) equivalent circuit for (a) during standby, $Q_1$ OFF, and $Q_2$ ON; (c) equivalent circuit for (a) during the one-shot period, $Q_1$ ON, and $Q_2$ OFF.

This change of state lasts for a brief period (determined by the value of circuit components), after which the circuit automatically reverts to standby. The trigger pulse therefore causes a single output pulse to be generated so that for one pulse in, we get one pulse out. The output may be taken at either collector; at the collector of $Q_2$ the pulse is positive-going, while at the collector of $Q_1$ it is negative-going. The triggering pulse can be a short, sharp spike, while the output pulse can have a flat top and be accurately controlled; its duration can either be shorter or longer than the input pulse.

To see how the circuit operates, we will assume ideal transistors. In standby, $Q_1$ is OFF, $Q_2$ is ON, and the equivalent circuit of Fig. 15.39b applies. The junction of $C_B$ and $R_3$ is at ground potential (for an accurate analysis the $V_{BE}$ drop of $Q_2$ would have to be taken into account). $R_3$ is therefore connected between $V_{CC}$ and ground; it does not enter the picture here and is omitted from the diagram. The negative supply keeps the base of $Q_1$ at a negative potential determined by the voltage-divider action of $R_2$ and $R_1$. Note that one end of $R_1$ is grounded through the low ON resistance of $Q_2$. The negative voltage at the base of $Q_1$ keeps $Q_1$ OFF. Also note that during standby $C_B$ is charged to approximately $V_{CC}$. $C_A$ is also charged to whatever voltage appears at the base of $Q_1$. Waveforms for $v_{B2}$, $v_{C2}$, and $v_{C1}$ are shown in Fig. 15.40. The standby levels are indicated for $t < t_o$.

Now consider what happens at $t = t_o$, when a positive-going pulse is applied to $C_T$. Remember that unless such a pulse is applied, the circuit remains in standby indefinitely. The positive pulse is coupled through $C_T$ to the base of $Q_1$, which has been maintained at a negative potential by the negative supply. The positive trigger overrides the negative supply and actually makes the base positive, causing $Q_1$ to start conducting. At this point the input trigger may be removed; the one shot has been initiated, and the full transition is taken over by the circuit. Since $Q_1$ has begun conducting, its collector starts falling. This drop in voltage is coupled, through $C_B$, to the base of $Q_2$. $Q_2$ comes out of conduction, causing its collector current to drop and hence its collector voltage to rise toward

$$V_{CC} \frac{R_1}{R_1 + R_4} \tag{a}$$

with a time constant

$$\tau = C_A \frac{R_1 R_4}{R_1 + R_4} \tag{b}$$

The rising voltage at the collector of $Q_2$ is coupled through the $R_1$-$C_A$ combination to the base of $Q_1$, which turns $Q_1$ on even more. This regenerative action takes place almost instantaneously, and the collector of $Q_1$ drops to approximately 0 V in essentially zero time (see Fig. 15.39c). Since $C_B$ is large, it does not appreciably discharge during the turn-on transient of $Q_1$. Figure 15.39b shows how $C_B$ was charged during standby. During the one shot, however, the circuit switches to the state shown in Fig. 15.39c. Here $C_B$ again tends to charge toward the positive supply, but with a polarity exactly opposite that maintained during standby. The charging time constant is

$$\tau = R_3 C_B \tag{c}$$

Note that the voltage across $C_B$ is also the base voltage for $Q_2$. As long as $C_B$ remains charged with the same polarity as during standby, the potential at

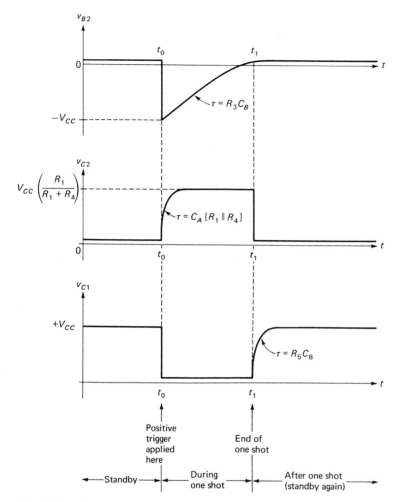

Fig. 15.40  One-shot waveforms for the circuit of Fig. 15.39a.

the base of $Q_2$ keeps $Q_2$ OFF. But as $C_B$ continues to charge toward $V_{CC}$, the voltage at the base of $Q_2$ keeps rising until it exceeds zero. When this happens ($t = t_1$), the one shot is over. The one-shot period is therefore equal to the time required for $V_{B2}$ to rise from $-V_{CC}$ to approximately 0 V. Since $V_{B2}$ actually aims to rise from $-V_{CC}$ to $+V_{CC}$ (a total swing of $2V_{CC}$), the swing from $-V_{CC}$ to 0 represents 50 percent of the usual exponential rise. From the graph of Fig. 15.4, it takes $0.69\tau$ to reach the 50 percent level; hence the one-shot period is

$$T = 0.69 R_3 C_B \tag{15.16}$$

$Q_2$ now starts conducting, and its collector potential begins to fall. This drop in $v_{C2}$ is coupled to the base of $Q_1$, making it less positive. Hence $Q_1$ conducts less, and its collector rises toward $V_{CC}$ with a time constant

$$\tau = R_5 C_B \tag{d}$$

The process continues until $Q_2$ comes fully ON and $Q_1$ goes fully OFF. When this happens, the negative supply holds $Q_1$ OFF indefinitely; we are now in standby once again. $Q_2$ is kept fully ON with base current drive from $V_{CC}$ through $R_3$. There is no current through $R_5$ or $C_B$, since $C_B$ is fully charged and $Q_1$ is OFF.

**Example 15.5**

For the circuit of Fig. 15.39a, assume ideal transistors (zero ON voltages, zero OFF currents), 12-V and $-12$ V supplies, $R_4 = R_5 = 1.5$ k$\Omega$, $R_1 = R_3 = 15$ k$\Omega$, $R_2 = 30$ k$\Omega$, $C_B = 0.1$ $\mu$F, and $C_A = 0.2$ $\mu$F. Determine

a. The one-shot period
b. $v_{C2}$ during the one-shot period
c. $v_{B1}$ during standby

**Solution**

a. The one-shot period is determined by $R_3$ and $C_B$:

$$T = 0.69 R_3 C_B \tag{15.16}$$
$$T = (0.69)(15 \times 10^3)(10^{-7})$$
$$T \simeq 1 \text{ ms}$$

b. During the one-shot period, $v_{C2}$ rises toward

$$V_{CC} \frac{R_1}{R_1 + R_4} = 12 \frac{15 \times 10^3}{16.5 \times 10^3} = 10.9 \text{ V}$$

c. During standby $Q_1$ is OFF, and the base draws negligible current. The base voltage is therefore determined by $R_2$ and $R_1$ as follows:

$$V_{B1} = -12 \frac{R_1}{R_1 + R_2}$$
$$V_{B1} = -12 \frac{15 \times 10^3}{45 \times 10^3} = -4 \text{ V}$$

## Problem Set 15.9

15.9.1. What does *monostable* mean?
15.9.2. Is the one-shot pulse present only while an input trigger is being applied?
15.9.3. Is it possible to generate a one-shot pulse that is longer than the input trigger pulse? Shorter?

15.9.4. For the circuit in Fig. 15.39a, assume ideal transistors, 6 V and $-6$ V supplies, $R_4 = 1$ k$\Omega$, and minimum $h_{FE}$'s $= 10$. Determine

   a. $I_{C2}$ during standby if $R_1 = 20$ k$\Omega$, $R_2 = 40$ k$\Omega$
   b. $R_3$ using the data from (a)
   c. $C_B$ for a 20-ms one-shot period using the data from (a) and (b)

15.9.5. If $R_1 = 10$ k$\Omega$ in Prob. 15.9.4, determine $R_2$ to yield 2 V reverse bias at the base of $Q_1$ during standby. What is $v_{C2}$ during the one-shot period?

## 15.3g The Schmitt Trigger

The Schmitt trigger is a bistable circuit. Its output can remain indefinitely at either of two possible levels. For now we will refer to these levels as *LO* and *HI*.

Consider the input and output waveforms of Fig. 15.41a and b. As long

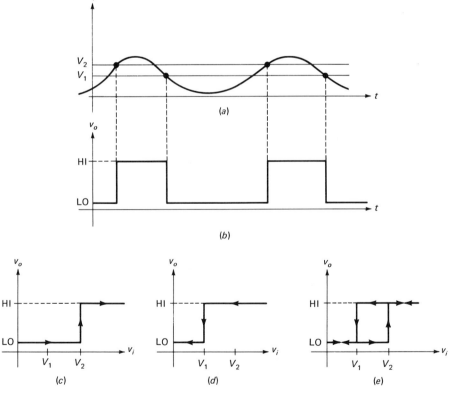

*Fig. 15.41* (a) *Typical Schmitt trigger input waveform;* (b) *output waveform;* (c) *Schmitt trigger transfer characteristic* ($v_i$ *increasing*); (d) *Schmitt trigger transfer characteristic* ($v_i$ *decreasing*); (e) *total characteristic (c and d).*

as $v_i < V_2$, $v_o$ is LO and can remain there indefinitely. As soon as $v_i$ exceeds $V_2$, however, $v_o$ switches to the opposite level (HI). The switching action from one level to the other is extremely fast, and as we shall soon see, it depends on regeneration (positive feedback). The level $V_2$ is called the upper trigger point. As long as $v_i$ remains above $V_2$, $v_o$ is HI.

As soon as $v_i$ drops below $V_1$ (lower trigger point), $v_o$ switches back to LO. Therefore $v_o$ is HI if $v_i \geq V_2$ and LO if $v_i \leq V_1$. The two trigger points are not usually the same, yielding a hysteresis effect. This can best be seen by observing the transfer characteristics in Fig. 15.41c, d, and e. Figure 15.41c applies to a rising $v_i$; note the switching action at $V_2$. Figure 15.41d applies when $v_i$ is decreasing; here the level change occurs at $V_1$. Figure 15.41e is a composite graph incorporating Fig. 15.41c and d.

Note that the output $v_o$ is rectangular; this means that the circuit can be used to reshape pulses to give them sharp leading and trailing edges. Clearly the output contains frequencies much higher than the input, and hence a nonlinear process is involved. When used this way, the Schmitt trigger is also called a "squarer" circuit, because it can generate a square wave from a sinusoidal or other gradually changing type of waveform. Because a change of state occurs whenever the input crosses a trigger point, the Schmitt trigger also finds applications as a level detector.

A Schmitt trigger circuit is shown in Fig. 15.42. This circuit is also called an *emitter-coupled binary*, because positive feedback occurs from coupling through emitter resistor $R_3$. The output, taken at the collector of $Q_2$, can be HI or LO. If $Q_2$ is OFF, $v_o \simeq V_{CC}$; if $Q_2$ is conducting, $v_o = V_{CE2} + V_E$. The LO and HI outputs are therefore characterized by $Q_2$ conducting and not conducting, respectively.

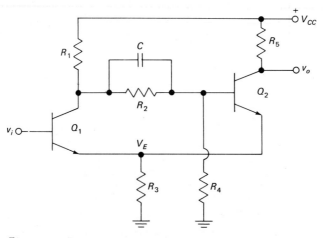

Fig. 15.42  Schmitt trigger circuit.

The circuit is discussed qualitatively first; a numerical example follows. Assume $v_i = 0$ and hence the base of $Q_1$ is grounded. $Q_1$ is cut off because its emitter is positive with respect to ground (the base-emitter junction is reverse-biased). With $Q_1$ OFF, $R_1$ and $R_2$ are in series ($C$ is fully charged in the steady state; therefore it does not conduct), and the equivalent circuit shown in Fig. 15.43a applies. Note that since $Q_1$ is cut off, there is no current through its emitter; therefore $R_3$ sees only the emitter of $Q_2$. The circuit is further simplified in Fig. 15.43b by replacing the base resistors with a Thevenin equivalent circuit in which

$$R_T = R_4 \| (R_1 + R_2) \tag{15.17}$$

$$V_T = V_{CC} \frac{R_4}{R_4 + (R_1 + R_2)} \tag{15.18}$$

As long as $Q_2$ is in the active region, we can write

$$I_{C2} = h_{FE2} I_{B2}$$

and

$$v'_o = V_{CC} - I_{C2} R_5 \tag{15.19}$$

and

$$V'_E = (I_{B2} + I_{C2})(R_3) \tag{15.20}$$

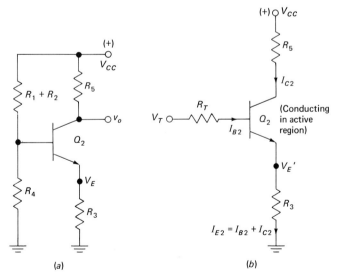

Fig. 15.43 (a) Equivalent circuit for Fig. 15.42 when $Q_1$ is OFF; (b) circuit of (a) with base Thevenin equivalent.

Normally $R_T$ and $V_T$ are chosen to operate $Q_2$ in the active region so that the above equations apply. Note that while $Q_2$ conducts, $Q_1$ is OFF.

Now suppose that $v_i$ begins to rise. Before $Q_1$ can start conducting, $v_i$ must exceed $V_E'$ and provide enough additional drive to exceed the base-emitter cut-in voltage for $Q_1$. The value of $v_i$ that causes $Q_1$ to start conducting is $V_2$, the upper trigger point. As soon as $Q_1$ begins to conduct, its collector voltage drops. This drop in $V_{C1}$ is coupled to the base of $Q_2$ so that $V_{B2}$ also drops, and $Q_2$ conducts less. As $Q_2$ conducts less, $I_{E2}$, which is also the current through $R_3$ due to $Q_2$, is reduced. Meanwhile $I_{E1}$, which also flows through $R_3$, has increased, but the increase in $I_{E1}$ is less than the decrease in $I_{E2}$. Therefore the net current through $R_3$ is reduced, and $V_E$ drops. Lowering $V_E$ increases the potential difference between base and emitter of $Q_1$, thus increasing all currents through $Q_1$. Now as $Q_1$ conducts more, its collector drops further, which causes $Q_2$ to conduct even less, therefore lowering $V_E$ still more. The process is obviously regenerative (positive feedback) and takes place almost instantaneously. $Q_1$ very quickly saturates, and $Q_2$ very quickly goes OFF. This "snap-action" switching is responsible for the almost instantaneous change in output level from LO to HI.

It should be pointed out that there will be no transition unless there is sufficient gain for regeneration to take place. Specifically, as $v_i$ rises, there must be at least an equal drop in $V_E$ so that the input is continuously regenerated. Another way of expressing this is to say that the voltage amplification of $v_i$, as it is processed through $Q_1$, the voltage divider $R_2$-$R_4$, and $Q_2$ back to the emitter of $Q_1$ must be at least unity.

With $Q_2$ cut off, $R_2$ and $R_4$ are in series. The equivalent circuit for $Q_1$ is shown in Fig. 15.44a and further simplified, using Thevenin's theorem, in Fig. 15.44b. Here we have

$$R_x = R_1 \| (R_2 + R_4) \tag{15.21}$$

$$V_x = V_{CC} \left[ \frac{R_2 + R_4}{(R_2 + R_4) + R_1} \right] \tag{15.22}$$

We can determine $V_E''$ as follows:

$$V_E'' = V_x - I_{C1} R_x - V_{\text{SAT}} \tag{15.23}$$

$Q_1$ remains saturated and $Q_2$ cut off as long as $v_i$ does not drop below $V_1$; $V_1$ is the sum of $V_E''$ plus the value of $V_{BE}$ required to saturate $Q_1$. But as soon as $v_i$ drops sufficiently to take $Q_1$ out of saturation, regenerative action is initiated that quickly cuts off $Q_1$ and forces $Q_2$ to conduct once more. The Schmitt trigger is now in its other stable state—that is, $v_o$ is LO.

$C$, in Fig. 15.42, is a speedup capacitor to couple voltage changes between the collector of $Q_1$ and the base of $Q_2$. $Q_2$ is not usually saturated; this produces faster switching action. With $Q_2$ operated in the active region, it can be more

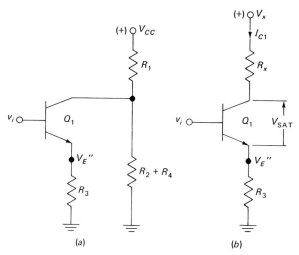

**Fig. 15.44** *Equivalent circuits for Fig. 15.42 when $Q_1$ is ON.*

quickly cut off, and $Q_1$ can therefore be saturated almost instantaneously, thus ensuring snap-action switching.

### Example 15.6

For the circuit in Fig. 15.42, assume the following values: $V_{CC} = 15$ V, $R_1 = 5$ kΩ, $R_2 = 3$ kΩ, $R_3 = 4$ kΩ, $R_4 = 8$ kΩ, $R_5 = 2$ kΩ. The active region $h_{FE}$ for $Q_2$ is 40. Both $Q_1$ and $Q_2$ are low-leakage silicon transistors. Determine LO and HI values for $v_o$, as well as upper and lower trigger points. Also determine the value of $V_{BE2}$ when $Q_2$ is OFF.

### Solution

As $v_i$ rises from zero, nothing happens until $v_i = V_2$, the upper trigger point. Until then, $Q_1$ is OFF and $Q_2$ conducts. The equivalent circuit of Fig. 15.43b applies. Here we have

$$V_T = V_{CC} \frac{R_4}{R_4 + R_1 + R_2} \tag{15.18}$$

$$V_T = 15 \frac{8 \times 10^3}{(8 + 5 + 3)(10^3)} = 7.5 \text{ V}$$

$$R_T = R_4 \| (R_1 + R_2) \tag{15.17}$$

$$R_T = (8 \times 10^3) \| (8 \times 10^3) = 4 \text{ kΩ}$$

Since $Q_2$ is in the active region, we can write

$$I_{B2} = \frac{V_T - V_{BE2}}{R_T + R_3(h_{FE2} + 1)}$$

A typical $V_{BE}$ is 0.6 V. This yields

$$I_{B2} = \frac{7.5 - 0.6}{4 \times 10^3 + (4 \times 10^3)(41)}$$
$$I_{B2} = 41 \ \mu\text{A}$$

But

$$I_{C2} = h_{FE2} I_{B2}$$
$$I_{C2} = 40(41 \times 10^{-6}) = 1.64 \text{ mA}$$
$$v'_o = V_{CC} - I_{C2} R_5 \tag{15.19}$$
$$v'_o = 15 - (1.64 \times 10^{-3})(2 \times 10^3)$$

The LO value of $v_o$ is

$$V'_o = 11.7 \text{ V}$$

The emitter voltage is

$$V'_E = (I_{B2} + I_{C2})(R_3) \tag{15.20}$$
$$V'_E = (41 \times 10^{-6} + 1.64 \times 10^{-3})(4 \times 10^3)$$
$$V'_E = 6.7 \text{ V}$$

To cut in $Q_1$, $v_i$ must rise beyond the upper trigger point:

$$V_2 = V_{BE1(\text{cut-in})} + V'_E$$
$$V_2 = 0.5 + 6.7 = 7.2 \text{ V}$$

where 0.5 V is the value of $V_{BE}$ required to cut in $Q_1$.

As soon as $v_i$ exceeds 7.2 V, there is a circuit transition in which $Q_2$ is cut off and $Q_1$ saturated. With $Q_2$ OFF, $v_o \simeq V_{CC}$; hence our HI output is

$$v''_o = 15 \text{ V}$$

Now the equivalent circuit of Fig. 15.44b applies. Using $I_{C1} \simeq I_{E1}$ and a saturation voltage of 0.3 V, we can estimate the emitter voltage as

$$V''_E = I_{E1} R_3$$
$$I_{E1} \simeq \frac{V_x - V_{\text{SAT}}}{R_3 + R_x}$$
$$V_x = V_{CC} \frac{R_2 + R_4}{R_2 + R_4 + R_1} \tag{15.22}$$
$$V_x = 15 \frac{(3 + 8)(10^3)}{(3 + 8 + 5)(10^3)} = 10.3 \text{ V}$$
$$R_x = R_1 \| (R_2 + R_4) \tag{15.21}$$

$$R_x = (5 \times 10^3) \| (11 \times 10^3) = 3.44 \text{ k}\Omega$$
$$I_{E1} \simeq \frac{10.3 - 0.3}{4 \times 10^3 + 3.44 \times 10^3}$$
$$I_{E1} \simeq 1.35 \text{ mA}$$
$$V''_E = I_{E1} R_3 = (1.35 \times 10^{-3})(4 \times 10^3)$$
$$V''_E = 5.4 \text{ V}$$

During this time, $Q_1$ is in saturation. A typical $V_{BE}$ to saturate $Q_1$ is 0.7 V; therefore $v_i$ must be at least $5.4 + 0.7 = 6.1$ V to keep $Q_1$ ON. If $v_i$ drops below 6.1 V, $Q_1$ comes out of saturation, and the switchover to the opposite state is initiated. The lower trigger point, $V_1$, is therefore 6.1 V. A complete transfer characteristic, including HI and LO levels as well as upper and lower trigger points, is shown in Fig. 15.45.

To determine $V_{BE2}$ when $Q_2$ is OFF, note that $R_2$ and $R_4$ are in series (Fig. 15.44a). The voltage at the collector of $Q_1$ therefore divides between $R_4$ and $R_2$ (Fig. 15.42), yielding

$$V_{B2} = V_{C1} \left[ \frac{R_4}{R_2 + R_4} \right]$$

where $V_{C1} = V_x - I_{C1} R_x$. But $V_x = 10.3$ V, $I_{C1} \simeq I_{E1} = 1.35$ mA, and $R_x = 3.44$ k$\Omega$; therefore

$$V_{C1} = 10.3 - (1.35 \times 10^{-3})(3.44 \times 10^3)$$
$$V_{C1} = 5.65 \text{ V}$$

We now have

$$V_{B2} = 5.65 \frac{8 \times 10^3}{11 \times 10^3} = 4.1 \text{ V}$$

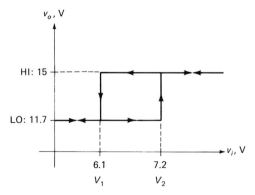

**Fig. 15.45** Transfer characteristic for the Schmitt trigger of Example 15.6.

But $V''_E = 5.4$ V, and $V_{BE2} = V_{B2} - V''_E$; therefore

$V_{BE2} = 4.1 - 5.4 = -1.3$ V

$Q_2$ is therefore kept OFF by 1.3 V reverse bias across its base-emitter junction.

### Problem Set 15.10

15.10.1. What are the outputs of a Schmitt trigger? What are the states of $Q_1$ and $Q_2$ in the circuit of Fig. 15.42 for each of these outputs?

15.10.2. What is meant by upper trigger point? Lower trigger point?

15.10.3. What is hysteresis? Can you think of other examples of hysteresis?

15.10.4. How is the snap-action switching of a Schmitt trigger accomplished?

15.10.5. What is meant by a squarer circuit?

15.10.6. What properties of a Schmitt trigger make it suitable as a level detector?

15.10.7. Estimate HI and LO values of $v_o$ in the circuit of Fig. 15.42 if $V_{CC} = 12$ V, $R_3 = 3$ k$\Omega$, and $R_5 = 3$ k$\Omega$. $Q_2$ is a low-leakage silicon transistor and operates in saturation.

15.10.8. Repeat Prob. 15.10.7 except that now $Q_2$ is not saturated but merely driven in the active region to generate the LO output. $Q_2$ has a saturation region $h_{FE} = 50$. The remaining components are $R_1 = 4$ k$\Omega$, $R_2 = 2$ k$\Omega$, and $R_4 = 6$ k$\Omega$.

15.10.9. Determine upper and lower trigger points for Prob. 15.10.8. Construct a transfer characteristic.

15.10.10. In the circuit of Prob. 15.10.8, what is the reverse base-emitter voltage of $Q_2$ when OFF?

## 15.3h  The Flip-flop

The flip-flop (FF) is a bistable circuit. It has two outputs and utilizes one or two inputs. Each output is the complement of the other; this means that if one output is HI, the other output is LO. Applications of the FF, also called a binary, include timing circuits (as a frequency divider) and counting circuits.

Each output of an FF cannot be determined entirely from a knowledge of the present inputs; instead both the present and previous inputs must be known. The FF "remembers" its previous state and responds to new inputs in a manner that takes this into account.

A basic FF circuit is shown in Fig. 15.46. The circuit is symmetrical; that is, $R_1 = R_4$, $R_2 = R_5$, and $R_3 = R_6$. An output is taken at the collector of each transistor. The circuit forces each transistor to be in a state opposite to the other's. For example, assume $Q_1$ is ON. This means that the junction of $R_1$ and $R_2$ is a virtual ground (except for the $V_{SAT}$ drop of $Q_1$) and that the base voltage of $Q_2$ is determined mainly by the negative supply in conjunction with the

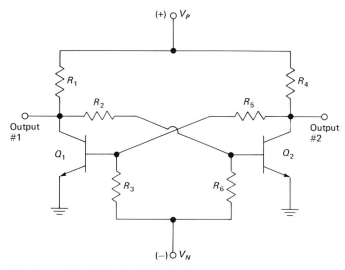

Fig. 15.46  Basic flip-flop (FF) circuit.

voltage-divider action of $R_6$ and $R_2$. This places the base of $Q_2$ at a negative potential with respect to ground, thus reverse-biasing the base-emitter junction and turning OFF $Q_2$.

The OFF state of $Q_2$ also ensures that $Q_1$ stays ON. The collector of $Q_2$ is at a positive voltage. Now $R_5$ is usually smaller than $R_3$, so that the base of $Q_1$ is closer to the positive level at the collector of $Q_2$ than to the negative supply. The net effect is to make the base sufficiently positive to keep $Q_1$ fully ON.

With $Q_1$ assumed ON and $Q_2$ OFF, these states will be maintained until a specific signal is applied to the circuit that forces it to change state. There are many ways in which the FF can be triggered to accomplish a reversal; each technique has certain applications. For now assume that we apply a positive trigger pulse at the base of $Q_2$; this starts $Q_2$ conducting. The circuit then completes a transition by itself, as follows: as $Q_2$ starts conducting, its collector voltage begins to drop; this drop is coupled by $R_5$ to the base of $Q_1$, which causes $Q_1$'s collector voltage to rise. The rising voltage at the collector of $Q_1$ is coupled through $R_2$ to the base of $Q_2$; now the positive trigger is removed, and $Q_2$ continues to conduct. Very quickly then, $Q_2$ comes ON and $Q_1$ goes OFF. Note that the trigger is applied only long enough to start the OFF transistor conducting; it is then removed, and the circuit completes the transition on its own.

## Example 15.7

Estimate the steady state base and collector voltages for each transistor in Fig. 15.46 if both $Q_1$ and $Q_2$ are silicon transistors with negligible OFF currents and $R_1 = R_4 = 2.7$ k$\Omega$; $R_2 = R_5 = 18$ k$\Omega$; $R_3 = R_6 = 100$ k$\Omega$; and $V_P = 15$ V, $V_N = -15$ V.

**Solution**

We will assume $Q_1$ to be ON and $Q_2$ OFF. Since the circuit is symmetrical, the answers we get also apply to opposite transistors when the FF state is reversed. An equivalent circuit is shown in Fig. 15.47. Here $V_{SAT}$ is assumed as 0.3 V and the base-emitter drive needed to saturate the ON transistor as 0.7 V. We therefore have

$V_{B1} = 0.7$ V
$V_{C1} = 0.3$ V

To estimate $V_{B2}$, note that there is a net potential difference across the series combination of $R_2$ and $R_6$ equal to $V_{C1} - V_N = 15.3$ V. $R_6$ develops a proportional amount of this voltage, that is, $15.3\ R_6/(R_2 + R_6) \simeq 13$ V. The base of $Q_2$ is therefore at $-15 + 13 = -2$ V relative to ground:

$V_{B2} = -2$ V

$V_{C2}$ is obtained by noting that there is a net voltage of $V_P - V_{B1} = 14.3$ V across the series combination of $R_4$ and $R_5$. $R_4$ develops $14.3R_4/(R_4 + R_5) \simeq 1.86$ V, yielding

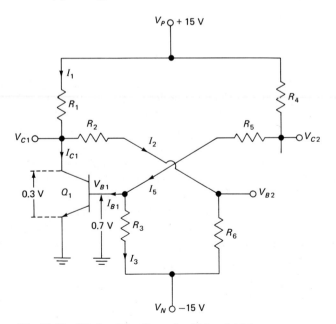

*Fig. 15.47  FF circuit for Examples 15.7 and 15.8.*

$V_{C2} = V_P - V_{R4}$
$V_{C2} = 15 - 1.86 = 13.14$ V

**Example 15.8**

Estimate the base and collector currents of $Q_1$ in Example 15.7. What is the minimum $h_{FE}$ required to saturate the ON transistor?

**Solution**

Using the circuit diagram of Fig. 15.47, we can write

$I_{B1} = I_5 - I_3$
$I_5 = \dfrac{V_{C2} - V_{B1}}{R_5} = \dfrac{13.14 - 0.7}{18 \times 10^3}$
$I_5 \simeq 0.69$ mA
$I_3 = \dfrac{V_{B1} - V_N}{R_3} = \dfrac{0.7 + 15}{100 \times 10^3}$
$I_3 = 0.157$ mA
$I_{B1} = (0.69 - 0.157)(10^{-3}) = 0.533$ mA
$I_{C1} = I_1 - I_2$
$I_1 = \dfrac{V_P - V_{C1}}{R_1} = \dfrac{15 - 0.3}{2.7 \times 10^3}$
$I_1 = 5.45$ mA
$I_2 = \dfrac{V_{C1} - V_{B2}}{R_2} = \dfrac{0.3 + 2}{18 \times 10^3}$
$I_2 = 0.128$ mA
$I_{C1} = (5.45 - 0.128)(10^{-3}) = 5.32$ mA

The minimum $h_{FE}$ required to saturate $Q_1$ is

$\underline{h_{FE}} = \dfrac{I_{C1}}{I_{B1}} = \dfrac{5.32 \times 10^{-3}}{0.533 \times 10^{-3}} \simeq 10$

We now consider some practical techniques of FF triggering. Shown in Fig. 15.48a is a modified FF to which speedup capacitors $C_3$ and $C_4$ have been added. The additional circuitry involving $C_1$, $C_2$, $D_1$, $D_2$, $R_7$, and $R_8$ makes up the triggering network. When $C_1$ and $C_2$ are tied together, as shown by the dashed lines, there is a single input called $T$—for toggle input.

The basic requirement of a triggering network is to ensure that whenever a trigger pulse is applied, the FF changes state quickly and reliably. The trigger pulse must only initiate the transition; after this each transistor base must be effectively disconnected from the trigger source to allow the circuit to carry on independently.

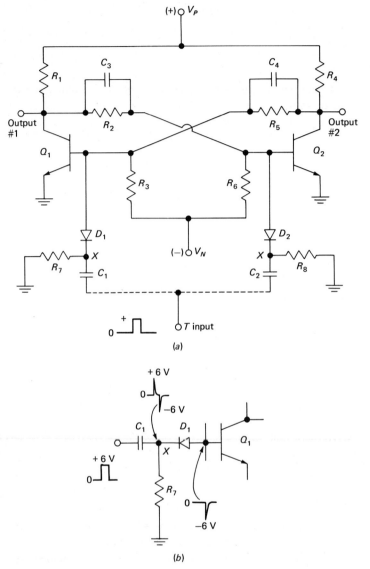

Fig. 15.48 *Modified FF with speedup capacitors ($C_3$ and $C_4$) and triggering circuitry.*

An equivalent circuit for the triggering network is shown in Fig. 15.48b. The positive-going pulse is differentiated by the $C_1$-$R_7$ combination (fast time constant). The output at point $X$ therefore consists of a positive 6-V spike when the input rises from 0 to 6 V and a negative 6-V spike when the input drops from 6 to 0 V. $D_1$ simply blocks the positive spike, allowing only a negative trigger to be developed at the base of $Q_1$. The same, of course, applies to $Q_2$—that is, both the base of $Q_1$ and the base of $Q_2$ receive negative-going spikes.

Now consider what happens if $Q_1$ is ON and $Q_2$ is OFF. At the instant that a positive trigger pulse is applied, neither base is affected, because $D_1$ and $D_2$ block the positive-going spikes. When the input pulse drops from 6 V to 0, though, a negative-going spike is generated by the differentiating action of $R_7$ and $C_1$ (as well as $R_8$ and $C_2$) and coupled by each diode to both bases. Since $Q_2$ is OFF, a negative spike simply keeps it OFF, but $Q_1$, which is now conducting, is driven in the OFF direction. As long as $Q_1$ begins to go OFF and the negative spikes disappear quickly enough, the circuit reverses state. If the spike does not decay soon enough, however, $Q_2$ is kept OFF, and the transition is delayed.

An improvement in FF triggering is possible using the modified circuit diagram of Fig. 15.49. Here $R_7$ and $R_8$ are returned to their respective collectors instead of ground as was done earlier. With $Q_1$ ON, its collector is close to ground

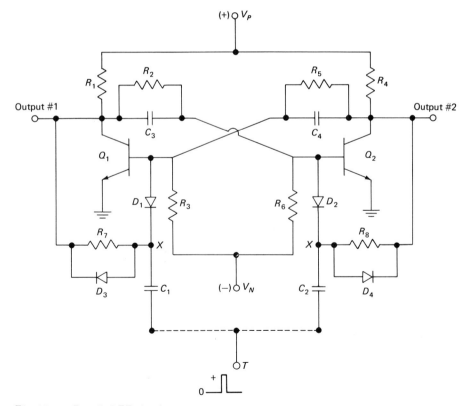

*Fig. 15.49  Practical FF circuit.*

anyway; therefore this half of the trigger circuit works in the same way as before. The addition of $D_3$ simply forces the positive-going spike developed at $X$ to decay very rapidly. The situation, however, is different for the OFF transistor (in this case $Q_2$). The collector of $Q_2$, which is positive, keeps point $X$ at a positive potential; this way, when the input trigger pulse drops, $X$ also drops, but from a positive level. This prevents $X$ from becoming negative. $D_2$ does not conduct, and the base of $Q_2$ is unaffected. This, of course, is precisely what is required. The negative-going trigger is applied only to $Q_1$, and as $Q_1$ turns OFF, there is nothing to prevent $Q_2$ from immediately responding by starting to conduct.

If the trigger input is a series of pulses, the waveforms of Fig. 15.50 apply. Note that a change of state occurs only when the trigger pulse is negative-going. This means that two "cycles" of the trigger input are needed to produce one "cycle" at either FF output. The FF is therefore a frequency divider, since two cycles in produce one cycle out.

In addition to bipolar transistor types, integrated circuit (IC) FFs are also quite common and simple to use. The basic principles are no different from those for the discrete circuits just discussed, but there are some variations designed to take advantage of IC manufacturing techniques. FETs also find application in FF design; some of their advantages over bipolars include greater noise immunity and lower power dissipation.

### Problem Set 15.11

15.11.1. How is each output of a FF related to the other?
15.11.2. What characteristic of the FF is responsible for its "memory"?
15.11.3. Estimate the ON base and collector voltages in the circuit of Fig. 15.46 if a germanium transistor is used.

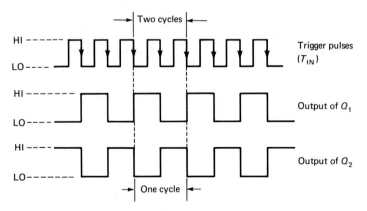

Fig. 15.50  Waveforms in the FF of Fig. 15.49.

15.11.4. Estimate the OFF base and collector voltages in the circuit of Fig. 15.46 if both transistors have zero OFF currents, zero ON voltages, and $V_P = 12$ V, $R_4 = R_1 = 2.2$ kΩ, $R_5 = R_2 = 15$ kΩ, $R_3 = R_6 = 100$ kΩ, and $V_N = -12$ V.
15.11.5. Repeat Prob. 15.11.4 by taking into account the ON voltages and assuming silicon transistors.
15.11.6. In Prob. 15.11.5, determine the ON collector and base currents as well as the minimum $h_{FE}$ required to saturate the ON transistor.
15.11.7. In the circuit of Fig. 15.49, indicate the function of each component below:

    a. $C_3$                 c. $C_1$
    b. $D_2$               d. $D_4$

15.11.8. What property of the FF makes it useful as a frequency divider?

## REFERENCES

1. Mulvey, J.: "Semiconductor Device Measurements," Tektronix, Inc., Beaverton, Ore., 1968.
2. "Silicon-controlled Rectifier Manual," General Electric, Syracuse, N.Y., 1967.
3. Phillips, A. B.: "Transistor Engineering and Introduction to Integrated Semiconductor Circuits," McGraw-Hill Book Company, New York, 1962.
4. Kiver, M. S.: "Transistors," McGraw-Hill Book Company, New York, 1962.
5. Amos, S. W.: "Principles of Transistor Circuits," Hayden Book Company, Inc., New York, 1965.
6. Cowles, L. G.: "Transistor Circuits and Applications," Prentice-Hall, Inc., Englewood Cliffs, N.J., 1968.
7. Joyce, M. V., and K. K. Clarke: "Transistor Circuit Analysis," Addison-Wesley Publishing Company, Inc., Reading, Mass., 1961.
8. Texas Instruments, Incorporated: "Transistor Circuit Design," McGraw-Hill Book Company, New York, 1963.
9. Angelo, E. J.: "Electronic Circuits," McGraw-Hill Book Company, New York, 1964.
10. Gibbons, J. F.: "Semiconductor Electronics," McGraw-Hill Book Company, New York, 1966.
11. "General Electric Transistor Manual," 7th ed., Syracuse, 1964.
12. "Motorola Switching Transistor Handbook," Phoenix, 1963.
13. Millman, J., and H. Tanb: "Pulse, Digital and Switching Waveforms," McGraw-Hill Book Company, New York, 1965.

14. Stern, L.: Silicon-controlled Rectifiers' New Applications, *Application Note* 141, Motorola Semiconductor Products Inc., Phoenix, 1966.
15. Zinder, D.: SCR Power Control Fundamentals, *Application Note* 240, Motorola Semiconductor Products Inc., Phoenix, 1968.
16. Zinder, D.: Thyristor Trigger Circuits for Power-control Applications, *Application Note* 227, Motorola Semiconductor Products Inc., Phoenix, 1966.
17. Smith, C.: SCR Switching Methods, *Bulletin* CA-66, Texas Instruments Incorporated, Dallas, 1969.
18. "General Electric Tunnel Diode Manual," 1st ed., Syracuse, 1961.
19. Kane, J.: The Field Effect Transistor in Digital Applications, *Application Note* 219, Motorola Semiconductor Products Inc., Phoenix, 1966.
20. Mitchell, B. B.: "Semiconductor Pulse Circuits with Experiments," Holt, Rinehart and Winston, Inc., New York, 1970.
21. Babb, D. S.: "Pulse Circuits: Switching and Shaping," Prentice-Hall, Inc.; Englewood Cliffs, N.J., 1964.

# appendix 1

**ABBREVIATIONS OF UNITS AND PREFIXES**

| Unit | Abbreviation |
|---|---|
| Centimeter | cm |
| Degree Celsius | °C |
| Degree Kelvin | °K |
| Coulomb | C |
| Ohm | Ω |
| Volt | V |
| Ampere | A |
| Watt | W |
| Hertz, cycles per second | Hz |
| Second | s |
| Henry | H |
| Farad | F |
| Decibel | dB |
| Radian | rad |
| Foot | ft |
| Joule | J |
| Pound | lb |
| Minute | min |
| Electron volt | eV |

## Prefixes

| | | | | | | |
|---|---|---|---|---|---|---|
| $10^{12}$ | tera- | T | | $10^{-6}$ | micro- | μ |
| $10^{9}$ | giga- | G | | $10^{-9}$ | nano- | n |
| $10^{6}$ | mega- | M | | $10^{-12}$ | pico- | p |
| $10^{3}$ | kilo- | k | | $10^{-15}$ | femto- | f |
| $10^{-3}$ | milli- | m | | $10^{-18}$ | atto- | a |

## Other

| | | | | |
|---|---|---|---|---|
| AC | alternating current | | emf | electromotive force |
| DC | direct current | | dc | direct-coupled |

# appendix 2

## DEFINITIONS OF SYMBOLS

Source symbols and conventions are shown in Fig. A.1. Current directions are defined in terms of conventional current flow. For example, in Fig. A.1d, current leaves the positive terminal of the DC voltage source, flows through $R$, and reenters at the negative terminal. There is a rise in voltage from point 2 to point 1, defined as $E_{12} = IR$. Similarly, current flowing through $R$ in Fig. A.1e produces a rise in voltage from point 2 to point 1, as shown. An example of correct symbol usage is given in Fig. A.2. This notation applies to current, voltage, and power in general; it has been adapted from IEEE (Institute of Electrical and Electronics Engineers) Standard No. 255.

$I_E$: DC component
$I_e$: RMS value of AC component
$I_m$: peak value of AC component
$i_e$: instantaneous value of AC component
$i_E$: instantaneous total (AC and DC) value
$I_{EM}$: peak total (AC and DC) value

*Note:*

$I_{EM} = I_E + I_{em}$
$I_e = 0.707 I_{em}$     for sinusoidal $i_e$

## 712  Solid State Electronic Circuits

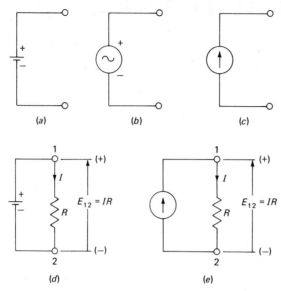

Fig. A.1  Source symbols and conventions: (a) *DC voltage source;* (b) *AC voltage source;* (c) *current source;* (d) *resistance across voltage source;* (e) *resistance across current source.*

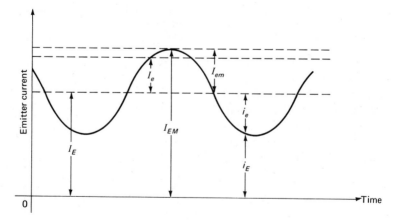

Fig. A.2  *Example of correct symbol usage.*

# appendix 3

## VOLTAGE AND CURRENT DIVIDERS

When voltage is applied to resistors in series, the total voltage will divide among the various resistors. In the circuit of Fig. A.3 the current is

$$I = \frac{E}{R_1 + R_2 + R_3 + \cdots + R_n}$$

This current is common to all resistors.

Let $E_1$ denote the voltage across $R_1$, $E_2$ the voltage across $R_2$, and so on. We then have

$$E_1 = IR_1 = \frac{ER_1}{R_1 + R_2 + R_3 + \cdots + R_n}$$

and in general,

$$E_n = IR_n = \frac{ER_n}{R_1 + R_2 + R_3 \cdots + R_n} \tag{A.1}$$

The voltage across any resistor $R_n$ is equal to the total applied voltage $E$ multiplied by $R_n$ and divided by the total resistance $R_1 + R_2 + R_3 + \cdots + R_n$.

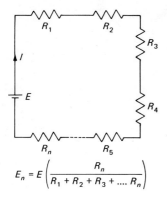

$$E_n = E\left(\frac{R_n}{R_1 + R_2 + R_3 + \ldots R_n}\right)$$

*Fig. A.3  Series circuit to study voltage division.*

Another useful relationship is the ratio of the voltage across any resistor $R_m$ to the voltage across any other resistor $R_n$, in the circuit of Fig. A.3. For example,

$$\frac{E_1}{E_2} = \frac{IR_1}{IR_2} = \frac{R_1}{R_2}$$

and in general,

$$\frac{E_m}{E_n} = \frac{R_m}{R_n} \tag{A.2}$$

Whenever current enters a junction, it will divide among the various available paths. In Fig. A.4, $I$ is assumed to enter junction $A$, while $I_1$ and $I_2$ are both assumed to be leaving. Direct application of Kirchhoff's current law and voltage law yields

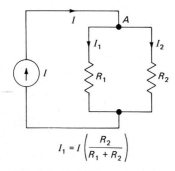

$$I_1 = I\left(\frac{R_2}{R_1 + R_2}\right)$$

*Fig. A.4  Parallel circuit to study current division.*

$$I = I_1 + I_2 \qquad (a)$$
$$I_1 R_1 = I_2 R_2 \qquad (b)$$

Manipulation of the above equations allows us to obtain some useful expressions. From Eq. (a):

$$I_1 = I - I_2$$

From Eq. (b):

$$(I - I_2)(R_1) = I_2 R_2$$
$$IR_1 = I_2 R_2 + I_2 R_1 = I_2(R_1 + R_2)$$
$$I_2 = I \frac{R_1}{R_1 + R_2} \qquad (A.3)$$

Following the same procedure, the student can show that

$$I_1 = I \frac{R_2}{R_1 + R_2} \qquad (A.3a)$$

Another useful relationship involving current division is now derived. The three resistors in Fig. A.5 are in parallel; the voltage between $A$ and $B$ is common to all three resistors. That is,

$$I_1 R_1 = I_2 R_2 = I_3 R_3$$
$$\frac{I_1}{I_2} = \frac{R_2}{R_1} \qquad \frac{I_3}{I_2} = \frac{R_2}{R_3}$$

and in general,

$$\frac{I_m}{I_n} = \frac{R_n}{R_m} = \frac{G_m}{G_n} \qquad (A.4)$$

The derivations also apply in the case of AC voltages and currents, provided impedances, instead of resistances, are used.

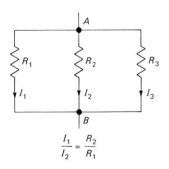

Fig. A.5 Parallel circuit to study current ratios.

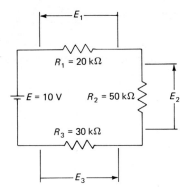

Fig. A.6 Circuit for Example A.1.

### Example A.1

Determine the voltage across each resistor in the circuit of Fig. A.6.

### Solution

$$E_1 = E \frac{R_1}{R_1 + R_2 + R_3} \tag{A.1}$$

$$E_1 = 10 \frac{20}{100} = 2 \text{ V}$$

$$\frac{E_2}{E_1} = \frac{R_2}{R_1} = \frac{50}{20} = 2.5$$

$$E_2 = 2.5 E_1 = 2.5(2) = 5 \text{ V}$$

$$\frac{E_3}{E_2} = \frac{R_3}{R_2} = \frac{30}{50} = 0.6$$

$$E_3 = 0.6 E_2 = 0.6(5) = 3 \text{ V}$$

The results are now checked:

$$E_1 + E_2 + E_3 = 2 + 5 + 3 = 10 \text{ V}$$

Hence Kirchhoff's voltage law is satisfied.

### Example A.2

Find $I_1$ and $I_2$ in the circuit of Fig. A.7.

### Solution

The total current supplied by the battery is

$$I = \frac{18}{2 + 5(20)/(5 + 20) + 3} = 2 \text{ A}$$

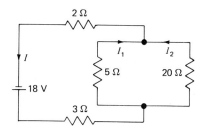

*Fig. A.7   Circuit for Example A.2.*

The 2-A current entering junction $A$ divides as follows:

$$I_1 = 2\,\frac{20}{25} = 1.6 \text{ A}$$

$$I_2 = 2\,\frac{5}{25} = 0.4 \text{ A}$$

# appendix 4

## DETERMINANTS

Determinants are useful for the solution of simultaneous equations. Only the essential features of determinants for solving such equations are given here.

A 2 × 2 (second-order) determinant takes on the following general form:

$$\begin{vmatrix} a_1 & b_1 \\ a_2 & b_2 \end{vmatrix}$$

The quantities $a_1$, $a_2$, $b_1$, and $b_2$ are called *elements*. The *value* of the determinant is

$$a_1 b_2 - a_2 b_1 \tag{A.5}$$

### Example A.3

Evaluate the following determinant:

$$\begin{vmatrix} 6 & 2 \\ 8 & 1 \end{vmatrix}$$

**Solution**

Using Eq. (A.5), we have

$6(1) - 8(2) = 6 - 16 = -10$

**720** Solid State Electronic Circuits

A 3 × 3 (third-order) determinant takes on the form

$$\begin{vmatrix} a_1 & b_1 & c_1 \\ a_2 & b_2 & c_2 \\ a_3 & b_3 & c_3 \end{vmatrix}$$

The above determinant may be evaluated using Cramer's rule, as follows:

$$a_1 \begin{vmatrix} b_2 & c_2 \\ b_3 & c_3 \end{vmatrix} - a_2 \begin{vmatrix} b_1 & c_1 \\ b_3 & c_3 \end{vmatrix} + a_3 \begin{vmatrix} b_1 & c_1 \\ b_2 & c_2 \end{vmatrix} \tag{A.6}$$

Here we have expanded the determinant about the first column. The elements $a_1$, $a_2$, and $a_3$ are called *cofactors*. The 2 × 2 determinants associated with each cofactor are called *minors*. The minor for any element is obtained by removing the row and column corresponding to that element. For example, to obtain the minor of $a_1$, we remove the first row ($a_1$, $b_1$, $c_1$) and the first column ($a_1$, $a_2$, $a_3$). The remaining determinant is the minor of $a_1$. The sign for each expansion is + if the sum of column number and row number is even, − if odd. For example, the minor of $a_2$ has a negative sign because $a_2$ is in the first column and second row (1 + 2 = 3—odd); on the other hand, $a_3$ is in the first column, third row (1 + 3 = 4—even), yielding a positive sign.

### Example A.4

Evaluate the following determinant:

$$\begin{vmatrix} 3 & 4 & 2 \\ 1 & 0 & 0 \\ 2 & 6 & 8 \end{vmatrix}$$

### Solution

Expanding about the first column, we have

$$3 \begin{vmatrix} 0 & 0 \\ 6 & 8 \end{vmatrix} - 1 \begin{vmatrix} 4 & 2 \\ 6 & 8 \end{vmatrix} + 2 \begin{vmatrix} 4 & 2 \\ 0 & 0 \end{vmatrix}$$
$$= 3[(0)(8) - (6)(0)] - 1[(4)(8) - (6)(2)] + 2[(4)(0) - (0)(2)]$$
$$= 3(0) - 1(20) + 2(0) = -20$$

It is not always desirable to expand about the first column; sometimes it is more advantageous to expand about another row or column. In Example A.4, the second row involves two elements whose value is 0; therefore a faster solution is to expand about this row, as follows:

$$-1 \begin{vmatrix} 4 & 2 \\ 6 & 8 \end{vmatrix} + 0 \begin{vmatrix} 3 & 2 \\ 2 & 8 \end{vmatrix} - 0 \begin{vmatrix} 3 & 4 \\ 2 & 6 \end{vmatrix}$$
$$= -1(32 - 12) = -20$$

The two minors whose cofactor is 0 are not usually written; they were included above only to illustrate the procedure for forming each minor.

The procedure developed above can be used to solve any order determinant. We now proceed to apply determinants to the solution of linear simultaneous equations. Suppose we have a set of equations of the following general form:

$$a_1 x + b_1 y + c_1 z = d_1$$
$$a_2 x + b_2 y + c_2 z = d_2$$
$$a_3 x + b_3 y + c_3 z = d_3$$

The $a$, $b$, $c$, and $d$ coefficients are numbers, while $x$, $y$, and $z$ are the unknowns we must find. To solve for $x$, we proceed as follows:

$$x = \frac{\begin{vmatrix} d_1 & b_1 & c_1 \\ d_2 & b_2 & c_2 \\ d_3 & b_3 & c_3 \end{vmatrix}}{\Delta} \tag{A.7}$$

where

$$\Delta = \begin{vmatrix} a_1 & b_1 & c_1 \\ a_2 & b_2 & c_2 \\ a_3 & b_3 & c_3 \end{vmatrix} \tag{A.8}$$

Note that the denominator $\Delta$ is formed by taking the coefficients of each term from the left side of the equations. The numerator determinant is similarly constructed, except that in place of the coefficients for $x$, the constants from the right side of the equation are substituted.

## Example A.5

Solve for the unknowns using determinants:

$$3x + 4y - z = 4 \tag{a}$$
$$-x - 3y + 10z = 34 \tag{b}$$
$$14x + 2y = 4 \tag{c}$$

### Solution

To solve for $x$, the following determinant is formed:

$$x = \frac{\begin{vmatrix} 4 & 4 & -1 \\ 34 & -3 & 10 \\ 4 & 2 & 0 \end{vmatrix}}{\Delta}$$

where

$$\Delta = \begin{vmatrix} 3 & 4 & -1 \\ -1 & -3 & 10 \\ 14 & 2 & 0 \end{vmatrix}$$

We now expand the numerator and denominator determinants about the third row to take advantage of the zero element:

$$x = \frac{4\begin{vmatrix} 4 & -1 \\ -3 & 10 \end{vmatrix} - 2\begin{vmatrix} 4 & -1 \\ 34 & 10 \end{vmatrix}}{14\begin{vmatrix} 4 & -1 \\ -3 & 10 \end{vmatrix} - 2\begin{vmatrix} 3 & -1 \\ -1 & 10 \end{vmatrix}}$$

$$x = \frac{4(40-3) - 2(40+34)}{14(40-3) - 2(30-1)}$$

$$x = \frac{4(37) - 2(74)}{14(37) - 2(29)} = \frac{0}{460} = 0$$

The remaining unknowns are similarly determined, yielding

$$y = \frac{\begin{vmatrix} 3 & 4 & -1 \\ -1 & 34 & 10 \\ 14 & 4 & 0 \end{vmatrix}}{\Delta} = \frac{920}{460} = 2$$

$$z = \frac{\begin{vmatrix} 3 & 4 & 4 \\ -1 & -3 & 34 \\ 14 & 2 & 4 \end{vmatrix}}{\Delta} = \frac{1,840}{460} = 4$$

# appendix 5

## LOOP AND NODE ANALYSIS

To solve for the various currents and voltages in circuits involving several sources, it is advisable to use a systematic approach, such as the writing of loop or node equations. Loop equations are written by applying Kirchhoff's voltage law, node[1] equations by applying Kirchhoff's current law at each principal node. At times the use of both loop and node equations may result in simpler algebra; the fundamental requirement is that one must write as many independent equations as there are unknowns. Two equations are independent of each other if one of them takes into account at least one element which is not taken into account by the other.

In the circuit of Fig. A.8a there are three different currents. To solve for these currents, three independent equations must be written. These may be loop equations or node equations or a combination of both. In this case we can write

Node equation at junction $A$: $\qquad I_1 + I_2 = I_3 \qquad$ (a)
Loop equation around $BDAB$: $\quad -26 + 8I_1 + 6I_3 = 0 \qquad$ (b)
Loop equation around $BACB$: $\quad -6I_3 - 4I_2 + 26 = 0 \qquad$ (c)

These equations are now solved using determinants:

---

[1] A node is a point where two or more components are connected.

$$I_1 + I_2 - I_3 = 0 \qquad (a)$$
$$8I_1 + 0I_2 + 6I_3 = 26 \qquad (b)$$
$$0I_1 - 4I_2 - 6I_3 = -26 \qquad (c)$$

$$I_1 = \frac{\begin{vmatrix} 0 & 1 & -1 \\ 26 & 0 & 6 \\ -26 & -4 & -6 \end{vmatrix}}{\begin{vmatrix} 1 & 1 & -1 \\ 8 & 0 & 6 \\ 0 & -4 & -6 \end{vmatrix}} = \frac{0 \begin{vmatrix} 0 & 6 \\ -4 & -6 \end{vmatrix} - 26 \begin{vmatrix} 1 & -1 \\ -4 & -6 \end{vmatrix} - 26 \begin{vmatrix} 1 & -1 \\ 0 & 6 \end{vmatrix}}{1 \begin{vmatrix} 0 & 6 \\ -4 & -6 \end{vmatrix} - 8 \begin{vmatrix} 1 & -1 \\ -4 & -6 \end{vmatrix} + 0 \begin{vmatrix} 1 & -1 \\ 0 & 6 \end{vmatrix}}$$

$$I_1 = \frac{-26(-6-4) - 26(6)}{1(24) - 8(-6-4)} = \frac{104}{104} = 1 \text{ A}$$

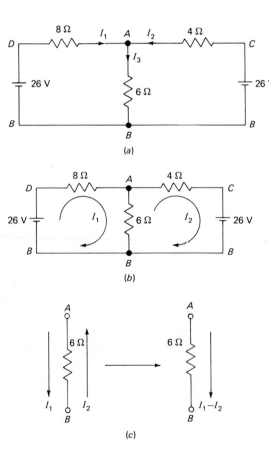

Fig. A.8 Analysis of DC circuit.

$$I_2 = \frac{\begin{vmatrix} 1 & 0 & -1 \\ 8 & 26 & 6 \\ 0 & -26 & -6 \end{vmatrix}}{104} = \frac{\begin{vmatrix} 26 & 6 \\ -26 & -6 \end{vmatrix} 1 \begin{vmatrix} & 6 \\ -8 & \\ -26 & -6 \end{vmatrix} \begin{vmatrix} 0 & -6 \\ -26 & -6 \end{vmatrix}}{104}$$

$$I_2 = \frac{1(-156+156) - 8(-26)}{104} = \frac{208}{104} = 2 \text{ A}$$

$$I_3 = \frac{\begin{vmatrix} 1 & 1 & 0 \\ 8 & 0 & 26 \\ 0 & -4 & -26 \end{vmatrix}}{104} = \frac{\begin{vmatrix} 0 & 26 \\ -4 & -26 \end{vmatrix} 1 \begin{vmatrix} 0 & 26 \\ -8 & \\ -4 & -26 \end{vmatrix} \begin{vmatrix} 1 & 0 \\ -4 & -26 \end{vmatrix}}{104}$$

$$I_3 = \frac{1(104) - 8(-26)}{104} = \frac{312}{104} = 3 \text{ A}$$

Another approach involves the writing of only two loop equations; here loop currents instead of currents through individual elements are used. In the circuit of Fig. A.8b currents $I_1$ and $I_2$ are assumed in loops $BDAB$ and $ACBA$, respectively. The net current through the 6-Ω resistor is the algebraic sum of $I_1$ and $I_2$, since this resistor is common to both loops, as shown in Fig. A.8c. The direction of both currents was assumed clockwise for convenience. Writing Kirchhoff's voltage law around both loops yields

$$-26 + 8I_1 + 6(I_1 - I_2) = 0 \qquad (d)$$
$$26 + 6(I_2 - I_1) + 4I_2 = 0 \qquad (e)$$

Rearranging, we get

$$14I_1 - 6I_2 = 26 \qquad (d)$$
$$-6I_1 + 10I_2 = -26 \qquad (e)$$

From Eq. (d):

$$I_1 = \frac{26 + 6I_2}{14}$$

From Eq. (e):

$$I_1 = \frac{-26 - 10I_2}{-6}$$

Therefore

$$\frac{26 + 6I_2}{14} = \frac{26 + 10I_2}{6}$$
$$156 + 36I_2 = 364 + 140I_2$$

$$104I_2 = -208$$
$$I_2 = -2 \text{ A}$$
$$I_1 = \frac{26 - 12}{14}$$
$$I_1 = 1 \text{ A}$$

The negative value for $I_2$ indicates that $I_2$ is in the opposite direction to that originally assumed. After the direction of $I_2$ has been corrected, it is found that both $I_1$ and $I_2$ flow through the 6-$\Omega$ resistor in the same direction; therefore the net current through this resistor is

$$I = 1 + 2 = 3 \text{ A}$$

The circuit of Fig. A.9a is one in which node analysis can be used. Generally, the unknown is a node voltage; in this case there is only one node, and the node voltage is measured between point $A$ and ground. This voltage is designated as $E_A$, while ground is just a convenient reference point. Before writing any equations, note that there are three currents in this circuit. These currents flow through resistors; for example, $I_1$ flows through $R_1$, $I_2$ flows through $R_2$, and so on. According to Ohm's law, the current through any resistor is given by the voltage across that resistor divided by the resistance. The bottom end of $R_1$ is at $+12$ V above ground; the top end is at $E_A$ above ground. The voltage across $R_1$ is simply the difference in potential between its two ends.

Since the direction of $I_1$ is assumed to be from $A$ to $B$ and since current through a resistor flows from plus to minus, we must assume point $A$ to be at a higher potential than point $B$ in writing our equation, even though the ultimate result might show this to be wrong. Thus the voltage across $R_1$ is assumed as in Fig. A.9b, and $I_1$ is

$$I_1 = \frac{E_A - E_B}{R_1} = \frac{E_A - 12}{2}$$

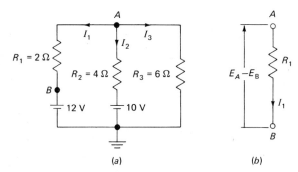

Fig. A.9  *Circuits for DC analysis.*

Writing a current equation at junction $A$, we get

$I_1 + I_2 + I_3 = 0$

or

$$\frac{E_A - 12}{2} + \frac{E_A - 10}{4} + \frac{E_A}{6} = 0$$

$$\frac{E_A}{2} - 6 + \frac{E_A}{4} - 2.5 + \frac{E_A}{6} = 0$$

$$E_A \frac{6 + 3 + 2}{12} = 8.5$$

$$E_A = 9.27 \text{ V}$$

The various currents can now be found:

$$I_1 = \frac{9.27 - 12}{2} = -1.365 \text{ A}$$

$$I_2 = \frac{9.27 - 10}{4} = -0.183 \text{ A}$$

$$I_3 = \frac{9.27}{6} = 1.548 \text{ A}$$

# appendix 6

## COMPLEX ALGEBRA

Complex algebra utilizes the concept of a $j$ operator to denote a phase angle of 90°. This is useful to describe AC voltages, currents, and impedances. A complex number can be written in the following general form:

$$A = a + jb \tag{A.9}$$

where $a$ is the real part and $b$ the imaginary part. The operator $j$ states that $b$ is shifted 90° relative to $a$. This is called the *rectangular form* because it describes the complex number in terms of a rectangular coordinate system. Such a system is shown in Fig. A.10. Here any quantity along the positive real axis has a phase of 0°; along the positive imaginary axis the phase is 90°, along the negative real axis it is 180°, and so on. Positive angles are always measured in a counterclockwise direction from the positive real axis.

Several complex numbers are graphically displayed in Fig. A.11. The rectangular form of these numbers is

a. $3 + j0$
b. $0 + j2$
c. $3 + j2$
d. $-2 + j0$
e. $-2 + j2$
f. $0 - j3$
g. $-3 - j2$
h. $2 - j2$

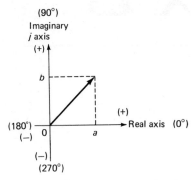

Fig. A.10 Graphic representation of a complex number.

Complex numbers may be added or subtracted by following the normal rules of algebra.

**Example A.6**

Perform the indicated operations:

a. $(3 + j8) + (-2 + j4)$
b. $(12 - j4) - (5 + j15)$

**Solution**

a. $3 - 2 + j8 + j4 = 1 + j12$
b. $12 - 5 - j4 - j15 = 7 - j19$

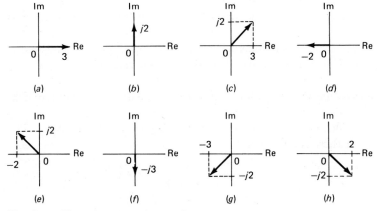

Fig. A.11 Examples of complex numbers.

To multiply or divide complex numbers, the rules of algebra are followed. Since the operator $j$ represents a phase shift of $90°$, then $jj$ must represent a phase shift of $180°$, which is equivalent to $-1$; therefore $j^2 = -1$. Similarly, $j^3 = -j$, while $j^4 = +1$.

**Example A.7**

Perform the indicated operations and simplify:
a. $(3 + j8)(-2 + j4)$
b. $(12 - j4)(-5 - j15)$

**Solution**

a. $-6 - j16 + j12 + j^2 32 = -6 - j4 - 32 = -38 - j4$
b. $-60 + j20 - j180 + j^2 60 = -120 - j160$

When dealing with ratios of complex numbers, the concept of the conjugate is useful. The conjugate of any complex number is obtained by changing the sign of its imaginary part. For example,

$$\text{Conjugate of } a + jb = (a + jb)^* = a - jb \tag{A.10}$$
$(3 + j5)^* = 3 - j5$

**Example A.8**

Divide $3 + j8$ by $-2 + j4$.

**Solution**

The quantity

$$\frac{3 + j8}{-2 + j4}$$

can be simplified by multiplying the numerator and denominator by the denominator's conjugate:

$$\frac{3 + j8}{-2 + j4} \cdot \frac{-2 - j4}{-2 - j4} = \frac{-6 - j16 - j12 + 32}{4 - j8 + j8 + 16}$$
$$= \frac{26 - j28}{20} = \frac{26}{20} - j\frac{28}{20}$$
$$= 1.3 - j1.4$$

### Example A.9

Simplify the following expression:

$$\frac{(3 + j5 - 14 + j18)(4 + j5)}{(6 - j2)(j4)}$$

### Solution

$$\frac{(-11 + j23)(4 + j5)}{j24 + 8}$$

$$= \frac{-44 + j92 - j55 - 115}{8 + j24} = \frac{-159 + j37}{8 + j24}$$

$$= \frac{-159 + j37}{8 + j24} \frac{8 - j24}{8 - j24}$$

$$= \frac{-1{,}272 + j296 + j3{,}816 + 88}{64 + 576} = \frac{-1{,}184 + j4{,}112}{640}$$

$$= -1.85 + j6.42$$

An alternative way of expressing complex numbers is the polar form. Here the magnitude and phase are given, as shown in Fig. A.12. The magnitude is

$$A = \sqrt{a^2 + b^2} \tag{A.11}$$

The phase is

$$\theta = \tan^{-1} \frac{b}{a} \tag{A.12}$$

The number is usually written $A\,\underline{/\theta}$. To convert from the polar to the rectangular form, the *phasor* $A$ in Fig. A.12 is broken down into rectangular (real) and vertical (imaginary) components, as follows:

$$A\,\underline{/\theta} = A(\cos\theta + j\sin\theta) \tag{A.13}$$

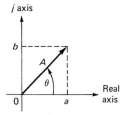

Fig. A.12 Polar form of complex number.

## Example A.10

Given the following complex numbers, convert to the polar form:

a. $3 + j4$
b. $8 - j6$
c. $0 + j10$

### Solution

a. $3 + j4 = \sqrt{3^2 + 4^2} \;\underline{/\tan^{-1} \dfrac{4}{3}}$
$= \sqrt{25}\;\underline{/\tan^{-1} 1.333} = 5\underline{/53°}$

b. $8 - j6 = \sqrt{64 + 36}\;\underline{/\tan^{-1} \dfrac{-6}{8}}$
$= 10\;\underline{/\tan^{-1} \dfrac{-0.75}{1}} = 10\;\underline{/-37°}$

c. $0 + j10 = 10\;\underline{/90°}$

## Example A.11

Convert the following complex numbers to the rectangular form:

a. $5\;\underline{/53°}$
b. $10\;\underline{/-37°}$
c. $10\;\underline{/90°}$

### Solution

a. $5\;\underline{/53°} = 5(\cos 53° + j \sin 53°)$
$= 5(0.6 + j0.8) = 3 + j4$
b. $10\;\underline{/-37°} = 10[\cos(-37°) + j \sin(-37°)]$
$= 10(0.8 - j0.6) = 8 - j6$
c. $10\;\underline{/90°} = 10(\cos 90° + j \sin 90°)$
$= 10(0 + j1) = 0 + j10$

The polar form is most useful when multiplying or dividing complex numbers. This is done as follows:

$$(A_1\;\underline{/\theta_1})(A_2\;\underline{/\theta_2}) = A_1 A_2\;\underline{/\theta_1 + \theta_2} \tag{A.14}$$

$$\dfrac{A_1\;\underline{/\theta_1}}{A_2\;\underline{/\theta_2}} = \dfrac{A_1}{A_2}\;\underline{/\theta_1 - \theta_2} \tag{A.15}$$

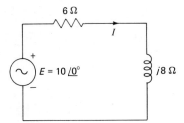

Fig. A.13  Circuit for Example A.23.

## Example A.12

Simplify the following:

$$\frac{(3\,/30°)(4\,/-45°)}{2\,/90°}$$

**Solution**

$$\frac{3(4)}{2}/30° - 45° - 90° = 6\,/-105°$$

## Example A.13

Determine $I$ and its phase relationship with respect to the applied voltage $E$ in the circuit of Fig. A.13.

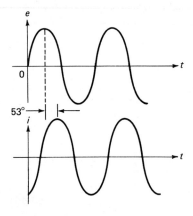

Fig. A.14  Voltage and current waveforms for the circuit of Fig. A.22.

## Solution

The impedance seen by the generator is

$Z = 6 + j8$

$Z = \sqrt{36 + 64} \; \underline{/\tan^{-1} \frac{8}{6}} = 10 \; \underline{/53°}$

The current is

$I = \dfrac{E}{Z} = \dfrac{10 \; \underline{/0°}}{10 \; \underline{/53°}}$

$I = 1 \; \underline{/-53°}$

Thus $I = 1$ A, lagging $E$ by 53°. The time graphs for both voltage and current are shown in Fig. A.14.

# appendix 7

## THEVENIN AND NORTON EQUIVALENT CIRCUITS

Any source of electric energy, DC or AC, may be simulated with an equivalent circuit of the type in Fig. A.15a or b. The Thevenin circuit of Fig. A.15a involves a constant voltage source and series internal resistance $R_T$. The Norton equivalent circuit of Fig. A.15b involves a constant current source and shunt internal resistance $R_N$. The load is connected between terminals $A$ and $B$. Both circuits of Fig. A.15 describe the same source; hence they behave electrically in the same way under all conditions. In general, we can write for the Thevenin circuit

$$V_L = E_T \frac{R_L}{R_T + R_L} \tag{A.16}$$

$$I_L = \frac{E_T}{R_T + R_L} \tag{A.17}$$

For the Norton circuit,

$$I_L = I_N \frac{R_N}{R_N + R_L} \tag{A.18}$$

$$V_L = I_L R_L = \frac{I_N R_N R_L}{R_N + R_L} \tag{A.19}$$

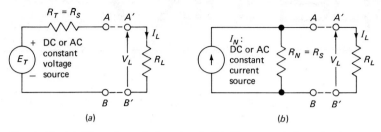

*Fig. A.15* (a) Thevenin equivalent circuit; (b) Norton equivalent circuit.

For the two circuits to be equivalent, they must behave electrically the same way under all operating conditions. When the load is an open circuit ($R_L = \infty$), the Thevenin circuit has zero current and hence no voltage drop across $R_T$; therefore all $E_T$ is developed across the load:

Open-circuit Thevenin voltage: $E_T$ (A.20)

At the same time, in the Norton circuit, all $I_N$ flows through $R_N$, yielding

Open-circuit Norton voltage: $I_N R_N$ (A.21)

When the load is short-circuited, the Thevenin circuit current is

Short-circuit Thevenin current: $\dfrac{E_T}{R_T}$ (A.22)

At the same time, in the Norton circuit, all $I_N$ bypasses $R_N$ in favor of the short circuit, yielding

Short-circuit Norton current: $I_N$ (A.23)

Equating Eq. (A.20) with Eq. (A.21) and Eq. (A.22) with Eq. (A.23), we have

$E_T = I_N R_N$ (A.24)
$E_T = I_N R_T$ (A.25)

from which we may conclude that

$R_N = R_T$ (A.26)

The Thevenin voltage is therefore the open-circuit voltage, while the Norton current is the short-circuit current.

The Norton or Thevenin resistance is obtained by removing the load, reducing the voltage or current source to zero, and looking into terminals $AB$. Voltage sources are reduced to zero by replacing them with a short circuit; current sources are reduced to zero by replacing them with an open circuit. Since

$R_N$ and $R_T$ are always the same (both represent the internal, or output, resistance of the source), from here on we will use $R_S$ to denote either $R_N$ or $R_T$.

### Example A.14

Find both Thevenin and Norton equivalent circuits for the circuit shown in Fig. A.16a.

### Solution

$E_T$ is the voltage between terminals $AB$ with no load. In this case no current flows through the 8-$\Omega$ resistor; hence no voltage is dropped across it. It follows that $V_{AB}$ is the same as the voltage across the 6-$\Omega$ resistor (see Fig. A.16b), that is,

$$E = V_{AB} = \frac{15(6)}{6+3} = 10 \text{ V}$$

$R_S$ is found by shorting the 15-V source and looking into terminals $AB$, as shown in Fig. A.16c:

$$R_{AB} = R_S = 8 + \frac{3(6)}{3+6} = 10 \text{ }\Omega$$

$I_N$ is the current through a short circuit connected from $A$ to $B$. Using the circuit of Fig. A.16d, we have

$$I = \frac{15}{3 + [6(8)/(6+8)]} = 2.33 \text{ A}$$

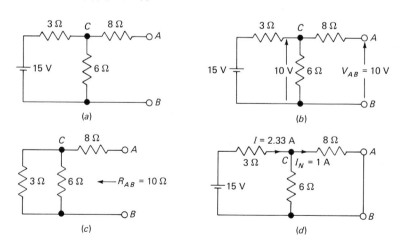

Fig. A.16  (a) Circuit for Example A.14; (b) obtaining the Thevenin voltage; (c) obtaining the internal resistance; (d) obtaining the Norton current.

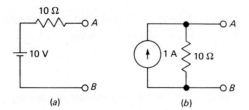

Fig. A.17 Thevenin (a) and Norton (b) equivalent circuits for the circuit of Fig. A.16a.

which divides between the two paths available at junction $C$. Hence the current through the short circuit is

$$I_N = \frac{2.33(6)}{6+8} = 1 \text{ A}$$

$I_N$ can also be calculated as the ratio of $E_T$ to $R_S$, yielding $I_N = 10/10 = 1$ A. Both equivalent circuits are shown in Fig. A.17.

# appendix 8

## POWER EXCHANGE BETWEEN SOURCE AND LOAD

Whether a source provides an approximately constant voltage (low internal resistance) or an approximately constant current (high internal resistance) to its load, energy is being transferred between source and load, and the rate at which this energy transfer occurs is the electric power. In the Thevenin equivalent circuit of Fig. A.15a the power delivered to the load is

$$P_L = \frac{V_L^2}{R_L} \tag{A.27}$$

$$P_L = \frac{E_T^2 R_L}{(R_S + R_L)^2} \tag{A.28}$$

The value of $R_L$ that will result in maximum power being delivered from the source to the load can be determined by differentiating $P_L$ with respect to $R_L$, setting the result equal to zero, and solving for $R_L$:

$$P_L = \frac{E_T^2 R_L}{(R_S + R_L)^2} \tag{A.28}$$

$$\frac{d(P_L)}{d(R_L)} = \frac{(R_S + R_L)^2 E_T^2 - (E_T^2 R_L)(2)(R_S + R_L)}{(R_S + R_L)^4} = 0$$

$$(R_S + R_L)^2(E_T^2) - (E_T^2 R_L)(2)(R_S + R_L) = 0$$

$$R_S + R_L = 2R_L$$

$$R_S = R_L \tag{A.29}$$

The student can verify that this is actually a maximum by substituting $R_S = R_L$ in the expression for $d^2P_L/dR_L^2$ and noting the result to be negative. Thus if we choose $R_L = R_S$, power to the load is maximized. The maximum power is obtained as follows:

$$P_L = \frac{E_T^2 R_L}{(R_S + R_L)^2} \tag{A.28}$$

If $R_S = R_L$:

$$P_L = \frac{E_T^2 R_S}{(R_S + R_S)^2}$$

$$P_{L(\max)} = \frac{E_T^2}{4R_S} \tag{A.30}$$

Maximum power is delivered to a load if the load is matched to the source, that is, if $R_S = R_L$. Under these matched conditions, the load voltage is

$$V_L = E_T \frac{R_L}{2R_L} = 1/2\ E_T \tag{A.31}$$

The conditions for maximum power transfer are the same whether one chooses a Thevenin or Norton equivalent circuit. The maximum power for a Norton circuit is

$$P_{L(\max)} = \frac{E_T^2}{4R_S} \tag{A.30}$$

But $E_T = I_N R_S$; therefore

$$P_{L(\max)} = \frac{I_N^2 R_S}{4} \tag{A.32}$$

### Example A.15

When the output terminals of a given network are short-circuited, the current through the short circuit is 2 A. With no load, 20 V is measured across the same output terminals. Determine the maximum available power from the source.

### Solution

The open-circuit voltage is the Thevenin voltage; therefore $E_T = 20$ V. If terminals $AB$ are short-circuited, the current is

$$I = \frac{20}{R_S} = 2\ \text{A}$$

$$R_S = \frac{20}{2} = 10\ \Omega$$

The maximum available power is

$$P_{L(max)} = \frac{E_T^2}{4R_S} = \frac{20^2}{4(10)} = 10 \text{ W} \quad (A.30)$$

With AC circuits, the source and load may involve reactive components so that, in general, an internal impedance $Z_S$ and a load $Z_L$ must be considered. In the circuit of Fig. A.18, we can write

$$I = \frac{E_T}{Z_S + Z_L} \quad (A.33)$$

In general both impedances may have resistive as well as reactive components, yielding

$$Z_S = R_S + jX_S \quad (A.34)$$
$$Z_L = R_L + jX_L \quad (A.35)$$

Only the resistive portion of $Z_L$ absorbs power so that

$$P_L = |I|^2 R_L$$
$$P_L = \left|\frac{E_T}{Z_S + Z_L}\right|^2 (R_L)$$
$$P_L = \frac{|E_T|^2 R_L}{|R_S + jX_S + R_L + jX_L|^2}$$

Now if $jX_S = -jX_L$, the denominator is minimized, therefore maximizing $P_L$. The load power expression now is

$$\frac{E_T^2 R_L}{(R_S + R_L)^2}$$

which was maximized earlier by making $R_S = R_L$:

$$P_{L(max)} = \frac{E_T^2}{4R_S} \quad (A.30)$$

To achieve maximum power, we must therefore have

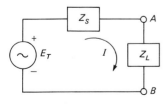

Fig. A.18  AC source and load.

$$R_S + jX_S = R_L - jX_L \tag{A.36}$$

or

$$Z_S = Z_L^* \tag{A.36a}$$

The above condition is appropriately called a conjugate power match. The same requirement, of course, applies to a Norton circuit.

**Example A.16**

Determine the load components in the circuit of Fig. A.19a for a conjugate match.

**Solution**

For a conjugate match, we require

$$Z_S = Z_L^* \tag{A.36a}$$

Since $Z_S = 1/Y_S$ and $Z_L = 1/Y_L$, we can also write

$$Y_S = Y_L^* \tag{A.36b}$$

which is more convenient to apply in parallel circuits. For the source

$$Y_S = \frac{1}{R_S} + j\omega C_S \tag{A.37}$$

$$Y_S = 0.01 + j(1{,}000 \times 10^{-5}) = 0.01 + j0.01 \quad \text{mho}$$

If $Y_S = Y_L^*$,

$$Y_L = 0.01 - j0.01 \quad \text{mho}$$

This involves a shunt conductance of 0.01 mho, from which $R_L = 1/0.01 =$

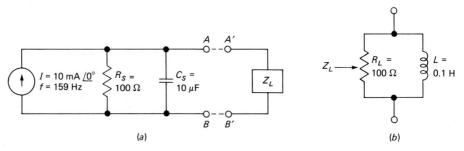

Fig. A.19 Circuits for Example A.16.

100 Ω. There is also a shunt susceptance of $-j0.01$ mho, which is equivalent to a reactance of $1/(-j0.01) = j100$ Ω. This requires an inductance $L$ as follows:

$$X_L = j100 = j2\pi f L$$

$$L = \frac{100}{2\pi(159)} = 0.1 \text{ H}$$

The load required for a conjugate match is shown in Fig. A.19b.

# appendix 9

## SUPERPOSITION

Superposition is a labor-saving tool that can be used when the relationship between physical variables is linear. A linear relationship means that the magnitudes of the physical variables are proportional to each other. More rigorously, if $y = f(x)$, the system is linear if

$$f(x_1) + f(x_2) = f(x_1 + x_2) \tag{A.38}$$

This statement will be illustrated with a numerical example. Take, for instance,

$$y = f(x) = 2x \tag{a}$$

When $x = 2$, $f(2) = 4$, and when $x = 6$, $f(6) = 12$. This yields

$$f(2) + f(6) = 4 + 12 = 16$$

If we compute $f(2 + 6)$, that is, $f(8)$, we get

$$f(2 + 6) = f(8) = 2(8) = 16$$

Since $f(2) + f(6) = f(2 + 6)$, the system described by the equation $y = 2x$ is linear. An example of a nonlinear system is given by

$$y = f(x) = x^2 \tag{b}$$

Repeating the previous procedure yields

$f(2) = 4$
$f(6) = 36$
$f(2 + 6) = 64$
$f(2) + f(6) = 4 + 36 = 40$

Since $f(2) + f(6) \neq f(2 + 6)$, the system is nonlinear.

Physically, when the value of a variable is doubled and the value of a related variable also doubles, the two variables are linearly related. Ohm's law is perhaps the simplest linear relationship in electricity; if we double the voltage, the current is also doubled, and so on. The power dissipated by a resistor, however, is not linearly related to the current, since doubling the current produces 4 times the power.

When a linear electric system is subjected to several excitations simultaneously, the total result can be obtained by superimposing the individual results for each excitation taken separately.

### Example A.17

For the circuit shown in Fig. A.20, find $I_x$, using superposition.

### Solution

Since there are two sources, we determine the contribution to $I_x$ from each source separately. The contribution from the 22-V supply is obtained in Fig. A.20b, where the current source has been removed:

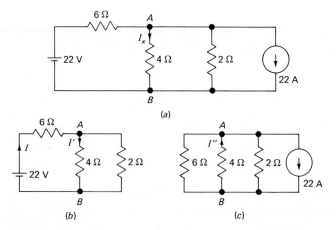

Fig. A.20 (a) *Circuit for Example A.17;* (b) *obtaining* $I'$ *by disregarding the current source;* (c) *obtaining* $I''$ *by disregarding the voltage source.*

$$I' = I\frac{2}{4+2} = \frac{I}{3}$$

$$I = \frac{22}{6 + [4(2)/(4+2)]} = \frac{22(3)}{22} = 3 \text{ A}$$

$$I' = \frac{3}{3} = 1 \text{ A}$$

To determine the contribution to $I_x$ from the current source only, the voltage source is replaced with a short circuit, as shown in Fig. A.20c. In this case we have

$$I'' = 22\frac{6\|2}{(6\|2) + 4}$$

$$I'' = 22\frac{12}{12 + 32} = 6 \text{ A}$$

Since $I''$ flows through the 4-Ω resistor in a direction opposite to that of $I'$, the total current $I_x$ when both sources are operating normally is

$$I_x = I' - I'' = -5 \text{ A}$$

that is, 5 A in the direction of $I''$.

# appendix 10

## SEMICONDUCTOR DATA SHEETS

| Number | Manufacturer | Type |
|---|---|---|
| 2N2137–46 | Motorola Semiconductor Products Inc. | Ge PNP |
| 2N2386 | Texas Instruments, Incorporated | P-Channel Si JFET |
| 2N3199–201 | Crystalonics | Si PNP |
| 2N3392–4 | General Electric | Si NPN |
| 2N3823 | Texas Instruments, Incorporated | N-Channel Si JFET |
| 2N3796–7 | Motorola Semiconductor Products Inc. | N-Channel Si MOSFET |

# MOTOROLA Semiconductors
BOX 20912 • PHOENIX, ARIZONA 85036

## 2N2137 thru 2N2141
## 2N2142 thru 2N2146

**3-AMP POWER TRANSISTORS**
GERMANIUM PNP
**30-90 VOLTS**
**62.5 WATTS**
OCTOBER, 1961
DS 3033

Actual Size

**RECOMMENDED for DRIVER APPLICATIONS REQUIRING $h_{FE}$ of 30-60 and 50-100 at $I_C = 0.5$ AMP**

- Low Collector Cutoff Current — 50 $\mu$A Max at $V_{CB} = 2V$, — 5 mA Max at $V_{CB(MAX)}$ @ +71°C
- Two-to-One Gain Ranges
- Low Saturation Voltage — 0.5 V Max at 2 Amps
- 125°C Stabilization Bake for 100 Hours
- Designed to Meet MIL-S-19500
- With or Without Solder Lugs

  Safe Operating Areas Specified
  Peak Power Derating Curves

## ABSOLUTE MAXIMUM RATINGS

| Characteristic | Symbol | 2N2137 2N2142 | 2N2138 2N2143 | 2N2139 2N2144 | 2N2140 2N2145 | 2N2141 2N2146 | Unit |
|---|---|---|---|---|---|---|---|
| Collector-Base Voltage | $BV_{CBO}$ | 30 | 45 | 60 | 75 | 90 | Volts |
| Collector-Emitter Voltage | $BV_{CES}$ | 30 | 45 | 60 | 75 | 90 | Volts |
| Collector-Emitter Voltage | $BV_{CEO}$ | 20 | 30 | 45 | 60 | 65 | Volts |
| Emitter-Base Voltage | $BV_{EBO}$ | 15 | 25 | 30 | 40 | 45 | Volts |
| Collector Current | $I_C$ | 3 | 3 | 3 | 3 | 3 | Amps |
| Power Dissipation at $T_C = 25°C$ | $P_C$ | 62.5 | 62.5 | 62.5 | 62.5 | 62.5 | Watts |
| Junction Temperature Range | $T_J$ | —65 to +100 | | | | | °C |

## THERMAL CHARACTERISTICS

| Characteristic | Typical | Maximum | Units |
|---|---|---|---|
| Thermal Resistance, Junction to Case, $\theta_{JC}$ | 1.0 | 1.2 | °C/W |

**QUALITY ASSURANCE:**

Every Motorola industrial power transistor receives a 125°C stabilization bake for 100 hours. Quality of each production lot is assured by storage life tests at 100°C, temperature-humidity tests, and seal tests. Samples from daily production are seal tested by using a helium leak detector and, after stabilization bake, lot samples are subjected to a detergent-pressure test. A third seal test is the 10-day moisture test per MIL-S-19500. These tests, and complete in-process quality control insure maximum reliability of Motorola power transistor performance.

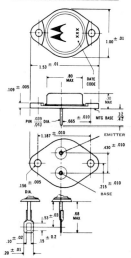

OUTLINE
(TO-3 without lugs; TO-41 with lugs. Collector connected internally to case. For units with solder lugs, specify devices MP2137 thru MP2146.)

**MOTOROLA Semiconductor Products Inc.**

# 2N2137 thru 2N2141 / 2N2142 thru 2N2146

**ELECTRICAL CHARACTERISTICS** (At 25°C case temperature unless otherwise specified)

| Characteristic | Types | Symbol | Minimum | Typical | Maximum | Unit |
|---|---|---|---|---|---|---|
| Collector-Base Cutoff Current ($V_{CB} = -2$ V, $I_E = 0$) | All Types | $I_{CBO}$ | — | 18 | 50 | μAdc |
| Collector-Base Cutoff Current | | $I_{CBO}$ | | | | mAdc |
| ($V_{CB} = -30$ V, $I_E = 0$) | 2N2137, 2N2142 | | — | 0.1 | 2 | |
| ($V_{CB} = -45$ V, $I_E = 0$) | 2N2138, 2N2143 | | — | 0.1 | 2 | |
| ($V_{CB} = -60$ V, $I_E = 0$) | 2N2139, 2N2144 | | — | 0.1 | 2 | |
| ($V_{CB} = -75$ V, $I_E = 0$) | 2N2140, 2N2145 | | — | 0.1 | 2 | |
| ($V_{CB} = -90$ V, $I_E = 0$) | 2N2141, 2N2146 | | — | 0.1 | 2 | |
| Collector-Base Cutoff Current ($V_{CB} = V_{CB\,max}$, $I_E = 0$, $T_C = +71°C$) | All Types | $I_{CBO}$ | — | 0.75 | 5 | mAdc |
| Emitter-Base Cutoff Current | | $I_{EBO}$ | | | | mAdc |
| ($V_{EB} = -15$ V, $I_C = 0$) | 2N2137, 2N2142 | | — | 0.08 | 2 | |
| ($V_{EB} = -25$ V, $I_C = 0$) | 2N2138, 2N2143 | | — | 0.08 | 2 | |
| ($V_{EB} = -30$ V, $I_C = 0$) | 2N2139, 2N2144 | | — | 0.08 | 2 | |
| ($V_{EB} = -40$ V, $I_C = 0$) | 2N2140, 2N2145 | | — | 0.08 | 2 | |
| ($V_{EB} = -45$ V, $I_C = 0$) | 2N2141, 2N2146 | | — | 0.08 | 2 | |
| Emitter-Base Cutoff Current ($V_{EB} = V_{EB\,max}$, $I_C = 0$, $T_C = +71°C$) | All Types | $I_{EBO}$ | — | 0.5 | 5 | mAdc |
| Collector-Emitter Breakdown Voltage* ($I_C = 300$ mA, $V_{EB} = 0$) | 2N2137, 2N2142 | $BV_{CES}$ | −30 | — | — | Vdc |
| | 2N2138, 2N2143 | | −45 | — | — | |
| | 2N2139, 2N2144 | | −60 | — | — | |
| | 2N2140, 2N2145 | | −75 | — | — | |
| | 2N2141, 2N2146 | | −90 | — | — | |
| Collector-Emitter Breakdown Voltage* ($I_C = 500$ mA, $I_B = 0$) | 2N2137, 2N2142 | $BV_{CEO}$ | −20 | — | — | Vdc |
| | 2N2138, 2N2143 | | −30 | — | — | |
| | 2N2139, 2N2144 | | −45 | — | — | |
| | 2N2140, 2N2145 | | −60 | — | — | |
| | 2N2141, 2N2146 | | −65 | — | — | |
| Floating Potential | | $V_{EBF}$ | | | | Vdc |
| ($V_{CB} = 30$ V, $I_E = 0$) | 2N2137, 2N2142 | | — | — | 1.0 | |
| ($V_{CB} = 45$ V, $I_E = 0$) | 2N2138, 2N2143 | | — | — | 1.0 | |
| ($V_{CB} = 60$ V, $I_E = 0$) | 2N2139, 2N2144 | | — | — | 1.0 | |
| ($V_{CB} = 75$ V, $I_E = 0$) | 2N2140, 2N2145 | | — | — | 1.0 | |
| ($V_{CB} = 90$ V, $I_E = 0$) | 2N2141, 2N2146 | | — | — | 1.0 | |
| DC Current Transfer Ratio | | $h_{FE}$ | | | | — |
| ($I_C = 0.5$ A, $V_{CE} = 2$ V) | 2N2137 - 2N2141 | | 30 | 45 | 60 | |
| | 2N2142 - 2N2146 | | 50 | 70 | 100 | |
| ($I_C = 2.0$ A, $V_{CE} = 2$ V) | 2N2137 - 2N2141 | | 15 | 25 | — | |
| | 2N2142 - 2N2146 | | 25 | 33 | — | |
| Collector-Emitter Saturation Voltage ($I_C = 2.0$ A, $I_B = 200$ mA) | All Types | $V_{CE(sat)}$ | — | 0.12 | 0.5 | Vdc |
| Base-Emitter Voltage ($I_C = 2.0$ A, $I_B = 200$ mA) | All Types | $V_{BE}$ | — | 0.75 | 1.2 | Vdc |
| Common Emitter Cutoff Frequency ($I_C = 2.0$ A, $V_{CE} = 6$ V) | All Types | $f_{ae}$ | 12 | 20 | — | kc |

*To avoid excessive heating of the collector junction, perform these tests with an oscilloscope

### POWER-TEMPERATURE DERATING CURVE

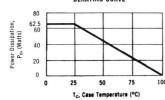

The maximum continuous power is related to maximum junction temperature by the thermal resistance factor.

This curve has a value of 62.5 Watts at case temperatures of 25°C and is 0 Watts at 100°C with a linear relation between the two temperatures such that:

$$\text{allowable } P_D = \frac{100° - T_C}{1.2}$$

**MOTOROLA Semiconductor Products Inc.**

## SAFE OPERATING AREAS

### 2N2137, 2N2142

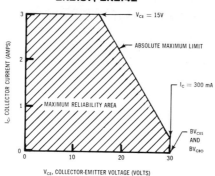

The maximum collector voltage and current are related by allowable d.c. power, transient conditions, and destructive voltage breakdown. The safe area of operation curves show a maximum reliable area of operation where destructive breakdown will not occur. The absolute limit lines can never be exceeded without possible damage to the transistor.

### 2N2138, 2N2143

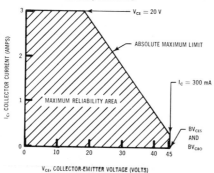

### 2N2140, 2N2145

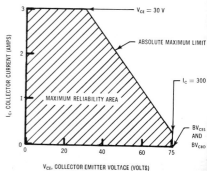

### 2N2139, 2N2144

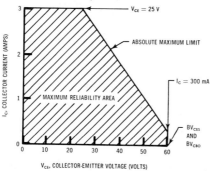

### 2N2141, 2N2146

*Appendix 10* 755

## 2N2137 thru 2N2141 / 2N2142 thru 2N2146

### COLLECTOR CHARACTERISTICS AT 25° C

**2N2137 thru 2N2141**
COLLECTOR CHARACTERISTICS
COMMON EMITTER CONFIGURATION

**2N2142 thru 2N2146**
COLLECTOR CHARACTERISTICS
COMMON EMITTER CONFIGURATION

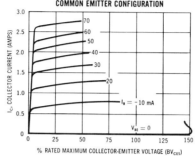

SATURATION REGION
(Constant base current; low voltage, high current)

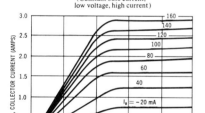

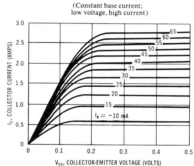

SATURATION REGION
(Constant base current; low voltage, low current)

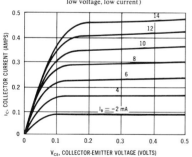

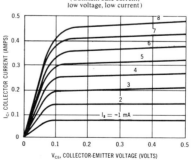

**MOTOROLA Semiconductor Products Inc.**

## INPUT & TRANSFER CHARACTERISTICS

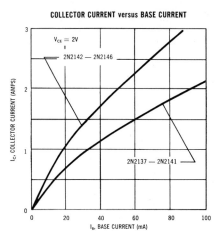

COLLECTOR CURRENT versus BASE CURRENT

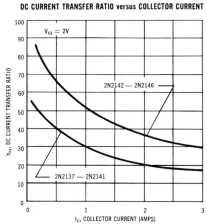

DC CURRENT TRANSFER RATIO versus COLLECTOR CURRENT

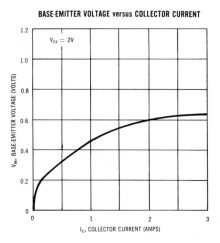

BASE-EMITTER VOLTAGE versus COLLECTOR CURRENT

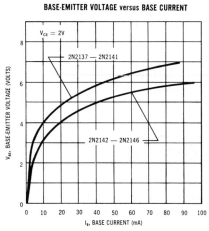

BASE-EMITTER VOLTAGE versus BASE CURRENT

Appendix 10   757

# 2N2137 thru 2N2141 / 2N2142 thru 2N2146

## TEMPERATURE CHARACTERISTICS

$I_{CBO}$ versus TEMPERATURE

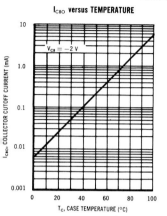

$h_{FE}$ versus TEMPERATURE

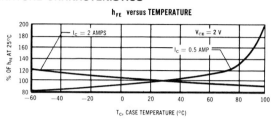

$g_{FE}$ versus TEMPERATURE

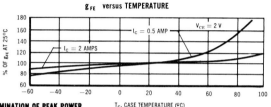

## DETERMINATION OF PEAK POWER

When a heat sink is used for increased heat transfer and greater thermal capacity, the following equation can be used to determine the allowable pulse power dissipation, designated as $P_P$. The allowable pulse power plus the steady state power, $P_{SS}$, gives the peak allowable power dissipation. Allowable pulse power is related by the following equation:

$$P_P = \frac{T_J - T_A - \theta_{JA} P_{SS}}{\theta_{JC}(1/C_P) + \theta_{CA}(t_1/\tau)},$$

where
$C_P$ = Coefficient of Power (from peak power derating curve),
$T_J$ = Maximum Junction Temperature
$T_A$ = Ambient Temperature (°C),
$\theta_{JC}$ = Junction-to-Case Thermal Resistance (°C/W),
$\theta_{CA}$ = Case-to-Ambient Thermal Resistance (°C/W),
$\theta_{JA} = \theta_{JC} + \theta_{CA}$,
$(t_1/\tau)$ = Duty Cycle = Pulse Width/Pulse Period,
$P_{SS}$ = Steady State Power Dissipation, and
$P_P$ = Allowable Pulse Power Dissipation Above $P_{SS}$.

The peak power derating curve is normalized with respect to the thermal time constant, $\tau$, which is about 50 milliseconds for these power transistors. The following example shows the application of this equation in conjunction with the peak power derating curve.

**EXAMPLE:**

Given:
$P_{SS} = 10$ W, $T_A = 40°C$, $t_1 = 1$ msec,
$(t_1/\tau) = 20\%$, $\theta_{CA} = 3°C/W$, $\theta_{JC} = 1.2°C/W$,
$T_J = 100°C$

**Solution:**
Enter the derating graph at $t_1/\tau = 1$ msec/50 msec, and duty cycle of 20%. Find $C_P = 5$. Substitute this value and the given parameters into the peak pulse power equation. This gives $P_P = 21.5$ watts. Thus the peak allowable power is $P_P + P_{SS}$, or 31.5 watts.

### PEAK POWER DERATING CURVE

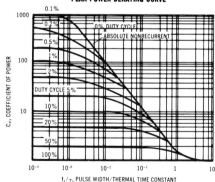

**Caution:** In all cases the peak pulse power should stay within the Safe Operating Area.

**MOTOROLA Semiconductor Products Inc.**
5005 EAST McDOWELL ROAD PHOENIX 8, ARIZONA   A SUBSIDIARY OF MOTOROLA INC

PRINTED IN USA 10-61   IMPERIAL LITHO                                 DS 3033

## TYPE 2N2386
## P-CHANNEL DIFFUSED PLANAR SILICON FIELD-EFFECT TRANSISTOR

### FOR INDUSTRIAL SMALL-SIGNAL APPLICATIONS
- High Input Impedance ( > 3 megohms at 1 kc)

\*mechanical data

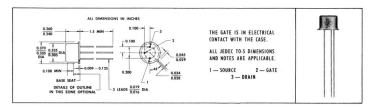

\*absolute maximum ratings at 25° free-air temperature (unless otherwise noted)

Gate Current . . . . . . . . . . . . . . . . . . . . . . . . . . . . . . 10 ma
Total Device Dissipation at (or below) 25°C Free-Air Temperature (See Note 1) . . . . 0.5 w
Total Device Dissipation at (or below) 25°C Case Temperature (See Note 2) . . . . . 1.5 w
Storage Temperature Range . . . . . . . . . . . . . . . . . . . . −65°C to +300°C

\*electrical characteristics at 25°C free-air temperature (unless otherwise noted)

| | PARAMETER | TEST CONDITIONS | MIN | MAX | UNIT |
|---|---|---|---|---|---|
| $BV_{DGO}$ | Drain-Gate Breakdown Voltage (See Note 3) | $I_D = -10\ \mu a$, $I_S = 0$ | −20 | | v |
| $I_{GSS}$ | Gate Cutoff Current | $V_{GS} = 10\ v$, $V_{DS} = 0$ | | 0.01 | $\mu a$ |
| | | $V_{GS} = 10\ v$, $V_{DS} = 0$, $T_A = 100°C$ | | 1.0 | $\mu a$ |
| $I_{D(off)}$ | Pinch-Off Drain Current | $V_{DS} = -12\ v$, $V_{GS} = 8\ v$ | | −10 | $\mu a$ |
| $|Y_{is}|$ | Small-Signal Common-Source Input Admittance | $V_{DS} = -10\ v$, $V_{GS} = 0$, $f = 1$ kc | | 0.3 | $\mu mho$ |
| $|Y_{fs}|$ | Small-Signal Common-Source Forward Transfer Admittance | $V_{DS} = -10\ v$, $V_{GS} = 0$, $f = 1$ kc | 1000 | | $\mu mho$ |
| $C_{iss}$ | Common-Source Short-Circuit Input Capacitance | $V_{GS} = 0$, $V_{DS} = -10\ v$, $f = 140$ kc | | 50 | pf |

NOTES: 1. Derate linearly to 175°C free-air temperature at the rate of 3.3 mw/C°.
2. Derate linearly to 175°C case temperature at the rate of 10 mw/C°.
3. This parameter corresponds closely to $BV_{DSS}$ (the Drain-Source Breakdown Voltage for $V_{GS} = 0$). $BV_{DSX}$ (the Drain-Source Breakdown Voltage for other values of $V_{GS}$) may be calculated from: $|BV_{DSX}| \cong |BV_{DGO}| - |V_{GS}|$.

\*Indicates JEDEC registered data.

Appendix 10   759

## TYPE 2N2386
## P-CHANNEL DIFFUSED PLANAR SILICON FIELD-EFFECT TRANSISTOR

**TYPICAL CHARACTERISTICS**

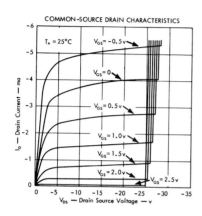
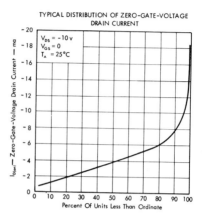

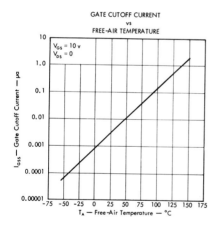
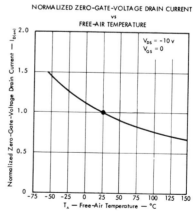

TEXAS INSTRUMENTS
INCORPORATED
SEMICONDUCTOR-COMPONENTS DIVISION
POST OFFICE BOX 5012 • DALLAS 22, TEXAS

## TYPE 2N2386
## P-CHANNEL DIFFUSED PLANAR SILICON FIELD-EFFECT TRANSISTOR

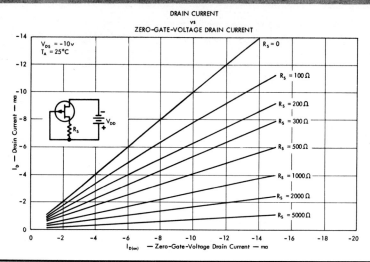

### TYPICAL APPLICATION DATA

HIGH INPUT IMPEDANCE AMPLIFIER

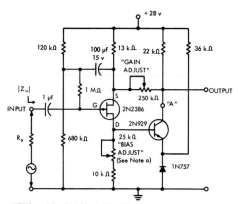

NOTES: a. Adjust for +18 v at Point "A".
b. All resistors ± 5% tolerance, 1/2 watt.

TYPICAL SMALL-SIGNAL CIRCUIT
PERFORMANCE CHARACTERISTICS

| FREQUENCY | $|Z_{in}|$ † |
|---|---|
| 10 cps | 70 MΩ |
| 100 cps | 70 MΩ |
| 1 kc | 50 MΩ |
| 10 kc | 10 MΩ |

| $R_g$ | 3 db BANDWIDTH † |
|---|---|
| 100 kΩ | 1 cps to 200 kc |
| 1 MΩ | 1 cps to 50 kc |
| 10 MΩ | 1 cps to 8 kc |

| VOLTAGE GAIN |
|---|
| Adjustable from 1 to 20 |

† $T_A$ = 25°C, "Gain Adjust" set for Gain of 10

TEXAS INSTRUMENTS
INCORPORATED
SEMICONDUCTOR-COMPONENTS DIVISION
POST OFFICE BOX 5012 • DALLAS 22, TEXAS

TI cannot assume any responsibility for any circuits shown
or represent that they are free from patent infringement.

PRINTED IN U.S.A.

TEXAS INSTRUMENTS RESERVES THE RIGHT TO MAKE CHANGES AT ANY TIME
IN ORDER TO IMPROVE DESIGN AND TO SUPPLY THE BEST PRODUCT POSSIBLE.

## TYPE 2N2386
## P-CHANNEL DIFFUSED PLANAR SILICON FIELD-EFFECT TRANSISTOR

**TYPICAL CHARACTERISTICS**

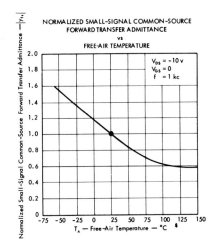

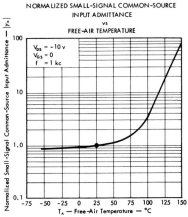

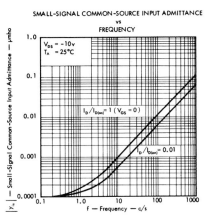

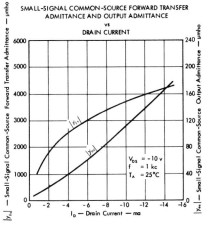

TEXAS INSTRUMENTS
INCORPORATED
SEMICONDUCTOR-COMPONENTS DIVISION
POST OFFICE BOX 5012 • DALLAS 22, TEXAS

**PNP POWER**
SILICON EPITAXIAL JUNCTION
TRANSISTOR

2N3199
2N3200
2N3201

- LOW $V_{CE}$ (sat)
- HIGH BETA
- HIGH $BV_{CEO}$

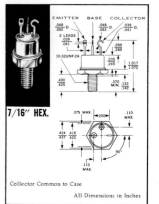

7/16" HEX.

Collector Common to Case
All Dimensions in Inches

ELECTRICAL DATA **ABSOLUTE MAXIMUM RATING**

| PARAMETER | SYMBOL | 2N3199 | 2N3200 | 2N3201 | UNITS |
|---|---|---|---|---|---|
| Collector to Base Voltage | $BV_{CBO}$ | 40 | 60 | 80 | Volts |
| Collector to Emitter Voltage | $BV_{CEO}$ | 40 | 60 | 80 | Volts |
| Emitter to Base Voltage | $BV_{EBO}$ | 10 | 10 | 10 | Volts |
| Collector Current | $I_C$ | -3.0 | -3.0 | -3.0 | Amps |
| Continuous Base Current | $I_{BC}$ | -1.5 | -1.5 | -1.5 | Amps |
| Continuous Power Diss. @ 25°C Case Temp. | $P_D$ | 40 Watts Derating 0.229 w/°C | | | |
| Junction Temp. Operating & Storage | $T_J$ | -65°C to +200°C | | | |

ELECTRICAL CHARACTERISTICS: $T_A = 25°C$ (UNLESS OTHERWISE STATED)

| PARAMETER | SYMBOL | CONDITION | 2N3199, 3200, 3201 Min. | Typ. | Max. | UNITS |
|---|---|---|---|---|---|---|
| Collector to Emitter Cutoff Current | $I_{CEX}$ | $V_{CE} = V_{CE}$ Max, $V_{BE} = 1.5V$ | — | 0.1 | -75 | μA |
| Collector to Emitter Cutoff Current | $I_{CEX}$ | $V_{CE} = 1/2\ V_{CE}$ Max, $V_{BE} = 1.5V$, $T_C = 150°C$ | — | 100 | -250 | μA |
| Collector to Emitter Cutoff Current | $I_{CEO}$ | $I_B = 0$, $V_{CE} = 1/2\ V_{CE}$ Max. | — | 1.0 | -100 | uA |
| Emitter Cutoff Current | $I_{EBO}$ | $V_{EB} = -10V$ | — | 1.0 | -50 | uA |
| Collector to Emitter Sustaining Voltage | $V_{CEO(SUS)}$ | $I_B = 0$, $I_C = -50mA$ | MAX. RATED | | | Volts |
| Forward Current Transfer Ratio | $h_{FE}$ | $I_C = -1.0A$, $V_{CE} = -2V$ | 20 | 45 | 60 | — |
| Forward Current Transfer Ratio | $h_{FE}$ | $I_C = -0.5A$, $V_{CE} = -2V$ | 30 | 80 | — | — |
| Forward Current Transfer Ratio | $h_{fe}$ | $I_C = -1.0A$, $V_{CE} = -2V$, $f = 1KC$ | 10 | 45 | 60 | — |
| Forward Current Transfer Ratio | $h_{fe}$ | $I_C = -1.0A$, $V_{CE} = -2V$, $f = 1MHZ$ | 1 | — | — | — |
| Base Emitter Voltage | $V_{BE}$ | $I_C = -1.0A$, $V_{CE} = -3.0V$ | — | — | -1.3 | Volts |
| Collector Emitter Saturation Voltage | $V_{CE(SAT)}$ | $I_B = 100mA$, $I_C = -1.0A$ | — | — | -0.3 | Volts |

**Crystalonics**
A TELEDYNE COMPANY

147 SHERMAN STREET, CAMBRIDGE, MASS. 02140
TELEPHONE: (617) 491-1670  •  TWX: 710-320-1196

Appendix 10   763

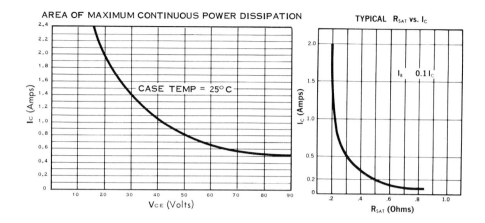

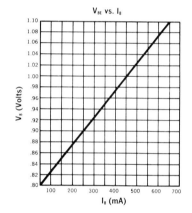

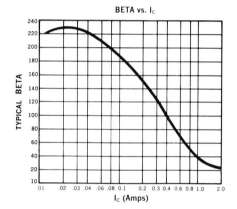

The 2N3199, 3200, and 3201 are **high voltage, high gain** silicon **Epitaxial Junction Power Transistors.** These transistors are manufactured by Crystalonics' exclusive **Epitaxial Junction Process,** which combines the advantages of alloyed, epitaxial, and planar techniques, and provides extreme ruggedness and parameter stability.

## Crystalonics
A TELEDYNE COMPANY

147 SHERMAN STREET, CAMBRIDGE, MASS. 02140

TELEPHONE: (617) 491-1670  •  TWX: 710-320-1196

764 Solid State Electronic Circuits

## Silicon Transistor
### PLANAR PASSIVATED

40.65

NPN

**2N3392**
**2N3393**
**2N3394**

The General Electric 2N3392, 2N3393 and 2N3394 are NPN silicon planar passivated transistors designed as small signal amplifiers. These devices feature tight beta control at an extremely low price.

### absolute maximum ratings: (25°C) (unless otherwise specified)

**Voltages**
| | | | |
|---|---|---|---|
| Collector to Emitter | $V_{CEO}$ | 25 | V |
| Emitter to Base | $V_{EBO}$ | 5 | V |
| Collector to Base | $V_{CBO}$ | 25 | V |

**Current**
| | | | |
|---|---|---|---|
| Collector (Steady State)[1] | $I_C$ | 100 | mA |

**Dissipation**
| | | | |
|---|---|---|---|
| Total Power (free air at 25°C)[2] | $P_T$ | 200 | mW |
| Total Power (free air at 55°C)[2] | $P_T$ | 120 | mW |

**Temperature**
| | | | |
|---|---|---|---|
| Storage | $T_{stg}$ | −55 to +125 | °C |
| Operating | $T_J$ | +100 | °C |
| Lead Temperature, 1/16″ ± 1/32″ from case for 10 seconds max. | $T_L$ | +260 | °C |

[1] Determined from power limitations due to saturation voltage at this current.
[2] Derate 2.67 mw/°C increase in ambient temperature above 25°C.

DIMENSIONS WITHIN JEDEC OUTLINE TO-98

NOTE 1: Lead diameter is controlled in the zone between .070 and .250 from the seating plane. Between .250 and end of lead a max. of .021 is held.

ALL DIMEN. IN INCHES AND ARE REFERENCE UNLESS TOLERANCED

3 LEADS

---

### electrical characteristics: (25°C) (unless otherwise specified)

**DC CHARACTERISTICS**

| | | Min. | Typ. | Max. | |
|---|---|---|---|---|---|
| **Collector Cutoff Current** ($V_{CB}=25V$, $I_E=0$) ($V_{CB}=25V$, $T_A=100°C$) | $I_{CBO}$ $I_{CBO}$ | | | 0.1 10 | μA μA |
| **Emitter Cutoff Current** ($V_{EB}=5V$, $I_C=0$) | $I_{EBO}$ | | | 0.1 | μA |
| **Collector to Emitter Voltage** ($I_C=1$ mA) | $V_{CEO}$ | 25 | | | volts |
| **Forward Current Transfer Ratio** ($V_{CE}=4.5V$, $I_C=2$ mA) 2N3392 2N3393 2N3394 | $h_{FE}$ $h_{FE}$ $h_{FE}$ | 150 90 55 | | 300 180 110 | |

**SMALL SIGNAL CHARACTERISTICS**

| | | Min. | Typ. | Max. | |
|---|---|---|---|---|---|
| **Output Capacitance** ($V_{CB}=10V$, $I_E=0$, $f=1$ MHz) | $C_{cb}$ | 4.5 | 7 | 10 | pf |
| **Input Impedance** ($V_{CE}=10V$; $I_C=2$ mA; $f=1$ KHz) | $h_{ib}$ | | 15 | | ohms |
| **Gain Bandwidth Product** ($I_C=2$ mA; $V_{CB}=5V$) | $f_t$ | | 120 | | MHz |
| **Forward Current Transfer Ratio** ($I_C=20$ mA; $V_{CB}=5V$, $f=20$ MHz) | $h_{fe}$ | | 15 | | |

GENERAL ELECTRIC

Appendix 10

# $V_{CE} = 10v; I_c = 1mA; f = 1KHz$

|  | | 2N3392 | 2N3393 | 2N3394 | |
|---|---|---|---|---|---|
| Forward Current Transfer Ratio | $h_{fe}$ | 208 | 150 | 100 | |
| Input Impedance | $h_{ie}$ | 6000 | 3400 | 2750 | ohms |
| Output Admittance | $h_{oe}$ | 14.0 | 10.0 | 7.7 | μmhos |
| Voltage Feedback Ratio | $h_{re}$ | .33 | .225 | .175 | $\times 10^{-4}$ |

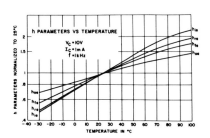

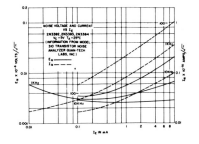

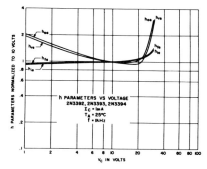

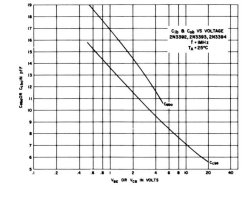

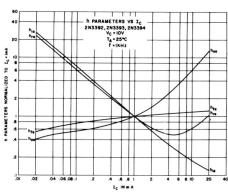

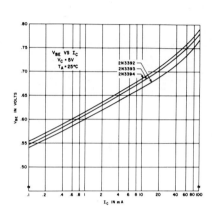
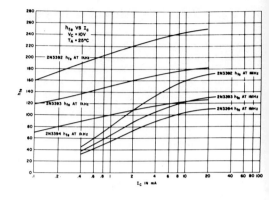
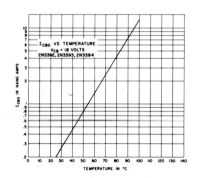
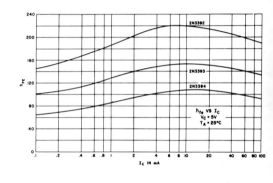
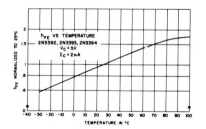

Appendix 10  767

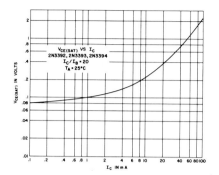

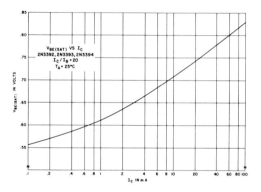

MOTOROLA Semiconductors
5005 EAST McDOWELL ROAD PHOENIX 8, ARIZONA

# 2N3796
# 2N3797

## SILICON N-CHANNEL MOS FIELD-EFFECT TRANSISTOR

N-CHANNEL MOS

FIELD-EFFECT TRANSISTORS

MARCH 1967 — DS 5106 R1
(Replaces DS 5106)

... designed for low-power applications in the audio frequency range

- Both Depletion and Enhancement-Mode Operation for a Broad Range of Applications
- Stable Performance at High Temperatures
  $I_{GSS} = 0.2$ nAdc max @ $T_A = 150°C$
- High Signal-Handling Capability at Low Drain Currents

**MAXIMUM RATINGS** ($T_A = 25°C$ unless otherwise noted)

| Rating | Symbol | Value | Unit |
|---|---|---|---|
| Drain-Source Voltage<br>　　2N3796<br>　　2N3797 | $V_{DS}$ | 25<br>20 | Vdc |
| Gate-Source Voltage | $V_{GS}$ | ±10 | Vdc |
| Drain Current | $I_D$ | 20 | mAdc |
| Power Dissipation at $T_A = 25°C$<br>Derate above 25°C | $P_D$ | 200<br>1.14 | mW<br>mW/°C |
| Operating Junction Temperature | $T_J$ | +200 | °C |
| Storage Temperature | $T_{stg}$ | -65 to +200 | °C |

## HANDLING CONSIDERATIONS:

MOS field-effect transistors, due to their extremely high input resistance, are subject to potential damage by the accumulation of excess static charge. To avoid possible damage to the devices while handling, testing, or in actual operation, the following procedure should be followed:

1. The leads of the devices should remain wrapped in the shipping lead washer or foil except when being tested or in actual operation to avoid the build-up of static charge.
2. Avoid unnecessary handling; when handled, the devices should be picked up by the can instead of the leads.
3. The devices should not be inserted or removed from circuits with the power on as transient voltages may cause permanent damage to the devices.

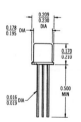

TO-18 PACKAGE

**MOTOROLA** Semiconductor Products Inc.  A SUBSIDIARY OF MOTOROLA INC.

# 2N3796 / 2N3797

**ELECTRICAL CHARACTERISTICS** ($T_A = 25°C$ unless otherwise noted)

| Characteristic | | Symbol | Min | Typ | Max | Unit |
|---|---|---|---|---|---|---|
| Drain-Source Breakdown Voltage | | $BV_{DSX}$ | | | | Vdc |
| ($V_{GS} = -4.0$ V, $I_D = 5.0 \mu A$) | 2N3796 | | 25 | 30 | — | |
| ($V_{GS} = -7.0$ V, $I_D = 5.0 \mu A$) | 2N3797 | | 20 | 25 | — | |
| Zero-Gate-Voltage Drain Current | | $I_{DSS}$ | | | | mAdc |
| ($V_{DS} = 10$ V, $V_{GS} = 0$) | 2N3796 | | 0.5 | 1.5 | 3.0 | |
| | 2N3797 | | 2.0 | 2.9 | 6.0 | |
| Gate-Source Voltage Cutoff | | $V_{GS(off)}$ | | | | Vdc |
| ($I_D = 0.5 \mu A$, $V_{DS} = 10$ V) | 2N3796 | | — | -3.0 | -4.0 | |
| ($I_D = 2.0 \mu A$, $V_{DS} = 10$ V) | 2N3797 | | — | -5.0 | -7.0 | |
| "On" Drain Current | | $I_{D(on)}$ | | | | mAdc |
| ($V_{DS} = 10$ V, $V_{GS} = +3.5$ V) | 2N3796 | | 7.0 | 8.3 | 14 | |
| | 2N3797 | | 9.0 | 14 | 18 | |
| Drain-Gate Reverse Current * | | $I_{DGO}$* | | | | pAdc |
| ($V_{DG} = 10$ V, $I_S = 0$) | | | — | — | 1.0 | |
| Gate-Reverse Current * | | $I_{GSS}$* | | | | pAdc |
| ($V_{GS} = -10$ V, $V_{DS} = 0$) | | | — | — | 1.0 | |
| ($V_{GS} = -10$ V, $V_{DS} = 0$, $T_A = 150°C$) | | | — | — | 200 | |
| Small-Signal, Common-Source Forward Transfer Admittance | | $|y_{fs}|$ | | | | $\mu$mhos |
| ($V_{DS} = 10$ V, $V_{GS} = 0$, $f = 1.0$ kHz) | 2N3796 | | 900 | 1200 | 1800 | |
| | 2N3797 | | 1500 | 2300 | 3000 | |
| ($V_{DS} = 10$ V, $V_{GS} = 0$, $f = 1.0$ MHz) | 2N3796 | | 900 | — | — | |
| | 2N3797 | | 1500 | — | — | |
| Small-Signal, Common-Source, Output Admittance | | $|y_{os}|$ | | | | $\mu$mhos |
| ($V_{DS} = 10$ V, $V_{GS} = 0$, $f = 1.0$ kHz) | 2N3796 | | — | 12 | 25 | |
| | 2N3797 | | — | 27 | 60 | |
| Small-Signal, Common-Source, Input Capacitance | | $C_{iss}$ | | | | pF |
| ($V_{DS} = 10$ V, $V_{GS} = 0$, $f = 1.0$ MHz) | 2N3796 | | — | 5.0 | 7.0 | |
| | 2N3797 | | — | 6.0 | 8.0 | |
| Small-Signal, Common-Source, Reverse Transfer Capacitance | | $C_{rss}$ | | | | pF |
| ($V_{DS} = 10$ V, $V_{GS} = 0$, $f = 1.0$ MHz) | | | — | 0.5 | 0.8 | |
| Noise Figure | | NF | | | | dB |
| ($V_{DS} = 10$ V, $V_{GS} = 0$, $f = 1.0$ kHz, $R_S = 3$ megohms) | | | — | 3.8 | — | |

*This value of current includes both the FET leakage current as well as the leakage current associated with the test socket and fixture when measured under best attainable conditions.

**MOTOROLA** Semiconductor Products Inc.

## 2N3796 / 2N3797

### TYPICAL DRAIN CHARACTERISTICS

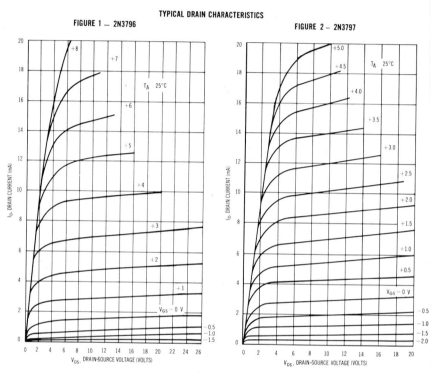

FIGURE 1 — 2N3796

FIGURE 2 — 2N3797

### COMMON SOURCE TRANSFER CHARACTERISTICS

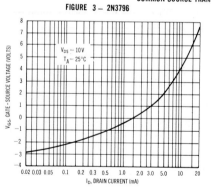

FIGURE 3 — 2N3796

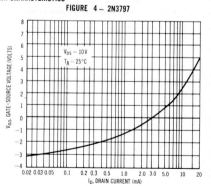

FIGURE 4 — 2N3797

**MOTOROLA** Semiconductor Products Inc.

## 2N3796 / 2N3797

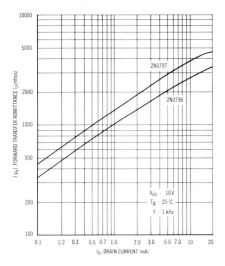

FIGURE 5 — FORWARD TRANSFER ADMITTANCE

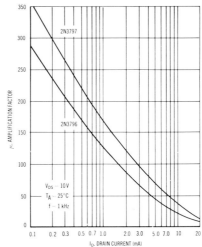

FIGURE 6 — AMPLIFICATION FACTOR

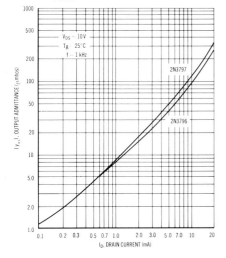

FIGURE 7 — OUTPUT ADMITTANCE

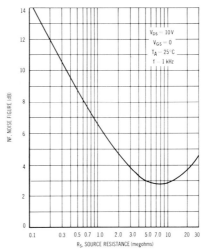

FIGURE 8 — NOISE FIGURE

 **MOTOROLA** *Semiconductor Products Inc.*

BOX 20912 • PHOENIX, ARIZONA 85036 • A SUBSIDIARY OF MOTOROLA INC.

## TYPE 2N3823
### N-CHANNEL EPITAXIAL PLANAR SILICON FIELD-EFFECT TRANSISTOR

**SYMMETRICAL N-CHANNEL FIELD-EFFECT TRANSISTOR FOR VHF AMPLIFIER AND MIXER APPLICATIONS**

- Low Noise Figure: $\leq$ 2.5 db at 100 Mc
- Low $C_{rss}$: $\leq$ 2 pf
- High $y_{fs}/C_{iss}$ Ratio (High-Frequency Figure-of-Merit)
- Cross Modulation Minimized by Square-Law Transfer Characteristic

BULLETIN NO. DL-S 657816, JULY 1965

\*mechanical data

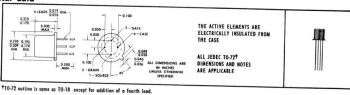

†TO-72 outline is same as TO-18 except for addition of a fourth lead.

\***absolute maximum ratings at 25°C free-air temperature** (unless otherwise noted)

| | |
|---|---|
| Drain-Gate Voltage | 30 v |
| Drain-Source Voltage | 30 v |
| Reverse Gate-Source Voltage | −30 v |
| Gate Current | 10 ma |
| Continuous Device Dissipation at (or below) 25°C Free-Air Temperature (See Note 1) | 300 mw |
| Storage Temperature Range | −65°C to +200°C |
| Lead Temperature 1/16 Inch from Case for 10 Seconds | 300°C |

\***electrical characteristics at 25°C free-air temperature** (unless otherwise noted)

| PARAMETER | | TEST CONDITIONS‡ | MIN | MAX | UNIT |
|---|---|---|---|---|---|
| $V_{(BR)GSS}$ | Gate-Source Breakdown Voltage | $I_G = -1\ \mu a$, $V_{DS} = 0$ | −30 | | v |
| $I_{GSS}$ | Gate Cutoff Current | $V_{GS} = -20\ v$, $V_{DS} = 0$ | | −0.5 | na |
| | | $V_{GS} = -20\ v$, $V_{DS} = 0$, $T_A = 150°C$ | | −0.5 | $\mu a$ |
| $I_{DSS}$ | Zero-Gate-Voltage Drain Current | $V_{DS} = 15\ v$, $V_{GS} = 0$, See Note 2 | 4 | 20 | ma |
| $V_{GS}$ | Gate-Source Voltage | $V_{DS} = 15\ v$, $I_D = 400\ \mu a$ | −1 | −7.5 | v |
| $V_{GS(off)}$ | Gate-Source Cutoff Voltage | $V_{DS} = 15\ v$, $I_D = 0.5\ na$ | | −8 | v |
| $|y_{fs}|$ | Small Signal Common-Source Forward Transfer Admittance | $V_{DS} = 15\ v$, $V_{GS} = 0$, $f = 1\ kc$, See Note 2 | 3500 | 6500 | $\mu mho$ |
| $|y_{os}|$ | Small-Signal Common-Source Output Admittance | $V_{DS} = 15\ v$, $V_{GS} = 0$, $f = 1\ kc$, See Note 2 | | 35 | $\mu mho$ |
| $C_{iss}$ | Common-Source Short-Circuit Input Capacitance | $V_{DS} = 15\ v$, $V_{GS} = 0$, $f = 1\ Mc$ | | 6 | pf |
| $C_{rss}$ | Common-Source Short-Circuit Reverse Transfer Capacitance | | | 2 | pf |
| $|y_{fs}|$ | Small-Signal Common-Source Forward Transfer Admittance | $V_{DS} = 15\ v$, $V_{GS} = 0$, $f = 200\ Mc$ | 3200 | | $\mu mho$ |
| $Re(y_{is})$ | Small-Signal Common-Source Input Conductance | | | 800 | $\mu mho$ |
| $Re(y_{os})$ | Small-Signal Common-Source Output Conductance | | | 200 | $\mu mho$ |

NOTES: 1. Derate linearly to 175°C free-air temperature at the rate of 2 mw/C°.
2. These parameters must be measured using pulse techniques. PW = 100 msec, Duty Cycle $\leq$ 10%.

\*Indicates JEDEC registered data.
‡The fourth lead (case) is connected to the source for all measurements.

**TEXAS INSTRUMENTS**
INCORPORATED
SEMICONDUCTOR-COMPONENTS DIVISION
POST OFFICE BOX 5012 • DALLAS 22, TEXAS

## TYPE 2N3823
## N-CHANNEL EPITAXIAL PLANAR SILICON FIELD-EFFECT TRANSISTOR

\* operating characteristics at 25°C free-air temperature

| PARAMETER | | TEST CONDITIONS‡ | MAX | UNIT |
|---|---|---|---|---|
| NF | Common-Source Spot Noise Figure | $V_{DS} = 15$ v, $V_{GS} = 0$, f = 100 Mc, $R_G = 1$ kΩ | 2.5 | db |

### TYPICAL CHARACTERISTICS‡

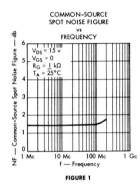

FIGURE 1

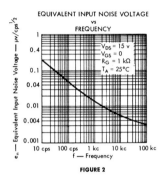

FIGURE 2

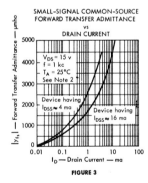

FIGURE 3

NOTE 2: These parameters must be measured using pulse techniques. PW = 100 msec, Duty Cycle ≤ 10%.

\*indicates JEDEC registered data.
‡The fourth lead (case) is connected to the source for all measurements.

TEXAS INSTRUMENTS
INCORPORATED
SEMICONDUCTOR-COMPONENTS DIVISION
POST OFFICE BOX 5012 • DALLAS 22, TEXAS

## TYPE 2N3823
## N-CHANNEL EPITAXIAL PLANAR SILICON FIELD-EFFECT TRANSISTOR

TYPICAL CHARACTERISTICS‡

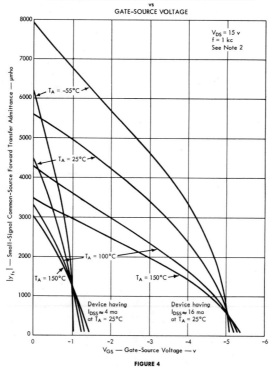

FIGURE 4

NOTE 2: These parameters must be measured using pulse techniques. PW = 100 msec, Duty Cycle ≤ 10%.

‡The fourth lead (case) is connected to the source for all measurements.

TEXAS INSTRUMENTS
INCORPORATED
SEMICONDUCTOR-COMPONENTS DIVISION
POST OFFICE BOX 5012 • DALLAS 22, TEXAS

Appendix 10  775

## TYPE 2N3823
## N-CHANNEL EPITAXIAL PLANAR SILICON FIELD-EFFECT TRANSISTOR

### TYPICAL CHARACTERISTICS‡

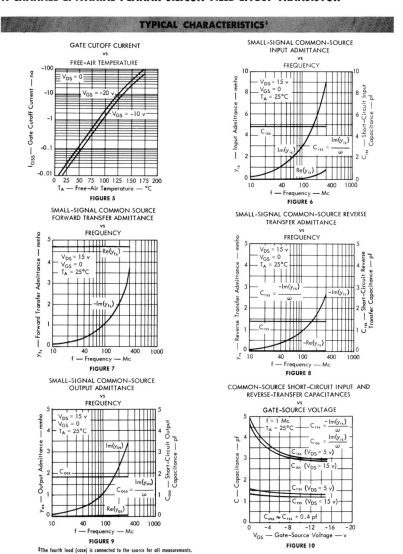

‡The fourth lead (case) is connected to the source for all measurements.

**TEXAS INSTRUMENTS**
INCORPORATED
SEMICONDUCTOR-COMPONENTS DIVISION
POST OFFICE BOX 5012 • DALLAS 22, TEXAS

PRINTED IN U.S.A.

TEXAS INSTRUMENTS RESERVES THE RIGHT TO MAKE CHANGES AT ANY TIME
IN ORDER TO IMPROVE DESIGN AND TO SUPPLY THE BEST PRODUCT POSSIBLE.

# appendix 11

## DERIVATIONS

### A.11.1 Average Value of a Waveform

The average value of any repetitive, time-varying waveform, also referred to as the DC component, may be obtained by algebraically summing the areas under the curve and dividing by the period. Mathematically, this can be stated as follows:

$$F_{AV} = \frac{1}{T} \int_0^T f(t)\, dt \tag{A.39}$$

**Example A.18**

Find the DC component for the waveforms shown in Fig. A.21.

**Solution**

a.  $T = 2$ s

   Area under curve:

from 0 to 1 s:  1(2) = 2
from 1 to 2 s:  1(−2) = −2

$$\text{DC component} = \frac{\text{net area}}{\text{period}} = \frac{2-2}{2} = 0$$

Hence, the waveform of Fig. A.21a has an average value (DC component) of 0 V.

b. $T = 4$ s

Area under curve:
from 0 to 2 s:  $2(10 \times 10^{-3}) = 2 \times 10^{-2}$
from 2 to 4 s:  $2(0) = 0$

$$\text{DC component} = \frac{\text{net area}}{\text{period}} = \frac{2 \times 10^{-2}}{4} = 5 \text{ mA}$$

c. $T = 2$ ms

Area under curve:
from 0 to 2 ms:  $(2 \times 10^{-3})(3.5) = 7 \times 10^{-3}$

$$\text{DC component} = \frac{\text{net area}}{\text{period}} = \frac{7 \times 10^{-3}}{2 \times 10^{-3}} = 3.5 \text{ V}$$

d. $T = 5$ μs

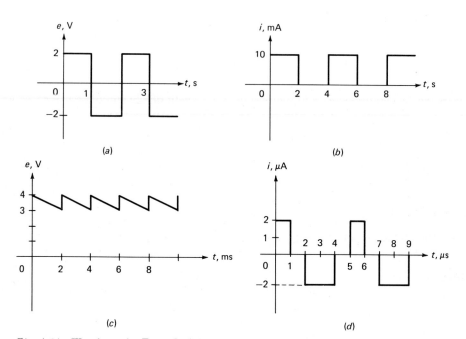

Fig. A.21  Waveforms for Example A.18.

Area under curve:
    from 0 to 1 μs: $(1 \times 10^{-6})(2 \times 10^{-6}) = 2 \times 10^{-12}$
    from 1 to 2 μs: 0
    from 2 to 4 μs: $(2 \times 10^{-6})(-2 \times 10^{-6}) = -4 \times 10^{-12}$
    from 4 to 5 μs: 0

$$\text{DC component} = \frac{\text{net area}}{\text{period}} = \frac{(2 \times 10^{-12}) - (4 \times 10^{-12})}{5 \times 10^{-6}} = -0.4 \ \mu A$$

**Example A.19**

Find the DC component for the waveforms shown in Fig. A.22.

**Solution**

Note that in all three cases, the horizontal axis has been labeled $\theta$, and the units are *radians*. This is done to facilitate the mathematical calculations that will have to be carried out. To determine the area under our curve, integration will have to be used. We have to do this only once, since the results can be used in other cases where the area under a sinusoidal curve is required.

a. The area under the curve from 0 to $\pi$ rad is found as follows:

$$\text{Area} = \int_0^\pi E_m \sin \theta \, d\theta \tag{A.40}$$
$$\text{Area} = [-E_m \cos \theta]_0^\pi = -E_m(\cos \pi - \cos 0)$$
$$\text{Area} = -E_m(-1 - 1) = 2E_m \text{ rad} \tag{A.40a}$$

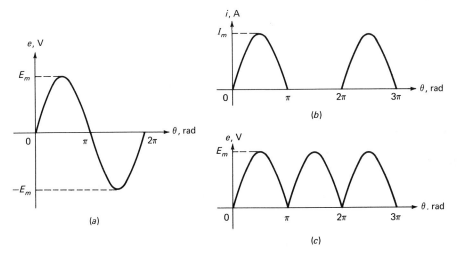

Fig. A.22  *Waveforms for Example A.19.*

The above indicates that the area under any half-cycle of a sinusoidal waveform is twice the peak value, as long as the horizontal axis units are radians, which results in the area being expressed in units of volt radians. Since the waveform between $\pi$ and $2\pi$ is identical to the first half (except for being negative), the area under it can be assumed equal to $-2E_m$ rad.

$$\text{DC component} = \frac{\text{total area for one period}}{T}$$

$$\text{DC component} = \frac{2E_m - 2E_m}{2\pi}$$

$$\text{DC component} = 0$$

as should be expected, since the negative area below the horizontal axis cancels the positive area above.

b. $T = 2\pi$ rad

Area from 0 to $\pi$:    $2I_m$ rad
Area from $\pi$ to $2\pi$:    0

$$\text{DC component} = \frac{2I_m}{2\pi} = 0.318 I_m$$

Hence, the average value of any half-wave rectified waveform, such as that in Fig. A.22b, is 0.318 of the peak value.

$$\text{DC} = 0.318 \text{ peak} \tag{A.41}$$

c. From the conclusions of (b), it follows that the average value for the full-wave rectified waveform of Fig. A.22c must be 2(0.318) of the peak value.

$$\text{DC} = 0.636 \text{ peak} \tag{A.42}$$

### A.11.2 RMS Value of a Waveform

The root mean square (RMS) value of a waveform, also known as the effective value, is obtained as follows:

1. Square the waveform.
2. Find the average value of the squared waveform.
3. Take the square root of the average value of the squared waveform.

Mathematically, this can be expressed as follows:

Let $F$ = RMS value of $f(t)$

$$F = \sqrt{\frac{1}{T} \int_0^T f^2(t)\, dt} \tag{A.43}$$

## Example A.20

Find the RMS value of all waveforms in Fig. A.23.

**Solution**

a. The first step is to square the waveform; this is shown in Fig. A.24a. The average value of the squared waveform is 100.

$$\text{RMS value} = \sqrt{100} = 10 \text{ V}$$

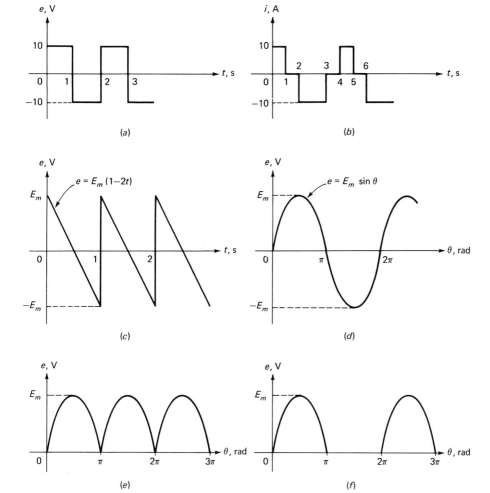

Fig. A.23  Waveforms for Example A.20.

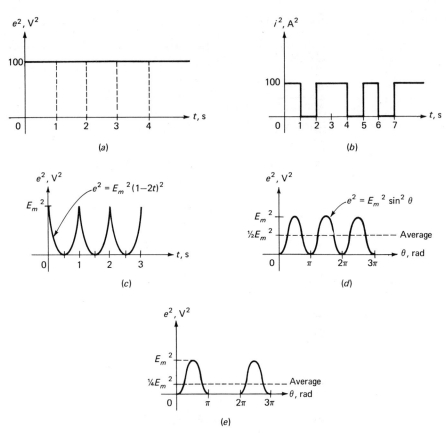

Fig. A.24  Waveforms for Example A.20.

b. The squared waveform is shown in Fig. A.24b.

Average value of squared waveform: $\dfrac{100 + 200}{5} = 60$

RMS value $= \sqrt{60} = 7.75$ A

c. The squared waveform is shown in Fig. A.24c. To compute the average value, we will need the area under the squared waveform. This can be done only by integration, as follows:

Equation of original curve (from 0 to 1): $e = E_m(1 - 2t)$
Equation of original curve squared: $e^2 = E_m^2(1 - 2t)^2$

$A$ = area under squared waveform: $\int_0^1 E_m^2(1-2t)^2\,dt$

$$A = \int_0^1 E_m^2(1 - 4t + 4t^2)\,dt = E_m^2\left(t - 2t^2 + \frac{4}{3}t^3\right)_0^1$$

Average value $= \dfrac{A}{T} = \dfrac{\frac{1}{3}E_m^2}{1} = \dfrac{E_m^2}{3}$

RMS value of sawtooth waveform $= \sqrt{\dfrac{E_m^2}{3}} = \dfrac{E_m}{\sqrt{3}}$ \hfill (A.44)

d. Equation of curve: $\qquad e = E_m \sin \theta$
Equation of curve squared: $\quad e^2 = E_m^2 \sin^2 \theta$

as shown in Fig. A.24d. The average value of $e^2$ can, because of symmetry, be located halfway between $E_m^2$ and 0. Hence the average value of squared waveform $= E_m^2/2$.

RMS value of sinusoidal waveform $= \sqrt{\dfrac{E_m^2}{2}} = \dfrac{E_m}{\sqrt{2}}$ \hfill (A.45)

e. The waveform of Fig. A.23e has the same RMS value as the waveform in Fig. A.23d; this is due to the fact that, when squared, these two waveforms are exactly the same. Hence

RMS value of full-wave rectified sinusoid $= \dfrac{E_m}{\sqrt{2}}$ \hfill (A.46)

f. The squared waveform is shown in Fig. A.24e. Because of the absence of any voltage for the second half of each cycle, it follows that the average value of this waveform is exactly one-half that of (d).

Average value: $\frac{1}{4}E_m^2$
RMS value of half-wave rectified sinusoid: $\sqrt{\frac{1}{4}E_m^2} = \frac{1}{2}E_m$ \hfill (A.47)

## A.11.3 Derivation of Eq. (9.15)

$I_D = I_{DSS}(1 - V_{GS}/V_P)^2$ \hfill (9.4a)

$y_{fs} = \left|\dfrac{d(I_D)}{d(V_{GS})}\right|$

$y_{fs} = |I_{DSS}(2)(1 - V_{GS}/V_P)(-1/V_P)|$

$y_{fs} = \left|\dfrac{2I_{DSS}}{V_P}(1 - V_{GS}/V_P)\right|$

But, from Eq. (9.4a),

$$1 - V_{GS}/V_P = \sqrt{\frac{I_D}{I_{DSS}}}$$

therefore

$$y_{fs} = \left|\frac{2I_{DSS}}{V_P}\sqrt{\frac{I_D}{I_{DSS}}}\right|$$

$$y_{fs} = \frac{2}{|V_P|}\sqrt{I_D I_{DSS}} \qquad (9.15)$$

## A.11.4 Calculation of Power Dissipation for Each Transistor in an Ideal Class B Amplifier

*Note:* Each transistor conducts for one half-cycle only.

- $p_d$: Instantaneous power dissipation, W
- $i_c$: Instantaneous collector current, A
- $v_{ce}$: Instantaneous collector-emitter voltage, V
- $V_{CC}$: Supply voltage, V
- $K$: $\dfrac{\text{Actual swing}}{\text{Maximum possible swing}}$
- $P_D$: Average value of $p_d$
- $R'_L$: Dynamic resistance seen by each transistor

$$p_d = v_{ce} i_c$$

$$i_c = K\frac{V_{CC}}{R'_L}\sin\theta \quad \text{from 0 to } \pi$$

$$i_c = 0 \quad \text{from } \pi \text{ to } 2\pi$$

$$v_{ce} = V_{CC} - KV_{CC}\sin\theta \quad \text{from 0 to } 2\pi$$

$$p_d = K\frac{V_{CC}}{R'_L}\sin\theta\, V_{CC}(1 - K\sin\theta) = \frac{KV_{CC}^2}{R'_L}(1 - K\sin\theta)(\sin\theta)$$

$$P_D = \frac{1}{2\pi}\int_0^\pi p_d\, d\theta$$

(There is no need to integrate between $\pi$ and $2\pi$ because $i_c = 0$ during this time.)

$$P_D = \frac{1}{2\pi}\int_0^\pi \frac{KV_{CC}^2}{R'_L}(\sin\theta - K\sin^2\theta)\, d\theta$$

$$P_D = \frac{KV_{CC}^2}{2\pi R'_L}\left[-\cos\theta - \frac{K\theta}{2} + \frac{K}{4}\sin 2\theta\right]_0^\pi$$

$$P_D = \frac{KV_{CC}^2}{R'_L}\left(\frac{1}{\pi} - \frac{K}{4}\right) \qquad (12.17)$$

When $K = 1$ (full swing),
$$P_D = \frac{V_{cc}^2}{R_L'}(0.318 - 0.25) = \frac{0.068 V_{cc}^2}{R_L'} \qquad (12.16)$$

The highest value of $P_D$ occurs when $\frac{dP_D}{dK} = 0$:

$$\frac{dP_D}{dK} = \frac{KV_{cc}^2}{R_L'}\left(-\frac{1}{4}\right) + \left(\frac{1}{\pi} - \frac{K}{4}\right)\left(\frac{V_{cc}^2}{R_L'}\right) = 0$$

$$-\frac{K}{4} + \frac{1}{\pi} - \frac{K}{4} = 0$$

$$K = \frac{2}{\pi} = 0.636$$

That this is a maximum (instead of a minimum) can be verified by evaluating $d^2P_D/dK^2$ at $K = 0.636$; the result will be negative.

$$\text{Highest } P_D = \frac{0.636 V_{cc}^2}{R_L'}\left(\frac{1}{\pi} - \frac{0.636}{4}\right)$$

$$\text{Highest } P_D = 0.1015 \frac{V_{cc}^2}{R_L'}$$

### A.11.5 Derivation of Eq. (13.13)

The current in a series resonant circuit of the type in Fig. 13.8a is

$$I = \frac{E}{(R_g + R_L) + j[\omega L - 1/(\omega C)]}$$

Let $R_g + R_L = R$, yielding

$$I = \frac{E}{R + j[\omega L - 1/(\omega C)]}$$

At resonance,

$$I_o = \frac{E}{R}$$

At $f_H$ and $f_L$, $|I| = 0.707 I_o$:

$$\frac{0.707 E}{R} = \left|\frac{E}{R + j[\omega L - 1/(\omega C)]}\right|$$

from which

$$R = \pm\left(\omega L - \frac{1}{\omega C}\right)$$

Above resonance ($f_H$),

$$R = \omega L - \frac{1}{\omega C} = 2\pi f_H L - \frac{1}{2\pi f_H C}$$

from which

$$f_H = \frac{R}{4\pi L} \pm \frac{1}{2}\sqrt{\frac{R^2}{4\pi^2 L^2} + \frac{1}{\pi^2 LC}}$$

Below resonance ($f_L$):

$$R = -\omega L + \frac{1}{\omega C} = -2\pi f_L L + \frac{1}{2\pi f_L C}$$

from which

$$f_L = \frac{-R}{4\pi L} \pm \frac{1}{2}\sqrt{\frac{R^2}{4\pi^2 L^2} + \frac{1}{\pi^2 LC}}$$

therefore

$$f_H - f_L = \frac{R}{4\pi L} + \frac{R}{4\pi L} = \frac{R}{2\pi L}$$

$$B = f_H - f_L = \frac{R}{2\pi L} = \frac{f_o}{Q} \qquad (13.13)$$

# answers to selected problems

**Problem Set 2.1**

2.1.3. $V$ = anything
$R = \infty$
$I = 0$

2.1.5.
a. OFF
b. ON
c. OFF
d. OFF
e. ON
f. $D_1$ (ON), $D_2$ (OFF)
g. OFF

**Problem Set 2.2**

2.2.1.
a. $V_T = 0.5$ V
b. $r_R = 100$ k$\Omega$
c. $BV = 13$ V
d. $r_F = 1$ $\Omega$
e. $r_Z \simeq 235$ $\Omega$

2.2.3.
a. $I = 0$
b. $I = 45.5$ mA
c. $I = 0.05$ mA
d. $I = 4$ mA
e. $I = 4$ mA
f. $I = 1.79$ A

**Problem Set 2.3**

2.3.1.
a. DC voltage = 0.17 V
DC current = 9 mA
AC voltage = 0.02 V peak
AC current = 3.33 mA peak

c. DC voltage $\simeq -12.2$ V
DC current $\simeq 20$ mA
AC voltage = 49 mV peak
AC current = 4.9 mA peak

## Problem Set 3.1

3.1.1.
  a. $E_s = 100$ V
  b. $PIV = 142$ V
  c. $I_{LM} = 14.2$ A
    $I_{L(RMS)} = 7.1$ A
  d. $P_L = 504$ W

## Problem Set 3.2

3.2.1.
  Peak $V_L = 163$ V
  Peak $I_L = 1.63$ A
  Peak $I_D = 1.63$ A
  Peak $E_D = 163$ V

3.2.3.
  $V_L = 81.5$ V
  $I_L = 0.815$ A
  $I_D = 0.815$ A

3.2.5. $RF = 1.21$

3.2.7. $P_{L(TOT)} = 66.4$ W

## Problem Set 3.3

3.3.1.
  $V_{LM} = 163$ V
  $V_{L(DC)} = 103.5$ V
  $V_L = 115$ V
  $V_{L(AC)} = 50$ V

3.3.3. 163 V

3.3.5. $P_L = 13.22$ W

## Problem Set 3.4

3.4.5.
  $V_{L(DC)} = 49.792$ V
  $V_{L(AC)} = 120$ mV

3.4.7.
  $C = 4{,}820$ $\mu$F
  $V_{L(DC)} = 50$ V
  $V_{L(AC)} = 50$ mV

## Problem Set 3.5

3.5.1.
  $C = 1{,}675$ $\mu$F
  $V_{L(DC)} = 70.35$ V
  $I_{L(DC)} = 140.7$ mA

## Problem Set 3.6

3.6.1.
  a. $\overline{P_{RS}} = 2.5$ W
  b. $\overline{P_Z} = 7.5$ W
  c. $I_Z = 200$ mA
  d. Regulation $\simeq 0.033$ percent
  e. $RRF = 200$
  f. $\overline{P_{RS}} = 40$ W

## Problem Set 3.7

3.7.5.
  a, f
  b, e
  c, d

3.7.7.
  $v_2 = 8$ V    for $v_1 \geq 8$ V
  $v_2 = v_1$    for $-15$ V $< v_1 < 8$ V
  $v_2 = -15$ V    for $v_1 \leq -15$ V

## Problem Set 3.8

3.8.1.
  a. DC = 100 V
  b. DC = 50 V
  c. DC = −60 V

## Problem Set 4.1

4.1.1.
  $A_v = 5$
  $A_i = 50$
  $G = 250$
4.1.3. $G = 400$

4.1.5. $R_i = 100\ \Omega$
4.1.7. $R_o = 11.1\ \text{k}\Omega$
4.1.9. $A_i = \infty$
4.1.11. $G_T = 400$

## Problem Set 4.2

4.2.1.
  a. 23.98 dB
  b. −16.98 dB
  c. 40 dB
  d. −75.46 dB
  e. −4.56 dB
  f. 12.04 dB

4.2.3. $P_o = 50\ \text{mW}$
4.2.5. 6.98 dB
4.2.7. $V_i = 1.58\ \text{V}$
4.2.9. 300 ft

## Problem Set 4.3

4.3.1. $P = 58.8 \times 10^{-9}\ \text{W}$
4.3.3. $R = 500\ \text{k}\Omega$
4.3.5. $R_L = 72\ \Omega$

## Problem Set 4.4

4.4.1.
  a. $R_i = \infty$
  b. $R_i = 0$
4.4.9. $A_v = k$
     $G = \infty$

## Problem Set 4.5

4.5.3.
  a. None
  b. None
  c. Degrades high-frequency response
  d. Degrades high-frequency response

## Problem Set 4.6

4.6.1. 3.32 octaves
4.6.3. $f_1 = 1.9\ \text{kHz}$
4.6.5. 2.7 decades
4.6.7.
  a. 0.043 dB
  b. 1 dB
  c. 0.043 dB
  d. 1 dB

4.6.11.
  a. $f_L = 1.59\ \text{Hz}$
  b. $f_H = 31.8\ \text{kHz}$
  c. $f_H \simeq 2\ \text{GHz}$
  d. $f_L = 530\ \text{Hz}$

### Problem Set 4.7

4.7.1. b. $f_L = 20$ Hz
     c. $f_H = 10$ kHz
     g. $f_H = 159$ Hz

4.7.3. 49 percent reduction

### Problem Set 4.9

4.9.3. $NF = 1.5$ dB

### Problem Set 5.1

5.1.11. $I_C = 3.5$ mA
      $I_E = 3.55$ mA
      $\alpha = 0.985$

### Problem Set 5.2

5.2.3. $h_{fe} = 130$;
     $h_{fe}$ drops with increasing $I_C$

### Problem Set 5.3

5.3.1. a. $R_B = 4.45$ kΩ
     b. $R_B = 149$ kΩ

### Problem Set 5.4

5.4.1. $R_C = 1,400$ Ω
5.4.3. $I_c = 0.35$ mA
     $V_{ce} = 0.53$ V
     $P_o = 185$ μW

### Problem Set 5.5

5.5.1. $V_{CE} = 12.6$ V

### Problem Set 5.6

5.6.1.

| | $I_B$, μA | $I_C$, mA | $I_E$, mA | $V_{CB}$ or $V_{CE}$, V |
|---|---|---|---|---|
| a. | 50 | 2.5 | 2.55 | 10 |
| b. | 100 | 4.9 | 5 | 5.2 |
| c. | 100 | 4.9 | 5 | 5.2 |
| d. | 0 | 0 | 0 | 15 |
| e. | 10 | 0.99 | 1 | −15.1 |
| f. | 10 | 0.49 | 0.5 | 5.1 |
| g. | 2.5 | 0.248 | 0.25 | 5 |
| h. | 2.5 | 0.248 | 0.25 | 5 |
| i. | 100 | 4.9 | 5 | 10.2 |
| j. | 50 | 2.45 | 2.5 | 10 |
| k. | 100 | 9.9 | 10 | −6.1 |

Answers to Selected Problems 791

|   | $I_B$, μA | $I_C$, mA | $I_E$, mA | $V_{CB}$ or $V_{CE}$, V |
|---|---|---|---|---|
| l. | 50 | 4.95 | 5 | $-15$ |
| m. | 50 | 5 | 5.05 | $-14.95$ |
| n. | 2.5 | 0.097 | 0.1 | 5 |
| o. | 100 | 1.9 | 2 | $-10$ |
| p. | 136 | 6.7 | 6.8 | $-2.73$ |
| q. | $I_{B1} = 100$ | $I_{C1} = 4.9$ | $I_{E1} = 5$ | 15 |
|   | $I_{B2} = 5{,}000$ | $I_{C2} = 245$ | $I_{E2} = 250$ |   |

5.6.3.  a.  $R_L = 200\ \Omega$         d.  $R_E = 2\ \text{k}\Omega$
            $R_B = 60\ \text{k}\Omega$              $R_B = 500\ \text{k}\Omega$
        b.  $R_L = 1\ \text{k}\Omega$
            $R_E = 392\ \Omega$
        c.  $R_L = 2\ \text{k}\Omega$
            $R_E = 990\ \Omega$

## Problem Set 5.7

5.7.1.

|   | A | B | C | D |
|---|---|---|---|---|
| a. | $I_{CEX}$ | $I_{CEO}$ | $I_{CER}$ | $I_{CES}$ |
| b. | 0.55 mA | 20 mA | 3 mA | 1.5 mA |

5.7.3.  a.  $I_{CEO} = 1\ \mu\text{A}$
        b.  $I_{CEX} = 100\ \mu\text{A}$

## Problem Set 5.8

5.8.1.  a.  $BV_{CBO} = 30\ \text{V}$       c.  $BV_{CEO} = 20\ \text{V}$
        b.  $BV_{CES} = 30\ \text{V}$       d.  $BV_{EBO} = 15\ \text{V}$

## Problem Set 5.9

5.9.5.  a.  Saturated         d.  Active
        b.  Cut in            e.  OFF
        c.  Saturated         f.  OFF
5.9.7.  $I_C$ rises from about 0 at A to maximum of 19.5 mA at B.
5.9.9.  $V = -2.18\ \text{V}$
        $R = 23.85\ \text{k}\Omega$
5.9.11. $A: I_C \simeq 0$
        $B: I_C = 0.88\ \text{mA}$
        $C: I_C = 2.94\ \text{mA}$

## Problem Set 6.1

6.1.5.  $G = 16.5$ dB

6.1.7.  $R_o = 1/h_o$, if $h_r = 0$ or $R_g = \infty$

## Problem Set 6.2

6.2.1.  $h_i = 10$ k$\Omega$, $h_f = 70$, $h_o = 2$ $\mu$mho, $h_r = 10^{-6}$

## Problem Set 6.3

6.3.1.  $h_{ib} = 25$ $\Omega$, $h_{fb} = -0.9$, $h_{ob} = 6.25$ $\mu$mho, $h_{rb} = 10^{-3}$

## Problem Set 6.4

6.4.1.  
a. $h_{ie} = 9{,}000$ $\Omega$  
b. $h_{fe} \simeq 190$ $\Omega$  
c. $h_{oe} = 84$ $\mu$mho  
d. $h_{re} = 0.66 \times 10^{-3}$

6.4.9.  $h_{fe} = 100$

## Problem Set 6.5

6.5.1.  $h_{ib} = 22.5$ $\Omega$, $h_{fb} = -0.993$, $h_{ob} = 0.0662$ $\mu$mho, $h_{rb} \simeq 0$

## Problem Set 7.1

7.1.5.  $A_i = -0.49$  
$R_i \simeq 25$ $\Omega$  
$R_o \simeq 10$ k$\Omega$  
$A_v = 196$

7.1.7.  $R_i = 1{,}180$ $\Omega$  
$R_o \simeq 3.3$ k$\Omega$  
$A_i \simeq 25.4$  
$A_v = -58.1$

7.1.9.  $R_i = 18.5$ k$\Omega$  
$R_o \simeq 100$ $\Omega$  
$A_i = -8.8$  
$A_v = 0.95$

7.1.11.  $R_i \simeq 25$ $\Omega$  
$R_o \simeq 2.7$ k$\Omega$  
$A_i \simeq -0.56$  
$A_v = 44.8$

## Problem Set 7.2

7.2.1.  
a. $A_v = -10$  
b. $A_v = 0.995$  
c. $A_v = 0.826$  
d. $A_v = -130$  
e. $A_v = 0.994$

7.2.3.  $R_o = 115$ $\Omega$  
7.2.5.  $A_v = 117.5$  
7.2.7.  $A_v = 102$, $R_i \simeq 38$ $\Omega$  
7.2.11.  $A_v \simeq -10$, $R_i = 187$ k$\Omega$

## Problem Set 7.3

7.3.1.  $f_L = 3.45$ Hz

7.3.5.  $n_1/n_2 = 10; L = 40$ H

## Problem Set 7.4

7.4.1.  $f_H = 262$ kHz

7.4.3.  $f_H = \dfrac{1}{1/f_\beta + 2\pi r_{b'e} C_{b'c}(1 + g_m R_L)}$

## Problem Set 8.1

8.1.3.  $I_C \simeq 2$ mA; $V_{CE} \simeq 0$ V

## Problem Set 8.2

8.2.1.
 a. $R_B = 72.5$ kΩ, $R_E = 4{,}150$ Ω
 b. $\Delta I_C = 0.95$ mA
 c. $\Delta V_{CE} = 4.5$ V

## Problem Set 8.3

8.3.1.
 a. Silicon
 b. $I_C = 8.25$ mA, $V_{CE} \simeq 22.5$ V
 c. $S_I \simeq 11.6$, $S_H = 16.8$ µA, $S_V = -0.423 \times 10^{-3}$ mho
 d. $V_{BE} = 0.41$ V, $I_{CBO} = 116$ µA, $h_{FE} = 150$
 e. $\Delta I_C$ due to $\Delta V_{BE} = 0.079$ mA
    $\Delta I_C$ due to $\Delta I_{CBO} = 1.345$ mA
    $\Delta I_C$ due to $\Delta h_{FE} = 1.26$ mA
    $\Delta I_{C(\text{TOT})} = 2.68$ mA
 f. $I_C = 10.93$ mA, $V_{CE} \simeq 17$ V

## Problem Set 8.4

8.4.3.
 a. $I_C \simeq 1$ mA, $V_{CE} \simeq 0$ V
 b. $S_I \simeq 1.2$
 c. $\Delta I_C \simeq 84$ µA
 d. $I_C = 1$ mA, $V_{CE} = 0$ V

## Problem Set 9.2

9.2.3.  1 pA

9.2.5.  $V_{GS(\text{OFF})} = -3$ V

## Problem Set 9.3

9.3.3.  $I_{DQ} \simeq 0.9$ mA,  $\overline{I_{DQ}} \simeq 2.6$ mA

9.3.5.  $I_{DQ} \simeq 1.8$ mA,  $\overline{I_{DQ}} \simeq 2.3$ mA

9.3.7.  $I_D \simeq -3.6$ mA,  $V_{DS} \simeq -8.2$ V, $R_i \simeq 20$ MΩ

## Problem Set 9.4

9.4.1. $y_{fs} \simeq 2{,}500 \ \mu\text{mho}, \ y_{os} \simeq 100 \ \mu\text{mho}$
9.4.3. $y_{fs}/C_{iss} = 128 \times 10^6$

## Problem Set 9.5

9.5.1.  a. $A_v = -20, \quad R_i \simeq 5 \ \text{M}\Omega$
  b. $A_v \simeq 0.79, \quad R_i \simeq 5 \ \text{M}\Omega$
  c. $A_v \simeq 0.82, \quad R_i = 25 \ \text{M}\Omega$
9.5.3. $R_{S1} = 1.3 \ \text{k}\Omega, \quad R_{S2} = 8.7 \ \text{k}\Omega$

## Problem Set 9.6

9.6.1. $f'_L = 0.0016 \ \text{Hz}, \ f''_L = 0.0795 \ \text{Hz}$

## Problem Set 10.1

10.1.3. $V_i = 0.91 \ \text{mV}, \ V_o = -909 \ \text{mV}, \ \beta V_o = -9.09 \ \text{mV}, \ A'_v = -90.9$, feedback is negative
10.1.5. $\beta_v = -0.01$
10.1.7. $\beta_v = -0.008, \ A'_v = 10{,}000$
10.1.9. $\beta_v = -0.000096$; change in $A'_v = 10.7$ percent

## Problem Set 10.2

10.2.1.  a. $A'_v = 83.3, \quad B' = 60 \ \text{kHz}$  10.2.5. No
  b. $A'_v = 9.8, \quad B' = 510 \ \text{kHz}$  10.2.7. $\beta_v = 0.04, \ V_i = -0.25 \ \text{V}$
10.2.3. $A_v f_H = 5 \times 10^6$, a constant

## Problem Set 10.4

10.4.3. a. $A_v = -883.6(10^3)$, b. $A'_v = -195$, c. $R'_i = 39.65 \ \text{h}\Omega$, d. $A'_v = -195$
10.4.5. $A'_i \simeq -67, \quad R'_i = 272 \ \text{h}\Omega$

## Problem Set 10.5

10.5.5. $\beta_v = 0.316$
10.5.7. DC attenuation: 0-dB, high-frequency attenuation = 6 dB, phase at 1,000 Hz = $-18.4°$

## Problem Set 11.1

11.1.7.  6 mV

## Problem Set 11.2

11.2.11.  a.  1 $\mu$V
        b.  18 mV

11.2.13.  86 dB

## Problem Set 11.4

11.4.5.  $V_o = 0$; circuit is adding
11.4.9.  $R_2 = 99R_1$
11.4.11.  $V_o = 21$ mV
11.4.13.  $10^7$ Hz

## Problem Set 11.5

11.5.3.  Lowest $V_i = 7.62$ V

## Problem Set 12.1

12.1.5.  $P_D = 11.25$ W
12.1.7.  a.  $h_{IB} = 15\ \Omega$,  $h_{FB} = -0.98$
       b.  $h_{IB} = 6\ \Omega$,  $h_{FB} = -0.974$
12.1.11.  $\eta = 33\frac{1}{3}$ percent

## Problem Set 12.2

12.2.3.  Minimum voltage: 1 V

## Problem Set 12.3

12.3.1.  a.  $\theta_{ca} = 12.5$ °C/W
       b.  $T_j = 61$°C
12.3.3.  $P_D = 187.5$ mW

12.3.5.  a.  Silicon
       b.  $\theta_{jc} = 1$ °C/W
       c.  $P_D = 21.6$ W
       d.  $T_j = 92$°C

## Problem Set 12.4

12.4.1.  $P_{o(\max)} = 28.5$ mW, $R_B = 100$ k$\Omega$, $V_{CE} = 7.5$ V, $I_C = 7.5$ mA
12.4.3.  $V_{CE} = 16.7$ V, $I_C = 1.32$ A, $R_L = 12.65\ \Omega$
12.4.5.  a.  When $h_{FE} = 15$:  $V_{CE} = 21.82$ V, $I_C = 1.03$ A
                                $P_{o(\max)} = 6.71$ W
       b.  When $h_{FE} = 30$:  $V_{CE} = 7.7$ V, $I_C = 1.83$ A
                                $P_{o(\max)} = 1.65$ W

12.4.7.  a. $P_{o(\text{max})} = 37.5$ W
b. $n_1/n_2 = 2.74$
c. $R'_L = 141$ Ω

## Problem Set 12.5

12.5.1.  a. 8 W
b. 2 W
12.5.3.  a. $R'_L = 10$ Ω
b. $V_{CE} = 40$ V, $I_C = 0$ A
c. 1. 40 V, 0 A
2. 0 V, 4 A
3. 80 V, 0 A
d. $P_{o(\text{max})} = 80$ W
e. 40 W
f. 80 V

## Problem Set 12.6

12.6.1.  a. $V_{CE} = 15$ V, $I_C = 0$ A    d. 3.75 W
b. 1. $V = 15$ V, $I = 0$    e. 1.02 W
2. $V = 0$, $I = 1$ A    f. 30 V
3. $V = 30$ V, $I = 0$    g. $60CT/4$
c. 7.5 W    h. 1.02 W
12.6.3.  $p_{d(\text{max})} = \tfrac{1}{2} P_o$ for class B
$p_{d(\text{max})} = 2 P_o$ for class A

## Problem Set 12.7

12.7.5.  a. $V = 20$ V, $I = 0$    d. $P_o = 16$ W
b. $p_{d(\text{max})} = 10$ W    e. $\overline{P_{DC}} = 25.4$ W
c. $\overline{P_D} = 2.72$ W    f. $40CT/5$

## Problem Set 13.2

13.2.1.  $R_S = 20.9$ Ω, $R_P = 470$ kΩ    13.2.17.  $L_1 = 75.3$ μH, $L_2 = 24.7$ μH,
13.2.7.  $Q = 125$              $C = 2.5$ pF
13.2.9.  If $Q \geq 5$    13.2.19.  $L_1 = 94.9$ μH, $L_2 = 5.1$ μH
13.2.13.  a. $C = 25$ pF    13.2.21.  $X = 7.77$ Ω inductive
b. $R_{\text{tot}} = 1{,}325$ Ω    13.2.23.  $n_2/n_1 = 14.14$
c. $I_o = 0.755$ mA    13.2.25.  $C = 17.8$ pF, $R_P = 33.6$ kΩ

d. $V_o = 453$ mV
e. $B = 211$ kHz
13.2.15. a. $C = 2.5$ pF
b. $Q = 6.3$
c. $B = 1.59$ MHz

13.2.27. $R = 64.6\ \Omega$
13.2.29. $C = 17.8$ pF, $n_3/n_2 = 6$,
$n_2/n_1 = 1.18$,
$R = 832\ \Omega$

## Problem Set 13.3

13.3.1. $R_1 = 158\ \Omega$, $C_1 = 42$ pF, $R_2 = 2{,}550\ \Omega$, $C_2 = 7.88$ pF, $R_3 = 30.2$ k$\Omega$, $C_3 = 4.55$ pF, $G_m = 16.5$ mmho
13.3.7. $n_2/n_1 = 31.6$
13.3.11. $L_n = 50\ \mu$H

## Problem Set 13.4

13.4.5. $y_{is} = 0.4 + j5.8$ mmho
$y_{fs} = 4.7 - j2.1$ mmho
$y_{rs} = 0 - j1.75$ mmho
$y_{os} = 0.1 + j2.4$ mmho
13.4.7. $y_i = 0.413 \times 10^{-4} + j4.75 \times 10^{-4}$
$y_f = 5 \times 10^{-3} + j0$
$y_r = 0 + j0.25 \times 10^{-3}$
$y_o = 0.645 \times 10^{-3} + j2.41 \times 10^{-3}$

13.4.9. $R_i = 29.3$ k$\Omega$, $C_i = 6.6$ pF
13.4.11. $R_o = 1{,}550\ \Omega$, $C_o = 37.4$ pF
13.4.13. $G_{pmn} = 23.7$ dB
13.4.15. $G_{pt} = 18.2$ dB

## Problem Set 13.6

13.6.7. Highest: 2,003 kHz
Lowest: 1,997 kHz
$B$: 6 kHz

13.6.9. 2, 4, 6, 8, 16 MHz

## Problem Set 14.2

14.2.7. Minimum $A_v = 29$
14.2.9. $C = 65$ pF

## Problem Set 14.3

14.3.7. $C_1 = 2{,}400$ pF, $C_2 = 230$ pF
14.3.15. $f_2 = 1{,}005$ kHz; $R_P = 635 \times 10^9\ \Omega$
14.3.19. $R_1 = 565\ \Omega$, $R_2 = 55\ \Omega$
14.3.21. a. $r = 50\ \Omega$
b. $r = \infty\ \Omega$
c. $r = -200\ \Omega$
d. $r = -600\ \Omega$
e. $r = \infty\ \Omega$
f. $r = 100\ \Omega$
14.3.23. $C = 56$ pF

### Problem Set 14.4

14.4.3. $f = 143$ Hz
14.4.7. $1{,}200\ \Omega < R < 1\ \text{M}\Omega$
14.4.9. $v_{B1} = 0.7$ V, $v_{C1} = 0.3$ V
$v_{B2}$ negative but rising, $v_{C2} = 15$ V
14.4.11. $\underline{h_{FE}} = 20$

### Problem Set 15.1

15.1.1.  
a. $-10$ V
b. 20 V
c. 0.4 $\mu$s
d. 1.3 $\mu$s
e. 770 kHz
f. 31 percent
g. 0.08 $\mu$s
h. 0.08 $\mu$s
i. 4.38 MHz

### Problem Set 15.2

15.2.1.  
a. $H = 4$ V, $t_r = t_f = 0$
b. $H = 5$ V, $t_r = t_f = 44\ \mu$s
c. $H = 1$ V, $t_r = t_f = 26.4\ \mu$s
d. $H = 5$ V, $t_r = t_f = 22$ ns
e. $v_{oo} = 4.16$ V, $v_{oss} = 4.5$ V, $t_r = t_f = 0$
f. $v_{oo} = 4.63$ V, $v_{oss} = 4.5$ V, $t_r = t_f = 0$
g. $H = 4.5$ V, $t_r = t_f = 0$

15.2.3. $C_1 = 5.55$ pF; voltage attenuation: 10
$R_i = 5\ \text{M}\Omega$, $C_i = 5$ pF

15.2.5. $C_1 = 4.5$ pF, $R_1 = 40\ \text{M}\Omega$, add a 20-M$\Omega$ resistor in parallel with $R_2$

### Problem Set 15.3

15.3.3. 2 percent

### Problem Set 15.4

15.4.1. Differentiation: a, c, d
Integration: b

### Problem Set 15.7

15.7.7. Amplitude reaches 5 V

### Problem Set 15.8

15.8.1. 9 steps: 0, 6, 11, 15.15, 18.6, 21.5, 23.92, 26, 27.67 V

## Problem Set 15.9

15.9.5. $R_2 = 20$ k$\Omega$, $V_{C2} = 5.45$ V

## Problem Set 15.10

15.10.7. HI = 12 V, LO = 6 V

15.10.9. Lower trigger point = 4.8 V, upper trigger point = 5.8 V

## Problem Set 15.11

15.11.3. Base: 0.3 V; Collector: 0.1 V

15.11.5. $v_{B1} = -1.3$ V
$v_{C1} = 10.55$ V

# index

Abbreviations of units and prefixes, 709–710
Acceptor impurities, 10
Active region, 160, 163, 193, 195, 666
Adding circuit, 435, 436
Admittance ($y$) parameters, 557–561
AGC (*see* Automatic gain control)
Alloy junction, 201
Alpha crowding, 461
Amplification:
    current, 95, 96
    voltage, 95, 96
Amplifiers:
    cascaded, 102, 136
    chopper-stabilized, 429–431
    class A, 474, 481
    class AB, 502
    class B, 488, 489, 502, 503
    class C, 479, 511, 512, 576, 578

Amplifiers:
    common base, 220, 249, 260, 261
    common collector, 221, 253, 262, 263
    common drain, 351
    common emitter, 168, 174, 217, 245, 258
    common gate, 354
    common source, 349
    complementary symmetry, 508
    current, 111
    Darlington, 416, 417, 452
    differential, 418–428
    direct-coupled, 168, 413, 415, 416, 475, 509, 510, 654
    emitter follower, 256
    feedback, 368–385
    ideal, 109–113
    intermediate-frequency (IF), 516
    large-signal, 457, 458

Amplifiers:
    operational, 431–433, 658, 661
    phase shift through, 94, 115, 138, 139
    push-pull, 488, 489, 502, 503, 508
    quasi complementary, 510
    radio-frequency (RF), 515
    RC-coupled, 116, 174, 258, 261, 263, 336, 340, 344, 350, 352, 355, 363, 654
    transconductance, 113
    transformer-coupled, 116, 291, 292, 470, 479, 481, 488, 489
    transresistance, 113
    voltage, 111
Amplitude modulation (AM), 516, 580, 581
Atomic theory, 1
Attenuation networks, 644
Automatic gain control (AGC), 571–573, 587
Average value of waveform, 759–762

Bandwidth, 115, 242, 517, 528, 540, 541, 653
Barkhausen criterion, 591, 610
Binary, emitter-coupled, 694
Bistable circuit, 693
Black box, 93
Bode plots, 118, 119, 121, 122
Bonding, 2
Bootstrapping, 277–281, 358, 359, 681
Breakdown voltage:
    avalanche, 18
    in diodes, 30, 31, 51
    in FETs, 329–331
    in junction (bipolar) transistors, 191
    zener, 18

Capacitance:
    diffusion, 12, 239, 667
    in FETs, 346, 671, 672
    in junction (bipolar) transistors, 239, 667
    space charge, 17
    transition, 17
Characteristic curves, 751–758
Charge, 1, 2
Charge generation, 8
Chopper-stabilized amplifier, 429–431
Clamping circuits, 85–88
Clipping circuits, 76, 77
Common-mode rejection ratio (CMRR), 420–426, 432, 433, 442–444
Compensated amplifiers, 655–657
Compensated probes, 647, 650
Compensated voltage dividers, 645–647, 650
Complex algebra, 729
Complex numbers, 729–733
    polar form, 732
    rectangular form, 729
Conduction band, 3, 4
Conductivity modulation, 626
Constant current source, 438
Constant voltage source, 438
Covalent bonds, 5, 6, 13, 18
Crystal (quartz), 611–614
Crystalline structure, 4, 5, 11
Current divider, 714, 715
Cutoff current, 177, 187–190
Cutoff region, 160, 163, 193, 195, 666

Darlington amplifier, 416, 417, 452
Data sheets, 751–758
Decibel, 100

Depletion region, 12, 13, 17, 18
Determinants, 719–722
Differential amplifier, 418–428
Differentiator, 658, 660, 661
Diffusion current, 10
Diode, 11
  ideal, 21, 22
  large-signal analysis, 26
  real, 26
  small-signal analysis, 36
  zener, 71
Distortion, 140–147, 377, 378, 461, 582
  crossover, 500
  harmonic, 142, 147
  intermodulation (IM), 147
  linear, 144–146
Donor impurities, 10
Doped semiconductors, 9
Drain current, 332
Drift current, 10
Droop, 641
Dynamic range, 517

Early effect, 239
Efficiency, 463, 475, 479, 481, 498, 499
Electron-hole pairs, 7, 8
Energy band, 2, 3
Energy gap, 6, 8
Energy level, 3, 4
Exponential graphs, 642

Fall time, 641
Feedback, 367–370
  effect of: on distortion, 377–379
    on frequency response, 375, 376
  local, 373, 374
  in oscillators, 590–595, 606

Feedback:
  overall (multistage), 373, 374, 396, 397
  stability, 401
Field effect transistor (FET), 319–325
  biasing, 333–345
  dynamic characteristics, 345
  graphic analysis, 325, 327
  static characteristics, 328
  switching characteristics, 668–671
Filtering, 49, 60, 115
  high-pass, 658
  low-pass, 661
Flip-flop, 701, 703–706
  as frequency divider, 706
  triggering, 703–706
Form factor, 53
Frequency modulation (FM), 516
Frequency multiplication, 578–580
Frequency response, 114–117, 362, 375, 444, 517, 540
  high, 296, 297, 656
  low, 285, 288, 654–656

Gain, power, 100, 563
  maximum available, 99
  maximum neutralized, 564
  maximum unneutralized, 564
  transducer, 99, 564, 565, 567
Gain-bandwidth product, 242
Gain margin, 403

Heat sinks, 469–473
Hybrid ($h$) parameters, 210–237
Hybrid-pi model, 237, 297

Impedance transformation, 533, 538
Integrated circuits, 204, 426, 427

Integrator, 658, 660–664, 706
Intermediate-frequency (IF) amplifier, 516
Intrinsic conduction, 6–8
Intrinsic standoff ratio, 624, 626, 630
Inverter, 435, 653, 669

$j$ operator, 729
Junction (bipolar) transistor, 151
  biasing: constant base current, 303
    nonlinear, 316–319
    series (emitter) feedback, 305
    shunt feedback, 314–316
  characteristics of: dynamic, 161
    static, 158–160
    switching, 192–195, 666, 667
  DC analysis, 176–184
  graphic analysis, 162–167
  models: $h$ parameter, 208
    hybrid-pi, 237, 545, 546
  parameters: $h$, 208
    large-signal, 460
  structure of: epitaxial, 203
    mesa, 202
    planar, 203

Kirchhoff's current law, 723
Kirchhoff's voltage law, 723

Leakage current, 328, 329
Lifetime, 12, 13, 154
Light-activated silicon-controlled rectifier (LASCR), 676
Loop analysis, 723

Majority carriers, 13–16, 153
Manufacturing techniques, 199

Metal oxide semiconductor field effect transistor (MOSFET), 322–325
  (*See also* Field effect transistor)
Miller effect, 272, 273, 383, 392, 664
Minority carriers, 13, 14, 16
Mismatch factor, 566
Mismatching, 555, 566
Mixing, 580, 582
Mobility, 7
MOSFET (*see* Metal oxide semiconductor field effect transistor)
Multivibrator:
  astable, 633–636
  bistable, 693
  monostable, 688–692

Neutralization, 547, 553–555
Node analysis, 723, 726
Noise, 148–150
Noninverter, 439
Norton equivalent circuit, 737, 738, 740
Nyquist stability criterion, 406, 408

Off region, 160, 163, 193, 195, 666
Offset current, 441, 442
Offset voltage, 440, 441
One shot circuit (*see* Multivibrator, monostable)
Operational amplifier, 431–433, 658, 661
Oscillators, 585
  astable multivibrator, 633, 636
  Clapp, 613, 614
  Colpitts, 602, 603, 605, 613, 614
  crystal, 611, 614
  feedback, 593
  harmonic, 586, 592, 593
  Hartley, 608–610

Oscillators:
   RC phase shift, 595–599
   relaxation, 586, 623
   tunnel diode, 614, 619
   unijunction transistor (UJT), 623, 631, 632, 636–639
Overshoot, 145, 641, 647

Peak rectifier, 88, 90
Phase margin, 403
Phase shift, 115
Piezoelectric effect, 103, 611
PN junction, 11, 14–16
   alloyed, 201
   diffused, 202
   etched, 201
Power match, 742, 743
   conjugate, 744, 745
Power supply, 47, 48
Prefixes, abbreviations of, 709–710

$Q$ factor, 521, 522, 587, 612

Radio-frequency (RF) amplifier, 515
Recombination, 6, 8, 12, 154
Rectification, 49
   bridge circuit, 58–60
   full-wave, 54, 67
   half-wave, 49–61
Resistance:
   dynamic, 40, 41
   forward, 28, 29
   input, 94, 113, 210–212, 265, 357, 381–387
   negative, 589, 590, 615, 616, 619, 620, 623
   output, 210–212, 388–395
   reverse, 29
Resonance, 528, 530, 531

Ripple, 49, 53, 63, 65, 68, 90
   factor, 53, 65
   reduction factor, 73, 74, 445
   in tuned circuits, 540
Rise time, 641, 643
Root mean square (RMS) value of a waveform, 762–765

Sag, 641
Saturation current, 16
Saturation region, 160, 163, 193, 195, 667
Sawtooth waveform, 64, 68, 680
Schmitt trigger, 693–700
Selectivity, 515
Shunt peaking, 656
Silicon-controlled rectifier (SCR), 672–678
Silicon-controlled switch (SCS), 676
Space charge potential, 13
Square-wave testing of amplifier, 145
Stability, 401–403, 406–408, 517, 566
Stability factors (operating point), 308, 309, 312
Staircase:
   generation, 684–686
   waveform, 684
Step voltage, 641, 684
Strain gage, 104
Subtracting circuit, 437
Superheterodyne receiver, 516
Superposition, 74, 747–748
Sweep:
   bootstrapping, 681, 682
   generation, 680, 681
   linearity, 681
   triggering, 681
   waveform, 680
Symbols, definitions of, 711, 712

Temperature, junction, 467–469, 493
Thermal resistance, 468–473
Thermal runaway, 467
Thermal time constant, 473
Thermistor, 316–318
Thevenin equivalent circuit, 737, 739
Threshold voltage, 15, 28
Tilt, 641, 654
Transadmittance, 347
Transconductance, 347
Transducers, 103
Transformer, isolation, 59
Transistors:
    field effect (*see* Field effect transistor)
    junction [*see* Junction (bipolar) transistor]
Tuned circuits, 520
    coupled, 539–541
    stagger, 541, 542
Tunnel diode, 614–617

Unijunction transistor (UJT), 623–632, 678, 679

Unilateralization, 547
Units, abbreviations of, 709

Voltage breakdown (*see* Breakdown voltage)
Voltage dividers, 714, 715
Voltage doubler, 88, 89
Voltage follower, 439, 440
Voltage regulation, 49, 70, 445
    percent, 73
Voltage regulator:
    with current preregulator, 453
    series feedback, 448, 449, 453
    shunt, 446
    zener, 71–76

$y$ parameters (*see* Admittance parameters)

Zener diode, 71